UTOPIANISM FOR A DYING PLANET

Utopianism for a Dying Planet

LIFE AFTER CONSUMERISM

GREGORY CLAEYS

PRINCETON UNIVERSITY PRESS
PRINCETON & OXFORD

Published by Princeton University Press
41 William Street, Princeton, New Jersey 08540
99 Banbury Road, Oxford OX2 6JX

press.princeton.edu

First paperback printing, 2024
Paper ISBN 978-0-691-23668-1
Cloth ISBN 978-0-691-17004-6
ISBN (e-book) 978-0-691-23669-8

British Library Cataloging-in-Publication Data is available

Editorial: Ben Tate and Josh Drake
Production Editorial: Jenny Wolkowicki
Jacket/Cover design: Lauren Smith
Production: Danielle Amatucci
Publicity: Alyssa Sanford and Charlotte Coyne
Copyeditor: Maia Vaswani

Jacket/Cover image: From woodcut in Thomas More's *Utopia*, 1516.
Lebrecht Music & Arts / Alamy Stock Photo

This book has been composed in Arno Pro

To Greta Thunberg,
who proves the power of one person with empathy

and to Cara Amal Claeys-Roberts,
b. 24 January 2022, the future

CONTENTS

PREFACE

IMAGINE: a shimmering, azure-blue sea, the waves gently lapping at the bow of our boat. Like divine breath, a tropical breeze wafts across our faces. On the horizon an island appears, palm trees drooping to kiss the billowing white surf, their green embracing the cloudless sky. As we land, smiling natives garland us with flowers, and extend shards of coconut. They are a people much like ourselves, we discover, only more content, and at peace with themselves and each other. We have reached utopia, have we not? So we stay on this island. We are happy. We proliferate.

After many generations, we have cut down the forests, and the soil is sterile. We are hungry. Our paradise is destroyed. Most of the indigenous population have died of our diseases and from enslavement. Their paradise too is gone. Then the water begins to rise, and our island, to sink . . .

ACKNOWLEDGEMENTS

FIRST AND FOREMOST I am thankful to my family for their patience and support, and to Christine especially. Secondly I am grateful to Lyman Tower Sargent, who generously shared his knowledge of the field, and to Artur Blaim, both of whom made many helpful suggestions. I owe further debts to many scholars past and present, and have benefited as always from the contagious enthusiasm of many groups, in schools, universities, and the public at large, about this topic and its relevance to where we are going. None bear any responsibility for the result.

In addition I would like to acknowledge: Sorin Antohi; Michael Brie; Anna, Chris, and Danny Claeys; Noa Cykman; Zsolt Czigányik; David Dwan; Dimitri El Murr; Aristides Agoglossakis Foley; Vita Fortunati; Jeff Frank; Golo Föllmer; Justyna Galant; Yaron Golan; Toon van Houdt; Masashi Izumo; Alan Kahane; Pierre Lurbe; Eduardo Marks de Marques; Andrew Milner; Marcello Musto; Mohammad Nasravi; Claudia Novak; Yoshifuri Ozawa; Juan Pro; Kohei Saito; Ryo Susato; the School of Marxism, Peking University; Ben Tate; Matthew Taylor (and the Royal Society of Arts); Philipp P. Thapa; the staff of the British Library, the London Library, and the Library of Royal Holloway, University of London, especially Inter-Library Loans; colleagues in the Utopian Studies Society and the Society for Utopian Studies; fellow protestors in the Extinction Rebellion movement; audiences at the Battle of Ideas, London; the University of St Andrews; the University of Greifswald; Southern Federal University, Rostov-on-Don; the Hungarian Academy of Sciences and Eötvös Loránd University, Budapest; the Académie Polonaise des Sciences; the University of Oxford; the University of Paris; Manchester Metropolitan University; Kyushu Sangyo University; Doshisha University; Chuo University; Osaka City University; Keio University; the Universidad Federal de Pelotas; the TEDx team, Goodenough College, London; the TEDx team at Linz, and especially Claudia Novak; members of the Royal Holloway University and College Union branch; Jon Erik Stubbings and the Camden School for Girls; and once again, the ever-admirable UK National Health Service, a utopia unto itself. Finally, I owe much to the fantastic team at Princeton University Press, from Ben Tate, who oversaw it from inception, to Maia Vaswani, a superb copy-editor.

Parts of the text have been previously published as "How Can We Imagine a Post-consumerist Character?" (*Tocqueville Review* 40 (November 2019): 313–22) and in "Marx and Environmental Catastrophe" (in *Rethinking Alternatives with Marx* (Palgrave Macmillan, 2021): 113–28).

PART I

Towards a Theory of Utopian Sociability

Community is the only heaven we can reasonably look forward to,—the only real and substantial salvation.

—HARRY HOWELLS HORTON, 1838[1]

Human nature is the true community of men.

—KARL MARX, 1844[2]

The love of possessions is a disease with them.

—SITTING BULL, 1877[3]

Forsooth, brothers, fellowship is heaven, and lack of fellowship is hell: fellowship is life, and lack of fellowship is death: and the deeds that ye do upon the earth, it is for fellowship's sake that ye do them, and the life that is in it, that shall live on and on for ever, and each one of you part of it, while many a man's life upon the earth from the earth shall wane.

—WILLIAM MORRIS, 1888[4]

The real objective of Socialism is human brotherhood.

—GEORGE ORWELL, 1943[5]

1. Harry Howells Horton, *Community the Only Salvation for Man: A Lecture* (A. Heywood, 1838), 14.

2. "Critical Marginal Notes on the Article 'The King of Prussia and Social Reform': By a Prussian", in *Collected Works*, by Karl Marx and Frederick Engels, 50 vols (Lawrence and Wishart, 1975–2005), 3: 204–5.

3. Quoted in John de Graaf, David Wann, and Thomas H. Naylor, *Affluenza: How Overconsumption is Killing Us—and How to Fight Back*, 3rd ed. (BK Currents Books, 2014), 115.

4. William Morris, *A Dream of John Ball; and, A King's Lesson* (Longmans, Green, 1912), 33.

5. George Orwell, *I Have Tried to Tell the Truth: 1943–1944* (Secker and Warburg, 1998), 42.

Utopianism is dead, and without it no radical philosophy can exist.

—JUDITH SHKLAR, 1957[6]

When all truth is found to be lies / And all the joy within you dies / Don't you want somebody to love . . . ?

—JEFFERSON AIRPLANE, 1966[7]

If Ecotopia is merely Utopia in the sense of a cheap castle in the air, then we have no future at all.

—RUDOLF BAHRO 1986[8]

6. Judith Shklar, *After Utopia: The Decline of Political Faith* (Princeton University Press, 1957), 208.

7. Jefferson Airplane, "Somebody to Love", recorded 1966 by RCA Victor; copyright Darby Slick.

8. Rudolf Bahro, *Building the Green Movement* (Heretic Books, 1986), 174.

1

Redefining Utopianism for a Post-consumer society

A SPECTRE IS haunting humanity—the ghastly dystopian image of its own extinction. Our world is burning up, and dying, nearly everywhere, from Indonesia and Australia to Brazil, the Arctic, Siberia, Greece, and California. We have at most a few short years to prevent its destruction, and our own extermination. To do so we need to rethink the very principles of our existence as a species from the ground up. This book proposes one way of doing this, derived from the tradition known as utopianism.

The despair sometimes called "doomerism" is one response to the environmental apocalypse. It often breeds depression, a sense of hopelessness, and a paralysing inactivity. Utopianism offers us a very different response.[1] The concept of utopia demands that we rise above the limits imposed by everyday reality and instead envision long-term futures in which humanity not only survives but flourishes. It can be used today both to help explain the history of humanity's aspirations to date and to project viable alternatives to the grim fate facing us. Every age has bred utopias in response its own specific crises. The foundational text of the literary tradition, Thomas More's *Utopia* (1516), was written in part to address the brutal displacement by greedy landowners of hundreds of thousands of agricultural labourers for sheep farming. Later phases of utopian thought reacted to the industrial revolution (the early socialists and Marx), then to the onset of monopoly capitalism (Edward Bellamy

1. The two main introductions to the wider concept are Krishan Kumar's *Utopianism* (Open University Press, 1991) and Lyman Tower Sargent's *Utopianism: A Very Short Introduction* (Oxford University Press, 2010).

and the later socialists), and more recently to late capitalism's evident failure to provide a satisfactory human life for the majority (the counterculture and political rebellions of the 1960s). At these points utopia served as what E. M. Cioran calls "a principle of renewal in both institutions and peoples", lamenting and sometimes satirising decay while mapping a way forward.[2] Our own unique crisis must in turn have its unique utopia, indebted to its predecessors but like no other. One such possible ideal society is sketched out here. But such visions cannot be a mere phantasm, a wish or dream; they must be historically and empirically grounded, and demonstrably practicable. What Immanuel Wallerstein calls "utopistics" involves "the serious assessment of historical alternatives, the exercise of our judgment as to the substantive rationality of alternative possible historical systems."[3] We need to know where we are going, not only that we do not want to be where we are.

Our own crisis is easily summarised. In the last decade we entered a period technically termed the *sixth mass extinction*. For the first time such destruction results from the actions of only one species, *homo sapiens*. The last was sixty-six million years ago, when 76 per cent of all species died. Now some 70 per cent of species have gone in the last half century, and insect loss is at 2.5 per cent per annum (p.a.). The world is hotter than at any point in the last twenty thousand years.[4] As early as 1912 assessments indicated that coal consumption was warming the earth. In 1953 the capacity of carbon emissions to raise temperatures by as much as 1°C were discussed in prominent publications like *Time* and the *New York Times*.[5] By the 1970s the implications were clear to those who sought to know; indeed, "Nearly everything we understand about global warming was understood in 1979".[6] By 2000 it was being asserted that "Many climate researchers believe we would need to cut greenhouse gas emissions immediately by 60 or 70 per cent, which would mean a complete halt in the use of petrol- and diesel-engined cars."[7]

2. E. M. Cioran, *History and Utopia* (1960; Quartet Books, 1996), 10. Cioran himself foresaw a new barbarism emerging from the 1930s.

3. Immanuel Wallerstein, *Utopistics; or, Historical Choices of the Twenty-First Century* (New Press, 1998), 1.

4. See Elizabeth Kolbert, *The Sixth Extinction: An Unnatural History* (Bloomsbury, 2014).

5. Nathaniel Rich, *Losing Earth: The Decade We Could Have Stopped Climate Change* (Picador, 2017), 189.

6. Rich, 3.

7. Jonathan Margolis, *A Brief History of Tomorrow* (Bloomsbury, 2000), 80.

A vast conspiracy of silence and denial, abetted by a flood of mis- and disinformation, and a natural inclination to hope for the best, to cling to "normality", and to fear bad news, has led us to ignore these warning signs and downplay their implications. Instead, we have cloaked reality in comforting euphemisms like "climate change", and the anodyne phrase "global warming", coined around 1975.[8] The more technical term "greenhouse effect" describes the trapping of emissions of water vapour, carbon dioxide (CO_2), methane, ozone, and nitrous oxide. It was identified by the 1930s and acknowledged secretly by fossil-fuel companies by the 1970s. With remarkable accuracy, Exxon scientists in 1982 estimated 1°C warming by 2019 and a CO_2 concentration of 415 parts per million (ppm), against a safe level of 350 ppm. In 2022 it reached 420 ppm. It could reach 427 by 2025, 450 by 2035, and 550 by 2050 or soon after. Around 1990, a rise in temperature of 1.5–4.5°C above a baseline some 250 years ago was predicted.[9] In 2021 the actual rise averaged about 1.25°C, but was 4°C in the Arctic, where unprecedented heatwaves are now common. It could be 1.5°C by 2024, and 2°C as early as 2035: the surge in emissions predicted for 2021 will be the highest in a decade. But warming will be uneven, with the United States and Russia, amongst others, likely to be hit worse than many countries.[10] And even at the current warming level, well below the limit we are supposedly aiming to achieve, tipping points are becoming imminent—in forest and ice loss, and other areas—which will increase the rate of degradation elsewhere and push temperatures higher more rapidly. Beyond any of these—and each may exacerbate others—there may be no return.

Most frighteningly, the worst-case scenarios suggested over the past thirty years or so have turned out both to have been the most accurate, and to be occurring much faster than anticipated. Prediction after prediction has been fulfilled decades before the dates estimated only a few years ago. Carbon emissions have grown steadily since the first major international discussions aimed at limiting them: they were 61 per cent higher in 2013 than in 1990.[11] Fifty per

8. For an introduction, see Emily Boyd and Emma L. Tompkins's *Climate Change* (Oneworld, 2009). CO_2 emissions are measured from a baseline early in the Industrial Revolution.

9. Bill McKibben, *The End of Nature* (Viking, 1990), ix, 18.

10. The more pessimistic view is offered in David Wallace-Wells's *The Uninhabitable Earth: A Story of the Future* (Allen Lane, 2019). A more optimistic estimate and critique of Wallace-Wells's "doomist" narrative is given in Michael Mann's *The New Climate War: The Fight to Take Back Our Planet* (Scribe, 2021, 211–17). Mann, however, says little about climate tipping points, and rather more about "tipping points" in public attitudes towards the problem. A good review of the movement is provided in Dale Jamieson's *Reason in a Dark Time: Why the Struggle against Climate Change Failed—and What It Means for Our Future* (Oxford University Press, 2014).

11. Naomi Klein, *This Changes Everything: Capitalism vs. the Climate* (Allen Lane, 2014), 11.

cent of all CO_2 emissions in history have been since 2000. Some 51 billion tons of greenhouse gases are now released annually, and the amount is rising by as much as 4 per cent p.a. It needs to fall by 7–15 per cent p.a. if we are to survive. But widely accepted projections indicate a 120 per cent increase in fossil-fuel consumption by 2030. For many years, at the climate-change summits from Kyoto (1997) to Paris (2015), 1.5–2°C of global warming was spoken of as "sustainable" and "liveable".[12] The most recent meeting, COP26 in Glasgow in 2021, continued to subscribe to this principle, while doing little to ensure that even 1.5°C would not be breached (1.8–2.8°C is estimated in the unlikely event that all pledges are met). But if the world is being destroyed at the current rate of 1.25°C, which is likely to bring us to 3–4°C or more, our problems are clearly worse than we think, and the enormity of what we face is simply not being recognised. Why should we settle on a ceiling higher than what is already destroying the planet? Can we really "compromise" with extinction, in order to retain "business as usual" for as long as possible? And the rate of degeneration will accelerate, since as the world gets warmer it will get hotter still if emissions are not drastically reduced. Nor are our distant "targets" viable. A notional zero-carbon goal by 2050 is now commonly touted—even in the proposed EU "Green New Deal", as of 2020. This is so improbably remote as to be meaningless. And the goals proposed for 2030 (as of 2021) would reduce emissions by only 0.5–1 per cent, when a 45 per cent reduction is necessary.[13] This is folly.

Such proposals still define our climate comfort zone. But they are seriously flawed. The non-threatening discourse which for the first decade of this century focused on a 1.5–2°C ceiling has now been replaced by forecasts of 4–5° or even more, and as early as 2060.[14] At 4°C, it is generally conceded, what we call civilisation will collapse, and most people will die. Even at the current rate of warming, the polar icecaps are melting in the Arctic (up to 20 metres thick, shrunken by 40 per cent since 1980, and melting at six times the rate of the 1990s). Some 40 per cent or more of the Siberian permafrost may vanish by 2100. Here and elsewhere, the permafrost could unleash as

12. But a recent account notes that "there is no specific scientific reason for picking these particular numbers": Eelco J. Rohling, *The Climate Question: Natural Cycles, Human Impact, Future Outlook* (Oxford University Press, 2019), 2. A 2°C ceiling is widely held to have originated with the economist William Nordhaus, who proposed it as early as the 1970s, but later described 4°C as "optimal", meaning that the costs and benefits of mitigating change balanced out. See Nobel Prize Organisation, "William D. Nordhaus: Facts", NobelPrize.org, Nobel Prize Outreach AB 2021, 8 November 2021, www.nobelprize.org/prizes/economic-sciences/2018/nordhaus/facts/.

13. *Guardian*, 7 May 2021, Journal, 1.

14. An early acute analysis of the problem assumed 2°C warming by 2030 as the "critical threshold": George Monbiot, *Heat: How to Stop the Planet Burning* (Allen Lane, 2006), 15.

much as a thousand million tons (a gigaton) of methane and 37 thousand million tons (37 gigatons) of organic carbon, more than has been released since the Industrial Revolution. In the Antarctic (up to 5 kilometres deep), which holds 90 per cent of the world's ice, 2.7 trillion tons have been lost, currently at the rate of 1.2 trillion a year, and this tipping point will likely be reached with only 2–3°C of warming, which the targets now agreed on will likely produce. The world's glaciers are disappearing—in Greenland, at nine times the rate of the 1990s, which will alone produce a 7-metre sea-level rise. These changes are pretty much irreversible. The last time CO_2 levels were near today's was the Pliocene period, some three million years ago, when sea levels were 12–32 metres above ours.[15] We can expect much worse.

As temperatures rise, deforestation, with increasingly devastating forest fires (now up to 25 per cent of carbon emissions), the degradation of agricultural land, desertification, and water shortages proceed apace. By 2030 demand for water may outstrip supply by 40 per cent. Vast areas will soon be too arid to cultivate or live in. Coastal regions will be inundated, and some islands are already endangered. Huge movements of population will occur, and war over habitable land, including the use of nuclear weapons, will become inevitable. Sea temperatures are rising and threaten dramatic alterations in currents. The oceans, which absorb most of the heat, are becoming more acidic, spelling the end of coral reefs (99 per cent of the Great Barrier Reef is already doomed) and much marine life. As the poles and glaciers melt they reflect less sunlight and thus assist greater warming. In total, oceans could rise 80 metres above today's levels—and 40 per cent of the world's population lives within 100 kilometres (60 miles) of the coast. To pessimists like David Wallace-Wells, there is "almost no chance" that we can avoid many of these scenarios.[16] Agreements come and go, and "greenwashing" with fine words and exuberant but often insincere promises becomes increasingly plentiful. Meanwhile, emissions just keep rising. For, despite rising anxieties, many of us are wedded to the opulent lifestyle the wealthy nations enjoy. We are living the good life, and are loath to give it up. Or we claim the right to "development" in order to get there. So we race towards the cliff edge at high speed in our flashy red sports car, music blaring, not even wearing our seat belts.

It is time to slam the brakes on. Not only would 2°C be catastrophic, even 1°C spells disaster. Following the 2016 Paris Agreement guidelines, the consensus at the United Nations Climate Change conference, COP25, in Madrid (2019) and then at COP26 (2021) in Glasgow still presumed that 1.5° was

15. Rohling, *Climate Question*, 127.
16. Wallace-Wells, *Uninhabitable Earth*, 9.

acceptable. It clearly is not: the Paris Agreement is far from adequate. It does not even dictate the need to cease fossil fuel use, and must be superseded by much more immediate and dramatic reductions in emissions. Even if existing goals were reached, warming of perhaps 3–4°C would still occur.[17] To Naomi Klein, 2°C of warming now "looks like a utopian dream."[18] A likely global temperature rise of 4–5°C by the mid-twenty-first century, rising to as high as 6°C by 2100, would mean summer temperatures in Europe and the Americas soon reaching 50–60°C.[19] At 3°C warming, trees will start to die, and few crops can be cultivated. Physical infrastructure (roads, electric wiring, window frames, etc.) then disintegrates rapidly. Rising sea levels would displace hundreds of millions of people, and the heat, billions more. By this point the process would be well-nigh unstoppable. Social and economic collapse would inevitably follow. A dramatic scramble for rapidly diminishing resources would result. Climate chaos breeds climate conflict. A "world where people shoot each other in the streets over a loaf of bread", in Bill McKibben's warning, may be just over the horizon.[20] A planet 4°C warmer could sustain only between 500 million and a billion people, out of a current population of some 7.8 billion, but as many as 10 billion in 2050. What will happen to the rest? And to the animals and natural world? No amount of posturing, virtue signalling, dithering, and delaying, or the announcing of distant goals, forever kicking the can further down the road, will save them. Action of an entirely different magnitude is required. This problem is greater than every other difficulty we face put together. It deserves our immediate, urgent, and undivided attention.

Here utopia enters into the equation. Defined as "no place", with "utopian" meaning "unrealistic" or "impossible", its relevance to the present is dubious. But it is much more. Utopianism allows us to project ideal societies or groups by imagining what might be but does not yet exist. Such maps of possibility take us beyond the thousand bubbles of everyday consciousness which envelop us, limiting our horizons to what our self-interest and desire for

17. Rohling, *Climate Question*, 95. Rohling regards 1.5°C as "too generous" (99). Wallace-Wells estimates 3.2°C of warming, provided all agreements are met (*Uninhabitable Earth*, 11), but also cites a UN report predicting 4.5°C by 2100 and an Intergovernmental Panel on Climate Change (IPCC) report suggesting over 4°C of warming is likely (p. 41).

18. Klein, *This Changes Everything*, 13. Such tipping points could also occur through nuclear war or volcanic eruption.

19. Mark Lynas, *Six Degrees: Our Future on a Hotter Planet* (Harper Perennial, 2008).

20. McKibben, *End of Nature*, 135.

happiness demand, and making vastly superior social arrangements and great future changes appear nearly inconceivable. Utopia is part of a family of concepts of imaginary spaces, including heaven and paradise, which have historically used images of ideal communities to offer a moral compass or rudder to steer us through the storms of life and restrain the excessive greed and selfishness which forever threaten to destroy us, or to lament the folly of failing to do so.[21] Now too it can function to warn of the extreme dangers of our present course. Without it, indeed, no real advancement can occur. It is, thus, far more than a merely interesting and provocative concept: it is vital to our progress. No plea to return to "normality" or the "everyday" is now worthwhile. Only the extraordinary can save us.

Utopian*ism* also involves putting utopian ideas into practice. Distinguishing the realistic and attainable aspects of utopia from those which are purely imaginary (which some necessarily are) is attempted in part I here. Utopianism is examined in terms of form, content, and function. Its forms, we will see, are three: utopian ideas, literary utopias and dystopias, and communitarianism. Its functions are two. One is to permit visionary social theory by hinting at possible futures on the basis of lost or imaginary pasts, or extrapolating present trends to their logical conclusions or outcomes. We can call this the *futurological function*. It allows us to reach beyond the horizons of everyday life and push back the boundaries of the possible. It may offer a blueprint or programme which can actually be reached—who would construct a building, much less a new society, without one? Utopia's second function is essentially psychological. Here a "desire" or "principle" is often viewed as the core or essence of utopian thinking. This may produce an image which serves as a critical standpoint on the present but necessarily recedes like a mirage as we approach it. So while we may realise past utopias we also constantly move the conceptual goalposts forward, somewhat on the "grass-is-always-greener" principle. We can term this the *alterity function*. Utopia's content is usually defined by equality and sociability, or "community", which is here narrowed and refined into a need for belonging. So the concept cannot be reduced to meaning nowhere, impossible, impractical, or perfect, or even merely better. Nor should we treat it as a substitute for or variant on religion. It does not seek a final, total, or permanent state of earthly perfection, bliss, holiness, blessedness, the complete abolition of alienation, or any other variation on salvation (which is what these are), though we need to consider claims that it does. It can succeed only where it has more modest and secular aims.

21. In the sense that they are spaces we imagine, which might of course be real, lack of evidence notwithstanding; the point is how we use what we imagine.

In a twenty-first-century utopia, it will be argued, these aims must include three key qualities: equality, sociability, and sustainability. Delving into earlier traditions reveals how they may be combined. Thomas More's humanist paradigm, here termed *utopian republicanism*, is defined by common property, relative social equality, and greater sociability, or a closer sense of community. More describes a form of polity or commonwealth, a mode of social organisation, and a sketch of the customs and manners appropriate to them. Suitably updated, and bolstered with suggestions drawn from later utopian thinkers, this tradition suggests clearer solutions for our unique problems than any other school of thought can offer. The examples drawn on here derive from all the emanations of utopianism; namely, literature, intentional communities, and utopian theory, which are introduced in a broadly chronological manner. They illustrate a rich and complex response to the central issues treated here, particularly that of needs, and of the desire for luxury goods and to emulate the wealthy, and suggest ways of releasing ourselves from the self-destructive mentality of consumerism and promoting what is now, and must remain, our own central principle: sustainability.

Equality and sociability are traditional utopian virtues, and indicate that the core of the ideal society is social relationships, not material plenty. This grates against many common images of utopia as universal abundance based on satisfying unlimited needs. For most of the last 150 years the "technological Utopia in which our descendants will live", as it was described in 1957, has evoked visions of gleaming skyscrapers and fantastic technologies which make everyone opulent and render onerous labour obsolete.[22] Yet in preceding centuries, both in fiction and in practice, utopia often stood for the image of a society defined much more by human interaction than by technology, physical infrastructure, or widespread luxury. Material needs have been satisfied in images of the ideal society, but they are not regarded as incessant or unrestricted. Utopia's key goal has been greater unity, camaraderie, and friendship, by contrast to dystopias, where fear and hatred of others predominate, and individual isolation and loneliness are deliberately promoted by rulers.[23] This gives us a spectrum of group types, with fear and alienation at the dystopian extreme, and friendship and belonging at the utopian end. People are nicer to each other in utopia, and less afraid. They have stronger feelings of what is here termed *belongingness*, or the need to feel accepted and to have a sense of home and place. This concept is at the heart of the aspiration for and experience of utopia, and is central to this

22. Harrison Brown, James Bonner, and John Weir, *The Next Hundred Years* (Weidenfeld and Nicolson, 1957), 160.

23. See my *Dystopia: A Natural History* (Oxford University Press, 2016), for this typology.

book. But it can flourish only where luxury, great social inequality, and an obsessive desire to possess and consume things and display our wealth are curtailed.

The need for sustainability requires re-examining utopian approaches to needs, consumption, luxury, simplicity, and nature, and tracing how the idea of unlimited consumption emerged from the eighteenth century onwards, and how utopians responded to it. This mentality has become universal, and, forgive the pun, all-consuming. We embrace it passionately like our one true love. Few utopians, we will see, have accepted it, however, and most have foreseen its dangers and limits.[24] But even here, as in the wider society, a tension developed between the growing appeal of luxury for all and the desirability on moral, then increasingly on ecological, grounds of a regime of greater simplicity and restraint which sustainably preserves the best that science and technology offer us, in the interests of the survival of all. The manifold forms this tension has exhibited are explored throughout part II of this book.

These three ideas—equality, sociability, and sustainability—are treated here as part of a single relationship. A central argument of this book is that the quid pro quo for exiting consumerism is greatly expanding opportunities for sociability to help compensate for diminished material consumption. People are unlikely to relinquish a consumer-oriented lifestyle unless they see their lives improved in other ways. The appeal of consumerism is otherwise simply overpowering. The chief problem here is the richest 15 per cent of humanity, who have decent housing, transportation, food, and clothing. If, and it is a very big *if*, we can wean these from their greedy obsession with luxury goods and conspicuous consumption, and dissuade the rest from emulating them, we stand some chance of preserving our planet. But if all we have to offer is gloom and hopeless pessimism, constant demands for self-sacrifice and austerity, the authoritarian prohibition of pleasurable activities, or a blind optimism based on ostrich-like denial and love of luxury, which in turn feeds the climate-change deniers, we will lose the battle and thence the war, and our planet. So the task before us involves not merely moving towards renewable fuels, drastically reducing carbon emissions, and the like. It involves fundamentally rethinking the society we live in, and altering our habits and behaviour as consumers, indeed our very identity as human beings, in the direction of a sustainable lifestyle. Science and technology alone simply cannot do the job. Our social relations must be rethought and restructured.

Utopia can help us meet humanity's greatest challenge, then, by illustrating how sustainability has been imagined in the past. Renouncing consumerism

24. For an exploration of these trends, see Rudolf Moos and Robert Brownstein's *Environment and Utopia: A Synthesis* (Plenum Press, 1977).

means acknowledging earth's incapacity to meet unlimited wants, and stressing conservation, repair, renewal, and autarky. It means ceasing to acquire products because we associate them with youth, sex appeal, beauty, or immortality, and prioritising instead their use value over their symbolic value. Following the sceptical tradition of Veblen, Packard, Galbraith, Potter, Riesman, and others, who throughout the twentieth century warned of the dangers of affluence, we must thus decouple our psychological wants from our physical needs.[25] This involves reversing a process developed over some three hundred years, which is here described as commodifying the self, "thinging ourselves", or identifying our personalities with objects. Decommodifying our lives allows us to view human progress in qualitative rather than quantitative terms, and to see this largely in terms of richer relationships with other people, or what is here called *enhanced sociability*, whose aim is belongingness, and whose antithesis is that type of alienation defined by a sense of lacking place. This compensatory sociability involves exchanging an unnecessarily wide range of consumer choices for a more nurturing, healthy, stimulating human environment. It implies constructing a set of institutions for promoting sociability, as well as nurturing attitudes towards other people which promote mutual assistance and conviviality. We must learn to interact more in order to shop less. And all this must be done within about a decade, or it will be too late.

This book thus rests on three premises:

1. That we value social relations more than anything else, which is easily illustrated, as we will see,
2. That the utopian tradition acknowledges this principle, and offers us a theory of the relationship between consumption and sociability, which is not too difficult to prove; and,
3. That a viable future is possible only if we relinquish a consumerist mentality in exchange for greater engagement with others. This is highly contentious, and not easily demonstrated.

This book, then, has three main aims: to defend a theory of "realistic" or realisable utopianism in order to describe the ideal we must aim for; to focus further on two main aspects of this theory—namely, sociability and restraining consumption, especially private luxury—by examining how the utopian tradition has treated these issues; and to make sense of what these imply for environmental degradation now. Some utopians have proposed ways of avoiding excessive private consumption by shifting our focus towards sustainable

25. On this tradition, see Daniel Horowitz's *The Anxieties of Affluence: Critiques of American Consumer Culture, 1939–1979* (University of Massachusetts Press, 2004).

luxuries in the public sphere. With appropriate modifications these proposals can be applied today, particularly by offering compensatory and often public sociability in exchange for the diminished private consumption of unsustainable goods. The process will not be painless, but at least we might survive.

We begin first by laying out the ground of the general argument and defining our key concept here, commencing with a brief overview of the tradition and some influential approaches to its analysis.

The History of Utopianism

It is sometimes claimed that utopianism has no history, but expresses a timeless desire for human improvement. This is only partially true at best, and is more misleading than not. From its invention in 1516, utopia has proven to be an organic concept which evolves to meet successive challenges. It has come to represent a constant, ongoing conversation about humanity's potential, and especially its capacity for moral amelioration. Every age projects idealised responses appropriate to its own problems and aspires to build on the past and, increasingly, to surpass the ambitions of its predecessors. Earlier ideals invariably disappoint later readers, because our expectations advance. Many older literary utopias fall far short of modern aspirations in being repressive, authoritarian, patriarchal, and undemocratic. Their austerity, harshness, and coercion jar against our sensibilities. Then there is the outlook of Europeans on the rest of the world. The utopian idea commenced from a less-than-universalist, usually Eurocentric if not downright imperialist perspective, with numerous nations putting themselves forward as the "elect". Slavery and war are sometimes retained. Until fairly recently utopia remained an imperial concept, denoting a "commonwealth of increase", in the seventeenth-century republican James Harrington's phrase, and a fantasy of conquest which implied dystopia for indigenous and non-European peoples. Decolonising utopia is a task only just begun, for the arrogance which empire breeds is very slowly dissipated.

Only in the late nineteenth century did a universalist utopia become possible—at the height of imperial expansion. And even then, notes Norbert Elias, the tendency remained that "almost all these political utopias express a wish for hegemony", or rule over others, "a wish for the preponderance of one's own class or one's own state over all other people," a prospect which could end only when "Present-day utopias . . . become utopias of humanity and not merely state utopias", through a world state and united international order.[26]

26. Norbert Elias, *Essays I: On the Sociology of Knowledge and the Sciences* (University College Dublin Press, 2009), 265, 287.

This trend began in the twentieth century, especially in the writings of H. G. Wells, when utopia evolved into an increasingly cosmopolitan, humanist idea centred on promoting world organisation for justice, peace, and plenty for all. Since then it has been marked by ever-greater aspirations to inclusivity, which demands recognising hitherto marginalised groups as entitled to equal respect.

Though it often harkened back to earlier, better times, from the eighteenth century onwards utopia can thus be understood as a subset of linear theories of indefinite progress, and the expectation that each generation would be better off than its parents, which became the most important modern social idea. Like rights theories, which it parallels in various ways but exceeds in ambition, utopian progress is, however, measured less by material than by moral improvement, defined by two key, related qualities, sociability and equality. Neither of these figures much in orthodox histories of ideas of progress, where, like some wayward relative arriving at the Christmas feast, introducing utopia often provokes an awkward silence in a celebration of humanity's glorious achievements, and a plea to have this somewhat unhinged guest ejected or sent to some distant table to avoid annoying the rest.[27] "Progress" has an orderly, bourgeois connotation. It results, we usually suppose, from gradual scientific and technological innovation combined with human ingenuity, efficiency, and ambition. It may emerge sedately out of an idea of Providence, or God's benign intention for the world, and plods relentlessly towards the future without spectacular frenzies of expectation, while always delivering something novel and delightful. By contrast, utopia often appears insistently disruptive and alarming. Utopians who embrace "an apocalyptic philosophy of history", in particular, are not seen as advocating "true progressivism" at all.[28] They, and especially Karl Marx, are instead derided as embodying what Robert Nisbet calls "a long and powerful tradition of Christian millennialist utopianism which could be, in some degree, secularized, with its apocalyptic intensity left undiminished."[29] Utopians are colourful cranks with bizarre and sometimes embarrassing or dangerous ideas. Some appear intoxicated by wild ideas, or to be downright mad. Utopia is the skeleton in the closet of progress.

But utopia's contribution to progress can indeed be measured by its aspiration for higher ethical ideals. It represents a process of continuous enlightenment, and a growth in maturity from humanity's childhood onwards, many

27. Robert Nisbet's *History of the Idea of Progress* (Heinemann, 1980) describes *Utopia* as "as far from the idea of progress as was More's beloved friend Erasmus's *In Praise of Folly*" (112).

28. John Baillie, *The Belief in Progress* (Oxford University Press, 1950), 91.

29. Nisbet, *Idea of Progress*, 67.

lapses notwithstanding. To Elias, "More's *Utopia* represents a specific stage in the development of social conscience".[30] In portraying communities where oppression and exploitation have been eliminated or greatly reduced, utopians demand more of us than mainstream proponents of progress do. They dare to imagine that we can ask more of ourselves. They insist that our moral capabilities have not yet met their limits and represent moral intelligence, not just instrumental reason or material gain. We have a better self, they say, if we would only aspire to realise it. From the Enlightenment onwards, most notably, their humanitarian agenda has constantly urged the limitation of cruelty, violence, pain, punishment, and coercion, and the exploitation of people, animals, and nature. This clearly implies a defence of the principle that no one's life should depend on another's suffering. Utopias have increasingly advanced an ideal of consent and condemned rule through tyranny and fear. Now that the blithely and blindly optimistic ideal of progress-as-material-advancement is dead, utopia can stand in its stead as an organising principle for our aspirations.

Despite some of the more extreme claims associated with cultural relativism and postmodernism, this moral philosophy implies a grand narrative of progress which makes the quest to minimise coercion, harm, and suffering the chief theme in our moral history and places utopia at the centre of the narrative.[31] It demands not just that we progress towards these goals but that all enjoy the results thereof. It insists not only that we formally end relationships of extreme domination, like slavery, but that we also attack their causes and justifications, like racism. It routinely asks a basic but radical question usually missing from popular discourse: how far has the progress of science and technology benefited the average person and made them happier and more at ease? So utopians have been found at the forefront of nearly every progressive movement for five hundred years, including sexual liberation, animal rights, vegetarianism, and many other practices. This alternative world view aims distinctly at our improvement as a species. So, far from being a marginal or eccentric idea, utopia is important to everyone. It sums up humanity's highest aspirations and defines our best selves.

After a phase which constitutes its prehistory, four major stages in the development of this agenda are evident, each defined, for our purposes here,

30. Elias, *Essays I*, 227.

31. The centrality of some variant on John Stuart Mill's famous harm principle is thus assumed here.

by a varying emphasis on three key qualities: equality, sociability, and sustainability.

The first stage is the concept's establishment in More's radical humanist text of 1516, against a background of profound social crisis, as the modern class war of the rich against the poor and the modern racial war of Europeans against the rest of humanity commenced. *Utopia*'s restatement of European conscience overlapped with the conquest of the "New World", to Europeans equally a destination ripe for plunder and embodying the lost virtues of the Garden of Eden or the Golden Age.[32] From the fifteenth through the twentieth centuries, the Americas in particular inherited aspirations drawn from ideas of the millennium, paradise, the chosen race and people, and the general salvation of humanity.[33] In a fundamental sense they became the chief modern spatial utopia, and the place to flee in order to escape the greater difficulty of promoting utopia at home, in old and corrupt nations.[34] To a lesser degree, Australia and the southern Pacific also came to embody ideals of primitive virtue and natural liberty, and the possibility of European conquest. And so to regain its own paradise, the west set out to conquer everyone else's.

The second phase occurred with the late eighteenth century's turn towards more forward-looking visions. The other place became the future time, replacing nostalgia for a lost Golden Age and the trope of discovering utopian places in distant lands with an imagined superior future. An air of profound expectation of imminent and dramatic improvement began to pervade the intellectual atmosphere. The wedding of *eutopia*, the imaginary good space, to *euchronia*, the future good time coming, sometimes implied the promise of being guaranteed by history, either rapidly, as in Marx, or more slowly, as in liberal theories of progress. As the west's main utopian project, Christianity, waned, a more secular utopia filled the void created by the loss of faith. Expectations of a millennium induced by God were now supplanted by visions of futures forged by human agency. A "religion of progress" saw faith in a future life replaced by "belief in human perfectibility and infinite progress".[35] In this phase utopia also represents a response to emerging crises, first of commercial

32. See Charles L. Sanford, *The Quest for Paradise: Europe and the American Moral Imagination* (University of Illinois Press, 1961).

33. Some of the spatialising process is traced in Jason H. Pearl's *Utopian Geographies and the Early English Novel* (University of Virginia Press, 2014).

34. See Ernest Lee Tuveson's *Redeemer Nation: The Idea of America's Millennial Role* (University of Chicago Press, 1968) and Frank Graziano's *The Millennial New World* (Oxford University Press, 1999).

35. E.g., Christopher Dawson, *Progress and Religion: An Historical Enquiry* (Sheed and Ward, 1929), xv, and generally 177–201.

society, then of urbanisation and industrialisation. As a form of satire it often lambasted the growing luxury, vanity, and egoism which commerce fuelled, and sometimes directly counselled a return to greater simplicity. The American and French revolutions revealed a powerful desire for greater social equality, soon defined as socialism. As the chief form of modern utopianism, it was quickly recognised as the progeny of Thomas More. This trend demonstrates a growing realism and universalism within the utopian tradition. It also coincided with the consumer revolution, which drove both to a greater extent than is usually appreciated.

The third major historical stage in utopianism occurred in the late nineteenth century, with the advent and dissemination of large-scale collectivist solutions driven by science, technology, and industry. This induced the first great explosion of utopian literature, sparked by the statist socialist proposals of the American Edward Bellamy's *Looking Backward 2000–1887* (1888). Then came the epoch of Karl Marx, the most influential non-religious utopian of all.[36] After 1917, spatially, anti-capitalist utopian hope was invested primarily in Russia's Bolshevik Revolution and, until the late twentieth century, in other societies inspired by or incorporated into the Soviet experiment. The idea of revolution inherited many assumptions formerly embedded in millenarian thinking, implying a moment of profound moral and psychological as well as social and political transformation, which overlapped with the language of conversion and redemption. Central to this expectation was the assumption that once private property was abolished, human behaviour would improve dramatically. If capitalism had betrayed the promise of universal happiness, this utopia seemingly offered a viable alternative, until, subverted by internal despotism and external pressures, the system finally collapsed in 1989–91. Meanwhile, capitalism also bred its own utopian alternatives, notably during the 1960s, in the form of a counterculture which often rejected consumerism.

The fourth historical turning took place in the late twentieth and early twenty-first centuries. This initially appeared as a claim of the ultimate victory of capitalism and liberal democracy over communism, and of the "end of history" and "end of utopia"—proposals which now look downright ridiculous. But now, as environmental catastrophe loomed, progress as such began to retreat rapidly. Dystopia came to overshadow utopia, whose prospects for revival now looked increasingly slender indeed.

36. In terms of readership as well as practical influence on ideas of a vastly improved future. Marx has easily outsold all the other practical utopists as well as the literary utopian and dystopian writers put together. He has thus a good claim to be the starting point for any history of the subject as a whole.

The Historiography of Utopianism

To comprehend these developments we need to clarify the meaning of our central concept. But what is the history of utopianism a history of? The answer depends greatly on our starting point. Yet there are many of these, and which if any is best is much disputed. In academia, petty jealousies often trump real issues of principle. Disciplinary rivalry is particularly disruptive in promoting claims for a monopoly on interpretation. For who would not aspire to own utopia? "Every discipline which concerns itself with this area would like to keep it to itself", writes Norbert Elias: "Literary scholars would like to define utopia as an exclusively literary genre, historians might perhaps wish it to be understood as a unique historical formation, philosophers as an eternal philosophical question and sociologists as a fact of society."[37] Amongst other things, literary scholars probe authorial intention, narrative strategies, the nature of the "canon", intertextuality, the formal variations of the utopian genre, and the shifting nature of its norms, and how these relate to their dystopian counterparts, and to science fiction and other genres.[38] Historians of communal societies detail individual experiments, map out long-term trends in communitarianism, and explain their relative success or failure, the dynamics of various groups, and their relations to the societies from which they emerge. Political theorists and intellectual historians trace the evolution of utopian ideas within broader trends in thought, and analyse and categorise their various causes, emanations, and interrelations. Then there are sociologists; architects and town planners; and students of religion, psychology, philosophy, science, and technology, of popular protest and revolutionary movements, and of imagined futures both positive and negative.[39] The utopian and dystopian dimensions of art, music, film, architecture, travel and exploration, empire, gender, nostalgia, and a hundred other themes also lend meanings to the subject. A dozen or more definitions of utopia, more or less plausible, meaningful, or helpful, emerge from these concerns. And the subject does not stand still but is ever in flux, making Mosaic efforts to set any one typology in stone difficult.[40] To muddy the waters still further, the most influential strand of the tradition by

37. Elias, *Essays I*, 212. More historians would in fact likely treat it as organic than "unique".

38. For reasons explained in my *Dystopia* (284–90), science fiction is separated from utopia/dystopia, and is also generally excluded from my account here, though a few such texts are mentioned. In practice, a strict separation of subgenres is sometimes extremely difficult.

39. A recent general inquiry of this type is Mark Featherstone's *Tocqueville's Virus: Utopia and Dystopia in Western Social and Political Thought* (Routledge, 2008).

40. For an extensive range of these, see Fred. L. Polak's *The Image of the Future* (2 vols (Oceana, 1961), 1: 379–438.

far, utopian theory, sometimes denies any affiliation with utopia. The most famous utopian of all, Karl Marx, is the key example here. But liberals too are often embarrassed by the label, which they associate with psychological weakness and political or religious extremism.

Many tensions exist between these approaches, and a lack of common ground causes persistent confusion in the field. Like the fabled elephant described by blind people, who touch its many parts and reason from the specific to the general without seeing the whole, utopia appears different to many who approach it. Much ink has been spilt over issues like whether or when texts should be privileged above context and history, or vice versa, and about what it is, ultimately, that we are trying to explain, or promote, in using the concept of utopia, or in subsuming it under or juxtaposing it to utopianism. Disciplinary jargon designed to exclude and intimidate sometimes functions as a substitute for thought. Some deploy the magical shield of various "isms", and invoke invisible armies of fellow "ists" who sprout like dragon's teeth every time the magic spell is uttered. (Shout "Marx" and imaginary millions, or at least real dozens, suddenly rally behind your banner.) At the opposite end of the spectrum, the mere word "utopia" may operate at the emotional level as a magical rescue concept, releasing us from our burdens, guilt, and feelings of inadequacy. Like back-page ads from 1950s comics for a Mr Atlas weightlifter's body so the bullies won't kick sand in your face on the beach, utopia can serve as a get-rich/strong-quick feel-good mantra, a wormhole whose fantasy releases us from oppressive reality, or a lifejacket which rescues us from drowning. (Say "utopia" and fly up into the sky away from it all.) Between these extremes lie a multiplicity of approaches and conflicting or overlapping meanings. This pluralism reminds us that utopia is too grand, brazen, complex, and epistemologically incisive a concept to be held captive for long by disciplinary imperialism. It cannot be owned: it is the common patrimony of humanity.

A loose agreement nonetheless exists that utopia consists in any ideal or imaginary society portrayed in any manner. This minimalist definition satisfies many and still portrays a discreet category. It overlaps with common-language conceptions of the term, where "fantastic" and "impossible" dominate. To Elias, utopia is a "fantasy image of a society which contains proposed solutions to specific unsolved problems of its society of origin".[41] Utopia thus sometimes posits an ideal which by definition is unattainable. Forever nowhere, like a mirage, it necessarily recedes as we approach it, because it is an aspiration based on something much better than the present. So as we reach that present, the ideal necessarily moves ahead of us and remains eternally in the future.

41. Elias, *Essays I*, 214. This would cover dystopia as well.

This process is nonetheless exceedingly useful. It allows us to pierce the bubbles of everyday life, the veils of ignorance which obscure the causes of exploitation and extreme inequality and allow them to be portrayed as "normal". Yet this conception must also be reconciled with the fact that utopia also envisions long-term futures for humanity by offering projections, both literary and historical, of where we are going, to give us alternative visions of much better places where we might go, and to suggest how we might get there, while sometimes also caustically mocking our hubris and our failure to reach the destination. This ability to rise above our chronic and debilitating tendency to short-termism and to challenge "normality" is crucial not just for visionary thinking but also for the anticipation and avoidance of disaster.

Some utopias, too, are realisable, and here utopianism must be deemed realistic: today's present, somewhere, was someone's ideal future in the past. This raises the issue of practical utopian experiments in living in a morally superior way. While fantasy and the imaginative projection of unrealisable ideals are part of the utopian narrative, so too are actual communities, and more collaborative and collectivist ways of life. Here utopia might in principle exist outside the imaginary no place and have actually been realised or still lie within reach. But how should we describe the common core shared by these dimensions of utopianism? What relationship exists between fantastic projections and utopian practice? While the heuristic utility of emphasising fantasy is crucial to understanding the psychology of utopianism, this book pushes at the purely imaginary sense of utopia and contends that it can also be understood as an achievable, if invariably imperfect, set of human relationships. The realist argument presented here thus bridges existing disciplinary and epistemological divisions, the gap between imagination and reality, the practical and theoretical aspects of the subject, and between *utopia* as "nowhere" and *eutopia* as the "good place".[42] It also permits a symmetry with the study of dystopia, where real societies are often, indeed increasingly, described as dystopian, meaning not that they have necessarily reached dystopia but that they lean alarmingly in that direction. Utopian*ism* then becomes the quest for the good place, (e)utopia, where these solutions are seemingly realised, if only temporarily and imperfectly, guided by the image of utopia, the no place which might be. Utopia and eutopia are easily confused, especially because they are pronounced the same way, so that "utopia" is often used interchangeably to mean both "no place" and "good place". Here, except where specified,

42. A useful discussion here, which moves in a similar direction, is provided by Wayne Hudson in *The Reform of Utopia* (Ashgate, 2003), which argues for a "non-totalising" account of utopia (p. 2).

"utopian" means "eutopian": the *e* is silently reinserted into *utopia*, without altering the spelling. But this too can be both real and imaginary.

———

The history of utopian*ism*, then, is the history both of its separate components and of what they share in common. The first use of the "ism" has been dated to 1649, though reflections on the meaning of utopia can certainly be traced back to the early reception of Thomas More's great work, which appeared in English in 1551.[43] On its broadest definition, the modern field of utopian studies commences with the revival of interest in More in the early nineteenth century. From the outset, this attention was not confined to literature, and the interdisciplinary nature of utopianism was recognised. In Britain, links between literary and communitarian utopias and trends in thought were already evident in the Pantisocracy proposals of the 1790s, and Plato, More, Robert Wallace, and others were connected by William Godwin as critics of private property.[44] A one-time disciple of Godwin, Robert Southey prominently associated the Welsh socialist Robert Owen with More in *Sir Thomas More; or, Colloquies on the Progress and Prospects of Society* (1829). To take only a few later examples, these links were also indicated in a series of studies of Owen, Charles Fourier, and Etienne Cabet; then in Marx's writings, which insisted on the now-indefensible distinction between "utopian socialism" and "scientific socialism",[45] and those of Louis Reybaud, who saw the "utopian movement" commencing with Plato and arriving at the "social utopias" of the period,[46] and others through the 1840s; through studies like William Lucas Sargant's *Social Innovators and Their Schemes* (1858), which associated socialism with More, Bacon, and Harrington.[47]

43. For the early period, see Robert Appelbaum's "Utopia and Utopianism", in *The Oxford Handbook of English Prose, 1500–1640*, edited by Andrew Hadfield (Oxford University Press, 2013, 253–66). For France, see Hans-Günter Funke's "Utopie, Utopiste", in *Handbuch politisch-sozialer Grundbegriffe in Frankreich, 1680–1820*, edited by Rolf Reichardt and Hans-Jürgen Lüsebrink (R. Oldebourg Verlag, 1991, 11: 5–104). A good account of the etymology of the various related terms is Lucien Hölscher's "Utopie", translated in *Utopian Studies* 7 (1996): 1–65.

44. William Godwin, *Enquiry Concerning Political Justice*, 2 vols (G.G.J. and J. Robinson, 1793), 2: 805. Unless otherwise noted, subsequent citations refer to this edition.

45. Marx only very rarely used the "ism". See Marx and Engels, *Collected Works*, 28: 71, 180; 17: 90.

46. Louis Reybaud, *Études sur les réformateurs ou socialistes modernes*, 2nd ed., 2 vols (Société belge de librairie, 1844), 1: 37.

47. William Lucas Sargant, *Social Innovators and Their Schemes* (Smith, Elder, 1858), 18. See Peter Fitting, "A Short History of Utopian Studies", *Science Fiction Studies* 36 (2009): 121–31.

Renewed stimulation of interest in the subject occurred during the second wave of European socialism, from the 1870s onwards, with works like Moritz Kaufmann's *Utopias; or, Schemes of Social Improvement* (1879), the best early study of this type, which uses "utopianism" once, and Karl Kautsky's *Thomas More and His Utopia* (1888) (also once).[48] In 1898 Vida Scudder used the term "social idealism" to describe More's inheritance amongst the "dreamers" of Victorian Britain.[49] In later studies, "utopianism" is introduced in Lewis Mumford's *The Story of Utopias* (1922), where thought and method are given pride of place.[50] Anticipating the psychological outlook now usually associated with Ernst Bloch, Joyce Oramel Hertzler's *The History of Utopian Thought* (1923) explicitly defines the "ism" as the "spirit of hope expressing itself in definite proposals and stimulating action, . . . meaning thereby the role of the conscious human will in suggesting a trend of development for society, or the unconscious alignment of society in conformity with some definite ideal."[51] Max Beer's *Social Struggles and Socialist Forerunners* (1924) adopts much of Marx and Engels's typology and does not discuss the "ism" as such, though Beer discusses More at length in his *History of British Socialism* (1929).[52] Much of the literature of this period and indeed subsequently relies on Marxist categories, and accepts, with Werner Sombart, the description that "All the older Socialists were Utopists because they mistook the real motive force in the life of society."[53] This insistence on proletarian class struggle as the dividing line between the utopians and more practical reformers, notably Marx, remained common through the 1980s.

Utopian*ism* as such became a serious object of study with the sociologist Karl Mannheim's *Ideology and Utopia* (1929), building partly on Bloch's early work, which did not use the "ism". Mannheim was chiefly concerned with the thought processes which motivate and structure knowledge, and with both wilful and unconscious desires for change, which in his interpretation ideology resists and utopia promotes. Of the utopians, he asserted, "Their thought is never a diagnosis of the situation; it can be used only as a direction for

48. For this period, see the incisive study by Toby Widdicombe "Early Histories of Utopian Thought (to 1950)" (*Utopian Studies* 3 (1992): 1–38). This terms Kaufmann's study "the first true history of utopianism" (6), while acknowledging that it does little to disentangle utopianism from socialism.

49. Vida D. Scudder, *Social Ideals in English Letters* (Houghton, Mifflin, 1898), 291.

50. Lewis Mumford, *The Story of Utopias* (1922; Viking, 1962), 9.

51. Joyce Oramel Hertzler, *The History of Utopian Thought* (George Allen and Unwin, 1923), 268. This implies that planning is, however, every bit as important as hope in defining utopianism.

52. Max Beer, *History of British Socialism*, 2 vols, (G. Bell and Sons, 1929), 1: 34–43.

53. Werner Sombart, *Socialism and the Social Movement* (J. M. Dent, 1909), 38–39.

action."[54] The "ism" is here thus more a habit of mind or sentiment than a catch-all description of the forms of thought and action which result. But the term did not catch on for many years, and, "scientific" Marxism having exempted itself from the label, was still often identified chiefly with literature. Paul Bloomfield's *Imaginary Worlds; or, The Evolution of Utopia* (1932), for example, associates utopianism with imaginative projection but conspicuously omits practical reformers like Owen, Fourier, and Marx. Frances Theresa Russell's *Touring Utopia* (1932) also links the "ism" to literature and decries the "erroneous identification" of socialism with utopianism.[55] Harry Ross's *Utopias Old and New* (1938) touches on socialism as having provided "pictures of a more pleasant future state", while again giving precedence to literary form, now often called the "utopian romance".[56] Marie Louise Berneri's *Journey through Utopia* (1950) abjures the "ism" but discusses socialist communitarianism and tries to rescue the utopians from some of Marxism's declamations.[57] The socialist movement continued to be the particular focus of most writers until after World War Two, when a shift towards literature is discernible, with dystopia looming centrally from the 1950s onwards.

Some early histories of communitarianism, like Charles Nordhoff's *The Communistic Societies of the United States* (1875), describe such experiments as "utopian", without delving further into definitions.[58] "Utopianist" and "utopianism" are used by George Jacob Holyoake in 1875 in relation to Owenism and the co-operative movement.[59] Morris Hillquit applied the label of "utopian socialist" to practical reformers in the "limited sphere" and at the national level.[60] Intellectual history did not arise as a separate discipline until the twentieth century, and treatments of utopian theory within mainstream ideas outside of Plato, Marx, anarchism, and socialism remain rare even today, especially with

54. Karl Mannheim, *Ideology and Utopia: An Introduction to the Sociology of Knowledge* (Kegan Paul, Trench, Trubner, 1936), 36. For a reconsideration of the main concepts here, see Lyman Tower Sargent's "Ideology and Utopia", in *The Oxford Handbook of Political Ideologies*, edited by Michael Freeden, Sargent, and Marc Stears (Oxford University Press, 2013), 439–51.

55. Frances Theresa Russell, *Touring Utopia* (Dial, 1932), 70.

56. Harry Ross, *Utopias Old and New* (Nicholson and Watson, 1938), 122.

57. Marie Louise Berneri, *Journey through Utopia* (Routledge and Kegan Paul, 1950), 207–9.

58. John Humphrey Noyes curiously uses the term "utopian" only once in his classic study, originally entitled *History of American Socialisms* (1870), in reference to Adin Ballou (*Strange Cults and Utopias of 19th-Century America* (Dover Publications, 1966, 131).

59. George Jacob Holyoake, *The History of Co-operation in England*, 2 vols (Trubner, 1875), 1: 22–51, 155.

60. Morris Hillquit, *History of Socialism in the United States*, 5th ed. (1903; Dover, 1971), 18–19.

respect to liberalism. The exception is where, as a "model of an ideal society" or "concept of a good life", as Barbara Goodwin insists, it is distinctively radical and totalistic but nonetheless seen as integral to virtually every world view.[61]

Modern utopian studies dates from the 1960s, and the revival of interest in the subject which the idealistic counterculture, politics, and communalism of that decade provoked. More recent scholarship has encouraged a tighter typology which acknowledges the different components of the tradition while avoiding reducing them to any one part. The quest for utopia is seen as a common endeavour which is expressed in various complementary ways, though the trend towards treating utopia as primarily a literary concept persists.[62]

In recent years two authors have most prominently addressed the problem of utopianism. Both adopt definitions which mix form and function, but hint only more obliquely at content.[63] To Lyman Tower Sargent the three "faces" or forms of utopianism are utopian social theory, literary utopias and dystopias, and utopian practice. The "ism" as a whole is best captured in the phrase "social dreaming", a functional description. Attempts to transform everyday life in utopian directions are seen as "essential for the improvement of the human condition", though what this advance consists in, the content of utopia, is implied rather than explicit.[64] Krishan Kumar adopts a similar typology in distinguishing among utopian social theory, including Marx; utopian fiction; and utopian communal experiments. The last represent "the practice of utopia", which is in tension with the idea of utopia as "nowhere", though it is aligned with Sargent's description of communities as "attempting to create a better society". Kumar pleads for "a direct connection between the expressions of the social or literary imagination and the practical life of society", to clarify relations between theory

61. Barbara Goodwin, *Social Science and Utopia: Nineteenth Century Models of Social Harmony* (Harvester, 1978), 4.

62. Some recent trends in literature are examined by David M. Bell in *Rethinking Utopia: Power, Place, Affect* (Routledge, 2017).

63. Another major contribution to this debate, Ruth Levitas's *The Concept of Utopia* (Syracuse University Press, 1990), offers the best survey of definitions based on form, content, and function, though its conclusions vary considerably from those defended here.

64. Sargent, *Utopianism*, 3, 5, 8–9. Sargent's "The Three Faces of Utopianism Revisited" (*Utopian Studies* 5 (1994): 1–37), first presented in 1975, is the most comprehensive attempt to clarify the definitional problem and provide a taxonomy of the various types of each "face". On current definitional disputes, see further Gregory Claeys and Lyman Tower Sargent's *The Utopia Reader* (2nd ed., New York University Press, 2016, 1–15).

and practice. He concludes in a Marxian vein that "All thought is shot through with practical elements; there is, likewise, no practice that is theory free, not governed by some sort of understanding that is essentially theoretical". So it may "be better to think of 'thought' and 'practice' not as in any way opposed to each other but as abstractions from a unified human activity." Practice produces thought, which induces further practice.[65]

The broadly accepted threefold typology of utopian forms does not indicate a specific *content* shared by them, only a definitive relationship among them. Clearly, ideal societies in literature and theory have much in common. Both are imaginary and textual, and they are separated only by the sometimes-thin veneer of fiction. The chief definitional problem arises here from including the third, practical component. How should we categorise the content of utopian practice? That is, how do we describe what happens when people think their way of life actually approximates to utopia, rather than merely aspiring to it or dreaming of the benefits thereof?[66] And how does this relate to the fictional and theoretical forms of utopianism? We cannot include communities in any typology simply because they *aim at* or *dream about* utopia and thus mirror a key utopian function. Like a mirage, this implies that they are condemned never to reach their goal, and by definition must fail or fall short, because their aspirations are unattainable in principle. In the meantime they must endure the Sisyphean frustration of seeing their goal inevitably receding as they approach it, remaining forever nowhere, at best a noble myth intended to guide aspiration. And they must swallow the mockery of those who condemn their claims of glimpsing utopia as deluded. But intentional communities and some similar spaces, discussed below, are clearly real, and are often called

65. Kumar, *Utopianism*, 64, 70; Lyman Tower Sargent, "Utopianism", in *Routledge Encyclopedia of Philosophy* (online) (1998), https://www.rep.routledge.com/articles/thematic/utopianism/v-1. Another useful account is Barbara Goodwin and Keith Taylor's *The Politics of Utopia* (Hutchinson, 1982), which argues that "utopianism as a tendency is a key ingredient of the whole process of modern politics" (9) and that "Utopianism depicts an ideal form of social life which, by definition, does not currently exist" (17). The authors also contend that "the Utopian enterprise is an attempt to depict the complete and concrete instantiation of these ideals in their society and the social consequences which this entails: this differs significantly from the approach which uses ideals as the light at the end of the tunnel, incapable of realization but useful to guide us in our choice of political means" (70). This incapacity is further modified later by the assertion that "Utopias are not realizable as totalities" (221) and the concession that many aspects of various aims and programmes both can and have been realised.

66. A few communes have also been called Utopia, like that founded by a group of Fourierists in Ohio in 1844.

"utopian societies".[67] When they work well, their inhabitants do usually regard them as "good places", or eutopias. Can we really deny them the right to make this claim, on the grounds of the superiority of theory to practice? We cannot. This sentiment needs to be fitted into any definition of utopianism.

Defining Utopianism: Some Components

Clarifying the approach to the forms, functions, and content of utopianism proposed here requires briefly introducing some common conceptualisations of the subject. Utopia is often identified with five things, to which it should not, however, be exclusively reduced.[68] When proposed as definitions, none of these aspects are "wrong". To identify any of them *exclusively* as "utopia" tends to ignore too many phenomena which arise in other parts of the field, making it impossible to define the "ism". What we need is a composite definition which brings every relevant aspect under its rubric while not suppressing the legitimate claim of any component to be included.

Utopia as Literary Text

Firstly, despite its close association with More's *Utopia* and the thousands of books it has inspired, utopia is not solely a literary tradition.[69] Most utopian texts describe ideal societies, and can be classed as "political novels" or "novels of ideas", where the message is more important than the means used to deliver it. Such works of fantasy, now including dystopias, might indeed be termed the fictional subset of social and political thought and are sometimes as rich in historical analysis and suggestive blueprints. But utopian literature is more than this, and cannot be reduced to its programmatic aspects or content. It is often the public face of the idea of utopia, as well as a leading focus of scholarship. Compared to academic studies or political tracts, novels possess a power and capacity for penetration other writing usually lacks. They generally

67. See Lucy Sargisson and Lyman Tower Sargent, *Living in Utopia: New Zealand's Intentional Communities* (Ashgate, 2004), 1–3.

68. Earlier and cruder versions of this argument are outlined in my "The Five Languages of Utopia", in *Spectres of Utopia*, edited by Artur Blaim and Ludmilla Gruszewska-Blaim (Peter Lang, 2012, 26–31), and in "News from Somewhere: Enhanced Sociability and the Composite Definition of Utopia and Dystopia" (*History* 98 (2013): 145–73).

69. The literature here is enormous. A good guide is John C. Olin's *Interpreting Thomas More's "Utopia"* (Fordham University Press, 1989). Recent trends are summarised in George M. Logan's *The Cambridge Companion to Thomas More* (Cambridge University Press, 2011). At last count, about 10,000 English-language utopian literary texts had appeared.

succeed better at portraying the richness and complexity of life, capturing the imagination, personalising experience, reaching our emotions, and conveying the force and majesty of utopian ideas. They make utopia human. To Sargent, utopian literature may have as many as seven purposes: it may "be simply a fantasy, it can be a description of a desirable or an undesirable society, an extrapolation, a warning, an alternative to the present, or a model to be achieved", or present an image of an attainable intentional community.[70]

As even a cursory glance at the genre reveals, the production of utopian and dystopian literary texts is a cumulative and self-reflexive process from the outset. Most such works have a self-consciously critical engagement with the tradition as a whole, from Thomas More (Plato, Sparta, Christianity) through Bellamy (Carlyle, Comte, socialism, Christianity), Wells (Plato, Bellamy), Huxley (Bolshevism, eugenics), and Orwell (socialism, Wells).[71] Depending on definition, utopian and dystopian elements are also often found interwoven, with some utopias being dystopias or containing dystopian elements, on some definitions, and vice versa, or both at the same time. No specific "critical" subgenre of utopia or dystopia thus emerges at any one point. Despite the definitive status of works like Francis Bacon's *New Atlantis* (1626) or H. G. Wells's *A Modern Utopia* (1905), moreover, the literary genre is not a subset of that related and immensely more popular genre science fiction, or of speculative fiction.[72] Some utopian novels are centred on science and technology, but many others are not. Utopian fiction is a form of fantasy literature but is closer to the "realistic" or realisable end of the spectrum, compared with the more extreme fantasy of science fiction. The latter indeed sometimes has anti-utopian implications insofar as it posits that technological solutions, like escaping to space, rather than moral and political responses can save earth or underpin the ideal society. Space invites greater displacement, not more intense engagement with terrestrial problems. Attaining utopia requires human effort in real life.

Utopia as Religion

Secondly, neither the study nor the content of utopia is a branch of theology. Utopia, if sometimes murky, generally relies on empirical premises about the possibilities of human happiness. Theology rests on unproven and unproveable

70. Sargent, *Utopianism*, 8.

71. These writers of course confronted other utopian traditions as well.

72. This is the substance of the argument in Fredric Jameson's *Archaeologies of the Future: The Desire Called Utopia and Other Science Fictions* (Verso, 2005). For elaboration, see my *Dystopia*, 284–90.

hypotheses which are accepted on "faith", without passing minimal standards of scientific evidence. Utopia and religion intermingle at many levels, however, and their relationship is sufficiently important and complex to require disentangling here. Both have idealised groups of believers, with group membership, and the desire to belong, often being much more important than dogma. In the western tradition, Christianity is one of utopia's parents, and its effort to diffuse an ethos of universal love is itself a utopian enterprise and a prototype for many which followed. Many lapsed Christians have found the concept of utopia appealing. There are many Christian literary utopias. The longest-lived forms of western communitarianism have also been religious: the Amish, Shakers, Moravians, Hutterites, and other groups. So have some of the more spectacularly unsuccessful, such as Thomas Müntzer's millenarians.

But in seeking the ideal society utopians have mostly been more concerned with promoting citizenship than sainthood, and order rather than salvation. Their aim is more often civic virtue than an inner "politics of the spirit", in J.G.A. Pocock's phrase, and their virtues do not rest on the piety of an ersatz holiness.[73] This makes utopia a moral and political, not a religious, category. Utopia is not reducible to Christ's reign on earth, or to the search for paradise or heaven in this life, or to secular versions of millenarianism. To Krishan Kumar, it is not a state of grace or the antechamber to salvation, because most utopians have been Pelagians who deny original sin and want to change people's behaviour, not human nature.[74] By contrast, the theologian Reinhold Niebuhr writes, Christian reformers reject the fundamental utopian precept that "human ills are due to bad institutions, that a fresh start with good institutions will result in a perfect commonwealth".[75]

But this too is a misconception. Utopia does not generically portray the "perfect" society, except in the very limited sense of "best possible".[76] Lewis Mumford reminds us that "The student of utopias knows the weakness that lies in perfectionism".[77] Utopia is often about *perfectibility*, in the sense of

73. *The Political Works of James Harrington*, ed. J.G.A. Pocock (Cambridge University Press, 1977), 72. The classic analysis here is Norman Cohn's *The Pursuit of the Millennium* (Secker and Warburg, 1947), after Mannheim the starting point in modern utopian studies.

74. Krishan Kumar, *Utopia and Anti-utopia in Modern Times* (Basil Blackwell, 1987), 100.

75. Reinhold Niebuhr, *The Kingdom of God in America* (Harper and Bros, 1937), 49.

76. Thus Condorcet, for instance, was working within a utopian tradition in insisting that human progress aims at "the true perfection of mankind" in his 1794 *Sketch for a Historical Picture of the Progress of the Human Mind* (Greenwood, 1979, 173). This is closely linked here with the growth of inequality within and between nations.

77. Lewis Mumford, *The Culture of Cities* (Secker and Warburg, 1940), 485.

searching for indefinite improvement. But this is a process, not an end: we never reach "perfection", which is essentially a theological category inapplicable to the "crooked wood" of humanity, from which, in Kant's famous formulation, "nothing perfectly straight can be built".[78] In More's *Utopia*, crime, imperialism, and many other evils persist. Most intentional communities, too, acknowledge that behavioural improprieties will exist and seek to regulate them, rather than assuming they can be eradicated. Disorder disturbs more utopias than is usually presumed. It is just better managed than elsewhere. Attempts to remove it completely thus result from confusing utopia and religion. Identifying the two invites making perfection, often described in terms of some form of salvation, the aim of utopia. Once we sever utopia from "perfection", however, its attainability, as "the good" place, eutopia, also becomes plausible. And so by aiming for less we gain more.

Disentangling utopia from religion, and especially millenarianism, given their long and intimate historical relationship from at least the fourteenth century, is a daunting if crucial task, which is attempted in greater detail below.[79] Here we can briefly unpack the definitional implications of its affinity to utopia. Millenarianism is a concept rich in meaning but subject to much confusion. In early modern Europe, utopia and the millennium are often found closely linked. To John Passmore, instancing the Ranters in the English Revolution, "Utopian communism and the doctrine of sinless perfection were . . . conjoined in a single stream".[80] But we must be wary of bald generalisations such as that, by the nineteenth century, "'Utopias and dystopias' are, in modern times, secularized heavens and hells, metaphors of eternity rather than mortality", and that "The ideas of 'progress' and 'utopia' had emerged to supplant the sacred myths of heaven and the millennium".[81] Other writers allege that "millenarians are generally thought of as utopian thinkers or political radicals."[82] The hypothesis, as J.F.C. Harrison presented it, "that Utopia may be a secular equivalent of the millennium or, alternatively, that the millennium may be a religious form of Utopia", turns out to be a minefield of

78. Immanuel Kant, *On History* (Bobbs-Merrill, 1963), 17–18.

79. A good account of these languages in relation to utopia is Alfred Braunthal's *Salvation and the Perfect Society: The Eternal Quest* (University of Massachusetts Press, 1979).

80. John Passmore, *The Perfectibility of Man* (Duckworth, 1970), 143.

81. W. Warren Wagar, *Terminal Visions: The Literature of Last Things* (Indiana University Press, 1982), 6, 61.

82. Jean-Robert Armogathe, "*Per Annos Mille*: Cornelius a Lapide and the Interpretation of Revelation 20:2–8", in *Catholic Millenarianism: From Savonarola to the Abbé Gregoire*, ed. Karl A. Kottman (Kluwer, 2001), 53.

definitional and philosophical issues.[83] Kumar writes that "the millennium is not Utopia. Its ideal order is predetermined. It is brought in by divine intervention. Human agency remains questionably relevant. The millennium is not, as is Utopia, a scheme of perfection to be realized—if at all—by conscious rational human action."[84] To Miriam Eliav-Feldon, "Genuine utopists do not indulge in fantasies about unattainable Gardens of Eden, but propose practical, though sometimes very drastic remedies for the defects of their societies." Thus, Renaissance utopians were "'Realistic' because they were not dependent on any supernatural conditions or on any divine intervention which would change the cosmos, human nature, or the course of history."[85] To Chloe Houston, by the 1640s, "*Utopia* was read by some not as an ironic jest but as a realistic model for the ideal society, albeit one that needed improvement, and utopian plans seemed timely and appropriate."[86] J. C. Davis's influential fivefold typology of early modern utopianism, which includes the "modes" of arcadia and cockaygne, sharply distinguishes among utopia, which "accepts deficiencies in men and nature and strives to contain and condition them through organisational controls and sanctions"; the millennium, which "assumes a coming state of redeemed and perfected men, restored to their prelapsarian command over nature, but such things cannot come from fallen man himself, only from a *deus ex machina*"; and the perfect moral commonwealth, where people are idealised.[87]

83. J.F.C. Harrison, "Millennium and Utopia", in *Utopias*, eds Peter Alexander and Roger Gill (Duckworth, 1984), 61. Harrison argues that in the eighteenth century "The millennium was secularised into a Utopia or perfect state of society, to be attained through a gradual and steady march of improvement" (65).

84. Kumar, *Utopianism*, 36.

85. Miriam Eliav-Feldon, *Realistic Utopias: The Ideal Imaginary Societies of the Renaissance, 1516–1630* (Clarendon, 1982), 129: "In this study 'utopia' will be used in a rather narrow sense to mean only *a literary work describing an ideal society created by conscious human effort on this earth*. This definition excludes any vision of an ideal existence that is other-worldly, unattainable, or dependent either on wonders of nature or on divine intervention. It also excludes states of bliss of one individual or other forms of personal salvation, as well as recommendations for piecemeal reforms and schemes for the amelioration of the human condition in any one particular field" (2).

86. Chloë Houston, *The Renaissance Utopia: Dialogue, Travel and the Ideal Society* (Ashgate, 2014), 163.

87. J. C. Davis, *Utopia and the Ideal Society: A Study of English Utopian Writing 1516–1700* (Cambridge University Press, 1981), 370. The other forms explored by Davis are arcadia, "a world of natural beneficence and human benevolence, where the deficiencies of both man and nature are made good in an atmosphere of calm and gentle fulfilment", and cockaygne, which

Life is nonetheless messier than theory, and these themes have long intermingled. The leading historians of western utopianism, Frank and Fritzie Manuel, view images of Eden, heaven, and the millennium as "a constant presence—in multiple variations—in all subsequent Utopian thought" after the Renaissance. This is evident in both confidence about the new order's inevitability, "a carry-over of millenarian certainty", and a common core of longing or desire.[88] Many Christian communitarians have been chiliasts or millenarians, expecting Christ's imminent return to earth. The early nineteenth-century Rappites expected this in 1829, and asserted that "A harmonious and united society of men may be said to be a Kingdom of God". John Humphrey Noyes's "Bible Communism" or "Perfectionism" at Oneida later in the century provided another well-known communitarian success story.[89] Christian communalists have often sought what Niebuhr calls "the unqualified realization of the Kingdom of Christ in history, which usually means the reconstruction of human society into a commonwealth of perfect love or perfect equality or perfect liberty."[90] Some enthusiasts today still aim to yoke utopia to religion, and insist with Martin Buber that "socialism without religion is body emptied of spirit, hence also not genuine body."[91]

Explaining post- or non-religious utopianism in terms of an inheritance of religious content requires confronting the murky concept of secular millenarianism. Many millenarians have had explicitly secular goals; this earthly dimension separates them from attempts to attain salvation in an afterlife. To Richard Landes, "millennialism" describes a "perfect and just society on earth (however defined), and thereby, collective salvation for its inhabitants".[92] To Eric Hobsbawm, millenarianism's "essence" lies in "the hope of a complete and radical change in the world". This includes "total rejection of the present, evil

---

"presumes no deficiencies in nature but an abundance distinguished by the capacity to satisfy the grossest appetite, leaving all men replete" (370).

88. Frank E. Manuel and Fritzie P. Manuel, *Utopian Thought in the Western World* (Basil Blackwell, 1979), 6, 32–33.

89. Quoted in A. E. Bestor, *Backwoods Utopias: The Sectarian Origins and Owenite Phase of Communitarian Socialism in America, 1663–1829*, 2nd ed. (1950; University of Pennsylvania Press, 1970), 5.

90. Reinhold Niebuhr, *Faith and History: A Comparison of Christian and Modern Views of History* (Nisbet, 1949), 235.

91. Martin Buber, *Paths in Utopia* (Beacon, 1958), xxv. See further Krishan Kumar, "Religion and Utopia", in *The Canterbury Papers: Essays on Religion and Society*, ed. Dan Cohn-Sherbok (Bellew, 1990), 69–79.

92. Richard Landes, *Heaven on Earth: The Varieties of the Millennial Experience* (Oxford University Press, 2011), 339.

world," coupled with "a fundamental vagueness about the actual way in which the new society will be brought about" and "a fairly standardized 'ideology' of the chiliastic type."[93] Such descriptions might plausibly cover many nineteenth- and twentieth-century revolutionary movements, including some strands of Marxism and anarchism. To Ludwig von Mises, Marx's view that "Socialism is made to appear as the inevitable goal and end of historical evolution" expressed a "socialist chiliasm" which invoked ideas of both an original Golden Age and a future state of perfection and salvation.[94] Frederik Polak, too, writes that "Chiliastic and eschatological elements were . . . still present, in a secularized form, in the utopian, prophetic, and messianic thought of Karl Marx, though he considered himself a declared enemy of utopian socialism."[95]

Various writers then blame totalitarianism on the persistence of this mentality amongst twentieth-century revolutionaries. Here, attempts to realise "heaven" on earth are misbegotten, a tyrannical and procrustean imposition of the idea on poor humanity. To Landes, "Most forms of secular millennialism—especially communism—anticipate a world of perfection never before realized, indeed never before possible", and Marx is a "secular Joachite" (after Joachim of Fiore, whom we meet below) who promised "secular salvation in Communism". On this reading, totalitarian tyranny commences by assuming one true world view. Lenin's claim that the Party could exercise limitless coercion stems from the belief in its absolute correctness and the certainty of its providing unparalleled benefit to humanity. So pursuing a "final, perfect" solution necessarily implies totalitarianism.[96] To Nicolas Berdyaev, "A utopia always includes a project for the complete, totalitarian ordering of life", and "Utopia is always totalitarian, and totalitarianism, in the conditions of our world, is always 'Utopian.'"[97] To Leszek Kolakowski, too, utopias are "visions

93. Eric Hobsbawm. *Primitive Rebels: Studies in Archaic Forms of Social Movement in the 19th and 20th Centuries* (Manchester University Press, 1959), 57.

94. Ludwig von Mises, *Socialism: An Economic and Sociological Analysis* (Jonathan Cape, 1936), 280–81.

95. Frederik L. Polak, "Utopia and Cultural Revival", in *Utopias and Utopian Thought*, ed. Frank E. Manuel (Beacon, 1967), 289.

96. Landes, *Heaven on Earth*, 27–28, 292, 340. But this implies that some forms of communism do not make this promise.

97. Nicolas Berdyaev, *Slavery and Freedom* (Centenary, 1944), 206; Berdyaev, *The Realm of Spirit and the Realm of Caesar* (Victor Gollancz, 1952), 175. Cf. Berdyaev, *Slavery and Freedom*, 206: "A utopia is the distortion in human consciousness of the Kingdom of God." Further: "Utopia is nothing more than a perversion and distortion of the religious faith in the coming of the Kingdom of God on earth, the grotesque rationalization of an unconscious millenarianism" (*The Meaning of History* (Centenary, 1945), 191).

of a perfectly unified society . . . where there is nothing to correct any more". They embody "a desperate desire to attain absolute perfection", which "is a degraded remnant of the religious legacy in nonreligious minds".[98] Indeed, Kolakowski's rendering of this issue directly addresses a key theme here. In his view, utopians "want to institutionalize fraternity", but "an institutionally guaranteed friendship . . . is the surest way to totalitarian despotism." A "conflictless order" can exist only "by applying totalitarian coercion".[99] Identifying utopianism with millenarianism thus sanctions the most important anti-utopian argument of all: utopia represents an impossible search for perfection which necessarily eventuates in intolerance, violence, and totalitarianism. So to John Gray "the logic of utopia" produces "universal despotism".[100] This accurately describes some strands of the tradition, but not all. For many utopians accept disorder, and not only respect liberty, but want the many, not only the few, to possess it. And while it may be impossible to "guarantee" friendship and similar relationships, we may do much to nourish them.

Utopia as Mental State

Thirdly, utopia is not reducible to a state of mind, a psychological impulse or propensity, a personality type, a "desire" or "principle",[101] or a deviant or pathological form of extreme fantasy. It is not an "obsession by the idea of regulating the life of the world by reason", as Berdyaev insisted in 1937, foreseeing the main line of attack on all planned economies thereafter.[102] Utopians are not generically bewildered or demented. No "utopian mind" exists in the sense of a pathological disposition to heretical extremism or a refusal to accept the

98. Leszek Kolakowski, "Need of Utopia, Fear of Utopia", in *Radicalism in the Contemporary Age*, vol. 2, *Radical Visions of the Future*, ed. Seweryn Bialer (Westview, 1977), 11; Kolakowski, *Modernity on Endless Trial* (University of Chicago Press, 1990), 132; Kolakowski, *My Correct Views on Everything* (St. Augustine's, 2005), 43.

99. Kolakowski, *Modernity on Endless Trial*, 139, 143. The argument here turns largely on two assumptions, firstly that "human needs have no boundaries we could delineate; consequently, total satisfaction is incompatible with the variety and indefiniteness of human needs" (138), and secondly that opposition to "The utopian dogma stating that the evil in us has resulted from defective social institutions and will vanish with them is indeed not only puerile but dangerous; it amounts to the hope, just mentioned, for an institutionally guaranteed friendship" (143).

100. John Gray, *The Silence of Animals: On Progress and Other Modern Myths* (Allen Lane, 2013), 52.

101. F. Jameson, *Archaeologies of the Future*.

102. Nicolas Berdyaev, *The Destiny of Man* (Centenary, 1937), 232.

"normality" of the status quo.[103] Karl Mannheim does speak of a "utopian mentality", or predisposition to thinking in utopian terms. This meant being "guided by wishful representation and the will to action", a process which "hides certain aspects of reality. It turns its back on everything which would shake its belief or paralyse its desire to change things". This process is also identified with the "collective unconscious", a concept whose implications for ideas of free will and an uncomfortably limited knowledge of our inner mental processes often make us squirm today.[104]

We might readily concede that a desire for equality, perhaps instinctive in origin, has underpinned much utopian sentiment over the centuries. The same can be said, though perhaps to a lesser degree, for liberty, progress, and other concepts. An innate desire for utopia can also be equated with what Bernard Mandeville, whom we meet later, called the wish to "make Men completely Happy upon Earth", a variant on that "restless Desire of mending their Condition" he thought could "never be obtain'd whilst the World stands".[105] Some, like Martin Buber and Ernst Bloch, posit a "principle of hope" or "wish-picture" where utopia functions as a fantastic longing or desire.[106] (Bloch is examined in detail below.) This too implies that the subject rests on an emotional substratum or semi-conscious dreamworld which instigates and structures our ideas of personal and social improvement. Utopia has also been described as the "education of desire", in a phrase adopted by E. P. Thompson from Miguel Abensour, where, it is suggested, the aim is to "teach desire to desire, to desire better, to desire more, and above all to desire in a different way". But it is unclear what "better" or "different" mean here, or when more is indeed desirable at all.[107]

Let us give these propositions their due. Psychology is central to understanding utopia. The heightened senses of expectation, hope, dream, and fantasy endemic to utopian thought and action are complex and well worth

103. This is an inversion of Stalinism's insistence that dissent from the system can assume the form only of mental illness. See Aurel Kolnai's *The Utopian Mind* (Athlone, 1995), where the accusation again hinges on utopia's obsession with "perfection".

104. Mannheim, *Ideology and Utopia*, 36. Here "ideologies", which aim to defend a given social order, are contrasted with "utopias", which propose to alter it dramatically (173–236).

105. Bernard Mandeville, *Enquiry into the Origin of Honour* (J. Brotherton, 1732), 15–16.

106. See Buber, *Paths in Utopia*, and Ernst Bloch, *The Principle of Hope*, 3 vols (Basil Blackwell, 1986). Ludwig Feuerbach's idea of God as a projection of human desire, and of love as the essence of Christianity, formed the methodological starting point for Marx's theory of alienation in the "Paris Manuscripts" of 1844.

107. E. P. Thompson, *William Morris: Romantic to Revolutionary*, 2nd ed. (1955; Merlin, 1977), 791.

scrutiny. "Dreaming" is a word "utopian" is frequently yoked to. "Social dreaming" is a subset of those fantasies, future-oriented projections, and nostalgic reflections which populate our mental world. To H. G. Wells, "All life has something dreamlike in it. No percipient creature has ever yet lived in stark reality. Nature has equipped us with such conceptions and delusions as survival necessitated, and our experiences are at best but working interpretations."[108] We experience everyday reality simultaneously in anticipation, in experience, and in retrospect. At any one moment we may be confronting the present, projecting into the future, and recalling, rewriting, or idealising the past. We constantly carry within us images of some improved condition or moment—a promotion at work, kicking the boss in the rear, a blissful holiday, the transcendent landscape, the ideal partner, Christmas, the Revolution which sweeps evil away—as well as nostalgic recollections of better moments or mythical times gone past.

These images populate an inner keep in our brains we might call "Dreamland", an interior good place or eutopia, where reality is negated or inverted, and where we retreat whenever circumstances demand it, slamming the gate shut and pulling up the drawbridge. In this happy (or happier) space, which is also a kind of greenhouse of ideas, imagined improvements are conceived and nurtured. At some points in our lives we may spend most of our time in Dreamland. But there is nothing intrinsically utopian, in the social sense, about this space: quite the contrary. For here ego triumphs, and we are kings or queens of our own castle. And here, too, in our inner sanctum, we are never completely safe from our inner demons, who dwell beneath, or thus free from anxiety, or the qualms of conscience, or suppressed urges, which are unconscious. Nonetheless escapism is a wonderful thing, and sometimes the imaginary world in our heads is the only fortress withstanding the recurrent pressures of reality. In this sense, as an inner mental space where we flee from oppressive reality, and which we use as a critical reference point, utopia, where again we possess autonomy and vanquish our foes, is indeed a state of mind.

Psychologically, utopia is thus at least initially compensatory rather than emancipatory. Illusions are often preferable to cold truths. Mannheim asserts that "When the imagination finds no satisfaction in existing reality, it seeks refuge in wishfully-constructed places and periods. Myths, fairy tales, otherworldly promises of religion, humanistic fantasies, travel romances, have been continually changing expressions of that which was lacking in actual life."[109]

108. H. G. Wells, *The Autocracy of Mr. Parham: His Remarkable Adventures in This Changing World* (William Heinemann, 1930), 283.

109. Mannheim, *Ideology and Utopia*, 184.

But "utopia" has a further function, as "that type of orientation which transcends reality and which at the same time breaks the bonds of the existing order". Its "wish-image" has a potentially revolutionary function.[110] We imagine justice for ourselves. Then, realising that this can be achieved only collectively, we take the leap to extend it to everyone else. Now our personal Dreamland or eutopia becomes the social space, utopia.

When this space is emancipatory, its individual and social sides also meet in a desire to be wanted, to feel at home and freed from the existential angst of isolation and loneliness. The concept of "belongingness", which we will shortly explore, when conceived as a more powerful desire than almost any other, captures these meanings and is crucial in defining utopia. It is the goal for which sociability is the means. It is the chief antidote to that alienation so often identified with the moderns.[111] The "modern mind" has been described as typically a "homeless mind", a condition so "psychologically hard to bear" that it induces a "permanent identity crisis". It "has therefore engendered its own nostalgias—nostalgias, that is, for a condition of 'being at home' in society, with oneself and, ultimately, in the universe".[112] This sense of "homesickness" was the core of the term "nostalgia", which first appeared as early as 1688. Bewildered by the blizzard of impulses the modern world creates, we feel we have lost both a unity with our community and a wholeness in our inner selves, and long to retrieve both.[113] Here a *Heimat*—the German term evokes a richness, depth of feeling, and existential longing lacking in English—or home, now lost to some other group, or just to time, easily becomes the focus of imaginary virtues. Where homesickness or *Heimweh* lacks a definitive, objective past or place upon which to focus, it may be preferable to conceive our imaginary home as a future utopia, where *Heimatslosigkeit*, the feeling of angst or loss, is conquered.[114]

This desire may thus be projected outward into space, or forwards or backwards in time. Like a quest for enlightenment in the mystic East, or elsewhere,

110. Mannheim, 171.

111. Though not of course other forms of alienation defined by the exploitation of labour, the work process, and class.

112. Peter L. Berger, Brigitte Berger, and Hansfried Kellner, *The Homeless Mind: Modernization and Consciousness* (Pelican Books, 1974), 74, 77. Martin Heidegger is often quoted as saying, referring to Novalis, that all philosophy is actually homesickness, "an urge to be at home everywhere."

113. For a study of the loss of the sense of place, see K.D.M. Snell's *Spirits of Community: English Senses of Belonging and Loss, 1750–1900* (Bloomsbury, 2016).

114. *Heimatslosigkeit* has no exact English equivalent, since "homefulness", sadly, is not a word and "homelessness" means simply being forced through poverty to live outside of a dwelling. Hence the use here of "belongingness", despite its awkwardness.

it may represent a search for something which can be found within us, or nearby, but which it is convenient to displace elsewhere. (For just as it was easier to envision utopia in a new world than it is to create one in Europe, it is easier to travel to India than to face up to our inner psychic life and intellectual or emotional confusion.)[115] It may be essentially social and express our need for others. Unfortunately, dissimulation and self-deceit are common here. Memory is not independent of will and is often sadly subservient to it. We prefer to construct happy stories than to confront unpleasant truths. Individually we have a natural tendency to force bad memories out with good, even to manufacture the latter when necessary. We ignore the shadows on trips down memory lane and remember the sunshine streaming down on our faces in a state of what seemed like eternal youth. We suppress the tearful byways and spotlight our triumphal marches down the highways.

The rewriting of history we call nostalgia, which nationalism demands, plays a similar role, leaving us trapped in a superior imaginary past. Denial and displacement are again key here. Increasingly, suggests Juliet B. Schor, we "yearn for what we see as a simpler time, when people cared less about money and more about each other".[116] Susan Stewart sees such nostalgia as a "social disease" which seeks "an authenticity of being" through presenting a new narrative, while denying the present.[117] Locating utopia in the past is easy because memory is often faulty and selective, and we can project backwards to an ideal starting point without worrying about its accuracy. Whole nations feel a romantic nostalgia, "a painful yearning to return home", for their lost golden ages of innocence, virtue, and equality, and for their mythical places of origin. Many of the largest lies are thus told by nationalists. The glorious fictional history of the imagined nation is often preferable over its more likely inglorious and bloodstained real past, as modern debates over imperialism and the statues of heroic conquerors and defenders of slavery make abundantly clear. This is analogous to individuals recalling their youth as the best period of their lives, the "good old days", or "going back home" to a place one has long since left and which is irredeemably changed, almost invariably for the worse when viewed through rose-tinted lenses.[118]

115. My own odyssey is recounted in a TEDx talk available on YouTube: Gregory Claeys, "Why Are Utopias Important for Human Mankind?" TEDx Talks, YouTube, 28 June 2019, https://www.youtube.com/watch?v=ouDdiJ0ey00.

116. Juliet B. Schor, *The Overspent American: Upscaling, Downshifting, and the New Consumer* (Basic Books, 1998), 24.

117. Susan Stewart, *On Longing* (Johns Hopkins University Press, 1984), 23.

118. Fred Davis, *Yearning for Yesterday: A Sociology of Nostalgia* (Free Press, 1979), 1. A well-known twentieth-century literary example is Thomas Wolfe's *You Can't Go Home Again* (Harper and Row, 1940).

Such projections divulge some interesting psychological secrets. The idea of returning to a more primitive condition as some sort of original home or starting point may be viewed primarily as a wish to expunge our sins or mistakes, to cleanse our consciences, to simplify our lives, and to start anew. With Jean-Jacques Rousseau, particularly, in mind, we will see, we do not really wish to return to simplicity, only to use the image to gain forgiveness and forgetfulness, to escape the consequences of our actions, to erase guilt, and to avoid the burdens and clutter of complexity, while retaining the advantages of civilisation. We want momentarily to cast aside our artificiality, while keeping our place in the pecking order. This quest is an exercise in forgetting, and in that well-practised art of having your cake whilst eating it too. So at this level, amongst others, utopia demonstrates a vital facet of the psychological history of humanity.

This psychological component is so all-pervasive that it must be included in any definition. Often associated with utopianism, and even defined as its "essence", "desire" is the language of instincts and propensities, or human nature. But while it is a vital function of utopian thinking, utopia cannot be reduced to "desire". This dimension comprises both the collective unconscious "motivations and presuppositions" which concerned Mannheim, and what today we more loosely refer to as emotions. In terms of content, it is the mood which indicates how far we feel at ease in our surroundings. We can liken this emotional substratum to a soundtrack playing in the background of everyday experience, though not moments of exuberant celebration. In utopia this is pleasingly calm and melodious, and reflects our inner peace and harmony.[119] In dystopia it is a cacophonic clashing of symbols and of discordant notes, and mirrors our inner anxieties and loneliness.

Utopia as Progress

Fourthly, utopia is not synonymous with social improvement as such, no matter how far-sighted or exotic proposals for amelioration may be. The long tradition of thinking about ideal cities, in particular, often includes stunning depictions of beautiful, symmetrical, harmonious piazzas, towers, and squares conceived on a dramatic scale, ecological plans for self-sufficient environments, and so on.[120] Such pleasing visions might be practicable at some point. Here, as is detailed in the discussion of urban life in part III, they are defined

119. My personal favourite: Neil Young's "Everybody Knows This is Nowhere".

120. A fine portrayal of this tradition is Ruth Eaton's *Ideal Cities: Utopianism and the (Un) built Environment* (Thames and Hudson, 2002).

as utopian only insofar as they imply improving human behaviour by enhancing sociability *as a consequence of such designs.*[121] To Kolakowski the same is true for technological utopias, which remain mere fantasies unless they "suggest the idea of an ultimate solution of mankind's predicament, a perfect satisfaction of human needs, a final state."[122] Reorganising or beautifying space as such, as opposed to moralising it, no matter how extensive, sublime, or awe-inspiring, is not sufficient to trigger the "utopian" label.[123] So utopia is not synonymous with "progress", and the desire to make society "better" tells us nothing about the distinctiveness of the utopian.

Utopia as Pleasure

Fifthly, and finally, utopia is also not indefinite leisure, immersion in hedonism, or a complete escape from work and drudgery. Of course, utopia implies greater happiness for most people. But happiness involves a respite from challenges and difficulties, or pain and discomfort, not their disappearance. As George Orwell insisted, "the highest happiness does *not* lie in relaxing, resting, playing poker, drinking and making love simultaneously . . . For man only stays human by preserving large patches of simplicity in his life, while the tendency of many modern inventions—in particular the film, the radio and the aeroplane—is to weaken his consciousness, dull his curiosity, and, in general, drive him nearer to the animals."[124] Endless sun-filled carefree days at the beach under tropical palms or at the Mediterranean coast, free of all responsibilities, with the related association of sex, relaxation, intoxication, and oblivion, may constitute one modern holiday ideal, as we maximise our distance from the woes of everyday life. But this is not utopia.[125] Though it too has ecstatic moments, and ideally a fair modicum of pleasure-seeking, utopia is a condition of social stability in which happiness derives from various sources, and not merely bodily pleasure. It rests on a principle of organisation which prioritises equality and close association. It requires complex organisation and confronting many challenges, not fleeing from them or dumping them on others. Making utopia work is actually itself hard work. Indeed, obsessively

121. The same argument holds against current proclamations that the internet is somehow "utopia".

122. Kolakowski, *Modernity on Endless Trial*, 132.

123. On the distinction, see Matthew Wilson's *Moralising Space: The Utopian Urbanism of the British Positivists, 1855–1920* (Routledge, 2018).

124. George Orwell, *Smothered Under Journalism: 1946* (Secker and Warburg, 1998), 32.

125. For this image, see Lena Lenček and Gideon Bosker's *The Beach: The History of Paradise on Earth* (Secker and Warburg, 1998).

satisfying bodily pleasures may well interfere in its attainment—and certainly does where sustainability is a key goal. Ecstasy is not definitive of the experience. Even happiness does not define utopia as much as contentment, freedom from fear, and the satisfaction of the sense of belonging.

Two Further Problems

Luxury, Consumerism, and Sustainability

Some knowledge of these components of utopianism is essential to understanding its many meanings. This book also explores two other approaches to the subject. Firstly, it reconstructs narratives of utopian thought and action in terms of the key principle urgently demanded now, sustainability, particularly by addressing both theoretical and practical utopian treatments of luxury and consumption. Secondly, portraying utopianism as realistic requires defining a realisable form of enhanced sociability centred on the concept of belongingness. This entails analysing types of association, including family, friendship, and groups, to see what utopian sociability might consist in. Let us briefly introduce each of these propositions.

The problem of luxury is often ignored in accounts of utopianism but is central to the narrative presented here.[126] Sociability and equality are valuable ends in themselves and in less critical times have served as such. But they must now be wedded to a higher priority, sustainability, which means abandoning lifestyles defined by consumerism. The lust for luxury is a leading cause of our current environmental debacle, because it fosters excessive and unnecessary consumption. The word "luxury" derives from two Latin sources, one (*luxus*) meaning splendour, pomp, and sensuality, the other (*luxuria*) implying riot, excess, and extravagance.[127] The term hints at illicit pleasures, self-indulgent idleness, and the existence of rigid class inequality. It is a relationship with people as well as things. It implies wanting things just because others have them, or because they cannot have them, and expending money and consuming goods and services to excess, often just to prove we can afford to waste things. This is the mentality we most need to reject, and it is explored in detail below.[128] Luxury is usually linked to avarice, vanity, envy, pretentiousness, and one-upmanship, to

126. It is not discussed in detail by Kumar, Levitas, or Sargent, though there are many references to the issue in Frank and Fritzie Manuel's *Utopian Thought in the Western World*, and in many accounts of early modern utopias.

127. Peter McNeil and Giorgio Riello, *Luxury: A Rich History* (Oxford University Press, 2016), 17.

128. E. J. Urwick, *Luxury and Waste of Life* (J. M. Dent, 1908), 213.

the desire to flaunt our superiority and to celebrate social inequality. Luxury by definition is what only the few possess, and indeed defines class more than any other concept. But everyone loves it. So here we have a problem.

At first glance, utopianism and luxury might seem to be prima facie incompatible. Utopians like Lycurgus and More portray the desire for luxury as subversive of order, peace, justice, and equality. Its restraint, we will see, was central to the tradition from the sixteenth to the later nineteenth century. Private luxury and the idea of unlimited material abundance were usually deemed incompatible with utopian order. Then a new paradigm emerged which proposed universalising luxury instead of abolishing it, primarily through a shift from private to public luxury. Nonetheless, sociability and equality still played key roles in the new understanding of luxury.

These facets of utopianism make it most relevant to our future, not merely its portrayal of "better" alternatives, which is too imprecise to be helpful. Renouncing the image of unlimited plenitude which has dominated modernity and adopting a sustainable, anti-consumerist lifestyle involves embracing the idea of a "stationary state" or "steady-state" economy. Moving beyond "fairytales of economic growth", in Greta Thunberg's apt phrase,[129] will be vastly more attractive, however, if this state is presented not as a regime of abstinence and painful withdrawal but as a condition, "degrowth" notwithstanding, in which the majority are happier than they are now, or than they will be if we do not dramatically reform our societies in the near future. This book aims to show how this might be possible without destroying much that we value in modernity.

Enhanced Sociability and Belongingness

Much of this prospect lies in our ability to promote a stronger sense of collective identity and sociability, and to make these pleasurable and meaningful to most people without compromising our need for privacy and individuality. So a second major aim here is to portray sociability and belongingness, which are often marginal to or even absent from accounts of the subject, as central goals of utopianism, and as the generic content linking most of its expressions.[130] Utopianism is understood here as both projecting and practising enhanced sociability, where our experience as social beings greatly exceeds the norm in

129. At a United Nations speech, 23 September 2019.

130. The concept of sociability is not treated by Kumar, Levitas, or Sargent. Its importance is most commonly noted in approaches rooted in intellectual history, like that of the Manuels, who discuss it in relation to Kant, Morelly, and others. Belongingness is rarely discussed in the secondary literature.

our own societies, and so, to brand the experience, is *utopianised*.[131] "Enhanced" means greater kindness, mutual affection, amicability, togetherness, solidarity, decency, respect for individuality and difference, a willingness to play by the rules, and much more. The role played by groups, and the desire to belong to them, or "groupism", as the French socialist Charles Fourier (discussed below) termed it, is central to this process. Different types of community and association are capable of sustaining diverse levels of sociability and types of relationship for varying lengths of time. Social interactions are as vital to our identity and stability as our prized sense of our unique self and its individuality. We need others to support us. We bask in the glow of their warmth when they do and wither when they do not. Money does not buy us happiness: rich people often feel friendless because they buy what others barter or give as love or services, and the suspicion that what they receive is mere pretence soon becomes evident. Sociability, which must be paid in kind, does the job much better. So psychologists tell us that "happiness depends on relationships more than anything else".[132] Social interactions are "the single greatest cause" of our feeling of well-being.[133] Market relations are "cold", in Robert E. Lane's phrasing, because they are calculating and commodified. Social relations are "warm" when they involve trust, openness, acceptance, and minimal exploitation.[134] They anchor us in a stormy world and are our best emotional safeguard against illness, stress, loneliness, and anxiety.

BELONGINGNESS

The feeling such sociability induces can be best understood through the concept of belongingness, a subcategory of group psychology which focuses on the desire to feel included. This reflects a deep-seated need first detailed by the psychologist Abraham Maslow, who believed it was superseded in importance only by the requirement of safety, and that when unsatisfied, and manifested as loneliness, a want of "at-homeness" caused "maladjustment and more severe

131. Credit for the verb is uncertain, but thanks to @Karroota and Philipp Thapa anyway.

132. Steve Duck, *Friends, for Life: The Psychology of Personal Relationships*, 2nd ed. (Harvester Wheatsheaf, 1991), 1.

133. Michael Argyle, *The Psychology of Happiness*, 2nd ed. (1987; Routledge, 2001), 71. On this theme see, e.g., Edward W. Soja's *Postmodern Geographies: The Reassertion of Space in Critical Social Theory* (Verso, 1989) and *Thirdspace: Journeys to Los Angeles and Other Real-and-Imagined Spaces* (Blackwell, 1996).

134. Robert E. Lane, "The Road Not Taken", in *Ethics of Consumption: The Good Life, Justice, and Global Stewardship*, eds David A. Crocker and Toby Linden (Rowman and Littlefield, 1998), 225–29.

pathology".[135] As elemental as "our need for water", to Kelly-Ann Allen, it is so fundamental that its manifestations often passed unnoticed.[136] Some see the need to belong as the primordial source for our desire for power, intimacy, approval, and much else. It commences in infancy, drives our willingness to conform through life, and may haunt us in our dotage. It is reflected in our attachment to places as well as people, and in our attitudes towards objects, and extends by association to all our senses, including smell and taste. The sense of belonging or connectedness is a crucial component in solidarity and has even been portrayed as the basis of morality as such.[137] Everyone has experienced the anxiety of feeling lonely, abandoned, ignored, friendless, rejected, shunned, dispossessed, foreign, alien, and alienated. Not being part of a group we aspire to join, and having our "I" become "we", can be devastating. Exclusion cuts us to the bone. No, you can't sit with us. No, we are not talking to you. No, you are not invited to our party. No, you cannot join our club. A word spoken or unspoken, an innuendo or snicker, the cold shoulder, a glance askance, the door closed as we approach, the seat not offered to us at the table, may make us feel unwanted. Why? we ask ourselves. Is it our accent, our body, our clothes, our hairstyle, our gender, our race, our opinions? Fear of rejection may become a phobia, like "impostor syndrome", which disables our social skills. It may paralyse our intellect, as in writer's block. So desperate is the need to be liked, and acknowledged, that the pain which the punishment of rejection entails may cause some of the deepest anxiety we ever feel. Unease may occur even in the best of circumstances, and may poison every other part of our lives. Status-consciousness and status-obsession, with which the need for belonging is wrapped up, intensify these fears greatly and produce a deficit of solidarity, as most intellectuals are acutely aware.

The feeling of belongingness is the main antidote to these fears and can be nurtured with relatively little effort. The utopian question is how to maximise this quality while heeding its limits, since exceeding them can be destructive.

135. Abraham H. Maslow, *Motivation and Personality*, 3rd ed. (Longman, 1987), 20–21, 43. The German term is even more precise: *Zusammengehörigkeitsgefühl*, which gives a sense of "togetherness" as well as belonging. A summary of the psychological literature is given in R. F. Baumeister and M. R. Leary's "The Need to Belong" (*Psychological Bulletin* 117 (1995): 497–529), which suggests that "much of what human beings do is done in the service of belongingness" (498).

136. Kelly-Ann Allen, *The Psychology of Belonging* (Routledge, 2021), 1.

137. B. F. Skinner insists that "A person does not act for the good of others because of a feeling of belongingness or refuse to act because of feelings of alienation. His behaviour depends upon the control exerted by the social environment" (*Beyond Freedom and Dignity* (Penguin Books, 1973), 110).

The utopian answer lies as much with public as private states of being.[138] Friends and family may supply a soothing, supportive, reassuring sense of belonging, and so may groups we identify with. We all recognise the sense of relief and the feeling of warmth which follows as we immerse ourselves in a friendly crowd and relax securely enveloped in the protective cocoon it provides. A sense of gratifyingly close, shared communality often emerges in ritual celebrations like festivals, or other forms of conviviality, which heighten the excitement of intimate contact, and in demonstrations, strikes, and many other forms of collective activity and organisation. Where it exists, the utopian component in our interactions lies in an at least temporary suspension of the unequal, alienating, and repressive social structures which otherwise define everyday life much of the time. As we foster unity, harmony, peace, trust, and togetherness, if only by shouting slogans together, we glimpse utopia in momentarily freeing ourselves from oppression and coercion. Some examples of the process are considered below.

The argument with respect to sustainability presented here requires recognising that this need for belonging can also be satisfied, at least partly, through the ego-reinforcing aspects of property and commodity ownership. Modernity has entailed increasingly defining ourselves by and through things, which we view as decorative and as moral and status extensions of our ego or sense of self. As an empire swells a national ego, property magnifies the individual's. "Look at me!" we say, as we admire our array of possessions, "I am all these things." The wealthier we are, the more we acquire things to display our grandeur and success. Commodities provide a sense of home akin to that which people and place may offer. Possession defines a domain of our sovereignty and psychological territory, and our safety zone is bounded by things we own, instead of our belonging to a place or to a group. Here, the scaffolding which props up our ego is commodified and interpreted in terms of power over material objects, rather than constructed from our human relationships. Our human essence thus comes to be expressed as a relationship with things rather than with people either individually or collectively, as "community". We are externalised in them, and our human attributes are rendered lifeless, or rather, they have been devoured, vampire-like, by objects. This is still a form of self-realisation, of course, but the "self" realised is also dead from a human viewpoint, as our devotion to objects drains us of our sociability. This imperious attitude towards things fuels consumerism. "Commodification", object overload, or "stuffication" naturally occurs as a result of growing individualism and the decline of family bonds.

138. For an exploration of this theme, see Richard Sennett's *Families against the City: Middle Class Homes of Industrial Chicago, 1872–1890* (Harvard University Press, 1970).

Under constant pressure to see ourselves as validated by consuming, we persuade ourselves that things can do what people seemingly cannot, or at any rate less successfully—namely, satisfy our emotional needs.

Superficially this is not as foolish as it may appear. But at a deeper level the assumption is mistaken. Commodification promotes inequality, while what we most prize about sociability is equality and the recognition of mutual dignity. Once we realise this, the corrosive effects of consumerism upon our sense of self-worth can be halted, and we can begin to move towards sustainability. Commodification provides an inferior form of belonging, whose emotional return is limited compared with what human interaction offers. Indeed, it may well further isolate us from others. The Covid pandemic has brought home our need for human closeness, warmth, touch, banter, smiles, and the immediately resonating sound of voices. Think of Zoom, or its equivalent, which we are all now much more intimately familiar with than anyone hitherto imagined, or desired. In Zoom our feelings are permanently on mute. Faces are blurred and reactions delayed or lost. Voices are echoey and garbled. We lack the reassuring sense of togetherness and fellow-feeling which only three-dimensional, close, direct, *real* human contact provides. We cannot weigh and measure other people's reactions on the screen: we need to face them. This need is profound and instinctual and goes to the root of what it is to be human. It is the core of what utopia, and most clearly intentional communities, represents. It can be attained through enhanced sociability, the deliberate effort to pursue greater closeness to others.

The case is presented here, then, for seeing an intimate relationship between sociability, our relations with other people and the sense of belonging we seek to achieve in these, and our attitude towards objects and consumption. This is where approaches to luxury, consumerism, sociability, and groups intersect.

Achieving a heightened sociability entails recognising a spectrum of behavioural possibilities, ranging from benign neutrality to effusive affection. Minimally, we should aim to be receptive and open to others, and bear them no ill-will, while of course still constructively criticising shortcomings of approach. Here a utopian outlook may be a weak or thin moral category: we abstain from threatening harm to others, or using them instrumentally, rather than endeavouring to help them. Just as the absence of pain is the most elementary form of happiness, or the feeling of well-being, the absence of hostility is the default for sociability. An outlook of benign neutrality is perhaps the weakest of such forms of acquaintance. Given the prevalence of hostility in everyday life, however, this still counts as enhanced sociability, for its basic demeanour is not aggressive or even competitive. From this it is a small step to a stronger moral category, kindness—that is, actually assisting others, out of a deeper sense of solidarity. Few do this from pure altruism, or love of humanity. Helping strangers is rare, and we should not build a moral philosophy on the possibility. We are more likely to do so when we feel a bond of identity with those who benefit from our actions. Utopian interaction does not require this stronger moral obligation, though it implies its desirability. At its strongest, it supplies those intense bonds of affection we call love. In the public sphere this is very rare. But even at its weakest, it provides greater sustenance than modern everyday life often does.

Utopia is not thus reducible to either love or hope, both of which are usually, in the west, identified with Christian moral imperatives and faith. Intense bonds of personal love provide a unique symbiosis and sense of being needed, though other forms of association can sometimes achieve this. Outside the family, love of more than a few might just be attainable on a very small scale, often bolstered by religion. But a wider ethos of love is too ambitious for any utopia which seeks to embrace larger numbers. Here we must settle for less in order to gain more, and renounce the perfect in order to gain the good. As a default position in our dealings with others, openness and some kindness are not morally strenuous or overly demanding. This also frees utopia from any need to be associated with religion, which is a decided advantage in a pluralistic world, is more conducive to gender equality, and is epistemologically crucial in any case, since all "faith" subverts a scientific outlook and weakens our ability to resist propaganda, as the Covid pandemic has abundantly demonstrated. But even warmth and openness are infectious, and will spark a response from others. This is how utopia is constructed, one emotional step at a time. It can, in this sense, as Ruth Levitas argues, be characterised as a method, or way of living, as much as an end destination, a plan, a blueprint, or an ideal state of being, though with these ends being complementary rather than, as is too often assumed, mutually exclusive.[139]

This stress on our need for others militates against the common presumption that increasing autonomy and individuality are the central goals of modern life. Optimally we aim not to maximise our "sovereignty" or independence from others but to balance our freedom with sociability and solidarity, which requires ever-greater collaboration. (This is true for states as well as individuals.) Since freedom is *a* if not *the* leading modern concept, at least in the west, utopia is at a clear disadvantage in weighing it against other first principles. Utopianism has often prioritised order, harmony, and equality over liberty. But

139. Ruth Levitas, *Utopia as Method: The Imaginary Reconstitution of Society* (Palgrave Macmillan, 2013).

it also acknowledges that solidarity, the willingness to lend our support to others based on a feeling and ideal of fellow humanity, and of sameness and common interest, is crucial to our well-being. It implies an ideal of loyalty like that promoted by Josiah Royce, where ethical obligations are the basis of a community.[140] It also acknowledges that solidarity requires a degree of social equality which has been in retreat in most societies for some five hundred years, in the face of the breakdown of traditional forms of community, of organised religion, and of extended family and tribal groupings, the most elementary forms of association throughout history, and even the great trades unions created in the nineteenth and twentieth centuries to safeguard workers' rights.[141]

Degrees of Association

The feeling of belongingness, and of more closely linking our identity with others, can result from various relationships. But which can intensify it without arousing antagonism towards others? How can we feel "in" without making others feel "out"? Once we move beyond mere neutrality towards others, we encounter a spectrum of sociable relations, from weak and shallow, perhaps merely amiable and jocular, to deep and intense bonds defined by trust, loyalty, and intimacy. Our closest tie to others is personal love, the most intense form of belongingness, which is sometimes so deep and all-consuming that we would willingly die for our beloved, our family, or our mates. Beyond this come varying degrees of affection for our kin, then the tribe, village, race, ethnic, language, or religious group, and the nation, forms of association based on work, like co-operation and solidarity, bonds based on leisure, as with sports teams, and more distant relations with strangers. This complex web constitutes our social network. How far we can extend the bonds between family and friends to wider groups without increasing exclusiveness is a key question for any concept of utopia. So let us briefly assess the utopian potential of each type of relationship to see which seems most suitable.

Family

We are today prone to idealise the family as a place where we retreat from the pressures of the wider society in search of reassurance, acceptance, and security. Families may provide a respite where we are accepted despite our behaviour.

140. Josiah Royce, *The Philosophy of Loyalty* (Macmillan, 1908), 15.

141. For this definition, and on variants of the concept, see Steinar Stjerno's *Solidarity in Europe: The History of an Idea* (Cambridge University Press, 2005).

But families can also be dysfunctional, and prisons from which we long to escape as fast as possible. While we are young they are necessarily hierarchical, and as we grow older new constraints may arise. Though Plato's Guardians rejected the nuclear family, the idea of a wider family has nonetheless often been put to utopian uses. Fictional utopias and intentional communities alike have often been described in terms of extended families of brothers and sisters, parents and children, often with patriarchal rule. Christianity prescribes loving our neighbours as "brethren". Religion has often tested this limit. And, as Zygmunt Bauman notes, the difficulty of "loving our neighbour" is almost insurmountable.[142] Luther believed it impossible "to rule a country, let alone the entire world, by the gospel".[143] The nuclear or extended family might still be an optimal basis for society, as some early modern utopias, like Tommaso Campanella's *The City of the Sun* (1623), suggest. Many communal groupings, however, recognise the competitive threat nuclear families pose to wider forms of identity, and discourage excessive intimacy or raise children in common to compensate. So Robert Owen condemned "single-family arrangements" as "hostile to the cultivation in children of any of the superior and ennobling qualities of human nature."[144] In the kibbutz, too, "familistic" attachment was generally scorned, as well as close friendships within peer groups.[145] Private affection may be frowned on by the group, which is why privacy—Orwell's "ownlife"—is often suppressed in dystopias. But a weakening of family bonds often induces greater loneliness unless it is compensated for by other forms of association.

Friendship

Friendship seemingly offers a better prospective model for utopian association. Outside the family, friendships, with their varying degrees of intimacy, offer the most valuable bonds we achieve. They may well surpass family affection: we can't choose our parents, but we do pick our friends. Friends value us, we imagine, because of who we are. They may be critical, but in a well-meaning way, without aiming to hurt or belittle us. We grow with them, and their progress marks our path as well as their own, as we calibrate our own

142. Zygmunt Bauman, *Liquid Love: On the Frailty of Human Bonds* (Polity, 2003), 77–118.

143. Quoted in Colleen McDannell and Bernhard Lang, *Heaven: A History* (Vintage Books, 1990), 151.

144. *New Moral World*, 1, no. 9 (27 December 1834), 67.

145. Yonina Talmon, *Family and Community in the Kibbutz* (Harvard University Press, 1972), 12, 154. Neither in the USSR nor in the Israeli kibbutz did removal of children for enforced community-centred group education prove particularly advantageous, however.

trajectories against theirs, anchoring our identity, often by generation, as we mark our progress through time. Even ordinarily we rely on friends for our sense of well-being. We laugh with them. They humour us, forgive us our foibles, and boost our confidence. Andrew M. Greeley calls them "a promise to ecstasy, a dream of pleasure and joy, a utopian vision".[146] To Horst Hutter, "Feelings of liking, cherishing, or being well-disposed are some of the chief defining characteristics of friendship".[147] To Cicero, friendship, "the most valuable of all human possessions", consisted in "a perfect conformity of opinions upon all religious and civil subjects, united with the highest degree of mutual esteem and affection".[148] To make a new friend is always a delight. To sustain a friendship over decades adds immensely to our experience of life. The betrayal of a friendship often occasions great suffering.

Most cultures mark this choice carefully. Though the British are now less formal, the use of a first name indicates an accepted familiarity, since the contrast of "you" and "thou" no longer exists.[149] Germans ask "Wollen wir uns duzen?"—shall we address each other in a familiar way?—to establish intimacy, while the equivalent French request is "On peut se tutoyer?" In many societies inviting someone to one's home to share food bestows this honour. The closeness such acceptance brings has many tangible benefits, not least the feelings of equality and acceptance which follow. The attractiveness of friendship as a model for utopia, as symbolised by personal greetings, is thus obvious. We recall George Orwell's comment, on reaching Barcelona during the Spanish Civil War, that "Nobody said '*Senor*' or '*Don*' or even '*Usted*'; everyone called everyone else 'Comrade' and 'Thou', and said '*Salud*!' instead of '*Buenos dias*'". This he immediately recognised was "a state of affairs worth fighting for". At this moment he first felt that socialism (and anarchism) meant an identity of belonging based on equality. It touched him deeply, and the memory, and the sense of value of the relationship, remained with him the rest of his life.[150]

Conditions of extreme duress, like imprisonment, despotism, or war, make these relationships all-important. Under Stalin's dictatorship in the USSR,

146. Andrew M. Greeley, *The Friendship Game* (Doubleday, 1971), 31.

147. Horst Hutter, *Politics as Friendship: The Origins of Classical Notions of Politics in the Theory and Practice of Friendship* (Wilfrid Laurier University Press, 1978), 5–6.

148. Cicero, *The Offices, Essays on Friendship and Old Age and Select Letters* (J. M. Dent, 1909), 176–77.

149. "Thou", the second person singular pronoun, is still used in some parts of northern England, but has been archaic elsewhere for several centuries outside of a few religious sects.

150. George Orwell, *Homage to Catalonia* (Macmillan, 2021), 3. Its inversion is portrayed in *Nineteen Eighty-Four*.

friendship was "an ultimate value".[151] In the terrible Pacific combat during World War Two, a marine private, E. B. Sledge, remembered that "Friendship was the only comfort a man had". He recalled that when veterans returned to the United States they sometimes expressed "a feeling of alienation from everyone but their old comrades", realising that "all the good life and luxury didn't seem to take the place of old friendships forged in combat". These bonds no one who had not experienced combat could understand, for here life itself depended on their resilience.[152] Their strength is doubtless one reason some men like war.

The relevance of friendship for any ideal of utopia lies centrally in the role played by equality. Where family is necessarily hierarchical, friendship from Aristotle onwards has been understood as based on equality: "Friendship implies the recognition by Self of Other as an equal in his humanity".[153] This has been interpreted as implying that there is an "irreducible, ubiquitous human desire for affection, companionship, and a sense of belonging", and even that "belonging as such is the truly irreducible root of human affection."[154] So too John Stuart Mill would argue that "the art of living with others consists first & chiefly in treating & being treated by them as equals."[155] Just how we conceive of this equality, however, is crucial and is arguably one of the main dividing lines between utopia and dystopia. Equality cannot entail a rigid uniformity, or what Kolakowski derides as "the aesthetics of impeccable symmetry and ultimate identity . . . in which all variety, all distinction, all dissatisfaction and therefore all development have been done away with forever."[156] Opposing extreme inequality need not mean suppressing difference and individuality, or neglecting our need for a degree of autonomy.

Different forms of friendship may thus exhibit varying degrees of utopian potential. John Reisman distinguishes between three main kinds of friendship,

151. Oleg Kharkhordin, *The Collective and the Individual in Russia: A Study of Practices* (University of California Press, 1999), 319.

152. E. B. Sledge, *With the Old Breed: At Peleliu and Okinawa* (Presidio, 1990), 218, 266. To James Michener, "The sense of belonging is one of the great gifts men get in battle" (*Tales of the South Pacific* (Fawcett Crest, 1947), 352).

153. Aristotle, *Nicomachean Ethics*, 1161b. It might be noted that Aristotle also regarded the passion for equality as "the root of sedition" (*Politics*, 1301b12).

154. Lorraine Smith Pangle, *Aristotle and the Philosophy of Friendship* (Cambridge University Press, 2003), 87, 125.

155. John Stuart Mill, *Collected Works*, 33 vols (Routledge and Kegan Paul, 1963–91), 17: 2001.

156. Kolakowski, *Modernity on Endless Trial*, 140–41. He adds: "The dream of a consistently egalitarian utopia is to abolish everything that could distinguish one person from another; a world in which people live in identical houses, identical towns, identical geographical conditions, wearing identical clothes and sharing, of course, identical ideas is a familiar utopian picture."

based on reciprocity, on receptivity, and on association. In the first, "each friend gives to, takes from, respects, and likes the other". Affection, loyalty, and admiration are all reciprocated. Friendships of receptivity are more imbalanced and include inequality, one perhaps giving more and receiving less affection, or good offices. In friendships of association, by contrast, little or no affection exists, and mutual interest predominates.[157] Clearly we want friendships of reciprocity to flourish, while recognising that those based on receptivity and association will always exist. We also want a multiplicity of group affiliations of the better sort, and active resistance to "othering" and exclusiveness.

The relationship between property and friendship is germane here. Sharing food and common dining have traditionally been used to initiate and sustain friendship. Reciprocity of obligation has often been built on exchanging gifts, which binds us by duty rather than proximity, and which forms the original basis of exchange economies.[158] In Marcel Mauss's famous account of "exchange courtesies", the "most important of these spiritual mechanisms is clearly the one which obliges us to make a return gift for a gift received."[159] To Marshall Sahlins, a "spectrum of reciprocities" exists, originating in mutuality and self-interest, and ranging from the most voluntary to the virtually coerced.[160]

Though it is a home-centred ideal of hospitality, the Greek concept of *xenia*, or guest-friendship, extending courtesies to all visitors, is sometimes held out as an ideal. To the Greeks, dining together expressed the equality of the (male) group, women and children being usually excluded. The banquet called a *symposion* could include eating, drinking, poetry, music, and more. In Greece, generosity in the household was counterbalanced by public banquets, and in Athens *c.*480 BCE the governing group of the fifty-eight most prominent citizens had a public building constructed where they could eat together daily. Wealthy Romans, too, regarded the *convivium*, or "living together", of the common meal as an essential part of civilised life, and by the late Republic special clothes were adopted for its performance, and frequent changes of costume became common. Public banquets given by the emperor could involve a thousand guests, and under Elogabulus such exotic fare as camels' feet and peacocks' heads were served. Displays of status through conspicuous consumption were already a norm. Medieval courts struggled to surpass such excess, but the burden

157. John M. Reisman, *Anatomy of Friendship* (Irvington, 1979), 2.

158. For the example of Greece, see Gabriel Herman's *Ritualised Friendship and the Greek City* (Cambridge University Press, 1987).

159. Marcel Mauss, *The Gift: Forms and Functions of Exchange in Archaic Societies* (Cohen and West, 1954), 5. For case studies, and an updating of Mauss's arguments, see Natalie Zemon Davis's *The Gift in Sixteenth-Century France* (Oxford University Press, 2000).

160. Marshall Sahlins, *Stone-Age Economics* (Tavistock, 1974), 193.

of serving sometimes thousands of retainers was an essential part of the social bond, and the banqueting hall was a centre of political power. Here too the association of food with social rank remained universal, and waste was a proof of power.[161] As we approach the modern period, *Ancien Régime* France stands out in conspicuous consumption; one financier had a room constructed solely for his guests to eat their deserts, where trees were planted with wired-on fruits and nightingales sang.[162]

Some utopians thus explicitly reject the subversive potential of involving property, and thus inequality, in any wider form of gift giving. Campanella's *The City of the Sun* insists that "it is worth the trouble to see that no one can receive gifts from another. Whatever is necessary they have, they receive it from the community, and the magistrate takes care that no one receives more than he deserves."[163] Edward Bellamy's *Looking Backward* (1888) describes a future where "People nowadays interchange gifts and favors out of friendship, but buying and selling is considered absolutely inconsistent with the mutual benevolence and disinterestedness which should prevail between citizens and the sense of community of interest which supports our social system."[164] We can readily concede that an ethos of helpfulness, and an exchange of services as well as goods, without any expectation of immediate return, certainly promotes a sense of social well-being.

Exchanging benefits, then, does not necessarily imply placing economic values on what is given. With friends we may err on the side of generosity. With strangers we demand a more exact reckoning. Clearly there is an important if invisible line between altruistic assistance and that based on reciprocal obligation or mutual interest. To Robert Lane, family life and friendship are key components in our subjective sense of well-being, and "Solidarity implies a degree of emotion, of warmth; reciprocal favors do not".[165] Amitai Etzioni also insists that community can be based only upon mutuality, "a form of

161. Roy Strong, *Feast: A History of Grand Eating* (Jonathan Cape, 2002), 14, 24, 37, 104.

162. Robert L. Heilbroner, *The Quest for Wealth: A Study of Acquisitive Man* (Eyre and Spottiswoode, 1958), 129.

163. Francis Bacon and Tomasso Campanella, *"The New Atlantis" and "The City of the Sun": Two Classic Utopias*, introd. G. Claeys (Dover, 2018), 51.

164. Edward Bellamy, *Looking Backward 2000–1887* [1888] (Oxford University Press, 2007), 52. Bellamy's essay "The Religion of Solidarity" linked love between individuals with love of the universe and saw the former as providing the basis for "unselfishness" and self-sacrifice as "the essence of morality" (*Selected Writings on Religion and Society*, ed. Joseph Schiffman (Liberal Arts, 1955), 3–26).

165. Robert E. Lane, *The Loss of Happiness in Market Democracies* (Yale University Press, 2000), 112.

community relationship in which people help each other rather than merely helping those in need", which is "undermined when treated like an economic exchange of services".[166] This implies that problems begin when money acts as an intermediary in such exchanges. But we may weigh benefits in non-material terms. To Georg Simmel, "the principle of sociability" is "the axiom that each individual should offer the maximum of sociable values (of joy, relief, liveliness, etc.) that is compatible with the maximum of values he himself receives." These values also include the spirit of the occasion, the motive of those who exchange, and the reciprocity of exchanged "personal qualities as amiability, refinement, cordiality, and many other sources of attraction." Even here Simmel thought that "the democracy of sociability even among social equals is only something played."[167] But the appearance of equality may suffice for our purposes: sometimes simulation and pretence are enough, and we pretend along.

These forms of sociability can all be contrasted to competitive behaviour, defined by Margaret Mead as "the act of seeking or endeavouring to gain what another is endeavouring to gain at the same time", where co-operation is "the act of working together to one end".[168] Like personal ambition, competition has its uses. But when it greatly undermines social solidarity and our sense of mutual humanity, its limits are reached. Unfortunately, extreme and often brutal competition became the dominant ideal of modernity, legitimated to many by ill-conceived social interpretations of the discoveries of Darwin, ensconced too in Marx's idea of class struggle, but the product chiefly of liberal ideas of market society, which rested loosely on a Hobbesian ideal of the war of all against all, and on European imperialism and racism.[169] Now the chief locus of competition is the workplace, a primary source of identity for many.

Here too, however, the advantages of mutual assistance are also evident. Solidarity at work necessarily plays an important role in defining utopian association. Originating in analyses of the division of labour and class structure, and possibly first defined in this context by Pierre Leroux, the term came by the 1840s to be closely linked with the championing of socialism, conceived as a new form of fraternity or brother- (and sometimes sister-) hood.[170]

166. Amitai Etzioni, *The Third Way to a Good Society* (Demos, 2000), 19–20.

167. Georg Simmel, *The Sociology of Georg Simmel* (Free Press, 1950), 45, 48.

168. Margaret Mead, ed., *Cooperation and Competition among Primitive Peoples* (Beacon, 1937), 8.

169. See my "The Origins and Development of Social Darwinism", in *The Cambridge Companion to Nineteenth-Century Thought*, ed. Claeys (Cambridge University Press, 2019), 165–83.

170. On its meanings, see Kurt Bayertz's *Solidarity* (Kluwer Academic, 1999, 3–28), and J.E.S. Hayward's "Solidarity: The Social History of an Idea in Nineteenth-Century France" (*International Review of Social History* 4 (1959): 261–84).

To Marxism this became a key cosmopolitan principle. Friedrich Engels, certainly, thought that "the simple feeling of solidarity based on the understanding of the identity of class position suffices to create and to hold together one and the same great party of the proletariat among the workers of all countries and tongues."[171] The "anarchist prince", Peter Kropotkin, construed solidarity in terms of a principle of mutual aid which could be traced throughout history, identifying its practice with tribal and village communities, and in modern society on mutual assistance amongst the working classes.[172] Another anarchist, Michael Bakunin, also regarded solidarity as "the only basis of all morality".[173] Other later nineteenth-century utopians also promoted the idea of solidarity as a central social value.

The effects of modern technology on friendship will be examined below. We should here, however, briefly note in passing two other relevant forms of friendship, with animals and machines. Friendships with animals, especially pets, can surpass in meaning those we have with people, and may exhibit what H. S. Salt called a "creed of kinship" with other species as well as a desire for unconditional loyalty (except from cats).[174] Loving animals can also help us to connect with other human beings. Machines are less cuddly and endearing. But we may also soon begin to relate emotionally with some types of mechanism—some already invite us to do so, though, so far, voice-controlled "personal assistants" like Alexa or Siri notwithstanding, they are mostly poor conversationalists. (Try talking to the ATM or food checkout machine.) Each of these relations implies an exchange of mutual good offices, perhaps the affirmation of mutual dignity (more doubtful for a machine), and the security that we will not be stabbed in the back (artificial intelligence (AI) dystopias like *I, Robot* notwithstanding). A utopian theory of friendship would certainly want to promote more of both. So we can hope that there will be more pets in utopia, not fewer, as H. G. Wells suggested, in the hope of eliminating diseases spread by them. Amiable cyborgs and robots with a sense of humour (often sadly lacking in utopia) and something approaching feelings would also be a bonus.[175]

If friendship, acquaintance, and solidarity on any scale require equality, we need to know whether this has existed beforehand on a large scale, and thus presents a useable precedent, or whether it must be invented. Until around the

171. Frederick Engels, "On the History of the Communist League", in Marx and Engels, *Collected Works*, 26: 330.

172. Peter Kropotkin, *Mutual Aid* (William Heinemann, 1903), 287.

173. Michael Bakunin, "Three Letters to Swiss Members of the International", in Bakunin, *From Out of the Dustbin: Bakunin's Basic Writings, 1869–1871* (Ardis, 1985), 57.

174. Henry S. Salt, *The Creed of Kinship* (Constable, 1935).

175. H. G. Wells, *A Modern Utopia* (Chapman and Hall, 1905), 230–34. AI can of course be programmed to tell jokes.

eighteenth century it was still plausible to claim that an original and still enviable equality had existed in some Golden Age or paradise, which might also persist in faraway primitive societies or remote settlements. Either could thus serve as a model for imagining current and future equality. These powerful myths—but they are more than this—are treated in greater detail below. But equality can also be conceived as an artificially constructed new relationship, like solidarity, which is contrived and created rather than natural and organic, regressive, atavistic, or remembered.

The language of rights and dignity is germane here. Where we once recalled or imagined original natural rights, we now innovate upon what are obviously conventional claims, which are more easily extended than any supposedly original rights can be. The same is true of the invention of the concept of dignity, which is closely aligned to rights concepts. To Richard Sennett and Jonathan Cobb, dignity "is as compelling a human need as food or sex", and this implies a need for equality.[176] This was also Ernst Bloch's conclusion, as we will see. Every ideal of friendship, indeed, requires such a reciprocal acknowledgement of dignity, which we also term "rights".

Both family and friendship require a degree of emotional intensity and investment of time which is extremely difficult to extend on a considerably large scale. Public bonds so intense cannot be normally expected, except in brief emergencies. An emotional limit, a cap on our inner resources, restricts our feeling of closeness to others. This is the boundary of the religious interpretation of utopia. Love might unite families or members of small, tightly knit units. Friendship can furnish a sense of belonging for more, but still not many, and still falls short of an ethos of public-mindedness. A cult-like association may instil such bonds, but the price is often too high for most. So never mind mutual love: even "liking" (except where only a click is involved) requires a degree of familiarity which is often lacking. You cannot *like* someone you don't know, and who is essentially a stranger to you. Even tolerating the alien seems difficult at times, particularly when fear and anxiety mount, for they generate hatred. So how might more extensive utopian forms of sociability be possible? The answer lies in another form of association: the group.

Groups

Utopias can be conceived as idealised groups, and understanding how groups work is central to the entire utopian project. But what kind of group is implied here? To extend the limited sense of belonging which family or friends

176. Richard Sennett and Jonathan Cobb, *The Hidden Injuries of Class* (Cambridge University Press, 1972), 191.

provide, and to reach larger numbers, especially in modern cities, a weaker form of friendship, something like acquaintance, best describes a potentially much wider set of relationships. Strangers we usually regard as people to whom we feel no obligation. Those we regard as "friends of friends", psychologists remind us, "are required to treat each other as equals."[177] Acquaintance, like neighbourliness, implies being open, helpful, and respectful. Though we do not have to, we help neighbours out of acquaintance and a sense of mutual need and a desire for reciprocity and trust building. The warmth of familial or friendly relations may be lacking, but this is far superior to mere neutrality.

Groups shape our actions in many ways. Often we behave appropriately, or in an expected manner, just because people know us, because we belong. But we gain much from membership. Groups make possible various levels of acquaintance. They can facilitate friendship, because belonging to a group implies the first requirements of trust have already been met. We know we have something in common and that people we engage with can be held responsible for their actions because we know where to find them. Groups usually provide feelings of togetherness, mutual respect, and trust. Often we solicit patronage from them. Amongst the members of clubs or societies, or in a college or military unit or professional association, membership may also entail a kind of social contract with tacit obligations based on honour as well as explicit rules. It may be denoted by a uniform or badge demarcating similarity and loyalty. In part III we will explore the possibilities of promoting the proliferation of such associations to increase this sense of acquaintance and mutuality, with a view to increasing belongingness, particularly in cities.

Here we must touch on group psychology, the main portal to the study of utopianism, and on the emotions central to it. The primary function of groups is to provide safety in numbers, and to civilise us by restraining our desire for natural liberty, to do simply what we want without interference. In isolation we are insignificant and fragile. When groups huddle, and envelope us, they generally lend us a sense of strength proportionate to their size, though the bond of fellow-feeling with humanity as such is often weak, the nation being the largest group with which we can readily identify. At the religious end of the communitarian spectrum, where it may be conceived as divinely sanctioned, groups may provide love, conceived as "a surplus condition of affective relationships; a luxurious state of social exchange" where "separate identities are (partially) merged in a relationship of desire, the consequence of which is

177. George McCall et al., *Social Relationships* (Aldine, 1970), 97. See, generally, Jeremy Boissevain, *Friends of Friends: Networks, Manipulators and Coalitions* (Basil Blackwell, 1974).

to infuse their (separate) lives with a common meaning."[178] More commonly they give us the acceptance we crave, and membership provides a position in the scheme of things by which to orient ourselves and validate our existence. As children we already strive to belong to exclusive groups or cliques, define our well-being by our success in so doing, and feel "equally a sense of rejection if not part of a group."[179]

Collectively, as a species, groups are our starting point. The earliest "primary groups", as E. H. Cooley termed them, were defined by face-to-face contact, and a relationship wherein "one's very self, for many purposes at least, is the common life and purpose of the group."[180] Groups, from the extended family to the village, tribe, nation, race, gender, or religion, define our immediate environment. They may also be largely imaginary and function as mental projections of ideal communities, like "the saved", "the righteous", or the "true patriots". Groups draw us out from ourselves, and inhibit excessive introspection, self-absorption, and narcissism. They guide and order our behaviour and subordinate our desires to their norms, permitting or assigning us roles within specific contexts. There are hierarchies of groups, and as we ascend the ladder, particularly towards whatever "establishment" we aspire to join, we may renounce membership of lesser groups when belonging to the more important group demands it. A sense of failing to belong brings loneliness, isolation, withdrawal, and depression. From the viewpoint of sociability, estrangement is our great enemy: belonging is a key antidote.

The "group utopia", as Elias terms it, embodies this aspiration.[181] Our desire for group identity is relatively constant, and we seem to seek a certain equilibrium and intensity in wanting to belong. That is to say, we reinforce one form of identity, like nationalism, when another weakens or fails, like religion or locality. But strengthening one form may also function to compensate for a perceived weakness in that form, and thus of our own inferiority, as in Germany's increased nationalism following its defeat in World War One. Either way, the desire is constant and must be sated.[182] The focal point of the group may also be the charismatic leader, to whom acolytes kowtow and whose word is sacred.

178. Bryan S. Turner and Chris Rojek, *Society & Culture: Principles of Scarcity and Solidarity* (Sage, 2001), 132.

179. Adrian Furnham, "Friendship and Personal Development", in *The Dialectics of Friendship*, eds Roy Porter and Sylvana Tomaselli (Routledge, 1989), 94.

180. Quoted in Jack Goody, *The Domestication of the Savage Mind* (Cambridge University Press, 1977), 15.

181. Elias, *Essays I*, 214, 217–18. "Unbelonging" would be a useful term here.

182. Maslow suggests that if our desire for love is satisfied in our early years we become "*more* independent than average" in our later desire for belongingness (*Motivation and Personality*, 31).

He or she may indeed define the group, and we ourselves by the degree of our homage to the leader. Recognising their susceptibility to flattery, careerist toadies will slavishly endeavour to emulate the leader's every foible, down to mannerisms and hairstyle. The ascendancy of such acolytes generally marks the end of any vibrant or progressive element in any institution.

Secondly, groups furnish us with key components of our identity, based on ritual, history, language, and shared symbols. Consciousness of being a group is central to its success, and to conquering individual isolation and alienation. "The emergence of an operational communal group", writes John R. Hall, "is found to be predicated on the mobilization of people who invoke one or another myth of 'who we are as a group.'"[183] While valuing our individuality, we are also group-defined creatures who gain much of our identity from associating with others. As Albert Camus put it relative to Hegel, "All consciousness is, basically, the desire to be recognized and proclaimed as such by other consciousnesses. It is others who beget us. Only in association do we receive a human value, as distinct from an animal value."[184] This identity is forged when groups extend us meaning through shared experience, which at an emotional level in particular bonds us with others. A beautiful sunset, or a moving piece of music, which might inspire in us an inward sensation of the sublime, becomes magnified and more exquisite when shared with others. (And so musicians will tell you that the audience "makes" the performance.) Such impressions are meaningful *because* they are shared, and because in their being shared we have been treated as equals, on human terms, without financial transactions intervening, perhaps even without the requirement of any emotional quid pro quo.

Cities are obviously places where groups most easily proliferate, and where we can encourage their growth with the greatest effect. Before we let this particular genie out of the bottle, however, by promoting their greater success in this area, we should briefly introduce two types of reservations about proposals to extend group identification here. The first rests on the nature of groups, and our recent experience in reinforcing them. The second relates to trends which are seemingly inherent in modernisation and urbanisation.

183. John R. Hall, *The Ways Out: Utopian Communal Groups in an Age of Babylon* (Routledge and Kegan Paul, 1978), 19. Here "groupness", the "feeling of belonging", is the key quality sought (28). Hall describes as "worldly utopian" groups which do not "struggle for control of the larger social order so as to implement change", nor "retreat from an intolerable world historical situation into a heavenly sanctuary", but instead "live out a model of an ideal world" (91).

184. Albert Camus, *The Rebel* (Alfred A. Knopf, 1954), 110.

Firstly, we recognise that groups divide as well as unite, and the intensity of our bonds within them may proportionately discourage relationships with non-members. This presents us with a distinctly utopian problem. Utopia is an idealised group characterised by enhanced sociability. But every "in" group or clique implies exclusion, or an "out-group". The attractiveness of the group may well be proportionate to its delineation of and antagonism towards its enemies. Germans, for instance, have been described as achieving a "grand utopia of belonging" in their united enmity to the Jews under Hitler.[185] So every utopia is a potential dystopia. Affection for and loyalty to some may be defined by hostility or friendly antagonism towards others: neighbouring countries, rival football clubs, other races, religions, and nations. The last are particularly guilty here. The most ambitious and universalistic theories of sociability demand a tempering of the cruder forms of national loyalty by cosmopolitanism to emphasise our common essential humanity. But often they fail, and national enmity prevails instead. Dystopian groups are notably defined by strict rules of inclusion/exclusion, by "othering" and separation, even hatred. Cults and gangs extend this tendency, often by encouraging submission to a charismatic leader rather than the group as such. We may develop groupophobia, a subset of sociophobia, in reaction to such prospects. It is also confusing that each group invites schizophrenia in insisting that our specific group personality displayed there is our "true" or "real" self. This hints at a constant multiplicity of selves, and at the triumph of appearance over "authenticity", and of insincerity over a principled "core" self.

But to some degree such repression is in the nature of groups as such. Taking a leaf from Orwell at his most anarchistic, we can concede that all groups seek power for its own sake.[186] They nurture, but their dynamic as groups also compels obedience and demands the sacrifice of our individuality and capacity for critical judgment, sometimes excessively.[187] They may ask us to lie and deceive as the price of admission and the test of loyalty. They may command deference to, even worship of, charismatic leaders. Enforced collectivism or compulsory public sociability of this sort produced dystopian ends during much of the twentieth century. Here utopia may degenerate into dystopia, or

185. T. Kuhne, *Belonging and Genocide: Hitler's Community, 1918–1945* (Yale University Press, 2010), 1, quoted in Allen, *Psychology of Belonging*, 81.

186. "The Party seeks power entirely for its own sake. We are not interested in the good of others; we are interested solely in power. Not wealth or luxury or long life or happiness: only power, pure power": *The Penguin Complete Novels of George Orwell* (Penguin Books, 1967), 895. We recall that Orwell self-identified as both an anarchist and a socialist.

187. This was one of Orwell's more prominent claims in his later writings. A more extended discussion of this process is in my *Dystopia*, 3–57.

our utopia consist of others' dystopia. The trick to avoiding the stifling conformity and mental degradation which encourages this transformation is to recognise how groups function and to anticipate and avoid these sacrifices of our integrity. We need to ensure a proliferation of groups, such that none swallows all our loyalties, as totalitarian parties attempt to do. We need also constantly to be aware of the emotional and often unconscious element in utopian thinking, or rather feeling, and to an underlying dimension of collective un- or semi-consciousness which underpins much of our public behaviour. Avoiding its manipulation is key to avoiding dystopia. From Mannheim to Orwell this has been a constant element in utopian criticism.

The demand for conformity may also impinge on our private relationships. At the best of times this does not occur. To Michael A. Hogg, the more cohesive the group is, the more friendships there are within it, and "group cohesiveness is widely treated as equivalent to interpersonal attraction". Here the outer reinforces more individual bonds by creating a reciprocal solidarity which strengthens both.[188] But more dystopian groups demand the sacrifice of individuality to social order, usually by subsuming the self within the group, the "I" within the "We" (hence Zamyatin's famous title), in order to promote the common good.[189] Hutter acknowledges that:

> The more numerous these relationships become, the better it is for societies, but the less satisfying they are to the individual on the assumption that the amount of emotional energy available to an individual is limited. Friendship and love present the lure of deeply-satisfying relationships at the expense of the networks of casual involvements, so needed by society, from which the individual necessarily has to withdraw some power of feeling for the purpose of investing this power in love and friendship.[190]

Over-conformity is a danger in all societies. A sense of growing individuality, uniqueness, and autonomy is a key if fragile and exaggerated achievement of modernity.[191] "Casting members as individuals is the trademark of modern society", writes Bauman. The "individualizing process" is not something we should willingly abandon or weaken: our still much too limited respect for the value of individual life is perhaps the greatest achievement of "civilisation". But neither should we pay the earth for it. The trick is rather to wed it successfully

188. Michael A. Hogg, *The Social Psychology of Group Cohesiveness* (Harvester Wheatsheaf, 1992), 24.

189. Yevgeny Zamyatin, *We* (1924; Jonathan Cape, 1970).

190. Hutter, *Politics as Friendship*, 5–6.

191. On groups and social identity theory, see my *Dystopia*, pt. 1.

to the need for group or social identities and not allow it to undermine the bonds the latter provide.[192] Emile Durkheim captured the problem well in asserting that "The human personality is a sacred thing; one dare not violate it nor infringe its bounds, while at the same time the greatest good is in communion with others."[193]

Tension between individualism and collectivism will always exist, and we cannot seek in principle to eliminate it. Though the progress of individualism may seem to subvert utopianism more than any other development, the more extreme the individualism in a given society is, and the more this is exacerbated by fear, the more powerful groups may become. To Michael Hechter, group solidarity "will be maximal in situations where individuals face limited sources of benefit, where their opportunities for multiple group affiliations are minimal, and where their social isolation is extreme."[194] Nations are one of the most dangerous forms of group—the growth of populist and racist nationalism, in particular, where the nation functions as an "imagined political community"—often promoting belongingness by emphasising exclusion rather than inclusion, showing why a proliferation of superior forms of community ("civil society") is desirable where groupism becomes an increasingly common response to alienation, fear, and isolation.[195] When we feel weak, we compensate by embracing national myths which play up our glory and greatness and ignore the rest. Here, as elsewhere, groups may magnify our vices as well as our virtues. Any nostalgia for primary groups where we lose ourselves in the collective needs to bear this in mind.

A second set of objections to proposals for group proliferation is rooted in modern life generally, and the problem of scale. Trust is central to friendship, belongingness, and neighbourliness.[196] It requires relative equality: "the roots of generalized trust lie in a more equitable distribution of resources and opportunities in a society", concludes a major study.[197] But we cannot trust everyone. Moreover, trust seems to be declining as we move more frequently and more commonly live in cities. "Magna civitas, magna solitudo," Francis

192. Zygmunt Bauman, *The Individualized Society* (Polity, 2001), 45.

193. Emile Durkheim, "The Determination of Moral Facts", in *Sociology and Philosophy*, trans. D. F. Pocock (Cohen and West, 1953), 37.

194. Michael Hechter, *Principles of Group Solidarity* (University of California Press, 1987), 54.

195. Benedict Anderson, *Imagined Communities: Reflections on the Origin and Spread of Nationalism* (Verso, 1991), 6.

196. Robert R. Bell, *Worlds of Friendship* (Sage, 1981), 16.

197. Bo Rothstein and Eric M. Ushaner, "All for All: Equality, Corruption, and Social Trust", *World Politics* 58 (2005): 41–72, quoted in Daniel J. Fiorino, *A Good Life on a Finite Earth: The Political Economy of Green Growth* (Oxford University Press, 2018), 118.

Bacon wrote five centuries ago: friends scattered in towns mean less fellowship, "so that there is not that fellowship, for the most part, which is in less [smaller] neighbourhoods".[198] Trust requires a sense that others are responsible and accountable for their actions. This depends on acquaintance and mutual recognition: we know that we will see people again and that when they know this, any propensity to misbehave is restrained.

So transparency and proximity are central to trust. But there are limits on the number of individuals proximate to us whom we can know well enough by sight to trust. Robin Dunbar argues that we can remember at most perhaps five thousand faces, and far fewer names, and are capable of at most 150 friendships, of which friends perhaps twelve to fifteen constitute a "sympathy group", and only three to five are intimate. The limit of friendships recurs as a common number in many forms of association. Dunbar cites as partial evidence the fact that Hutterite communities split when they reach 150 members, the reason given being that "They find that when there are more than about 150 individuals, they cannot control the behaviour of the members by peer pressure alone", which would imply a need for "hierarchies and police forces".[199] This is a central principle of group dynamics, as well as of utopian organisation. Tribal groupings of 500–2500 are the largest association in most traditional societies, where language and cultural identity are closely linked. Within these, clan groups typically average about 150 members, and these usually constitute the limit for ceremonies, as well as village sizes. A similar number, however, corresponds in later societies to the number of people we might send Christmas cards to, and the number present in the workplace before more formal hierarchies are required. This sense of a natural or organic group size restricts our horizons. Trust clearly requires familiarity and is eroded by estrangement and anonymity, not to mention competition and animosity.[200] Yet most of us now live in large cities, where anonymity prevails. Sometimes we do not even know our neighbours. Trust is also eroded by constant bombardment of mis- and disinformation and downright lies. These factors seemingly invalidate, or at least mitigate and challenge, the very possibility of any utopian sociability being widely extended. How can we establish trust in cities, where most of the people we encounter daily are strangers whose casual gaze we avoid, and our workplace too may be a pit of spiteful bullies consumed by petty jealousies?

198. Francis Bacon, *The Essayes; or, Counsels Civill and Morall* (1597; J. M. Dent, 1906), 80.

199. Robin Dunbar, *Friends: Understanding the Power of Our Most Important Relationships* (Little, Brown, 2020); Robin Dunbar, *How Many Friends Does One Person Need?* (Faber and Faber, 2010), 24–28. The figure of 150 is now commonly referred to as "Dunbar's number".

200. Jacques Derrida, *Politics of Friendship* (Verso, 1997), 88–89.

Finally, there is the problem that all sociability demands time and effort, and its value may indeed be judged by what has been expended. Friendship has declined in later modern societies because less free time is available for its cultivation. We also have less need for mutual assistance, which is one of its traditional sources. We would rather pay someone to fix something than incur an obligation to someone else by allowing them to do it for free, on the barter system. This is why money was invented, we reason. But barter, or lending or giving things and services to people, is actually better for sociability, because it demands reciprocity in kind. ("Assume *man* to be *man* and his relationship to the world to be a human one", Marx wrote, "then you can exchange love only for love, trust for trust, etc.")[201] Money dilutes or disguises this human relationship. Specialisation of function also segregates our lives in ways unknown to more traditional communities, and offers a sharp distinction between the workplace and home.

These factors "reduce friendliness and mutual concern in society as a whole as it becomes richer in material goods and ever more pressed for time."[202] They pose substantial challenges to the general hypothesis offered here. Indeed they may fatally undermine it. It may be that, once it is lost, we cannot recreate "community", whose varied meanings we must now briefly assess. We may have to make do with inferior substitutes which relegate "home" and personal closeness to family and the private sphere, and "society" to our own networks of associates, who are mostly bonded by self-interest rather than any sense of solidarity or mutual respect. This may simply be the price to be paid for maximising individualism and independence, and for being modern. This was a common enough conclusion, we will now see, amongst those who first scrutinised the process of modernisation.

The Sociology of Community

The search for "community", to Robert Nisbet the "most fundamental and far-reaching of sociology's unit-ideas", is a central theme in the history of utopianism and is often linked to ideas of belonging.[203] But calling utopia a "community" tells us little. A prison is also a community, and might sometimes be better than desperate poverty outside, but is hardly utopian. Voluntary or intentional communities formed on the basis of shared belief, however, have

201. Marx and Engels, *Collected Works*, 3: 326.

202. Fred Hirsch, *Social Limits to Growth* (Routledge and Kegan Paul, 1976), 77–78.

203. Robert A. Nisbet, *The Sociological Tradition* (Heinemann, 1966), 47; see also Robert A. Nisbet, *The Quest for Community* (Oxford University Press, 1953), 25.

produced many enduring egalitarian societies throughout history, with some lasting centuries. More modern groups have often attempted to recreate in modified form the advantages of rural village life, and the values of autarky, close co-operation, and mutuality. Early nineteenth-century Owenites, Fourierists, and other socialists, notably, juxtaposed their ideal associations to industrial and urban degradation and commercial excess. Throughout this period there was a growing sense that the moderns had lost something essential in leaving the countryside for crowded and unhealthy cities. A "pastoral impulse" led some "back to the land" seeking a simpler and more "authentic" life.[204] Renewed sociability was central to this appeal. Hamlets, villages, or small towns embodied familiarity. Cities were rootless, anonymous, restless, and heartless. Particularly in the United States, nostalgia for small-town life survives to the present day, even as rural communities decline steeply into drug abuse and poverty, and the small-farm ideal disappears.

This nostalgia coincided with an essentially Romantic revolt against the more sordid aspects of bourgeois life, and the suffocating, repressive, formal conventionalism of the age. Critiques of money-grubbing Mammon could come from religious, moral, artistic, or humanitarian viewpoints. They did not require retreating to the country. Already in the 1840s an urban counterculture of "Bohemians" haunted the cafés of the Quartier Latin in Paris seeking creativity, personal style, and emotional fulfilment. At least one artist, Maurice Quay, established a communal group "dedicated to vegetarianism, illuminist philosophy, and the rediscovery of the Golden Age."[205] A later generation of Aesthetes championed the primacy of beauty and of "art for art's sake". The appreciation of nature was encouraged by further movements, including nudism, conservation, and hiking, which in the German context have been collectively termed "naturism".[206] The Simple Life movement, and later still the Beat Generation and the Hippies, who are discussed below, championed similar values. All rejected the bourgeois work ethic, and in the twentieth century, the middle-class corporate "executive" lifestyle and compulsory "groupthink" of what William Whyte called the "organization man".[207] All gravitated towards the culture, music, and condition of the poor. All celebrated art, music, poetry, and the

204. For Britain, see Jan Marsh's *Back to the Land: The Pastoral Impulse in England, from 1880 to 1914* (Quartet Books, 1982), and more generally, Raymond Williams's *The Country and the City* (Chatto and Windus, 1973).

205. As discussed in Neil McWilliam's *Dreams of Happiness: Social Art and the French Left, 1830–1850* (Princeton University Press, 1993), 29.

206. John Alexander Williams, *Turning to Nature in Germany: Hiking, Nudism, and Conservation, 1900–1940* (Stanford University Press, 2007).

207. William H. Whyte Jr, *The Organization Man* (Jonathan Cape, 1957). This ethos is a key target in Ray Bradbury's *Fahrenheit 451* (Ballantine Books, 1953).

erotic, the spontaneous over the structured. All rejected the expectation that human value should be measured by the standards of machines, or money, and dismissed ideals of the good life dominated by wealth and materialism.

These trends all implied a new theory of sociability, which by the late nineteenth century was addressed by the emerging academic discipline of sociology under the rubric of "community". Emile Durkheim, Ferdinand Tönnies, and others described the dramatic disruption of family and kinship bonds, and their replacement by connections defined by commerce and cities. They pointed to isolation, alienation, or anomie as resulting from large-scale urban, mass, or democratic societies. Later sociologists would add an affinity between the "age of consumption" and the acceleration of this process.[208] At a time of rapid imperial expansion, the question of the relative value of "savage" customs also entered into play. "Primitive society" usually came off badly in the comparison. But sometimes challenges were made to the much-vaunted claims of the superiority of "civilisation" which were central to justifying the suppression of indigenous peoples. For there was sometimes a lingering sense that something valuable, a kind of primal bond akin to what we have termed belongingness, and perhaps also an elemental freedom, had been lost in this transition.

The most enduring account of this contrast was Tönnies' juxtaposition of *Gemeinschaft*, or traditional community, to *Gesellschaft*, or modern society, in *Community and Civil Society* (1887). Tönnies assumed that "in the original or natural state there is a complete unity of human wills" based on kinship. Memory, emotional ties, and habits augmented ties of blood, which also forged a similarity of character. Communities of blood gave rise to those based on place, and in spirit, again rooted in "close proximity". Communities "bound together in an organic fashion by their inclination and common consent" could be described in terms of kinship, neighbourhood, and friendship or comradeship. A brotherly spirit in the family was matched by the exercise of paternalistic authority in the village. Communal religious worship was central to maintaining these bonds. In towns, friendship or comradeship grew through frequent interactions, notably in the workplace. Consensus, a reciprocal binding sentiment or peculiar will of the community, might arise here. The sense of community life derived particularly from "*mutual* possession and enjoyment, and possession and enjoyment of goods held *in common*". "Common goods" included "common evils; common friends—common enemies".[209]

208. The classic study is David Riesman's *The Lonely Crowd: A Study of the Changing American Character* (Yale University Press, 1952).

209. Ferdinand Tönnies, *Community and Civil Society*, ed. José Harris (1887; Cambridge University Press, 2001), 22–62, 170, 253, 255.

This description owed something to the eighteenth-century Scottish Highlander Adam Ferguson's republican critique of the threat of what was coming to be called commercial society to civic virtue and military preparedness.[210]

What replaced traditional communities were more formal and less organic associations based on self-interest and commercial exchange, the "cash nexus", typically in cities. Their underlying principle, "individualism", encouraged people to be "essentially detached" rather than united, "so that everyone resists contact with others and excludes them from his own spheres". Selfishness predominated over any sense of common endeavour or outlook, and dependency and subjection resulted. Capitalism embodied and promoted this disunity. Now all "conventional sociability may be understood as analogous to the exchange of material goods", with each seeking advantage over others.[211] Tönnies hinted that renewing religious ties might promote *Gemeinschaft* feelings in modern society. Obviously these were no substitute for kinship association. But equally he suggested that big cities threatened much that defined *Gemeinschaft*, including the family. The city and "community" might well be prima facie incompatible. This challenge to any future possibility of urban enhanced sociability, and to any idea of urban utopianism at all, is addressed below.

Many other writers took up these themes. Emile Durkheim assessed group solidarity in terms of mechanical or organic association, based upon likeness or interdependence. Max Weber examined cohesion in armies, corporations, religious brotherhoods, neighbourhoods, and similar associations. "Associative" relations involved "a rationally motivated adjustment of interests or a similarly motivated agreement". "Communal" relationships, including the family, erotic bonds, religious brotherhoods, military units, and the nation, were based on "a subjective feeling of the parties, whether affectual or traditional, that they belong together". They did not exclude conflict, "coercion of all sorts" being "a very common thing in even the most intimate of such communal relationships if one party is weaker in character than another."[212]

Twentieth-century sociologists and political theorists continued to understand community in terms of *Gemeinschaft*, while usually lamenting its "loss" or "death". Many suggested this left a void to be filled by forging new ideas of unity and identity in large-scale societies. Some thought the loss of extended family and small group associations encouraged identification with mass movements, and linked crowd behaviour with religious sectarianism and mass

210. See Ferguson's *Essay on the History of Civil Society* (1767; Edinburgh University Press, 1966), a text which, amongst other things, evidently offered the first use of the word "civilisation" in English.

211. Tönnies, *Community and Civil Society*, 52, 65.

212. Max Weber, *Economy and Society* (University of California Press, 1978), 40–42.

politics, especially fascism. Bernard Bell and David Riesman echoed Tocqueville's warnings of the "tyranny of the majority" in democracies and associated "mass society" with political populism, or pandering to the worst instincts of the majority.[213] Some saw the unitary sovereignty of the "general will" as extending from politics to culture.[214] As "mass culture" gained conceptual ground from the mid-twentieth century onwards, however, the disparaging associations it possessed for earlier psychologists declined. Its threat to nonconformity, central to John Stuart Mill, for example, seemingly receded. The mass or "group mind" now became to sociologists and historians a more neutral phenomenon. The Le Bonian tradition now mostly disappeared.[215] Crowd behaviour became less pathological and more conventional. Crowds were rarely seen as sinister, at least until the last few years and the revived threat of populism and now, again, fascism.[216] Others adapted Tönnies' analysis to give what the American sociologist Louis Wirth called "urbanism" more positive qualities, noting that while kinship and family bonds declined, a "levelling influence" occurred in cities as ways of life converged.[217] The city had its allies too. It produced a different form of sociability, less close, perhaps, but richer.

Yet few thought urban life was on balance superior to what it replaced, except in terms of its sheer variety of life, which for some implied greater sophistication and toleration of difference. In America in the 1940s the claim was still made that it is only in the "small, primary community" that people can "find the way to live well."[218] By the late twentieth century, nostalgia dominated many of these debates. Particularly in the United States, both in common parlance and in some scholarly literature, "community" gained the conceptual consistency of a marshmallow: warm, gooey, sweet tasting, pastel tinged. Its Disneyland-like ideas of belonging, safety, security, and acceptance still evoked the rapidly declining aura of small-town American life, which the middle classes now attempted to recreate in the suburbs, and later in private gated communities, whose very existence increases fear amongst their residents.[219]

213. See Bernard Iddings Bell, *Crowd Culture: An Examination of the American Way of Life* (Gateway, 1956); Nisbet, *Quest for Community*; Riesman, *Lonely Crowd*.

214. Robert E. Park, *The Crowd and the Public and Other Essays* (University of Chicago Press, 1972), 63–84.

215. A good review of this process is Christian Borch's *The Politics of Crowds: An Alternative History of Sociology* (Cambridge University Press, 2012).

216. Erich Goode, *Collective Behavior* (Saunders College Publishing, 1992), 5–17.

217. Louis Wirth, *Louis Wirth on Cities and Social Life* (University of Chicago Press, 1964), 35.

218. Baker Brownell, introducing Arthur E. Morgan's *The Small Community: Foundation of Democratic Life* (Harper and Bros, 1942, vii).

219. See Anna Minton, *Ground Control: Fear and Happiness in the Twenty-First-Century City* (Penguin Books, 2009), 61–82.

Nisbet's definition of community as encompassing "all forms of relationship which are characterized by a high degree of personal intimacy, emotional depth, moral commitment, social cohesion and continuity in time", in particular, has been criticised as "loaded with positivity", and akin to "an idealized description of marriage, or close friendship".[220]

Clearly, the more hostile urban environments became, as crime rates rose in the 1960s and city centres became impoverished, the more attractive were such images and the stronger became the tendency to idealise them. So "community" refused to go away. Contrasted to the cold, alien anonymity, crass individualism, and competitiveness of the urban beehive, where millions pass each other daily, eyes averted, hostility scarcely muted, stepping over ever-increasing numbers of beggars and rough sleepers, "communities" were still seen as "closely-knit", and possessing "frequent and intense interaction." The concept "feels good", Zygmunt Bauman writes, because it "stands for the kind of world which is not available to us—but which we would dearly like to inhabit and which we hope to repossess".[221] But increasingly, by the late twentieth century, this sense of closeness was no longer a remembered experience. Slipping into the realm of the imaginary, it had become an alterity unreachable to most, a marker of what we have lost but cannot regain, the other which we require to contrast with daily life but which we know is unattainable. It functioned, in other words, very much like its distant cousin, utopia.

Utopia, the City, and Belongingness

The tantalising mirage of *Gemeinschaft* remains today as a ghostly reminder of what we have, or at any rate imagine we have, lost, for most of us have no real memory upon which to base a contrast. Any account of utopia today must confront the argument that the peripatetic and isolating nature of modern urban life and our inability to build the bonds of trust in cities seem to preclude *Gemeinschaft*. By the mid-twentieth century cities provided a stark contrast between what Lewis Mumford described as necropolis and utopia.[222] The new ideal city for many, and especially the technological utopians, was epitomised by the gleaming skyscrapers of Manhattan, whose scale and size, however, often overwhelmed the individual. This was a spatial, not a sociable,

220. Nisbet, *Sociological Tradition*, 47, criticised by Graham Day in *Community and Everyday Life* (Routledge, 2006), 9. On some distinctions between community and intimacy, see Lynn Jamieson's *Intimacy: Personal Relationships in Modern Society* (Polity, 1988), 80–89.

221. Zygmunt Bauman, *Community* (Polity, 2001), 1–2, 48.

222. Lewis Mumford, *The City in History* (Secker and Warburg, 1961), 3.

vision, designed to dazzle and impress us with the marvels of human ingenuity and our Babel-like capacity to reach into the skies, rather than to humanise us. The profound loss of sociability and sense of place and belonging which the often-overwhelming spaces and heights of modern great cities have promoted was not appreciated. We were dwarfed by our creations, and our smallness was worrying. But we remained impressed by our accomplishments.

Let us give the city its due. We should recall how liberating the urban experience has been and continues to be. Its essence is energy, stimulation, temptation, variety, entertainment, bustle, anonymity, and luxury. Most of all, excitement epitomises the modern. We relish many of the city's most temporary qualities, the rapid interchange with people we will never see again, alongside the possibility of meeting like-minded others who may be associates for life. The city also cloaks us in its anonymity. Millions have left the land, village, or small town fleeing petty tyranny, over-proximity, superstition, intolerance, gossip, jealousy, suffocating conformity, and downright boredom. To be known is a good thing when we are good but a bad thing when we are bad, unless those who surround us are bad too. Familiarity breeds contempt as well as friendship. To millions the appeal of anonymity is obvious: it brings freedom. Cities are more tolerant and liberal. Encounters with different people erode our prejudices. So the city has as many friends as enemies—probably more. In it a plethora of subgroups and subcultures can furnish an antidote to alienating anonymity by bringing together similar people. Such groups permit much more specialised interests to develop. Virtual communities function similarly, and in many cases an imaginary virtual community may well be more satisfactory than face-to-face reality, in terms of the freedom we have to express our ideas, and the protection which impersonal contact offers.[223]

These still do not give us a sufficient sense of belonging to satisfy our psychological needs, however. Loneliness remains a great creeping plague amongst us. To counter it, a new ideal of the city is required. Utopia and the city can coalesce, we will see, only by creating smaller-scale life and a proliferation of group identities within an urban context. This became clear by the late nineteenth century. From Ebenezer Howard's planned garden cities, which aimed to combine the advantages of rural life with suburban proximity, to the ambitious schemes of Frank Lloyd Wright, Le Corbusier, and others, a sense of the loss not only of the aesthetic value of rural existence but of the sociability of smaller spaces often came to the fore. A century later it was becoming

223. See Majid Yar, *The Cultural Imaginary of the Internet: Virtual Utopias and Dystopias* (Palgrave Macmillan, 2014). Some recent studies show the young would rather interact online than person to person, but Covid indicates the limit to this claim.

evident that much of the city's potential for sociability was being lost to the suburbs, as the wealthy fled inner cities, taking their spending and tax base with them. But suburbs often did not create real neighbourhoods, and produced instead an epidemic of loneliness and, at least in the United States, a severe diazepam crisis to compensate, exceeded in danger only by the later methamphetamine and synthetic opioid epidemics. When people began to return to the cities, too, via gentrification, they brought a new set of problems.

Critics like Patrick Geddes, Lewis Mumford, Jane Jacobs, Henri Lefebvre, and David Harvey have resisted this trend, and exemplify a confidence that the city can yet express the higher, humane and life-enhancing aspirations of utopianism. To Lefebvre and Harvey, a radical approach to the city involves rethinking a "heterotopian" conception of urban space in a manner very different from Foucault's.[224] Many, like Mumford, insist that the building block of any such ambition must be the "neighbourhood unit", a term coined by Clarence Perry in 1929. As a space, the *quartier* or *arrondissement*, even the medieval walled town, corresponds to the outer limits of our psychological sense of immediate personal identity, and of the number of faces we can recognise. Here we can practise a kind of local citizenship or patriotism. Besides the solidarity of family and work, where these exist, only these forms of association can provide "so close, so intense, so narrow" a feeling of belonging, by establishing both "a civic nuclear to draw people together", in the form of some kind of local centre, and "an outer boundary to give them the sense of belonging together".[225] Small must balance large, neighbourhoodness cosmopolitanism, the decentralised the centralised, is the message here. The new *Gemeinschaft*, "belongingness", links the city and utopia. The challenge of recreating it, we will see in part III, remains even more urgent today.

Utopianism Restated

Let us summarise this definitional discussion, and some potential objections to the approach adopted here. Utopia will necessarily remain a contested concept in any event, since no one definition can satisfy all perspectives on the

224. David Harvey, *Rebel Cities: From the Right to the City to the Urban Revolution* (Verso, 2012), xvii. See Henri Lefebvre, *The Urban Revolution* (University of Minnesota Press, 2003), 9–11, 33, 128–31. To Lefebvre, "heterotopia" remains "the other", rather than a specific subset of utopia. See further Nathaniel Coleman's *Lefebvre for Architects* (Routledge, 2015) and *Utopias and Architecture* (Routledge, 2005).

225. Lewis Mumford, *The Urban Prospect* (Secker and Warburg, 1968), 56–78, quotes at 57 and 68.

subject. But the interpretation defended here offers one means of extracting from this tradition something valuable for the present and future, by presenting the prospect of a realisable utopia which can serve as an organising category for the social and other transformations now demanded of us.

Utopias are ideal groups defined by their satisfaction of our needs for security, equality, sociability, and belongingness. The term "utopia" normally denotes "eutopia", or the good place, which may be imaginary or real, while retaining the ambiguity of being "no place".

Utopianism is both the imagining of these groups and the quest to make them real. It can be defined by form (of which there are three), function (two), and content (one).

Formally, its three domains are utopian theory, literary utopias and dystopias, and communitarianism.

Its two functions are, first, psychologically, to provide an interior mental space which distances us critically from the present, and which is in principle always ideal and unattainable, since whenever we reach what we have dreamt about, we move the goalposts of our aspiration. This space may be compensatory, displacing our anxieties with reassuring images, or emancipatory, suggesting change. It may be nostalgic or future oriented. Here the idea of utopia functions to provide alterity. Utopianism's second function is futurological, where, in "social dreaming", we imaginatively project visions of dramatically improved societies, or eutopias.

The normative content of utopianism represents the quest for an ideal sociability defined by belongingness, underpinned by a supportive infrastructure whose particulars can vary considerably. To George Kateb, utopia's "positive identifiable content" lies in the core themes which define much of the tradition from Thomas More onwards.[226] Given *Utopia*'s central influence for five centuries, this can be termed a Morean paradigm of utopian republicanism, depending on how much of More's agenda is accepted. Not all utopias express this goal explicitly, or in the same way, but many do. Where such content is realised, and a sense of belongingness, identity with others, equality, and human fulfilment are attained, even temporarily, utopia exists. Their realisation is what makes utopia "good" or "better" or eutopian. But the content of these terms also evolves. While it has sometimes rested historically on conquest, slavery, and exploitation, the history of the tradition indicates a movement towards ever-greater inclusivity, equality, and non-exploitation, increasing resistance to inflicting pain on any species, and the replacement of minority coercion by democratic

226. George Kateb, *Utopia and Its Enemies*, 2nd ed. (Schocken Books, 1976), vi.

consent.[227] Utopia's explicit moral content thus demands ever more progress towards these ends.

Utopianism thus retains both a fantastic and imaginary and a real content, and functions which remain discrete and indeed irreconcilably separated from one another. Confusion between these, and between utopia and eutopia, accounts for many misunderstandings about the subject. But utopia does not aspire to salvation, or any secular variation thereof. Except in satire and science fiction, it does not consist in living forever, possessing magical powers, or being infinitely happy, or of unceasing immersion in pleasure, or bliss, grace, or perfection of any kind. Instead, utopian ideas, literature, and communities all aim to enhance human sociability and encourage belongingness. They aim to bring us closer together and to minimise conflict, coercion, and violence. Utopia is their attainment, temporary or prolonged, imagined or real.

The sense of bonding or cohesion here termed "belongingness" may, however, be only temporary or intermittent, or conditional or dual edged, hinting at exclusion. Now we begin to see just how tricky the proposition offered here is. Clearly we can be on our best behaviour, obliging, and respectful of others, beyond what society normally demands of us, for short, often formal and ritualistic, periods, in the spirit of mutual embrace, putting on our party clothes and being at our politest, laughing at our relatives' off-colour jokes and dubious political opinions while biting our tongues. The utopian question is how to prolong these episodes, inscribe them in everyday life, and practise them widely—extending utopian groups, in other words, and moving from neutrality towards acquaintance and deeper forms of friendship, by making people happier and more fulfilled by belonging to them.

Here, however, a substantial problem arises: we cannot *make* people more sociable, or make them feel that they belong, any more than we can demand that someone be happy. Attempts to do so are prone to backfire. Feelings are too anarchic, volatile, and unpredictable to be susceptible to such guidance. We can only furnish the means by which people can engage more amicably with others and identify with them and with their surroundings. We can invite them to join, and make them feel good in joining, but we cannot force them to accept and should not force them to fake their willingness or act in bad faith. By definition, enhanced sociability and belongingness are voluntary, willing, and consensual, not coercive. So we need to define more closely how this sociability works, and what kinds of typologies have been offered to explain its flourishing or decay. Then we can promote it.

227. Bearing in mind that minorities in democracies who bow to the will of the majority are in a sense "coerced".

So we now have a definition of utopianism which accommodates form, function, and content:

> Utopianism is the projection of both imagined and real ideal groups which embody the feeling of belongingness. Formally it is expressed as literature, theory, or intentional community. The functions of utopia are to represent a necessarily unattainable state of betterment, which always recedes before us but provides us with critical alternatives to the present, and to describe ideal past or projected future societies. In its content it promotes an enhanced sociability defined by friendship, neighbourliness, acquaintance, communality, and solidarity, commencing with an attitude of benign neutrality but aspiring to stronger and more egalitarian, but still consensual, bonds. These goals are summarised in the concept of belongingness, which is the opposite of that form of alienation that is defined by the sense of not fitting into and feeling a part of our environment. In the degree to which we achieve it, the dominant principles of dystopia, loneliness and fear, are reduced.

The basic question this book examines, then, is whether a society framed around utopian social relations can help to offset the inevitable decline in consumerism which sustainability demands in the coming decades. To answer this question (with a firm but qualified *yes*) we must consider how utopian sociability has been approached in the past, as well as how, in voluntary communities, it has actually been practised, and how it has been projected or imagined in literary form. These responses are treated here historically and contextually, in light of the commercial revolution which in so many ways defines the modern period, from the eighteenth century to the present. Initially hostile to luxury, utopians gradually came to adopt counter-strategies of more just but equally plentiful technologically based futures, while often insisting that luxury enjoyed in public could supplant excessive private consumption and even assist in fostering greater sociability thereby.

We commence with the prehistory of utopia in the myths of an original human equality which date from the earliest period. More than anything else, the sense of common identity and purpose utopianism seeks can result only by creating a more equal society. Equality was utopia's original premise. We now turn to its origins.

2

The Mythical Background

REMEMBERING ORIGINAL EQUALITY

IT IS MUCH easier to project a utopian society that has never abandoned primitive simplicity than to imagine one corrupted by commerce and luxury which then renounces them. It is also easier to envision exiting a society where luxury has infected only the upper and middle but not the lower classes, as was the case in most modern revolutions. In this chapter we will consider some memories and practices which reflect a recurrent belief in an original human equality, prior to such degeneration. Historically these were central to the image of what might be returned to during much of the utopian tradition. They are the prehistory of utopia. We may treat them today as the legacy of the childhood of humanity, and their relevance may be challenged. But they illustrate much about the core utopian theme, equality, whose relationship to needs, as treated in literature, we will then examine.

We should first reiterate the progressive unfolding of the utopian concept. In the western tradition, the image of what became known as utopia has undergone five major historical transformations. All societies and peoples require grand narratives to situate themselves and to provide a sense of structure, order, and meaning. Utopia too performs this function. It began as an ideal original stage, with Greek images of the Golden Age and Christian myths of Eden and heaven. Secondly, Christian millenarianism implied that this stage, or something like it, might be realised again, and in the sixteenth century began acting on the idea. Thirdly, the Renaissance suggested that utopia might still exist somewhere, awaiting rediscovery, as More's *Utopia* and some fantastic travel literature implied. More described this as a humanist ideal republic or city-state, and left the hint that it might be recreated. Fourthly, utopia was transformed in the late eighteenth century into the future good time, or *euchronia*, to emerge through scientific and technological advance. As divine providence weakened in the age of revolutions, human agency was strengthened. Utopia now became a vastly improved state of society capable of realisation. Fifthly, an ideal of unlimited abundance was added to many visions of the future utopia by the late nineteenth century. This chapter will survey the first two of these stages, and some of their later modifications.

The Golden Age

Western culture rests on two contradictory myths: the first of foundation and degeneration from a condition of original harmony and purity, as represented in Greek notions of the Golden Age and Christian ideas of the Garden of Eden; the second, of ascent from a condition of primitivism and poverty to one of progressive complexity, enlightenment, and opulence. Since myths spring from needs, both can also be read as psychological glosses on humanity's history.[1] Arthur Lovejoy, George Boas, and Hiram Haydn distinguish between a "hard" primitivism, in Haydn's words, "a simple, rigorous life of natural righteousness", and a "soft" "age of ease and indulgence—even sometimes of unbridled individualistic licence".[2] Intermingled with these were variants on the "hard" ideal which did not romanticise the earliest stages but sought to avoid the degenerative aspects of civilisation for military reasons in particular. The tough guy in the utopian family, Sparta, long epitomised this ideal.

Possibly the folk memory of the Minoan or Mycenaean peoples of the second millennium BCE, the Greek myth of the Golden Age long functioned as a ghostly memory of humanity's imagined original state, and provided one basis for both Christian millenarianism and secular conceptions of ideal societies. To Lovejoy, Boas, and Haydn, the Greeks espoused "chronological primitivism" in portraying a degenerative sequence of four ages of Gold, Silver, Bronze, and Iron, in which peace, plenty, virtue, or at least innocence, and freedom were successively lost as history progressed.[3] The appeal of this myth of corruption is obvious. Those who desire equality can claim a vital precedent. So can those who wish to blame degeneration and existing inequalities on human nature. The desire for equality clearly links utopia to this myth,

1. See, e.g., H. J. Massingham, *The Golden Age: The Story of Human Nature* (Gerald Howe, 1927).

2. Hiram Haydn, *The Counter-Renaissance* (Charles Scribner's Sons, 1950), 501. See Arthur O. Lovejoy and George Boas, *Primitivism and Related Ideas in Antiquity* (Johns Hopkins Press, 1935), 10–11, 28.

3. Lovejoy and George Boas, *Primitivism and Related Ideas*, 1; Haydn, *Counter-Renaissance*, 465.

which can also function as a fantasy of an original anarchic freedom to do as we choose, following our instincts in natural liberty rather than being hampered by law and convention. This fantasy gave rise to an antinomian strand in utopianism, which we will also encounter later.

The myth of the Golden Age rests upon a few key sources.[4] The Greek term for paradise was adapted from the Persian word for garden. Hesiod's *Works and Days* (*c.*700 BCE) relates that the gods who dwelt on Olympus created a golden race of men who knew neither toil nor grief and enjoyed peace and plenty without labour. In the present age of iron, however, "men never rest from labour and sorrow by day". Plato's account of the great island empire of Atlantis (*c.*340 BCE) drew on similar themes: its inhabitants lacked private property, renounced gold and silver, and "thought scorn of all things save virtue and counted their present prosperity a little thing"—themes very close, we will see, to Thomas More's ideal.[5]

In Plato's *Republic* (*c.*380 BCE), generally taken as the starting point of western utopianism, Sparta and Crete, where many communal institutions existed, are an influence, as are some Pythagorean settlements in Italy, where community of property was supposedly practised. Here the allegory of degeneration shifts from time to character. The ruling class is described as golden, by contrast to the silver, brass, and iron temperaments of lesser mortals.[6] These "guardians" hold property, wives, and children in common to bond their group, while remodelling the entire society on the familial ideal.[7] The dangers of the "luxurious city" whose inhabitants were prone to a "fevered state" and fond of "relishes and myrrh and incense and girls and cake" (*Republic*, 373e) are clearly spelt out. Plato established the problematic that great disparities in wealth threaten to undermine any polity fatally, and especially republics, and that the greatest danger lies with the ruling class itself, as writers down to

4. See, generally, John Ferguson, *Utopias of the Classical World* (Thames and Hudson, 1975); Doyne Dawson, *Cities of the Gods: Communist Utopias in Greek Thought* (Oxford University Press, 1992); Rhiannon Evans, *Utopia Antiqua: Readings of the Golden Age and Decline at Rome* (Routledge, 2008).

5. Hesiod, quoted in Claeys and Sargent, *Utopia Reader*, 18; Plato, *Critias*, 120d, in *The Collected Dialogues of Plato*, eds Edith Hamilton and Huntington Cairns, trans. Hugh Tredennick (Pantheon Books, 1961), 1224.

6. Plato, "Critias", in *The Collected Dialogues of Plato*, eds Edith Hamilton and Huntington Cairns, transl. Hugh Tredennick (Pantheon Books, 1961), 1214–24.

7. A. W. Price, *Love and Friendship in Plato and Aristotle* (Clarendon, 1989), 189. See Plato, *Republic*, 457d. This arrangement Aristotle rejected in due course as destructive of affection and property alike, and Plato himself seems later to have abandoned the scheme, though in the later *Laws* extremes of property are avoided (J. Ferguson, *Utopias of the Classical World*, 73).

at least H. G. Wells in the early twentieth century would reiterate. Only with Thomas More, however, perhaps following a hint from Joachim of Fiore, would Plato's communism be universalised to an extensive population.

There are other variants on the classical myth of original purity. The Roman historian Plutarch linked Homer's Elysian Fields to the Islands of the Blessed, whose inhabitants enjoyed "all things without trouble or labour".[8] There is a hint in some authors, notably in Virgil's *Fourth Eclogue* (*c.*42 BCE), that the lost Golden Age might yet be regained, war ended, and the soil once again spontaneously yield its fruits. Ovid's *Metamorphoses* (*c.*8 CE) stressed that war, law, and punishment were absent in the Golden Age, a condition of everlasting spring where the untilled earth yielded plenty without labour. After the fall of Cronos, the father of Zeus, however, this was succeeded by a Silver, a Bronze, and then finally an Iron Age, defined by the peoples Zeus had created, and characterised by their increasing violence, foolishness, and savagery. In the last stage, greed and strife, conquest, and inequality became overwhelming, and Jove finally inflicts a great inundation upon the world. The Christian Flood myth is a variant on this allegory.

Sparta

Sparta is the earliest society in the western tradition to renounce luxury to ensure equality. Aristotle thought its communal institutions might well have been copied from Crete, where the public was fed from state-owned lands.[9] Its founder, the possibly mythical ninth-century-BCE lawgiver Lycurgus, supposedly appeared at a moment of great crisis in the state, and instituted a system of education called the *Agoge*, or "raising", to create a mighty army from the nation's male youth.[10] These lived communally in barracks from the age of seven.[11] Xenophon (*c.*430–354 BCE) tells us that Lycurgus also introduced several other peculiar customs. These included giving great stress to producing healthy children, regulating the age of marriage, whipping the negligent, and forcing the young to wear the same cloak all year and to go barefoot to harden their feet. The diet of young men was limited, but Lycurgus permitted them to

8. Plutarch, *The Lives of the Noble Grecians and Romans* (Modern Library, n.d.), 683.

9. Aristotle, *Politics*, 1271c, 1272a. "Moderation in eating" was the subject of legislation here.

10. Lycurgus may have imitated Cretan custom, but most of what is associated with him is swathed in obscurity. What is chiefly important for the argument here, however, is that people *believed* the account to be founded in fact. See Elizabeth Rawson, *The Spartan Tradition in European Thought* (Clarendon, 1960), and Stephen Hodkinson and Ian Macgregor Morris, eds, *Sparta in Modern Thought* (Classical Press of Wales, 2012).

11. Paul Cartledge, *The Spartans* (Macmillan, 2002), 28, 155.

steal food when hungry (their cuisine was notoriously poor). He "imposed on them much labour and contrived that they should have very little leisure". In addition, "since he wanted them to be imbued with a strong sense of respect, he ordered that even in the streets they should keep their hands under their cloaks, walk silently, turn around nowhere, and keep their eyes fixed [on the ground] in front of their feet." At their common meals, extravagance and drunkenness were avoided and they spoke only when addressed. Hunting and sports hardened them for military service. Free men were prohibited from "having anything to do with the acquisition of wealth", their "only appropriate activities" being "those that promote freedom for cities." So owning gold and silver was forbidden. Lycurgus also "made it compulsory to practise every civic virtue without excuse", on penalty of losing the rights of full citizenship. Weakness and cowardice were despised, and deformed infants were abandoned to the elements. By these means, avarice and weakness were conquered. By Xenophon's time, however, standards had fallen. Some Spartans "even prided themselves" on possessing money, and their leaders were said to "take much more trouble to be rulers than to be worthy of rule".[12] Both Plato and Aristotle (*Politics*, 1269b) also commented that Spartan women were prone to indulge in luxury.

The best-known account of Sparta, by Plutarch (*c.*46–120 CE), added detail to this portrait. He describes Lycurgus as dividing the lands of Laconia, as Sparta was also called, into some thirty-nine thousand shares, allowing each family "about seventy bushels of grain for the master of the family, and twelve for his wife, with a suitable proportion of oil and wine. And this he thought sufficient to keep their bodies in good health and strength; superfluities they were better without." He then set about expelling "luxury and crime" by allowing only heavy iron money, which could not be accumulated or used for foreign trade, outlawing all "needless and superfluous arts", and despising personal vanity. Thus luxury, "deprived little by little of that which fed and fomented it, wasted to nothing and died away of itself." Dining in common was enjoined to avoid any inequality in private consumption.[13] Barracks-like communal sleeping arrangements, partly designed to ensure constant military alertness, were required for those under thirty.[14] Male Spartan citizens were forbidden any occupation except being soldiers. Sparta's victory over Athens in the fifth-century-BCE Peloponnesian Wars seemingly proved her command of the civic and military virtues. But if the Spartans were models in self-restraint and

12. *Xenophon's Spartan Constitution*, ed. Michael Lipka (Walter de Gruyter, 2002), 67, 69, 73, 79, 85, 93, 95.

13. Plutarch, *Noble Greeks and Romans*, 55–56.

14. Paul Cartledge, *Spartan Reflections* (Duckworth, 2001), 14–15.

self-control, embracing modesty and simplicity in attire, diet, sexual desire, and much else, they were not equal. The *helots*, or servile/slave class, seven-eighths of the population, did most of the menial labour. They were treated with great brutality and once a year could even be killed with impunity.

In the ancient world, the Spartans were often contrasted, sometimes favourably and sometimes not, to the Athenians, whose extensive foreign trade brought them great opulence, with all its attendant vices. Sparta enjoyed a fitful reputation for many centuries as a synonym for civic virtue, autarky, military hardness, and personal asceticism, as well as, to some, militarism and proto-totalitarianism.[15] It represents the most extreme form of anti-luxury utopia. As with the Golden Age, if this version of Sparta had not existed, it would have been necessary to invent it, to describe one alter ego of the more-commercial peoples. To critics a key problem, as later commentators like the Marquis de Chastellux observed, was that what the Spartans did not expend psychologically in consumption became instead, by way of zealous overcompensation, "too extravagant a search for glory, and ambition."[16] Private indulgence was exchanged for public glory in the form of a lust for conquest: a poor trade for the rest of humanity, and the helots. David Hume too later described "ancient policy" as "violent, and contrary to the more natural and usual course of things", which ought to be to augment the well-being of a state's citizens.[17] But neo-classical republicans often invoked Spartan virtues, especially a contempt for opulence. Some Renaissance writers, like Francesco Guicciardini, admired Lycurgus' "most remarkable achievement, bringing about in one day in his city such moderation in living and such zeal for virtue and such low esteem for wealth".[18] From Plato onwards, indeed, Spartan ideals remained central to utopianism until at least the early twentieth century. They also came to overlap with the noble savage motif. For there was a pronounced tendency in western travellers and observers to "transfer the moral qualities of the Greeks, particularly those of the Spartans", and especially "Spartan endurance", to some indigenous peoples.[19]

15. On its progress, see Rawson, *Spartan Tradition*. The totalitarian charge is usually associated with Karl Popper's *The Open Society and Its Enemies* (2 vols, Routledge and Kegan Paul, 1945).

16. Jean François, Marquis de Chastellux, *An Essay on Public Happiness*, 2 vols (T. Cadell, 1774), 1: 50.

17. David Hume, *Philosophical Works*, 4 vols (Adam and Charles Black, 1854), 3: 283.

18. Quoted in Eric Nelson, *The Greek Tradition in Republican Thought* (Cambridge University Press, 2004), 72.

19. Benjamin Bissell, *The American Indian in English Literature of the Eighteenth Century* (Yale University Press, 1925), 5, 24.

Central to the Spartan myth was the assertion that luxury caused moral and political corruption. Plato, Xenophon, and other classical thinkers warned that luxury made men physically "soft", "effeminate", and cowardly. In Greece, Rome, and eighteenth-century Britain, luxury was associated with the conquest of Asia, and came itself to seem like a victorious army, sweeping all the virtues beside it, or a contagious disease, infecting all who touched it.[20] Thus Alexander the Great was described by the philosopher Thomas Reid as "conquered by oriental luxury".[21] The ancient world had many symbols of extravagant consumption. The wealth of the Persian kings, concentrated in Persepolis, which Alexander conquered, was legendary. Babylon, with its famous hanging gardens, symbolised great wealth and extravagant consumption. Rome under Nero (54–68 CE) was long renowned for its dissipation.

Greece had a counterpart. The epitome of luxury and "a life of loose indulgence beyond all measure" was the small Greek colony in Italy called Sybaris, in modern Calabria, which lasted some two centuries (*c.*720–510 BCE). Its sole legacy today is the term "sybaritic". To the later moderns its lifestyle was vastly more agreeable than Sparta's. It was renowned for its splendorous state banquets, feasts, and public entertainments. Invitations to these were sent out a year early to allow citizens to prepare their adornments. Here, it is reputed, the cookbook was invented, and distinguished chefs were adorned with golden crowns and allowed to patent their recipes for a year. The scent of flowers and perfumes pervaded the proceedings. The Sybarites taught their horses to dance to flute music. They invented chamber pots, in order not to detract from their celebrations, and possibly steam baths as well, doubtless to help recover from their excesses. One Sybarite found himself in Sparta and, appalled by the barracks life of its citizens, vowed he would rather die a coward's death than to endure such conditions.[22]

Between Sparta and Sybaris lay various less extreme compromises. The Greek Cynics and the Roman Stoics both championed a life guided by nature. The Cynics, spoilsports who regularly gave speeches on the subject at banquets and festivals, warned of "the gold that is so sought after, the silver, the luxurious houses, the elaborate clothing—and then remember through how much toil and trouble and danger these have been acquired—yes, and through

20. Emanuela Zanda, *Fighting Hydra-like Luxury: Sumptuary Regulation in the Roman Republic* (Bristol Classical, 2011), 9.

21. Thomas Reid, *Essays on the Active Powers of the Human Mind* (1788; MIT Press, 1969), 133.

22. Joseph Sevier Callaway, *Sybaris* (Johns Hopkins Press, 1930), 35, 86–87, here 17; William D. Desmond, *The Greek Praise of Poverty: Origins of Ancient Cynicism* (University of Notre Dame Press, 2006), 135. For a later invocation of these themes, see Edward Everett Hale's *Sybaris and Other Homes* (Fields, Osgood, 1869).

how many men's blood and death and ruin."[23] Their founder, Diogenes, was renowned for his tirades against bankers and corrupt politicians and was even known to spit in rich men's faces.[24] The Cynics also had their own utopia, the city of Pera, described by Crates, whose peaceful inhabitants were slaves to neither sensual appetite nor gold. They evidently championed the abolition of marriage, property, and government. Some specifically blamed luxury and gluttony for mankind's descent from the Golden Age.[25] To the Roman Stoic Seneca, luxury

> turned her back on nature, daily urging herself on and growing through all the centuries, pressing men's intelligence into the development of the vices. First she began to hanker after things that were inessential, and then after things that were injurious, and finally she handed the mind over to the body and commanded it to be the out and out slave of the body's whim and pleasure.[26]

Another Stoic, Epictetus, echoed this sentiment in urging readers to limit their bodily wants, to "take just so much as your bare need requires", and to "cut down all that tends to luxury and outward show."[27]

The romantic appeal of a more primitive condition has, however, also long had its critics. In the first century CE the Roman writer Juvenal suggested that the moral simplicity of the Age of Saturn had hardly been enviable compared with Roman civilisation.[28] In the Renaissance, the Golden Age came increasingly to be viewed as a kind of Stone Age of dubious achievement; the luxury of trading societies was already indicating a bias in favour of the moderns. In the late seventeenth century, the idea of a state of nature which was not only primitive but savage and wild, "ruled by the principle of tooth and claw", became commonly associated with Thomas Hobbes.[29] Now such virtue as could ever be attained was often assumed to be contingent on politics and social organisation, and to be artificial rather than natural. To Rousseau there was "no natural society among men."[30] Sociability had to be achieved, and consciously fabricated. It was always fragile and easily subverted, especially by

23. Lovejoy and Boas, *Primitivism and Related Ideas*, 143.

24. Desmond, *Greek Praise of Poverty*, 53.

25. Donald R. Dudley, *A History of Cynicism* (Methuen, 1937), 245, 44, 219.

26. Seneca, *Letters from a Stoic* (Penguin Books, 2014), 192.

27. Epictetus, *The Discourses and Manual*, 2 vols (Clarendon, 1916), 2: 229.

28. Lovejoy and Boas, *Primitivism and Related Ideas*, 71.

29. Haydn, *Counter-Renaissance*, 519.

30. Quoted in Robert Wokler, *Rousseau, the Age of Enlightenment, and Their Legacies* (Princeton University Press, 2012), 97.

vanity, egoism, and selfishness, all of which flourish most readily in commercial societies. But the appeal of an ideal of original equality was never entirely lost. It stood for independence, and for virtue. In the Enlightenment historian Constantin Volney's phrase, "As the earth was free, and its possession easy and secure, every man was a proprietor, and the division of property, by rendering luxury impossible, preserved the purity of manners."[31]

The Christian Paradise

The Christian creation myth described in Genesis (*c.*1000–500 BCE) drew on accounts as old as Sumer (*c.*4000 BCE), where a land called Dilmun was described as without fear or terror, where "the lion mangled not. The wolf ravaged not the lambs."[32] Like the Golden Age narrative, the biblical account offers an allegory to explain the origins of evil, and reconciles this with an idea of original goodness or virtue which flatters divine intentions while damning humanity. The Eden story is fraught with ambiguity. Some later believed that Adam and Eve spent only a day in paradise—Luther proposed a mere two hours, an alarmingly short period of innocence. From the outset the presence of the snake, Satan, mars the purity of the location, making Eve's succumbing to temptation and seduction by Satan seemingly inevitable.[33] So Eden is not paradise. Nor is it utopia, for virtue requires knowledge of good and evil, the desire for which provokes Adam and Eve's expulsion. Yet as an ideal place, or moral postulate, paradise, like heaven, persisted notwithstanding. Such narratives provide a necessary and reassuring sense of fixity and belonging, for we can belong to a narrative as much as to a place or group, and in any case we have to start somewhere. Their absurdities and inconsistencies merely reveal the depth of our desire to believe in them, and the ability of priesthoods to exploit our gullibility to the full.

Eden long retained a sense of spatial as well as temporal reality. During the Renaissance, the notion of the actual existence of the terrestrial paradise was

31. M. Volney, *The Ruins; or, A Survey of the Revolutions of Empires* (1791; H. D. Symonds, 1801), 25.

32. Damian Thompson, *The End of Time: Faith and Fear in the Shadow of the Millennium* (Sinclair-Stevenson, 1996), 10.

33. Morris Jastrow, "Adam and Eve in Babylonian Literature", *American Journal of Semitic Languages and Literatures* 15 (1899): 193–214, quote at 209. See, generally, George Huntston Williams, *Wilderness and Paradise in Christian Thought: The Biblical Experience of the Desert in the History of Christianity & the Paradise Theme in the Theological Idea of the University* (Harper and Bros, 1962). The two roles are even confused: the Hebrew name for Eve was also a common name for serpents.

revived, and Eden was variously located in Ethiopia, Sri Lanka, or the Middle East. Mysterious islands reached by intrepid voyagers were also reputed to enjoy Edenic conditions. In the eighth-century Irish *Imramm Brain* (Voyage of Bran), one such group engaged in

> A beautiful game, most delightful,
> They play sitting at the luxurious wine,
> Men and gentle women under a bush,
> Without sin, without crime.[34]

Such images raised the tantalising possibility that paradise might be rediscovered, regained, or recreated. This ran parallel to the notion that the whole earth might be restored in such an image, an idea encouraged, we will see, by millenarians. As faith in an actual earthly paradise located in this world declined, this ideal in turn gained in strength, nurtured by the need to believe in some form of rescue scenario for humanity.

The most important shift in these expectations came in the sixteenth century, when the search for paradise tilted westwards. What was once deemed fantastic now became credible, as the Americas were invested with dramatic hopes of human transformation. This process involved reinventing an imaginary utopian space defined by ideas of virtue, innocence, novelty, grace, equality, and infinite abundance of land and resources.[35] Here, a constant intermingling of fantasy, curiosity, wishful thinking, and unparalleled greed and aggression occurred. With the bestselling fictional voyage narrative Sir John Mandeville's *Travels* (*c*.1356) in his pocket, Christopher Columbus sought in the Americas, perhaps inconsistently but entirely conveniently, both the long-lost earthly paradise and the fabled city of gold, El Dorado, waiting to be pillaged. Long on confidence and short on modesty, he proclaimed that "God made me the messenger of the new heaven and the new earth, of which He spoke in the Apocalypse by St. John, after having spoken of it by the mouth of Isaiah; and He showed me the spot where to find it."[36] God may have expelled humanity from Eden, but Columbus was now ready to retake it by storm.

The reconquest of paradise took some five hundred years. Some seventeenth-century explorers, inspired by Sir Walter Raleigh, placed Eden further north, at precisely 35° north latitude, around modern Tennessee, which

34. Aisling Byrne, *Otherworlds: Fantasy and History in Medieval Literature* (Oxford University Press, 2016), 37. This is a reputed model for the semi-mythical paradise, St Brendan's isle.

35. Edmundo O'Gorman, *The Invention of America* (Indiana University Press, 1961), 4.

36. Kirkpatrick Sale, *The Conquest of Paradise: Christopher Columbus and the Columbian Legacy* (Hodder and Stoughton, 1991), 190.

it was claimed had a perfect climate.[37] There was much speculation about the American natives, whom some thought descended from the ten lost tribes of Israel. In 1530 the historian Peter Martyr D'Anghiera thought they had never been "civilised nor had any intercourse with any other races of men. They live, so it is said, as people did in the golden age, without fixed homes or crops or culture".[38] Other early accounts described wives, land, and property alike as held in common, sometimes hinting at the association of virtue with equality.[39] John Locke wrote in 1689 that "In the beginning all the world was *America*, and more so than that is now; for no such thing as *Money* was any where known."[40] As Henri Baudet indicates, the idea of the "noble savage" soon emerged.[41] To the eighteenth-century French explorer Lahontan, the Hurons were "strangers to . . . *Meum* and *Tuum*", and lacked the vices resulting from private property in Europe.[42] The Jesuit missionary Father Joseph Lafitau, too, identified them with, in Gilbert Chinard's words, an admirable "combination of idyllic simplicity and spartan severity", which he attributed partly to their being "ignorant of all the refinements of vice which luxury and abundance have introduced".[43] By the Enlightenment, ideas of a life lived according to "nature" overlapped with what many people imagined by "utopia". Here, as Lois Whitney puts it, the primitive was associated with the "good, wise, benevolent", and civilisation with corruption and degeneracy.[44] Sparta had been reinvented for a new age.

37. Louis B. Wright, *The Colonial Search for a Southern Eden* (University of Alabama Press, 1953), 42.

38. Peter Martyr, *De Orbe Novo: The Eight Decades of Peter Martyr D'Anghera*, 2 vols (1530; G. P. Putnam's Sons, 1912), 1: 380.

39. Hugh Honour, *The New Golden Land: European Images of America from the Discoveries to the Present Time* (Allen Lane, 1975), 12.

40. John Locke, *Two Treatises of Government*, 2nd ed., ed. Peter Laslett (Cambridge University Press, 1970), 319.

41. Henri Baudet, *Paradise on Earth: Some Thoughts on European Images of Non-European Man* (Yale University Press, 1965), 28. This remains the best brief introduction to this theme.

42. Tzvetan Todorov, *On Human Diversity: Nationalism, Racism, and Exoticism in French Thought* (Harvard University Press, 1993), 273.

43. Father Joseph François Lafitau, *Customs of the American Indians Compared with the Customs of Primitive Times*, 2 vols (Champlain Society, 1974), 1: xcix, 90. The editor continues: "therefore, Rousseau became Lafitau's main continuator". Lafitau also commented that Lycurgus had insisted that Spartan houses were made only of wood, since he "feared emulation, which, opening the way for luxury and magnificence, would have brought them out of that state of frugality and equality" which underpinned their success (2: 22–23).

44. Lois Whitney, *Primitivism and the Idea of Progress in English Popular Literature of the Eighteenth Century* (Johns Hopkins Press, 1934), 7. See also Adam Kuper's *The Invention of Primitive Society: Transformations of an Illusion* (Routledge, 1988).

The discovery of the Americas presented Europeans with the intriguing and potentially exhilarating possibility of regressing to primitive virtue. So America, first generically and then as the United States, that most island-like of nations, became the utopian topos upon which many of the Old World's most fervent hopes were focused from the fifteenth until the late twentieth century. By the early 1800s the view was common, as the French pamphleteer François Barbé-Marbois put it, that "The golden age, a fiction of the Old World, is realized in the New."[45] The attraction of the New World lay in the possibility of self-remaking, and literally in the discovery of medicines and "wonderfull cures of sundrie greate deseases" for healing hitherto incurable European ailments, as a 1577 text put it.[46] America's conquerors saw themselves as "new men". Indeed, the apostolic fervour of Christianity, which, John Leddy Phelan reminds us, had "long since died out in the Old World", was now reborn.[47] Yet extreme virtue and genocide were never far asunder in the New World. For those conquerors whose ambitions defined the new golden age, the pillage and exploitation of the alter ego of the noble savage, the pagan cannibal, proceeded at a pace and on a scale virtually unprecedented in human history. Some thought the natives anxiously awaited conversion to Christianity, and hence salvation, and were better off dying as Christians than living as pagans. To its indigenous inhabitants the Americas were soon a dystopia on the greatest scale ever conceived. Eventually about 90 per cent—perhaps fifty-five million people, a tenth of the world's population—succumbed to violence or disease. And so, by 1820, did some 75 per cent of those who had been moved there forcibly, mostly as African slaves. The price of Europe's utopia, which depended on this racial annihilation, was high indeed.

Utopia and Millenarianism

A longing for Christ's return to earth to re-establish divine rule has long been a source of utopian thought. It has antecedents in Judaism, where the Messiah transforms this world assisted by divine intervention.[48] Known as millenarianism, after the thousand-year period of rule which would supposedly occur, such beliefs are also described as chiliasm, or eschatology, the doctrine of the

45. Quoted in Harry Levin, *The Myth of the Golden Age in the Renaissance* (Oxford University Press, 1969), 68.

46. Howard Mumford Jones, *O Strange New World: American Culture; The Formative Years* (Chatto and Windus, 1965), 37.

47. John Leddy Phelan, *The Millennial Kingdom of the Franciscans in the New World* (University of California Press, 1956), 45.

48. Joseph Klausner, *The Messianic Idea in Israel* (George Allen and Unwin, 1956), 10.

end of the world.[49] They are sometimes represented as bringing "heaven" to earth. The analogy is not entirely apt, however. Many images of heaven are quite austere, hierarchical, and dictatorial, and resemble an ascetic monastery more than a happy society, though there are utopias like this too. Augustine (354–430 CE) thought there would be no friendships, household, or family there, but equally no strangers, only "the society of the angels and the heavenly community" united in undifferentiated love.[50] Some later images of heaven, however, are more mundane. Emmanuel Swedenborg's, in the eighteenth century, included considerable sensual gratification, spiritual development, and communal and family love, while the Mormon image has marriage and procreation too.[51]

Expectations of Christ's return commenced in the first few centuries CE, and are often associated with Roman persecution of Christianity. They assumed two main forms. Pre-millennialists believed that Christ appears before the millennium, while to post-millennialists he arrives after it. Either way, an apocalypse occurs. Amongst various accounts of its arrival, a key text here is the Book of Revelations (especially chapter 20), written *c.*64–93 CE. Lactantius (d. *c.*320) forecast Christ's return within two hundred years, when "all evil will be swept off the earth and justice will reign for a thousand years", after which the devil would be released and defeated in final battle. Then "After God's coming the just will gather from all over the world, and after his judgment the holy city will be set up at the centre of earth, and God himself will dwell in it with the just in control." Classical sources are freely intermixed here, giving a Cockaigne-ish twist to the narrative. The millennium implied returning to the Golden Age of Saturn, where "streams of wine shall run down, and rivers flow with milk". For God "had given the land for all to share, so that life should be lived in common, not so that a ravenous, raging greed should claim everything for itself; what was produced for all should not be denied to any . . . but all were equally well off because abundant and generous giving was done by those with to those without."[52]

For many centuries the interpenetration of visions of the ideal life makes a strict separation between utopianism and millenarianism difficult.[53] Like the

49. For an introduction, see D. Rudolf Bultmann's *History and Eschatology* (Edinburgh University Press, 1957).

50. Quoted in McDannell and Lang, *Heaven*, 64.

51. McDannell and Lang, 183.

52. Lactantius, *Divine Institutes* (Liverpool University Press, 2003), 291.

53. For an introduction, see Ernest Lee Tuveson's *Millennium and Utopia: A Study in the Background of the Idea of Progress* (University of California Press, 1949) and Theodore Olson's *Millennialism, Utopianism, and Progress* (University of Toronto Press, 1982), which indicates how

magical world view it notionally replaced but really incorporated, "religion" was not a separate domain, or an occasional dalliance with the lip service of obligation, but a primary way of understanding and belonging to the world. To Sylvia Thrupp, the term "millennium" "may be applied figuratively to any conception of a perfect age to come, or of a perfect land to be made accessible."[54] This implies a considerable overlap with perfectionist definitions of utopianism, and allows "millennium" to be associated with secular images of the good life. But it is unwise to blur this distinction unduly. To Norman Cohn, the quest for salvation is the main dividing line here.[55] As defined by More's humanism, utopianism does not aim at salvation, for the term has no precise secular equivalent. "Salvation" refers to eternity. For More, "utopia", by contrast, makes this life good, without necessarily implying forgiveness of "sin" or complete moral regeneration. Divine intervention also plays no role in descriptions of the founding of most subsequent utopias. Though Christian utopians often insist on the identity of the millennium and utopia, fusing the two terms and the phenomena they describe has some unfortunate consequences. As we have seen, assuming that utopia involves "perfection" or living in a "state of grace" "free of sin", or as a quest for redemption, in the sense of forgiving sin, or salvation, merges the utopian into the religious and subordinates the institutional and programmatic to the mystical.[56] These are attributes of millenarianism,

---

serious this confusion is (5–6n). Olson concludes that "Millennialism and utopianism, while almost antithetical in character, are combined in the notion of progress" (7), an ideal which presupposes that "*there is a blind force in history, a force uncontaminated by historical contingency, yet dedicated to the continued improvement of man*" (9, italics in original). Millennialism by contrast relies on three assumptions: "history is conceived of as 1) the locus of divine action and, accordingly, the only source of salvation or historical vindication; 2) a strongly periodized drama; and 3) as proceeding to its conclusion in a deeply dialectical fashion." Utopianism is described as resting on the view that "man's proper focus is on transcendent truth derived from a higher realm in a hierarchical universe. History is not a source, or the source, of value or truth in the task of embodying what is necessarily and eternally true" (7–9).

54. Sylvia Thrupp, ed., *Millennial Dreams in Action: Studies in Revolutionary Religious Movements* (Schocken Books, 1970), 12.

55. Norman Cohn, "Medieval Millenarism: Its Bearing on the Comparative Study of Millenarian Movements", in Thrupp, *Millennial Dreams in Action*, 31–43.

56. Levitas, in *Utopia as Method* (12–13), attempts to bring utopia closer to theology, based on Bloch and Tillich. Some utopians have made this claim. The founder and paterfamilias of the Oneida community, John Humphrey Noyes, declared himself free from sin in 1834, and followed Paul's proclamation that "Ye are not under law, but under grace" (George Wallingford Noyes, ed., *Religious Experience of John Humphrey Noyes* (Macmillan, 1923), 120). My point is twofold: that the concept does not fit the genre as a whole, and that it is not identical to that

not utopianism. They are theological categories, not social and political postulates.

Nonetheless the two ideals are difficult to prise apart, and confusion between them persists. Their meeting ground is secular millenarianism, where we get many of the promises of the millennium without divine assistance. Here, the entire burden of dramatic moral improvement is shifted onto humanity. Here, utopianism mimics the logical structure of religious ideals, and portrays human behaviour as capable of perfection rather than constrained by the more-limited malleability which secular humanism presupposes.[57] Central here is the assumption that degenerate or unregenerate humanity becomes capable of conversion and purification through something like a process of secular baptism, where our sins are washed away and we are "renewed". Once "saved", people are expected to possess saint-like qualities, especially where controlling the passions is concerned.

Secular millenarianism thus hints at profound, dramatic, total, and rapid moral change towards a hitherto unrealised state of virtue. This may be portrayed as freedom from sin in an antinomian state of grace, completely superseding "alienation" and returning to a condition of psychological wholeness, or the anticipation of a "Novum", a completely unknown state of being, as Bloch, discussed below, proposes. To Theodor Adorno, utopia hints at a world "without hunger and probably without anxiety" where people "could also live as free human beings."[58] But completely eliminating anxiety is a dubious proposition, indeed impossible, and totally abolishing alienation smacks of millenarianism. Such assumptions are often linked to socialism, the main form of utopianism in the nineteenth and twentieth centuries. To Landes, the "*secular* dimension of millennialism—its insistence that redemption occurs in the world of time and history, in the *saeculum*—makes it possible for non-theistic versions to emerge, like utopianism and communism", making "redemption", not salvation, the shared ground of the secular and religious concepts.[59] "Redemption" here may have a more secular meaning. It implies a release from sin, thus, and the curse of labour, in this life, rather than any promise of an eternal afterlife.

The charge of secular millenarianism has been levelled against Marx and his followers, in particular, and a host of crimes linked to its adoption. Marx's ideas have been described by Michael Löwy as "typically a secularized version

---

pre- or post-legal internalisation of norms which is an ideal in both the anarchist and Marxist traditions, though the latter is clearly indebted to it.

57. See my *Dystopia*, 243–63.

58. Quoted in Ernst Bloch, *The Utopian Function of Art and Literature* (MIT Press, 1988), 4.

59. Landes, *Heaven on Earth*, 21.

of biblical messianism", and by Nicolas Campion as "the most powerful and enduring form of modern millenarianism".[60] Jacob Talmon uses "political messianism" to describe these trends, giving greater stress to "the religion of Revolution" than in his earlier description of "totalitarian democracy".[61] Nicolas Berdyaev suggests that for Marx "collectivity takes the place of the lost God". Here "the old Hebrew millenarism has come to life in a secular shape", with the proletariat as "the new Israel, and all the attributes of God's chosen people are transferred to it."[62] Bloch builds on the same tradition, citing the twelfth-century prophet Joachim of Fiore (*c.*1135–1202) and Thomas Müntzer's quest for a "pure community of love, without judicial and state institutions" as the ideal condition for humanity, and seeing utopia as manifesting apocalyptic consciousness.[63]

Often seen as one of More's sources, Joachim is the starting point for many of these discussions.[64] The greatest early medieval millenarian prophet, he proposed extending monastic common life and communal ownership to a wider population. His grand analogy projected three great stages of human progress, corresponding to the Father, Son, and Holy Spirit. This hinted at a potential process of perfectibility for mankind towards a state of universal love. Such notions were revived periodically thereafter, partially in response to the decay of feudalism and the insecure life of the landless peasantry in particular, as well as the new anxieties of oppression and exploitation in the towns. We can trace the allegory at least as far forward as Auguste Comte in the nineteenth century.

At least in the west, utopianism from the Renaissance onwards thus commences from a religious basis, as the dominant metaphor available for imagining the ideal society, while also drawing on the revival of classical republicanism. Millenarianism implied a symmetry between the ideal future and the ideal past, with history coming around full circle to return mankind to its virtuous origins and divine order. The crucial shift here came when humanity rather than

60. Michael Löwy, *Redemption and Utopia: Jewish Libertarian Thought in Central Europe* (Athlone, 1992), 14–15, 18; Nicolas Campion, *The Great Year: Astrology, Millenarianism and History in the Western Tradition* (Penguin Books, 1994), 427. For clarification see my *Marx and Marxism* (Penguin Books, 2018), 108–9. On Marx's approach to an ideal based on love see *Collected Works*, 3: 326.

61. J. L. Talmon, *Political Messianism: The Romantic Phase* (Secker and Warburg, 1960), 19.

62. Nicolas Berdyaev, *The End of Our Time* (Sheed and Ward, 1933), 39.

63. Ernst Bloch, *Thomas Münzer Als Theologe der Revolution* (Kurt Wolff Verlag, 1921), 179. See also Bloch, *Man On His Own: Essays in the Philosophy of Religion* (Herder and Herder, 1970), 133–41, and Klaus Vondung, *The Apocalypse in Germany* (University of Missouri Press, 2000), 195.

64. Abraham Friesen, *Reformation and Utopia: The Marxist Interpretation of the Reformation and Its Antecedents* (Franz Steiner Verlag, 1974), 23–24.

God became the agent of this transformation. For Cohn this commenced around 1380, in a series of peasant uprisings across Europe partly triggered by the undermining of medieval order by the Black Death.[65] Equality was a central theme here. In Britain, peasant leaders like John Ball began to preach as early as 1360 against the class system as such, using a language of "quasi-millenarian prophecy", and claiming that "things will never be well in England so long as there be villeins and gentlefolk".[66] The Peasants' Revolt of 1381 began by insisting that serfdom be abolished and land rents fixed. Then demands were made for "the abolition of lordship and the division of property between all men", though kingship remained untouched. Thus commenced the secular, and also the revolutionary, road to utopia.[67] Here one figure was central: Thomas Müntzer.

The Origins of Secular Millenarianism: Thomas Müntzer, Revolution, and Republicanism

Thomas More's direct influence on communist ideas in central Europe in the early sixteenth century was limited.[68] It was the Anabaptist Thomas Müntzer (*c.*1489–1525), rather, who donned the mantle of Joachitism, and converted millenarianism into a passionate revolutionary struggle.[69] To Abraham Friesen, the proposed "transformation . . . of the Christian life to a new morality" was not entirely invented on the spot.[70] The Black Death, which reached

65. Cohn, "Medieval Millenarism", 40. To Cohn, millenarianism is a "religious movement inspired by the phantasy of a salvation" in terms of five components. It is collective, in the sense that it is to be enjoyed by the faithful as a group; terrestrial, in the sense that it is to be realised on this earth and not in some otherworldly heaven; imminent, in the sense that it is to come both soon and suddenly; total, in the sense that it is utterly to transform life on earth, so that the new dispensation will be no mere improvement on the present but perfection itself, accomplished by agencies which are consciously regarded as supernatural. (31)

66. H. de B. Gibbins, *English Social Reformers* (Methuen, 1902), 15.

67. Alastair Dunn, *The Great Rising of 1381* (Tempus, 2002), 60, 68.

68. On the history of its translation and reception see Terence Cave's *Thomas More's Utopia in Early Modern Europe* (Manchester University Press, 2008), which discusses some fifty editions. The editor notes that while "most of the paratexts concede that Utopia is an ideal, an imaginary construction", there are cases which "claim unequivocally that the book is a practical guide to governance" (10).

69. Claus-Peter Clasen, *Anabaptism: A Social History, 1525–1618* (Cornell University Press, 1972), 184.

70. Joachim's influence came mostly through the Taborites: Friesen, *Reformation and Utopia*, 21–26.

Europe in 1348 and eventually killed half the population, naturally generated immense paranoia, hysteria, a profound sense of sin, and anticipation of the end of the world. In the 1420s groups like the Brethren of the Free Spirit, the Hussites, the Waldensians, and the Taborites anticipated an imminent Judgment Day, and asked the poor to pool their goods in common in anticipation thereof, claiming that they need no longer "pay rents to your lords any more, nor be subject to them". Some had libertarian and antinomian tendencies. A chiliast group called the Adamites lived on an island in the Luznic River in Bohemia (now in the Czech Republic). Recalling humanity's original condition, they went about naked, at least in warm weather, and preached sexual liberation as well as murder and plunder, claiming that all was permitted to those in a state of grace beyond sin.[71] Müntzer himself advocated community of goods to achieve apostolic purity and pave the way for a community of the "elect" on earth, who alone would be saved. He had read Plato's *Republic*, and displays a Joachite influence too. Both Joachim's monastic vision and Plato's elite communism were thus compatible with his vision.

To twentieth-century commentators like Bloch and Mannheim, Müntzer was the central figure who first brought heaven to earth, and demanded a millenarian transformation of behaviour in the image of Christian purity. To Cohn, Müntzer aimed to lead "the poor who were potentially the Elect, charged with the mission of inaugurating the egalitarian Millennium". In a final contest of the righteous, or elect, against the unrighteous, Christian peoples would overthrow their godless rulers, and "put away all luxuriousness, intemperance and come to know the advent of faith".[72] Here, however, the mask fell to reveal the true face of religious fanaticism. "The sword", Müntzer declared, was "necessary to exterminate" all the godless, "For a godless man has no right to live if he hinders the godly." The wages of sin were death, to be paid promptly. Indeed, to Cohn, "To judge by his writings he certainly showed far less interest in the nature of the future society than in the mass extermination which was supposed to precede it."[73] It would not be the last time zeal to reach heaven would result in trampling on anyone standing in the way.

Some of Müntzer's followers took his message to mean that all should be equal, share goods in common, and "be brothers and love one another like brothers". As one historian notes, "Relief from the tremendous conviction of

71. Howard Kaminsky, *A History of the Hussite Rebellion* (University of California Press, 1967), 340.

72. Cohn, *Pursuit of the Millennium*, 260; Thomas Müntzer, *The Works of Thomas Müntzer*, eds T. Matheson and T. Clark (T. and T. Clark, 1988), 301.

73. Cohn, *Pursuit of the Millennium*, 239.

sin and the yearning for a purity of life prompted them to share all things."[74] A unity of belief was also presupposed. Some suggested eliminating all pagan knowledge. At Anabaptist Münster in 1534 all books but the Bible were banned, and some demanded that the rest be burnt. There was, however, some ambiguity as to whether the injunction that "all things be one and communal, alike in the bodily gifts of their Father in heaven" applied only to "Saints", or to the entire congregation. Some assumed that an obligation to assist their neighbours was sufficient, and that apostolic poverty, usually defined by reference to Acts 2:44–45, was unnecessary.[75] The core of the "medieval millennial promise", as Landes terms it, was the assumption, "abolish private property and evil will vanish from the earth."[76] At various places through the 1570s, property was indeed shared, and sometimes wives. The luxury and avarice of rich and corrupt priests were also explicitly targeted.

Yet paradoxes abounded amidst these militant proclamations of virtue and holiness. Most Anabaptists denounced splendour in dress, and in Moravia (now in the Czech Republic) they ceased making sumptuous garments altogether. But at Münster the clothing of the wealthy was appropriated and displayed to a "theatrical" extreme, occasioning "a good deal of grumbling".[77] (Distinction in dress, fashion, and the expense of clothing recur constantly as utopian issues from the sixteenth century even to the present.) In Bohemia the rebel Taborites sought to punish such sins by death, though here too some of those Cohn calls the "elite of amoral supermen" "made themselves garments of truly regal magnificence, which they wore under their tunics." These "Free Spirits" or "adepts" openly proclaimed that in a state of grace every degree of pomp and luxury was accessible to them.[78] Their antinomianism, the assumption of being above all laws, or possessing attributes associated with divinity, included the idea that "man in his present life can acquire such a degree of perfection as to render him wholly sinless"; that in this condition "a man can freely grant his body whatever he likes"; that "those who are in this degree of perfection and in this spirit of liberty are not subject to human obedience . . .

74. Cohn, 258; George Huntston Williams, *The Radical Reformation* (Weidenfeld and Nicolson, 1962), 124.

75. George Huntston Williams, ed., *Spiritual and Anabaptist Writers* (SGM, 1957), 277. See, generally, James M. Stayer, *The German Peasants' War and Anabaptist Community of Goods* (McGill-Queen's University Press, 1991), 95–159.

76. Landes, *Heaven on Earth*, 291.

77. Karl Kautsky, *Communism in Central Europe in the Time of the Reformation* (T. Fisher Unwin, 1897), 253; Friedrich Reck-Malleczewen Bockelson, *A History of the Münster Anabaptists*, eds George B. von der Lippe and Viktoria M. Reck-Malleczewen (Palgrave-Macmillan, 2008), 92.

78. Cohn, *Pursuit of the Millennium*, 149, 231.

for, as they assert, where the spirit of the Lord is, there is liberty"; "that man can attain the same perfection of beatitude in the present as he will obtain in the blessed life to come; that the perfect soul does not need to practice acts of virtue"; that "the carnal act is not a sin, since nature inclines one to it"; and that the members of the sect "should not stand up when the Body of Jesus Christ is elevated, nor show reverence to it".[79] Some possibly even thought such a state of freedom "releases all from servitude including those who had been previously bound to a king or other lord".[80] A debt to some images of the Golden Age is obvious.

Such ideas indicate a transfer of the projected heaven to earth, and were shared by other Protestants at the time. Luther too assumed that in heaven "As the world will have an end, so also will government, magistry, laws, the distinction of ranks, the different orders of dignity, and everything of that nature. There will be no more any distinction between servant and master, king and peasant, magistrate and private citizen."[81] This may look like utopian equality, but it is not. It is a state of grace. One of its last incarnations would be with Nietzsche, or at least the vulgar adaptation of the "superman" ideal. Those who know Hitchcock's *Rope* (1948), which explores the theme that murder is permitted to those beyond good and evil, will recognise the delusion. It also has interesting parallels with a state of war, where soldiers justify what would otherwise be crimes by the higher authority of following orders, or dealing with a less-than-human enemy. This line of thought is incompatible with a universalist utopianism.

Müntzer has long been assumed to foreshadow the emergence of modern revolutionism, where utopian ideals are clad in sometimes sumptuous millenarian garb. To Friedrich Engels, writing in 1850, Müntzer was a "magnificent figure" to whom the "kingdom of God" meant "a society with no class differences, no private property and no state authority independent of, and foreign to, the members of society." Though "the class struggles of those days were clothed in religious shibboleths, and though the interests, requirements, and demands of the various classes were concealed behind a religious screen", "the chiliastic dream-visions of early Christianity offered a very convenient starting

79. Howard Kaminsky, "The Free Spirit in the Hussite Rebellion", in Thrupp, *Millennial Dreams in Action*, 166. For English parallels, see also Gertrude Huehns's *Antinomianism in English History* (Cresset, 1951).

80. Robert Lerner, *The Heresy of the Free Spirit in the Later Middle Ages* (University of California Press, 1972), 87–88. Feudalism had also been a target in Johann Eberlein von Günzburg's *Wolfaria* (1521), where all live by agriculture and wear similar plain clothing.

81. Quoted in McDannell and Lang, *Heaven*, 154.

point", if not one capable of modern application.[82] Following Bloch, Mannheim sees Müntzer's "orgiastic chiliasm" as the first expression of the utopian mentality: "The decisive turning-point in modern history was, from the point of view of our problem, the moment in which 'Chiliasm' joined forces with the active demands of the oppressed strata of society." Now, "Longings which up to that time had either been unattached to a specific goal or concentrated on other-worldly objectives suddenly took on a mundane complexion." The resulting "spiritualization of politics", demanding on earth what had previously been imagined only in heaven, marks a key turning point in history: "It is at this point that politics in the modern sense of the term begins, if we here understand by politics a more or less conscious participation of all strata of society in the achievement of some mundane purpose." For Mannheim this was, however, driven not by flux in or a fundamental alteration of ideas, but rather "had its roots in much deeper-lying vital and elemental levels of the psyche", namely the "collective unconscious", a concept indebted to Le Bon, Freud, and Jung.[83] To Cohn, what Müntzer specifically understood was that:

> A social struggle is seen not as a struggle for specific, limited objectives, but as an event of unique importance, different in kind from all other struggles known to history, a cataclysm from which the world is to emerge totally transformed and redeemed. This is the essence of the recurrent phenomenon—or, if one will, the persistent tradition—that we have called "revolutionary millenarianism".[84]

The leading Protestant sects roundly condemned such trends throughout the sixteenth century, with the Helvetic Confession in 1566 deriding "Jewish dreams that there will be a golden age before the Day of Judgment, and that the pious, having subdued all their godless enemies, will possess all the kingdoms of the earth."[85] The following century witnessed many variants on millenarian and utopian thought, where returning to original equality or creating a New Jerusalem sometimes loomed large. Seventeenth-century millenarianism was by no means always radical.[86] But the English Revolution witnessed an intense revival of millenarian sentiments, amongst movements associated with the Fifth Monarchy men, Ranters, Diggers, and Quakers, who saw renovation in this world as a key goal: "We look for a new earth as well as a new heaven", wrote

82. Marx and Engels, *Collected Works*, 10: 409, 412, 415, 422.

83. Mannheim, *Ideology and Utopia*, 190–92, 36. See my *Dystopia*, pt. 1.

84. Cohn, *Pursuit of the Millennium*, 281.

85. Quoted in Jeffrey K. Jue, *Heaven upon Earth: Joseph Mede (1586–1638) and the Legacy of Millenarianism* (Springer, 2006), 126.

86. As in the case of Joseph Mede, who supported the Church of England: Jue, 35.

the Quaker Edward Burrough in 1659.[87] The language of the millennium and utopia were now interwoven in an ideal associated with the dispossessed. Through the English Civil War (1642–51) the idea of an imminent millennium grew in strength and influence. One Fifth Monarchist, Peter Chamberlen, said property only divided humanity into factions, and until "the world return to its first simplicity, or . . . to a Christian Utopia . . . covetousness will be the root of all evil."[88] Such views had political dimensions: some Fifth Monarchists challenged the legitimacy of the existing system of kingship as repugnant to God, and in the 1670s some were involved in a plot to kill the king.[89] But others, it has been asserted, focused, at least in England, more on godly behaviour than godly rule.[90]

The most direct inheritor of Müntzer's spirit in this period, and a resolute opponent of luxury, was the English millenarian republican Gerrard Winstanley (1609–76).[91] He established the fundamental postulate of exploitation in contending that "all rich men live at ease, feeding and clothing themselves by the labors of other men, not by their own". Originally the "whole mankinde walked in singlenesse and simplicity each to other". But once the stronger imagined that they might "have some larger part of the Earth then they, and be in some more esteeme then others, and that they should acknowledge me in some degree above them", this thought, "the Serpent that deceives the man", meant "Mankinde falls from single simplicity to be full of divisions, and one member of Mankinde is separated from another, which before were all one, and looked upon each other as all one." As with Thomas More, the key problem for Winstanley was that the rich were depriving the poor of the right to cultivate formerly common land. The former, he insisted, "tell the poor People that they must be content with their poverty, and they shall have their Heaven heareafter", adding, "But why may not we have our Heaven here, (that is, a comfortable livelihood in the earth.)" He urged ending all buying and selling, and money too, in his chief utopian work, *The Law of Freedom in a Platform;*

87. Christopher Hill, *Some Intellectual Consequences of the English Revolution* (Weidenfeld and Nicolson, 1980), 58.

88. Quoted in Christopher Hill, *The World Turned Upside Down: Radical Ideas during the English Revolution* (Temple Smith, 1972), 92.

89. B. S. Capp, "Extreme Millenarianism", in *Puritans, the Millennium and the Future of Israel: Puritan Eschatology, 1600 and 1660; A Collection of Essays*, ed. Peter Toon (James Clarke, 1970), 87, 89.

90. William M. Lamont, *Godly Rule: Politics and Religion, 1603–60* (Macmillan, 1969), 173.

91. See, generally, Timothy Kenyon, *Utopian Communism and Political Thought in Early Modern England* (Pinter, 1989); J. Davis, *Utopia*; James Holstun, *A Rational Millennium: Puritan Utopias of Seventeenth-Century England and America* (Oxford University Press, 1987).

or, True Magistracy Restored (1652). The revolution, he thought, had "made a free *Commonwealth*" where "the common Land belongs to the younger Brother, as the enclosures to the elder Brother, without restraint". Beggars and idleness would disappear once the poor, exercising the right given them by God at the creation, cultivated the common land, with families ruled patriarchally, as Winstanley imagined Adam had governed, and fathers beating their recalcitrant children. Plenty would result, and "*This Freedom in planting the common Land will prevent robbing, stealing, and murdering*".[92]

During the English Revolution, a few groups tried to practise such ideas. Most of the Levellers, in A. L. Morton's view, "envisaged a society of secure and independent small producers whose property needed protection from the great owners and rich trading monopolies."[93] Some possibly proposed an agrarian law limiting landed property.[94] The Diggers, certainly, attempted to farm common land and implement Winstanley's ideas with respect to equality, commencing on what they proclaimed as the first day of Christ's millennial kingship, 1 April 1649. The Familists, or Family of Love, led by Henry Niclaes, inspired ideas of communal marriage. Throughout the century there was also another, countervailing utopian trend which stressed the need to employ the poor with greater efficiency and to increase production. Gabriel Plattes's *A Description of the Famous Kingdom of Macaria* (1641), which cited More and Bacon, for instance, urged that landholdings be limited to what could be cultivated, and advised that those who failed to improve their lands should be banished, with a 5 per cent inheritance tax for national improvement.[95] This heralded the so-called "full-employment" utopias of the subsequent period.[96] Some of these sought to utilise the advantages of monastic-style communities. In *A Way Propounded to Make the Poor Happy* (1659), an early scheme for planned settlements of the poor, the Dutchman Peter Plockhoy insists that "when God by the dispensation of his secret Counsell, joyned some such together, as do agree with his divine will, and with the rules of nature, and they will not exchange their union or fellowship for all the riches in the World."

92. Gerrard Winstanley, *The Complete Works of Gerrard Winstanley*, 2 vols (Oxford University Press, 2009), 2: 289, 216–17, 244, 246, 248.

93. A. L. Morton, *The World of the Ranters: Social Radicalism in the English Revolution* (Lawrence and Wishart, 1970), 179.

94. Jonathan Scott, *Commonwealth Principles: Republican Writing of the English Revolution* (Cambridge University Press, 2004), 82–83.

95. Charles Webster, ed., *Samuel Hartlib and the Advancement of Learning* (Cambridge University Press, 1970), 82.

96. See J. K. Fuz, *Welfare Economics in English Utopias from Francis Bacon to Adam Smith* (Martinus Nijhoff, 1962), 18–62, and J. Davis, *Utopia*, 288–368.

Thus "in this way of amity God only is the bond, wherewith they are tyed together without being lyable to be unloosed, and upon which foundation being fixed, they resolve to withstand all assaults whatsoever". So he enjoined, "let us reduce our friendship and society to a few in number, and maintain it in such places as are separate from other men, where we may with less impediment or hinderance, love one another".[97]

Besides Christian millenarian sources, arguments favouring limiting property in the seventeenth century were also bolstered by the revival of classical republicanism. Various disciples of Machiavelli, in particular, warned of the dangers of luxury and gross inequality of property to social and political stability as well as civic and individual freedom.[98] Most republicans viewed inequality of property as the greatest threat to a free commonwealth. In Britain the central figure was James Harrington (1611–77), author of the leading republican and quasi-millenarian utopia of the period, *Oceana* (1656), whose immense influence on both sides of the Atlantic has been documented by J.G.A. Pocock. In his view, *Oceana* is less a work of fiction than an early example of the "historical utopia" "in which social forces supposedly operating at the time of writing are shown reaching culmination and resolution in an ideal future."[99] Its proposed constitution praised Sparta's "agrarian" as a means of avoiding the greatest danger to any state, an over-concentration of property, and warned that luxury accompanied any shift of power towards a monarchy or oligarchy, which would bring "such corruption of manners as should render them uncapable of a commonwealth." Harrington regarded covetousness as the root of all evil and proposing limiting landed estates for Britain to ensure that at least five thousand properties existed.[100]

Amongst Harrington's associates, Marchamont Nedham's *The Case of the Commonwealth of England* (1650) noted of an "absolute community" of property that "neither the Athenian nor Roman levelers ever arrived to this high pitch of madness, yet we see there is a new faction started up out of ours known by the name of Diggers" who offered "a new plea for a return of all men *ad tuguria* [in peasant cottages]; that like the old Parthians, Scythian nomads, and other wild barbarians, we might renounce towns and cities, live at rovers

97. Peter Plockhoy, *A Way Propounded to Make the Poor Happy* (1659), 16, 31.

98. See Z. S. Fink, *The Classical Republicans: An Essay in the Recovery of a Pattern of Thought in Seventeenth-Century England* (Northwestern University Press, 1962), and, generally, J.G.A. Pocock, *The Machiavellian Moment: Florentine Political Thought and the Atlantic Republican Tradition* (Princeton University Press, 1975).

99. Pocock, in Harrington, *Political Works*, 74.

100. Harrington, *Political Works*, 202. For commentary, see Rachel Hammersley's *James Harrington: An Intellectual Biography* (Oxford University Press, 2019, 102–7).

[without a definite aim], and enjoy all in common."[101] Nedham's *The Excellencie of a Free-State* (1656) stressed that avarice and luxury at Rome meant that "the love of their Country was changed into a Study of Ambition and Faction" and warned that "where Luxury takes place, there is as natural a tendency to Tyranny, as there is from the Cause to the Effect".[102] Later seventeenth- and early eighteenth-century so-called neo-Harringtonians or "Commonwealthmen" like Walter Moyle and Henry Neville even resurrected the Platonic image of Sparta under Lycurgus, and lamented its decline under the corrosive effects of luxury and growing inequality of property.[103] One neo-Harringtonian literary utopia, *The Free State of Noland* (1696), bans the "meer Luxury and Vanity" associated with the royal court.[104]

While religious enthusiasm abated somewhat in Britain during the eighteenth century, both republican and millenarian trends still continued. Anglican divines like Richard Clarke still expected a "*paradisiacal* world," where the curses induced by the Fall would "go away gradually, as the progressive restitution of all things comes on, and is ripening through this *Millennial* reign". John Wesley anticipated "an unmixed state of holiness and happiness, far superior to that which Adam enjoyed in Paradise . . . a more beautiful Paradise than Adam ever saw".[105] Awakened with a dramatic fervency, this spirit would later permeate the French Revolution period, through the writings of Godwin and others, and extend itself in the early socialism of Owen as well as the vision of communist society portrayed by Marx.

101. Marchamont Nedham, *The Case of the Commonwealth of England, Stated* (1650; University Press of Virginia, 1969), 109.

102. Marchamont Nedham, *The Excellencie of a Free-State* (1656; Liberty Fund, 2011), 12, 31.

103. In "An Essay upon the Lacedaemonian Government", discussed in Caroline Robbins's *Two English Republican Tracts* (Cambridge University Press, 1969, 30).

104. *The Free State of Noland* (J. Whitlock, 1696), 16.

105. Quoted in David Spadafora, *The Idea of Progress in Eighteenth-Century Britain* (Yale University Press, 1990), 125.

3

Theories of Realised Utopianism

THIS BRIEF OVERVIEW of various medieval and early modern concepts of the ideal society has stressed the confluence of religious and classical sources with respect to ideas of limiting property holding in particular, and emphasised the evolution from the early sixteenth century onwards of concepts of heaven, Eden, and the Golden Age into demands to transform the earth in their image. This implied that utopia might actually be created. We turn now briefly to consider some much later theoretical defences of such a proposition.

To describe utopianism as positing a realistic dimension, not necessarily definitive of everyday life but existing as an alternative to it, a eutopian island both imaginary and real within an often-shark-filled dystopian sea, we need to introduce three concepts: *heterotopia*, *liminality*, and the *concrete utopia*. All indicate how everyday life can be conceived as intersecting with, harbouring, or promoting utopia at various levels. Though problems exist with some applications or interpretations of these concepts, all suggest that daily life can be a rich and rewarding utopian milieu, if only locally and temporarily. Later we will consider some intentional communities, which are sometimes termed "practical utopias", insofar as they indicate ways of working together co-operatively.[1] Part III then explores how such proposals might be extended in both space and time in the coming decades.

Michel Foucault and Heterotopia

In a 1967 article entitled "Different Spaces", Michel Foucault examined how different social spaces exemplified contrasting sets of rules as "heterotopias".[2] While the concept appeals to some historians of utopianism, Foucault's main

1. E.g., by Dennis Hardy in *Alternative Communities in Nineteenth Century England* (Longman, 1979, 1–10).

2. Reprinted in Michel Foucault, *Aesthetics, Method, and Epistemology*, transl. Robert Hurley et al. (Allen Lane, 1998), 175–85. On the uses of the concept, see, e.g., Roland Ritter and Bernd Knaller-Vlay's *Other Spaces: The Affair of the Heterotopia* (Haus der Architektur, 1998), Eric C.

aim was more mundane; namely, to describe behavioural heterogeneity across social spaces.[3] Utopias are "emplacements having no real place" which "maintain a general relation of inverse or direct analogy with the real space of society": "They are society perfected or the reverse of society, but in any case these utopias are spaces that are fundamentally and essentially unreal." By contrast, other spaces exist that are "designed into the very institution of society, which are sorts of actually realized utopias in which the real emplacements" are "at the same time, represented, contested, and reversed".[4]

This suggests a theory of actually existing utopian spaces. Yet the locations described—brothels, theatres, cemeteries, museums, libraries, fairs, holiday camps, public baths, and, above all, ships—are primarily self-enclosed places removed, at least temporarily, from broader social mores and norms. They are defined by separation and isolation, not "utopian" norms as such, in the sense of being moralised spaces guided by higher or more humane principles. They are not readily differentiated into "sacred" and "profane" spaces. They are not "sorts of actually realized utopias", or oases in a cruel world, but vary from other spaces in querying the norm's "normality". They are just different modalities of sociability which indicate an often-temporary shared "otherness" defined by specific boundaries. But they are so multitudinous as to challenge the very idea of a "norm" as such.

Nonetheless, some such locations seemingly share utopian qualities. "Taboo" spaces in earlier societies have sometimes protected the "impure"; for instance, menstruating women. Various forms of sacred place, like churches, or spaces defined by pilgrimage, have embodied specific moral principles which are "higher" or more humane than those of the wider society, thus providing for instance the possibility of sanctuary, or mercy, for criminals. The traditional Persian garden, a key source for western ideas of paradise generally, could also function to order and refine wider outside spaces, later becoming "a sort of blissful and universalizing heterotopia". Planned colonies—Foucault mentions the Puritans in North America and the Jesuits in Paraguay—also represent "absolutely perfect places . . . in which human perfection was effectively achieved".[5]

Smith's *Foucault's Heterotopia in Christian Catacombs* (Palgrave Macmillan, 2014), and Michiel Dehaene and Lieven De Cauter's *Heterotopia and the City: Public Space in a Postcivil Society* (Routledge, 2008). Foucault had used the concept earlier, in *The Order of Things* (1966), without elaborating on it.

3. See Tobin Siebers, ed., *Heterotopia: Post-modern Utopia and the Body Politic* (University of Michigan Press, 1994); Dehaene and De Cauter, *Heterotopia and the City*.

4. Foucault, *Aesthetics, Method, and Epistemology*, 178.

5. Foucault, 184. The seventeenth-century Jesuit "reductions" in Paraguay were much discussed in eighteenth-century Europe as they involved community of property as well as an apparently humane approach to indigenous peoples.

Part of Foucault's point is thus simply that moral norms and practice are spatially differentiated, and norms of propriety vary enormously. We do not dress the same way for the beach as for the office, or talk the same way at funerals as in cafés. Singing along with the music in clubs is fine; in commuter trains it is not. Talking loudly on one's phone on the train is accepted in London, but prohibited in Tokyo. Some heterotopian spaces are clearly more "utopian" than others. These are our concern here. They are not defined solely by formally demarcating "other" or different spaces. Alterity is not the issue: we require a core definition of utopian content, of what is markedly "better"; namely, one rooted in sociability.

We can see why "heterotopia" might assist a theory of utopianism, and why a "utopian heterotopia" might describe a pocket of utopian activity in a non-utopian society. Yet Foucault's essay is suggestive rather than definitive. Whether literary or communal, utopias are typically spaces removed from the rest of the world. Here specific norms are adopted, institutions designed, and behaviour modified, to produce greater order, equality, and happiness, and less anxiety, hatred, fear, and oppression. Utopias are akin to sacred spaces where higher moral norms govern behaviour than are practised elsewhere, especially with respect to consent and coercion. "Heterotopia" allows us to imagine spaces where everyday utopianism flourishes, and which function critically to show up, satirise, or condemn wider social practices. Some types of festival, we will see, are a key example of this process.

Arnold van Gennep, Victor Turner, and Liminality

The suggestion that specifically utopian types of heterotopian space involve an atmosphere of expectation, anticipation, or optimism about the unfolding of the much better "not-yet" but still "possibly-coming", which will be discussed shortly, requires introducing another concept. The idea of liminality gives us a theory of time which corresponds to heterotopia's conception of space. Developed by the anthropologists Arnold van Gennep and Victor Turner, liminality describes a process of passing from one state or demarcation of time and social status to another, typically in rituals which involve the three stages of separation, transition, and incorporation or reaggregation—for example, from youth into puberty or marriage.[6] Such processes are assumed to define social life throughout history and have a wider collective function. Derived from the Latin *limen*, or "threshold", liminality describes a heightened state of expectation or anticipation, as well as an intensified reorientation of spatial boundaries, particularly between sacred and profane. Turner refers to

6. See Arnold van Gennep, *The Rites of Passage* (Routledge and Kegan Paul, 1960), 11.

"liminoid" spaces, which are neutral or privileged areas away from the mainstream, akin to heterotopias, like universities (as they once were). He speaks of liminal phases of tribal society which invert, without usually subverting, the status quo. Here "reversal" indicates to a community that "chaos is the alternative to cosmos", a process typically played out in "saturnalian or lupercalian revelry"; in other words, in specific types of festival. Liminality is further understood as fomenting "communitas", an intensified feeling of belonging in small, often non-literate societies, but capable of being expressed on a much larger scale: "Revolutions, whether violent or non-violent, may be the totalizing liminal phases for which the limina of tribal *rites de passage* were merely foreshadowings or premonitions."[7]

Liminality thus reorders time to produce "an instant of pure potentiality" when "the past is momentarily negated, suspended, or abrogated, and the future has not yet begun". Not all forms of liminality involve an intensified desire for equality or communitas, though all generate expectations of altered behaviour. But such moments entail experiences of social unity, which Turner sees as projected in utopian form, especially during religious revivals, and, in Europe, during early capitalism. Indeed, these theoretical constructs might form the basis for a utopian model of society, Turner proposes, if only we could move beyond merely positing ideal structures of "what the world or land or island would look like if everyone sought to live in communitas with his and her neighbor."[8]

Following the Polish sociologist Florian Znaniecki, Turner develops the idea of communitas, defined as "a bond . . . uniting people over and above any formal social bonds":

> In communitas there is a direct, total confrontation of human identities which is rather more than the casual camaraderie of ordinary social life. It may be found in the mutual relationships of neophytes in initiation, where communitas is sacred and serious, and in the festal ecstasies of great seasonal celebrations. In many initiations the participants undergo a leveling process. The distinctions of their previous status, sex, dress, and role disappear, and as they share common trials and eat and sleep in common, a group unity is experienced, a kind of generic bond outside the constraints of social structure, akin to Martin Buber's "flowing from I to Thou." Communitas,

7. Victor Turner, *From Ritual to Theatre: The Human Seriousness of Play* (Performing Arts Journal, 1982), 33, 41, 44–45, 48–49, 53. Paul Goodman uses "communitas" to denote the feeling of community, rather than the space a community occupies. See Percival Goodman and Paul Goodman, *Communitas: Means of Livelihood and Ways of Life* (1947; Vintage Books, 1960).

8. V. Turner, *From Ritual to Theatre*, 48–49, 53.

> however, does not merge identities; instead it liberates them from conformity to general norms, so that they experience one another concretely and not in terms of social structural (e.g., legal, political, or bureaucratic) abstractions.[9]

Here, "communitas" reflects Simmel's concept of sociability as "an ideal sociological world in which the pleasure of the individual is closely tied up with the pleasure of others."[10] Sociation merely acknowledges the need for interaction within existing social structures. For the dispossessed or oppressed, who benefit little from these structures, it does not validate our humanity, which requires asserting and recognising equality and is an essentially nutritive acknowledgment of mutual dignity. Temporarily, this process also functions as a vital escape valve, permitting rising social tensions to dissipate through short-lived fantasy, play-acting, and exuberant release. This mimics the compensatory function of the utopian idea. Our doubts, tensions, anxieties, and guilt allayed, we find it easier to maintain daily life in oppressive and exploitative societies, as virtually all are. Rebaptism and evangelical or political conversion to a "new" self perform the same function.

This gives us an anthropological perspective on much of what we call utopia, as well as a historical schema for understanding how these phenomena mutate in the modern period, and why "history itself seems to have discernible liminal periods", or waves of acute criticism and transition which precede and demand social transformation.[11] To Turner, Van Gennep understood that liminality applies "to all phases of decisive cultural change, in which previous orderings of thought and behavior are subject to revision and criticism, when hitherto unprecedented modes of ordering relations between ideas and people become possible and desirable". Liminality involves "not only transition but also potentiality", not only "going to be" but also "what may be".[12] It gives a historical grounding and meaning to ideas of intense group hope and expectation. It explains how continuous moments of expectation bind groups together, especially through festive celebration. The latter term, derived from the Latin *celeber*, "numerous, much-frequented", reflects the qualities of vivacity or, in Emile Durkheim's term, "effervescence", exhibited by crowds of people

9. Victor Turner, ed., *Celebration: Studies in Festivity and Ritual* (Smithsonian Institution Press, 1982), 205.

10. See Simmel, *Sociology*, 42–48.

11. Victor Turner, *Dramas, Fields, and Metaphors: Symbolic Action in Human Society* (Cornell University Press, 1974), 285.

12. Victor Turner and Edith Turner, *Image and Pilgrimage in Christian Culture* (Basil Blackwell, 1978), 2.

who share common purposes and values. These originally sacred moments revolved around "objects and activities believed to be charged with supernatural power".[13] As the sacred meanings fall away, reinforcing the group itself becomes a main function of the celebration. How fine it is to shout with the crowd, and how much louder we are together. "Look at us!" we proclaim, as our mutual exuberance reinforces our personal sense of well-being, strength, and unity when bonded with and protected by the multitude. Thus, after a fashion, it is now the group which becomes sacred, and the giver of norms. Temporarily we fuse with it, and belongingness overwhelms our egotism.

Turner thus adopts from Simmel a distinction between "sociation" and "sociability". The former means "being with and for others in that construction of society out of contending interests, duties, and purposes," while "sociability" is "the autonomous or play-form of sociation . . . the feeling . . . and satisfaction . . . of being sociated." Sociability, the pure joy of interacting with others, is evident specifically in carnival, with its joking, laughter, mocking, and humour.[14] It celebrates the moment, its special quality and alterity, and the joy of feeling acceptance. To Simmel, its "aim is nothing but the success of the sociable moment and, at most, a memory of it."[15] To Turner, its meaning lies primarily in an "experience of unity", or what we have here termed belongingness. But the "great difficulty is to keep this intuition alive".[16] Momentarily we affect equality, mocking the foibles and pretensions of our supposed social superiors. Then this moment passes, before the danger of repercussion commences. Temporarily achieving Tönnies' community of feeling is one thing. Maintaining it in an unequal society is quite another. So, too, Turner's communitas, "representing the desire for a total, unmediated relationship between person and person, a relationship which nevertheless does not submerge one in the other but safeguards their uniqueness in the very act of realizing their commonness", seems extraordinarily elusive, and verges on the religious.[17] This language is shared by the young Marx. It presumes a "totality" and unmediated intimacy or transcendental unity, which are at best extraordinarily rare, and cannot even in principle be long sustained.

In the modern period liminality is clearly associated with secularised millenarianism.[18] The millenarian quest for redemption and salvation exceeds

13. V. Turner, *Celebration*, 13, 16.

14. V. Turner, 28–29.

15. Simmel, *Sociology*, 45.

16. V. Turner, *From Ritual to Theatre*, 47.

17. V. Turner, *Dramas, Fields, and Metaphors*, 274.

18. Christopher Lasch, *The True and Only Heaven: Progress and Its Critics* (W. W. Norton, 1991), 40–41. A critical assessment of this view is offered in Hans Blumenberg's *The Legitimacy*

what the more secular and less perfectionist traditions of utopianism usually aim at, thus requiring a clear distinction between them. Only the religious forms of utopianism aim at "heaven on earth".[19] To Turner, millenarian movements seek communitas through promoting homogeneity, equality, the absence of property, anonymity, humility, and unselfishness, by abolishing ranks, and by maximising religion and piety.[20] Modern revolutionism mimics this process insofar as it shares with millennialism the expectation of novelty, even of complete moral renewal and rebirth, a wiping clean of our moral slates by purging guilt, and a condition of sublime peace as an end state. But a sense of communitas can also be developed in other religious forms, notably in pilgrimages, treated below, which represent a kind of moveable utopia. To Turner these "exhibit in their social relations the quality of communitas; and this quality of communitas in long-established pilgrimages becomes articulated in some measure with the environing social structure through their social organization."[21] Typically this is achieved by personal sacrifice, self-renunciation, group prayer, common dining, with all often eating and drinking from the same vessels, and fasting, all of which foster more intimate and equal group bonds. Pilgrimage thus represents a release from the ordinary world, which

> liberates the individual from the obligatory everyday constraints of status and role, defines him as an integral human being with a capacity for free choice, and within the limits of his religious orthodoxy presents for him a living model of human brotherhood and sisterhood. It also removes him from one type of time to another. He is no longer involved in that combination of historical and social structural time which constitutes the social process in his rural or urban home community, but kinetically reenacts the temporal sequences made sacred and permanent by the succession of events in the lives of incarnate gods, saints, gurus, prophets, and martyrs.[22]

To Turner, then, communitas "often appears culturally in the guise of an Edenic, paradisiacal, utopian, or millennial state of affairs, to the attainment of

---

of the Modern Age (MIT Press, 1983), pt. 1, which argues that the idea of a Last Judgment was not transferred into historical thought (32).

19. To Bloch, "religion seeks the utopian perfection in the totality, and salvation places the individual case completely within the universal, places it into 'I will make everything anew'" (*Utopian Function of Art*, 147).

20. Victor Turner, *The Ritual Process: Structure and Anti-structure* (Transaction, 2008), 111–12.

21. V. Turner, *Dramas, Fields, and Metaphors*, 166–67. On some limits of Turner's approach, see Maribel Dietz's *Wandering Monks, Virgins, and Pilgrims: Ascetic Travel in the Mediterranean World, A.D. 300–800* (Pennsylvania State University Press, 2005, 30–35).

22. V. Turner, *Dramas, Fields, and Metaphors*, 207.

which religious or political action, personal or collective, should be directed. Society is pictured as a communitas of free and equal comrades—of total persons".[23] It aims at a form of sociability like that ideal of friendship which Tönnies described as a *Gemeinschaft* community of feeling tied to neither blood nor locality.[24] Turner did not, however, think the hippies of his own time had the capacity to transform their "Edenic fantasy . . . the ecstasy of spontaneous communitas" into a sustained condition. Paraphrasing Rousseau, he adds that:

> Spontaneous communitas is a phase, a moment, not a permanent condition. The moment a digging stick is set in the earth, a colt broken in, a pack of wolves defended against, or a human enemy set by his heels, we have the germs of a social structure. This is not merely the set of chains in which men everywhere are, but the very cultural means that preserve the dignity and liberty, as well as the bodily existence, of every man, woman, and child.[25]

Clearly a theory of utopianism might profitably draw upon a concept of liminality. Utopia is the place we aim to go, and utopianism is the process which gets us there. Both seeking and arriving at utopia involve a "rite of passage" in leaving the inadequate, lapsed condition of unequal humanity, then passing through a transitional phase to a higher level of existence. Would that the passage from adolescence to adulthood, so sacred to many, could represent this transition—humanity could proclaim a genuine maturation! The physical space of utopia appears here as a variant of sacred spaces throughout history, where "higher" moral norms prevail—the occasional human sacrifice notwithstanding—than in the wider or previous society. This higher moral existence and the closer social bonds which define it is utopia's telos, or end. In utopianism, the concept of liminality applies particularly to transitions from old to new, and the conscious anticipation of the qualitative superiority of the new over the old. This the festival does for sociability, and the revolts of the 1960s, to take another example treated below, did for cultural transformation more widely. Liminality can also be understood psychologically as part of a periodic process by which we expunge guilt and responsibility and wipe our moral slates clean.

We must, however, eventually reach the stage of reaggregation or incorporation. We cannot remain forever in transit, in limbo, homeless, on the road, or, like millenarians, in a permanent state of expectation. Nor, conceptually,

23. V. Turner, 237–38.

24. Tönnies, *Community and Civil Society*.

25. V. Turner, *Ritual Process*, 139–40.

outside of theology, can we sustain an eternal state of "bliss": bliss exists only by contrast to experiencing non-bliss. The same is true of "grace", orgasm, intoxication, the "experience of unity", and most other forms of idealised and intensified temporality. If every day were Christmas, or our birthday, there would be nothing special about Christmas and birthdays; it is the contrast and their uniqueness which makes them so. Yet the reaggregative stage represents a "higher" form of normality, preserving something of the greater sociability of the liminal stage. The festival ends, and we return from the pilgrimage. What do we retain from these liminal experiences? Have we changed, personally or morally, and more to the point, is our society different as a result? If pilgrimage partly involves reconstructing our identity, how much of the new identity remains once the process ends?

To Turner, then, liminality is utopian in three main senses. Firstly, it denotes a period when "anti-structures" clash with dominant social structures. Secondly, the condition of liminality may posit communitas, or a more intense social bond than we usually experience, sometimes in overtly utopian form. Thirdly, this communitas may be "spontaneous, immediate, concrete", and come to define an entire way of experiencing sociability.[26] These three meanings correspond so closely to what we have defined as utopianism that liminality must be seen as one of its core components.

Ernst Bloch and the Concrete Utopia

So far we have introduced two concepts, heterotopia and liminality, to clarify potential intersections between utopianism and everyday life. The writings of the best-known theorist of utopia, the German philosopher Ernst Bloch (1885–1977), defend a third relevant idea, the "concrete utopia", freed from "ideology and abstraction".[27] Given the standard and, after 1918, official Marxist condemnation of all utopianism as inferior to the "scientific socialism" incarnated in Bolshevism, Bloch offers the foremost twentieth-century Marxist attempt to re-legitimise utopianism within socialism. Yet his approach too has limitations, which, given Bloch's prominence in utopian theory, require detailed exploration. In particular, Bloch never escaped that faith either in the historical destiny of the revolutionary proletariat which distinguishes Marxist socialism or in Marx's vision of the coming communist society. He also adopted a quasi-mystical idea of this transformation which hinders more

26. V. Turner, *Ritual Process*, 127.

27. Bloch, *Utopian Function of Art*, 43. We must necessarily pass over this epistemological claim here.

secular theories of utopia, while appealing to those who promote utopia as a new form of religion.[28] These weaknesses reduce, without eliminating, his usefulness today.

Like Victor Turner, Bloch studied under Georg Simmel. Exiled in the United States from 1938 to 1949, he was shunned by the Frankfurt School for defending the Stalinist show trials in the late 1930s. In East Germany between 1949 and 1961, Bloch resisted attempts to reform its political system, which he broadly supported. Suspected of heterodoxy, however, he was forced to retire after the 1956 Hungarian Uprising. When the Berlin Wall was erected he moved to West Germany. Here he was more critical of the USSR: his verdict on the Prague Spring's suppression in 1968 was that "The slogans and alibis circulating in the Soviet Union today are pure ideology, and the best that can come from them is the warning: This is not the way to act."[29] Focused on the writings of the young Marx, Bloch's Marxism was also intermixed with speculative idealism. Jürgen Habermas linked him to traditions of Jewish mysticism, and termed him "A Marxist Schelling", after the German philosopher who conceived the ideal as emerging from the real.[30]

Bloch's engagement with utopianism commenced with *The Spirit of Utopia* (1918), written in 1915–16 and revised in 1923 against the backdrop of the Russian Revolution. Contemporaries said Bloch saw himself as a new messiah poised at the brink of the apocalypse, heralding what Michael Löwy calls "a new Church—the 'space of a still-flowing tradition and a link with the End'—as one of the necessary outcomes of the socialist revolution".[31] Here utopia is the "waking dream", a fantasy expressed and fulfilled in art and the "utopian reality of transcendental opera" as well as politics. Two themes introduced here became central to Bloch's later thinking: the idea of the "not-yet-conscious", and Marx's "revolutionary mission absolutely inscribed in utopia".[32] Bolshevism appears as "a transitionally necessary evil" heralding the replacement of the state by "an international regulation of consumption and production". To Bloch, Marx presented "the very first true total revolution, the end of every class struggle", in which "an entirely new class steps forward, social nothingness, emancipation as such". Marx had surpassed "the all too Arcadian, the abstract-utopian kind of socialism—appearing since the Renaissance as the secular mode of the thousand-year Kingdom". Instead he offered an image of

28. For Bloch's religious views, see in particular *Atheism in Christianity* (Verso, 2009).

29. Bloch, *Utopian Function of Art*, xii.

30. Jürgen Habermas, *Philosophical-Political Profiles* (Heinemann, 1983), 61–78.

31. Löwy, *Redemption and Utopia*, 139–40.

32. Ernst Bloch, *The Spirit of Utopia* (1918; Stanford University Press, 2000), 3, 103, 187, 237.

"utopia's distant totality", "everyone producing according to his abilities, everybody consuming according to his needs, everyone openly 'comprehended' according to the degree of his assistance."[33] There is a sense, too, that this socialism solved a more existential problem, "the basic motif of socialist ideology" being that each person needs to "have things set right with himself, with his moral party membership"—the title of this section is, after all, "Karl Marx, Death, and the Apocalypse". Thus Marx is conceived as resolving German idealism and Christianity alike, as indeed he saw himself. So the fact "*that there can be a Heavenly Kingdom*" becomes "*simply necessary*".[34] Bloch later described this and his second work, a biography of Thomas Müntzer, as expositions of "revolutionary Romanticism" and as anticipating Bakunin's idea of free association beyond states. Both Müntzer and Bolshevism heralded the "end of all time", the "final return to paradise".[35] Such language, even if more allegorical than literal, is best described as millenarian rather than utopian.

Bloch's famous "principle of hope" is centrally identified with utopia in his chief work, *Das Prinzip Hoffnung*, written during 1938–47, and revised in 1953 and 1959.[36] Here he revived the idea of the "not-yet-conscious" as an aspect of emerging awareness of the new, while still insisting that Marx's version of communism was the inevitable and only legitimate destination of such hopes. To Bloch, "Everybody's life is pervaded by daydreams". Utopia commences as a daydream driven by a desire for wish fulfilment, rooted in a psychological need for hope and expectation which is intensified by dissatisfaction and unhappiness.[37] But such wishes assume many forms, including images of the past, religious conceptions of paradise, fairy tales, imaginary travel, film and theatre. All "either present a better life, as in the entertainment industry, or sketch out in real terms a life shown to be essential."[38] "Hope" is thus not identical with "utopia" or utopianism, which are only one subset of its results, or with a "dream", or even collective and social rather than individual fantasies.

This approach is thus suggestive, but hardly unproblematic. Images of a "better life" are not utopian as such. Indeed, they may reject utopia as overambitious

33. Bloch, 239–40, 244–46.

34. Bloch, 268, 276.

35. Bloch, 279. See Bloch, *Thomas Münzer*. Müntzer is, however, broadly described in terms of chiliasm rather than utopianism (67–85).

36. Its first title was "Dreams of a Better Life". Accounts of Bloch include Wayne Hudson's *The Marxist Philosophy of Ernst Bloch* (Macmillan, 1982), Vincent Geoghegan's *Ernst Bloch* (Routledge, 1996), and Peter Thompson and Slavo Žižek's *The Privatization of Hope: Ernst Bloch and the Future of Utopia* (Duke University Press, 2013).

37. Bloch, *Principle of Hope*, 1: 1.

38. Bloch, 1: 13.

or, more commonly, deflect us from utopian goals by mere entertainment, sucking us into unrealisable escapist fantasies like superhero films, which give vent to our psychological frustrations but confuse or obscure our sense of reality, for instance by pushing us back into a magical world view. Our aspirations may also demand other people's servitude: my glory means conquering you. Hope can be more subjective or wishful, and as "hopium" may be a mere distraction; or it may be more objective, and rooted in real possibility. Commonly it focuses on improving the present, more and better of the same, or a removal of current woes, rather than offering alternatives to it. Or it represents vague and unspecified improvements.

Bloch's acolytes sometimes propose that the "core of utopia is the desire for being otherwise, individually and collectively, subjectively and objectively."[39] But "otherwise" is too vague to be helpful. They also describe utopia as consisting in the "desire for a better way of being or of living", where the "essence of utopia seems to be desire". Besides being circuitous, this implies so many possible outcomes as to be unhelpful, since all life involves desires of many kinds.[40] This reductionist misinterpretation of Bloch shrinks utopia to a psychological principle, and a general desire for improvement, an impulse to "better our condition", or vaguely to hope in general. This dilutes the specificity of the utopian considerably. Whenever we are in pain, or mere discomfort, we seek the "otherwise" and the "better": winning the lottery would define most people's "otherwise". We constantly desire a multiplicity of different "other" things and feelings. But most of them are far from utopia. We need to know more precisely what is "good" about the "good place". Merely being "better" than our current condition, insists J. C. Davis, "is too imprecise, vague and subjective".[41] This approach merges utopia into a general theory of progress and leaves it without a distinctive content, as well as reducing it essentially to a psychological principle.[42]

If such narratives subsume utopia under the western idea of progress, Bloch himself is at points much clearer about the utopian "better". While broadly adopting Marx's vision of communism, Bloch early on expressed an admiration for Bakunin's idea of "the confederation, freed of all 'states' and of all authoritarian organizations."[43] He also projected a new ideal of the relationship between work and leisure. In a delightful neologism, Bloch described how "Sundayness" (*Sonntäglichkeit*) would mean that we would "no longer have

39. Levitas, *Utopia as Method*, xi.

40. Levitas, *Concept of Utopia*, 7, 181. Cf. the subtitle to Fredric Jameson's study of utopia *Archaeologies of the Future: The Desire Called Utopia and Other Science Fictions.*

41. J. Davis, *Utopia*, 13.

42. Levitas, *Concept of Utopia*, 7.

43. Quoted in Löwy, *Redemption and Utopia*, 141.

any separate Sundays and holidays, but just as it will have the hobby as a profession and the public festival as the finest manifestation of its community, so it will also be able, in a happy marriage with the mind, to experience with it its *festive weekday*."[44] To Hans Jonas this "earthly paradise of active leisure" would supersede differences between work and leisure as well as city and country.[45] We will return to this concept in part III. Nonetheless, the juxtaposition is, at least for the time being, overstated. Everyday cannot be Christmas, or even Sunday, for then the special quality of the day is lost. What Bloch implies more practically, rather, is merging work and play as far as we can, but without the distinction of challenge and effort by contrast to relaxation and celebration being lost. Moreover, this has also to be reconciled today, we will see, with diminished growth and consumption, and sustainability. This Bloch, as a Marxist-Leninist, does not allow for, any more than most others writing in this period. For it was not yet necessary to do so.

Bloch identifies two landmarks in the development of utopian hope, whose destination he sees in Marx's ideal of "concrete freedom", "the socially succeeding will which has become communally clear". Firstly, like Mannheim, he dates modern utopianism to Müntzer's millenarian revolt.[46] Here, for the first time, the "not-yet-conscious", Bloch's description of how the past prefigures possible futures, hinted at its modern form. This "prepared the way for socialism from the sixteenth century on", and for Friedrich Engels's "millennium of freedom", which Bloch describes "as a concrete one, history's finality or the final chapter of the history of the world." Thereafter, More's *Utopia* exhibited "the first modern portrait of democratic-communist wishful dreams".[47]

Secondly, socialism then became the specific modern "concrete utopia", the ideal future condition wherein the "total man" is realised. But its achievement is not inevitable. History contains many possibilities, including non-utopian forms of novelty, progress, and improvement. To Bloch the central figure here is Marx, whose "philosophy of the New" rests on "real humanism". Bloch applauds "the complete reversal of alienation", a goal usually associated with the young Marx.[48] Its abolition could be understood through the concept of *Heimat*, a place "without expropriation and alienation, in real democracy . . . and in

44. Bloch, *Principle of Hope*, 3: 819, 912; Bloch, *Das Prinzip Hoffnung*, 2 vols (Aufbau Verlag, 1955), 2: 394.

45. Hans Jonas, *The Imperative of Responsibility: In Search of an Ethics for the Technological Age* (University of Chicago Press, 1984), 194–95.

46. Bloch, *Principle of Hope*, 2: 533. For Bloch on Mannheim's limitations, see *Heritage of Our Times* (Polity, 1991), 264–65.

47. Bloch, *Principle of Hope*, 1: 40–41; 2: 515, 519; Bloch, *Utopian Function of Art*, 137.

48. Bloch, *Atheism in Christianity*, 254.

which no one has yet been: homeland." This implies that what we have here called belongingness, akin to Marx's "species" identity, is the main solution to the social aspects of Marx's theory of alienation, "the return of man from religion, family, state, etc., to his *human*, i.e., *social*, existence", provided we include within the concept an ideal of solidarity, and the recognition of mutual dignity and respect—a sense, in other words, of belonging to the group defined as humanity.[49] But a "complete reversal" comes close to the promise of salvation and the secular millenarian project.

At this point, however, which has no direct conceptual parallel in Marx, millenarian mysticism clouds Bloch's vision.[50] This perfectionist description, aligned with the language of "final", "total", and "complete"—a place "no-one has yet been" sounds curiously like "salvation" in "heaven"—accords with Bloch's use of the term "Novum" to indicate a radically new condition which has never existed. Such a conceptual shift is inherently suspicious, for it threatens to subsume the utopian under the mystical and theological. (It has, however, been interpreted as principally psychological and historical.)[51] Yet this is also wedded to a fairly orthodox account of socialism's transition from the "utopian" to the "scientific", a narrative few accept today. To Bloch, "Concrete utopia designates precisely the power and truth of Marxism . . . It is only Marxism, together with science, that discovered this topos—in fact with the development of socialism from utopia to science" (written in 1959). Thus "science superseded" all the utopias "from Thomas More to Weitling":

> The rescue of the good core of *utopia* is equally overdue (as a concept which at the most lay in mist, never in deception); the *concrete-dialectical utopia of Marxism, that grasped and alive in real tendency*, is such a rescue . . . The socialist revolution is distinguished from its predecessors by its scientific character and concreteness, by its proletarian mandate and classless goal.[52]

49. "Economic and Philosophic Manuscripts of 1844", in Marx and Engels, *Collected Works*, 3: 297. Marx's fourfold account of alienation comprises alienation from the process of work, from the produce of labour, from species-being, and from other people. "Solidarity" could cover all of these forms. But "belongingness" does not solve the exploitation of labour.

50. Marx uses "Heimat" loosely in the "Economic and Philosophic Manuscripts", in discussing the worker's sense of alienation from his humble dwelling, but does not elevate the term conceptually. See Marx and Engels, *Collected Works*, 3: 314.

51. David Gross writes that its two meanings are "the inner awareness of what man could be but has not yet become" and "an epiphany . . . of something radically new, something which literally 'breaks into' history and shatters its orderly progression": "Ernst Bloch: The Dialectics of Hope", in *The Unknown Dimension: European Marxism since Lenin*, eds Dick Howard and Karl. E. Klare (Basic Books, 1972), 120.

52. Bloch, *Principle of Hope*, 3: 1376; Bloch, *Utopian Function of Art*, 107; Bloch, *Heritage of Our Times*, 137. On "Heimat", see Bernhard Schlink's *Heimat als Utopia* (Suhrkamp, 2000),

This implies that out of all the competing utopian narratives, the "concrete utopia" can have only one true end. Humanity's destiny is socialism, "the age without the cocky, sharp, heel-clicking wolves", which implies the triumph of a particular form of sociability. "Only socialist society can fulfil the wish of old age for leisure", and only abolishing private property can provide socialist freedom. But "*concrete order is not opposed to concrete freedom*".[53] This for Bloch requires, in the Leninist formulation, "democratic centralism", "common organization of the processes of production, a common unified plan of human information and cultivation."[54] No spectre of cocky Stalinist wolves clouds this vision.

Like Marx, then, Bloch thought the state and classes would eventually die out, following that proletarian revolution which the earlier utopians did not anticipate, but which remains the key step towards realising "dialectical materialist humanism". Marx had overcome the abstract character of previous utopianism. Since him, "mere utopianizing, apart from still having a partial active role in a few struggles for emancipation, has turned into reaction or superfluous playful forms." William Morris's "neo-Gothic Arcadia", the last form of the bourgeois utopia, is "redundant" by comparison, and later utopias are dismissed as "a diluted modernization of Thomas More". To Bloch, what separates Marx from the utopians is that his "Must", or moral imperative, "lies within the economically immanent manifestations of capitalist society itself and causes the latter to collapse only in immanent-dialectical terms." That is to say, jargon to one side, we cannot *impose* a new society upon reality based upon the standards of the ideal. What will be must emerge from what is—which is a tautology, since *whatever* happens can emerge only out of what exists. (But this also clearly reveals the mystical status of the "Novum" and of "Heimat".) Thus "Actual *descriptions* of the future", beyond the "sparse concept" of the classless society, are "deliberatively missing" in Marx, "precisely because Marx's whole work serves the future, and can only be comprehended and implemented at all in the horizon of the future, yet one which is not pictured in a Utopian-abstract way". For Marx, "the rejection of all fantasies of the so-called State of the future . . . occurred solely for the sake of the future, a comprehended future into which it was finally possible to travel with a map and compass . . . Marxism is *not no anticipation (utopian function), but the Novum of a processive-concrete anticipation*" and a "*unity of hope and knowledge of*

---

which discusses its post-1989 German meanings, and the difficulties of "Ortlosigkeit", the loss of a sense of place.

53. Bloch, *Principle of Hope*, 1: 41; 2: 533.

54. Bloch, 2: 534. Bloch does not mention Lenin here, but readers are doubtless aware of Lenin's modification of Marx's democratic theory, involving decisions emanating from the centre downwards via a tightly controlled centralised party mechanism.

process". Thus Marx, converted to a historical standpoint, was notionally a materialist first and foremost by 1846.[55]

This opposition to proposing or imposing grand, detailed blueprints of the future society is portrayed, as it is by virtually all Marxists, as a great advantage. There is some sleight of hand here, however. It can also be seen as a colossal ethical misunderstanding which was used to justify mass murder and the slaughter of whole classes and categories of individuals on the basis of the hypothesis that any means were appropriate to attain this supreme end.[56] "What counts" is only "actively conscious participation in the historical immanent process of the revolutionary re-organization of society". Aldous Huxley thought this end might justify "every atrocity on the part of true believers", and so it did.[57] Such a theory is clearly indebted to and parallels Christian and millenarian ideas of a state of grace beyond earthly morality, where all is permitted to the redeemed, since to "the pure all things are pure" (Titus 1:15). It exposes the limitations of Marx and Bloch alike vis-à-vis the possibilities of conceiving utopia. We never have to reveal what we have in mind, or plan, or seek. And when it all goes wrong we can blame "history", or inadequate revolutionary zeal, or a misinterpretation of dialectics, but never our own sacred theory, which permits us to do anything in the moment so long as we call it "imminent" and "dialectical".

Yet at the same time Bloch's achievement lies in his ability to use "utopia" in a positive sense, not conceived of by Marx, though implied in his early humanist writings in particular, to indicate the vital process of imagining the future. In *A Philosophy of the Future* (1963) Bloch emphasised that "*humanism has developed in utopia*", and renewed his assertion that "a real utopian critique" remained possible, that another "not-yet" lay ahead.[58] So utopia for Bloch is understood in two main senses: as one expression of the principle of hope, or

55. Bloch, 2: 530, 533, 547, 573, 581, 583, 615, 617, 620–22; 3: 1359; Bloch, *Utopian Function of Art*, 53. On Lenin's deviation from Marx, see my *Marx and Marxism*, 287–336. Whether a stress on rights mitigates the evils of democratic centralism is the key question here. Geoghegan describes Bloch's position here as one of "thoroughgoing Leninism" (*Ernst Bloch*, 127). Equally crucial is the fact that while Bloch centres his expectations on the narrative created by Marx's theory of alienation in 1844, the "transition from utopia to science" not only takes place after this (in 1845–46) but in key respects negates the humanist viewpoint which guided Marx in 1844. Indeed it can be contended that the abolition of all antagonism between humanity and the world, and of all alienation, is a theological concept of the type Marx clearly abandoned after 1845. Bloch has indeed often been seen as a quasi-theological thinker: see Geoghegan, *Ernst Bloch*, 102.

56. See my *Marx and Marxism*, 103–6, 455–56, and *Dystopia*, 128–76.

57. Aldous Huxley, *Prisons* (Trianon, 1949), 12.

58. Ernst Bloch, *A Philosophy of the Future* (Herder and Herder, 1963), 89.

expectation by the subject, and as a specific end or object created by historical development. Hope functions at the subjective level to inspire change; the modification of Marxism here challenges any economic determinism which guarantees and predicts, as Marx did, an inevitably victorious proletarian revolution emerging from capitalist crisis. And this revalidates not only the dream but the dreamers, especially the utopian intellectuals, who constantly posit the ideal by contrast to the real.

But what then is specifically "utopian" about what is expected or dreamt about? Why should dreams assume a utopian form or reach a utopian destination when they can posit many other end states, including mere entertainment, and fantasies about individual or group power, like nationalism? And what do we do when we realise, as Bloch observed wistfully whilst languishing in American exile in 1942, when considering the "Burial Ground and Commemoration of Utopias", that "It was better in the dream . . . The object, once manifested, is never the same as it was imagined to be, and the effect is all the more acute in relation to the vast number of badly, partially, or wrongly fulfilled wishes . . . so the dream remains better than the reality"?[59] Yet Bloch also acknowledged that "the waking dream of a thing is always nurtured by that which is missing". Even if "good dreams can go too far," as in the fairy tale where a "Wish becomes a command", the dream still functions as a critique of its object, ever spurring it onwards. But the object never becomes, and can never become, the realised dream.[60] This, as we have seen, is one of utopia's two key functions. In fact the dream and the fantasy are *always* better—that is the *point* of dreaming: we get what we want without the painful effort and compromises which reality invariably extorts from us. Utopia, then, is a mirage imagined by thirsty travellers, an oasis in the desert. But in aspiring to reach it we may encounter real water.

The substance of hope, Bloch asserted in his 1961 Tübingen inaugural lecture, was "real humanism".[61] This he believed was the essence of Marx's system. But this is in many respects an arbitrary destination, one wish within many dreams. Much hope aims at far lesser, and often antithetical, goals. Bloch himself was all too aware that bourgeois utopianism, particularly in the United States, had swallowed up much of modern expectation of change and improvement. Idealism had become fixated on the cornucopia of material prosperity, not the utopia of solidarity and social equality. And socialist dreams were hardly faultless either. Utopias, Bloch stressed, had long been "dreams of a

59. Ernst Bloch, *Literary Essays* (Stanford University Press, 1998), 464.

60. Bloch, *Utopian Function of Art*, 163, 171.

61. Bloch, xxv.

better life . . . a transformation of the world to the greatest possible realization of happiness, of social happiness." They needed too, however, to take into account art, religion, and much else in order to reach a "totality".[62]

In the realm of religion, in particular, this is a substantial innovation within Marxism. Gone is Marx's idea that religion is a mere opiate, a distracting delusion, an epistemological obstacle, a crude form of utopian projection, as for Feuerbach. Much hinges on whether we regard Bloch, who wrote of a "mythical God", as an atheist, for broaching a doctrine of secular messianism does not require God, only an account of charisma, which seemingly shifts much of the revolutionary burden from the proletariat to its leaders.[63] But this idea of atheism may mean only, in Peter Thompson's phrasing, "that one has to be against the Creator-God and the assumption of authority by the church and the state, who act as keepers of his word", leaving the possibility of adhering to a religion still searching for the "Holy Grail", and which still includes a "faith in Christ".[64] To Jack Zipes, Bloch aimed at "providing a religious basis to Marxism and uncovering atheism in Christianity." Religion was for Bloch a major source of utopianism, and thus of hope, and not just historically. But is "religion" here groupism, or theology, or both? To Zipes, Bloch presents the paradox of having God "eliminated in order for humankind to be liberated and seize the essence of religion", meaning reintegrated into "communality, fraternity, and solidarity", which is Marx's position, following Feuerbach. Religion to Bloch is "an evolutionary process of self-awareness" expressed through messianic figures like Moses and Jesus in particular. Its ultimate aim, in Bloch's words, was "a messianic kingdom of God without God."[65] But this religion without God smacks of mysticism. Wayne Hudson views Bloch's strategy with respect to religion as "open to serious criticism".[66] It is difficult not to agree that a "religion of hope" looks suspiciously like another variant on secular millenarianism, where "hope" merely substitutes for "faith", proposing faith where analysis belongs, and compensating for Marxism's failings by proffering a theory of charismatic messianic leadership, with Lenin as the latest prophet. (But this may be an improvement on Marx's notoriously skimpy theory of leadership.)

There is also the question as to how open-ended Bloch meant the category of "dream" to be. He distinguished between mere daydreaming and utopian

62. Bloch, 4.

63. Bloch, *Atheism in Christianity*, viii.

64. Bloch, xiv, xxiii.

65. Jack Zipes, *Ernst Bloch: The Pugnacious Philosopher of Hope* (Palgrave-Macmillan, 2019), 87, 94, 97, 135; Bloch, *Man On His Own*, 162. In Zipes's words, Bloch thought "Religion was a part of a process of human hope that can only be kept alive through Messianism" (94).

66. Hudson, *Marxist Philosophy*, 190.

daydreaming, contrasting "the goal of More and Campanella . . . always the realm of conscious dreaming, one that is more or less objectively founded or at least founded in the dream and not the completely senseless realm of daydreaming of a better life."[67] So defining utopia solely as the "desire for a better way of being or of living" was "completely senseless", and not accepted by Bloch himself, who crucially modified it by adding the goal of "social happiness" to give "dream" a specific social content. We need to separate this conception of dreaming from the anticipation which grows from the modern theory of progress. Bloch's "conscious" daydream is much more ambitious than "world-improving" or a dream of a "better life". Its social content resembles Turner's "communitas", or what is here described as the enhanced sociability which aims to promote belongingness. Social utopias aim at "the *eudaemonia of everyone*", happiness, and abolishing suffering and degradation.[68] This happiness includes a form of sociability or an ideal of friendship akin to the "kingdom of neighbourly love" sought by Christianity:

> The associative-federative utopias seek the birth of a new collective from among a circle of friends . . . this is the highly substitutive element in social utopias of this kind. Marxism, though radically opposed to the wolf-state, has no reason to expect salvation or even upheaval from friendship, the commune, and the small business resuscitated in it . . . And yet the old hope of friendship lives on, precisely in the image of a classless society it unfolds with new wishful and life dimensions.[69]

Sociability is part of the content of the future utopia, then, though we have already seen that friendship too is a complex concept. Natural rights is another component. By the late 1950s Bloch came to see the natural law tradition as key to moving beyond and against Stalinism. "Freedom as feeling does not appear in utopia but in natural law", he later wrote, in what many might regard as a remarkable concession to rights- and theologically based liberalism. Natural law had engendered the "heroic illusions" which instigated the French Revolution.[70] So neither Marxism nor any previous form of utopianism provides a complete image of the best possible future. In *Natural Law and Human Dignity* (1959), Bloch asserted that the essence of humanism, "human dignity", required reasserting the cause of "human rights". In the *German Ideology* (1845–46), Marx rejected this perspective as a bourgeois, idealist conception

67. Bloch, *Utopian Function of Art*, 5.

68. Ernst Bloch, *Natural Law and Human Dignity* (MIT Press, 1986), 205.

69. Bloch, *Principle of Hope*, 2: 964.

70. Bloch, *Utopian Function of Art*, 9, 15.

resting on a Feuerbachian, abstract, theologically rooted concept of human "essence". Marx also assumed that the need for such rights, which were rooted in private property, would end once property was held in common. Bloch agreed that virtually all crime would disappear in the future classless society; "fraud, theft, embezzlement, trespass, murder, and robbery all become more than outdated, they become as good as prehistoric." Yet natural law might provide what positive law did not, a claim for human dignity which "does not allow any authority (insofar as one is necessary) . . . to become cocky".[71] This elliptical allusion to Stalinism is unhelpful, however. It is constitutional checks and safeguards, legal protections, political pluralism, and a culture which upholds the rule of law and protects the individual which inhibit this "cockiness"—the apparatus of the state, and the mentality of its administrators, into which Marxists rarely delve in detail. Merely restating humanist principles will not do the trick, even if it aligns Bloch with Thomas More.

Bloch's "humanism", then, derives from Marx's "*categorical imperative to overthrow all relations* in which man is a debased, enslaved, forsaken, despicable being", an ambitious goal, which embellishes what we earlier described as a non-coercive ideal.[72] But this is of course the young Marx, before the breach which the *German Ideology* is usually conceived as representing.[73] Could the materialist conception of history still support such a "categorical imperative", which we normally associate with idealism, after morality had been relegated to being purely historical, relative, and class based? Bloch thought it could: "Socialism can raise the durable flag of the ancient fundamental rights . . . with illusions that have been seen through, corrected class ideology, perfected seriousness of the matter". Yet here he encounters an obstacle. Bloch claims that such rights could not be construed as "*innate* rights; they are all either acquired or must be acquired in battle", implying (as any secularist must) that all rights are merely conventional. Yet he insists that they could be derived from humanism, and rested on assuming an inherent sense of human dignity. This effectively makes them innate, and also at root theological—where else do we get "dignity" from?[74] For we are certainly not born displaying it. So again Bloch's "atheism" is questionable.

71. Bloch, *Natural Law*, 184, 187.

72. Marx and Engels, *Collected Works*, 3: 182.

73. It thus ignores the problems of any "epistemological break" between the texts. For the difficulties here, see my *Marx and Marxism*, 79–112, and further Louis Althusser's *For Marx* (Penguin Books, 1969). Strictly speaking, even such policies as communism itself would have to be excluded from being recommended if no ideals were being offered.

74. Bloch, *Natural Law*, 188. The concept is usually linked to Pico della Mirandola and then Kant. See Michael Rosen, *Dignity: Its History and Meaning* (Harvard University Press, 2012). In Mirandola it is intimately linked to humanity's unity with God (*Oration on the Dignity of Man*,

Bloch also supposed that such inalienable rights could not include a right to property, or rest on any idea of a "fixed generic essence of man". This is what the *German Ideology* rejects but, again, just what an innate sense of dignity implies. So dignity, once God is removed from the equation, cannot be innate, much less *the* definitively human quality. It can be only something we aspire to possess. It represents an "ought", indeed a utopian "ought": we *should* treat people with dignity. But Marx contended that "Communism is for us not a *state of affairs* which is to be established, an *ideal* to which reality [will] have to adjust itself", but rather "the *real* movement which abolishes the present state of things".[75] This statement announced his supposed break from the "utopians". It may be synonymous with the coming into being of proletarian class consciousness, and solidarity, the unity which will define the future society in formation. But Bloch's response is thus, like Marx's, somewhat convoluted and disingenuous: "the ideal looks entirely different when it is not immediately introduced into history from above, when it is concretely and utopianly extracted from its dialectical mediations". Thus the ideal still remains—the overthrow of inhumane relations—but it is "posited by the tendency" of the "development of the social being of the proletariat . . . and adjusted by the praxis of the tendency toward always deeper levels of reality".[76] Intellectually this is doubtless having your cake and eating it too. And once the proletariat fails to fulfil its historical mission, or even to understand its "social being", the theory is bankrupt, unless some messiah fortuitously intervenes to save it. When the ideal fails to be "imminent" it ceases to exist.

It is difficult to see Bloch's resolution of the problem of morality as more successful than Marx's. Bloch's is still a transhistorical humanist ideal even if it is realised only in practice. It is normative and prescriptive in introducing an abstract ideal of human nature based on dignity to bolster the theory. It is also ambiguous about restricting rights to those generated by work. Bloch describes the assertion of rights usually proscribed or merely formally recognised

eds Francesco Borghese, Michael Papio, and Massimo Riva (Cambridge University Press, 2012), 34). Whatever later variations emerge, its origins are irreducibly theological; that is, it is our divine status which gives us dignity, a quality we possess which animals do not, as Cicero had already argued. One could as well argue that humans are uniquely savage in their treatment of one another. The problem thus lies in a differentiation of the human from the animal, and the righteous privileges of murder and exploitation deduced for humanity from it. These issues are explored further in my "Socialism and the Language of Rights", in *Revisiting the Origins of Human Rights: Genealogy of a European Idea*, edited by Miia Halme-Tuomisaari and Pamela Slotte (Cambridge University Press, 2015, 206–36).

75. Marx and Engels, *Collected Works*, 5: 49.

76. Bloch, *Natural Law*, 192, 197.

in Marxism. Such rights are, however, often conceived of by socialists as those "of the working classes instead of the individual". They seem to be defined, thus, by labour, and are consequently subject to the same weaknesses as other claims of this productivist kind; namely, that they cannot project rights outside of work, which are rooted instead in humanity as such.[77] Here Bloch seemingly faults socialists. The "ultimate quintessence of classical natural law" requires that rights are grounded in "man, and not only his class": their basis is humanism, not the fact of labour. And so "the authentic inheritance of the natural law that was revolutionary" becomes "the abolition of all relations that have alienated man from things that have not only been reduced to being merchandise but are even stripped of their own value." Bloch thought that this assertion of rights would help balance the greater weight given to happiness by social utopias, by providing "juridical guarantees of human security or freedom as categories of human pride", hence promoting "*autonomy* and its *eunomia*" [right ordering]. This is an extremely significant innovation within Marxism, and scarcely one calculated to win official approval.[78]

As we have seen, however, even after leaving East Germany Bloch was rarely an outspoken critic of Leninism, and never proposed any substantial alternative to it. Some claim his Marxism never revealed "how an alternative society could be organised and maintained in purity over an extended period."[79] By 1968 he acknowledged that the USSR had generated a "personality cult, an extensive and absolutist centralisation, lack of room for any except a 'criminal' opposition, the terror and the police state, and an all-powerful state police". There had been a "comparatively weak emphasis on personal freedoms in Marx", and returning to natural rights could redress the imbalance.[80] But while Leninists complimented liberalism by adopting this language, its functioning within the Leninist state was negated by the absence of structures which balanced the power of the Party.

The connection between Bloch's two main propositions, the principle of hope and the concrete utopia, is moreover not as intimate as their proximity in his works suggests. Indeed, no necessary relation exists between them. The "principle of hope", the expectation of the "not yet", which refers to developing human nature as well as states of being, is not the "concrete utopia": the first is

77. On the origins of this perspective, see my *Machinery, Money and the Millennium: From Moral Economy to Socialism, 1815–60* (Princeton University Press, 1987, 31–32).

78. Bloch, *Natural Law*, 200, 203, 205. In 1968 Bloch noted that "There are men who toil and are burdened, those are the exploited. But in addition there are also men who are degraded and offended. Of course the exploited are also degraded and offended, but a distinction must be made between these two aspects" (interview at Korčula, in *Telos* 25 (1975): 171).

79. Hudson, *Marxist Philosophy*, 48.

80. Quoted in Geoghegan, *Ernst Bloch*, 128–29.

a mentality, the second its clearly delineated object, as well as the process of creating it. Hope can take us anywhere and everywhere. Mostly it does not take us towards utopia, or humanism, but rather towards ameliorating our own immediate condition, if only in compensatory fantasy. Exciting hope is thus not the goal of utopianism, since false or misleading or "bourgeois capitalist" hopes might be more damaging than a realistic assessment of our state of being. Anticipating the best future state we are capable of realising is, instead, what utopianism aims at. Envisioning the "not yet" is crucial to bringing it about. The concrete utopia thus grows out of the abstract utopia, not just out of Marxian praxis, which in any case must require subordinating any ideals to its own definition of the aim of classless society, as Bloch himself implies.[81] Hope helps to motivate and urge it on, but it must be utopian hope, not hope in the abstract, and not the false hope of proletarian revolution or pseudo-religious faith either. To Vincent Geoghegan, the concept of the concrete utopia is "problematic", given its dependence on Marx's theory of proletarian revolution and the immanent qualities utopia must possess as a result.[82] To Wayne Hudson, too,

> Bloch's concept of concrete utopia is thus more metaphysical than his identification of concrete utopia with Marx's "realm of freedom" would suggest . . . Clearly it would be helpful if Bloch distinguished more clearly between (1) concrete utopia as a content given now in human experience; (2) concrete utopia as the praxis of mediated anticipation of the good; and (3) concrete utopia as the correlate of such praxis with real not-yet-being.[83]

Yet at points Bloch does seem to indicate a clear preference for reinstating the role of ideals in Marxism, and of utopia as the chief amongst them:

> A merely empirically aimed Marxism eliminates two kinds of materials with utopian connotations which are actually inspirational in the end: firstly, *ideals* and secondly, the *finale utopica* in general . . . It is precisely these two elements which belong to Marxism, both today and tomorrow.[84]

Bloch's main achievements, then, lie firstly in having reinstated the role of utopianism within Marxism, and secondly in bolstering this with a more or less

81. As David Kaufmann, for example, acknowledges: see "Thanks for the Memory: Bloch, Benjamin and Philosophy of History", *Yale Journal of Criticism* 6 (1993): 143–62.

82. Geoghegan, *Ernst Bloch*, 152.

83. Hudson, *Marxist Philosophy*, 103.

84. Quoted in Hudson, 54.

idealist and religiously grounded natural-rights ideal. To Hudson, "Bloch attempts to supplement Marxism with a series of radical developments designed to counter the main *lacunae* of the Marxist tradition". He "contributed a new idea to the history of utopian thought. This was the idea that utopia was not 'no place,' but existed in the darkness of the lived moment and its transcending dynamic."[85] Yet Hudson adds that Bloch fails to "analyse the nature of Marx's utopianism in detail" and distinguish his own ideals from Marx's.[86] Undoubtedly Bloch remained unduly fixated on Marx, a great utopian no doubt, though not one, we will see, wholly adaptable to our own century's predicaments. But Bloch's adherence to an essentially messianic and apocalyptic tradition is still more problematic. His insistence that the "Novum" be unprecedented jars against the narrative offered here, where everyday utopianism has many precedents, parallels, and incarnations—indeed, is precisely an extension of what is most noble, generous, warm, and agreeable in humanity. The wish to break free of all human experience is an essentially religious, anti-historical strategy, a form of secular millenarianism, and a strategy and set of assumptions we have already rejected here as antithetical to secular utopianism. Here Kolakowski is harshest on Bloch, terming his expectation of "absolute perfection" "a speculative belief in the tendency of the universe towards a perfection about which we can predicate nothing". It remains entangled, if not incarcerated, within the prophetic mode, and lies beyond science, beyond materialism, and beyond empiricism. "As a philosopher," Kolakowski supposes, Bloch "must be termed a preacher of intellectual irresponsibility. He cannot be credited with inventing a Utopia, still less a 'concrete' one". The fetishism of "final perfection"—the secular millennium—is the rock on which his system splits.[87]

So if to some degree Bloch reconciles utopia and Marxism, he does not take utopia beyond either Marxism or theology, as we must obviously do today. Beyond the argument that it is unwise to continue to plump for noble lies or grand illusions, we will see that Marxism and, even more, Marxism-Leninism are wedded to a theory of growth, and, for Hans Jonas, to "the naïve Baconian idea of dominating nature". They are indeed "no less dedicated to the Baconian idea than its capitalist rival". For this reason, if no other, and despite the obvious merits of Marx's theory of alienation, as a world view Marxism-Leninism

85. Wayne Hudson, "Bloch and a Philosophy of the Proterior", in P. Thompson and Žižek, *Privatization of Hope*, 26; Hudson, *Marxist Philosophy*, 67.

86. Hudson, *Marxist Philosophy*, 57; and 59 for a delineation of Marx's utopianism.

87. Leszek Kolakowski, *Main Currents of Marxism*, 3 vols (Oxford University Press, 1978), 3: 428, 433, 445. And see, generally, 421–49 for one of the most sophisticated treatments of Bloch.

is today essentially conservative, if not reactionary.[88] It remains trapped within nineteenth- and twentieth-century assumptions about industrialisation in particular, and, if theoretically capable of embracing sustainability, practically lacks the will to do so. If the "concrete utopia" is limited to the parameters laid down by Marx, it offers the possibilities thrown up or permitted by history at any one time, but not the same range offered by the "abstract utopia".[89] It limits the concept to what Marxism permits, plus a supplemental theory of rights. So Bloch's two main deviations from Marx are in proposing a humanist account of rights, and in dressing utopia up once again, or rather disguising it, in religious and mystical garb.

To summarise this section, then. Firstly, suitably and critically adapted, Bloch enables us to understand existing practices, particularly of sociability, solidarity, co-operation, and comradeship as anticipating their own later generalisation, as an acorn to an oak, except that we must will the process of growth, which does not occur naturally. But such practices must be understood as ideals which we impose on and use to shape reality—exactly what Marx denied they were—and that is how we will treat them here. Bloch's main limitation as a thinker lies in conceiving utopia as a substitute for religion, so as a form of secularised millenarianism. This makes him popular amongst those who sanctimoniously view utopia as a halfway house between religion and unbelief, who see "faith" and "hope" as equivalents, and who promote "love" as *the* utopian virtue. But it is uncomfortable for secular thinkers, for whom utopia cannot serve as a fig leaf disguising religious, faith-based, anti-materialist, and anti-empiricist sentiments, and for whom religion encourages belief without foundation, which subversively spills over into social life generally. Thus the pious Bible salesmen of hope encourage an anti-rationalist approach to the subject, which in commending the quest for "totality" and complete newness, the "Novum", is millenarian rather than utopian according to the distinctions proposed here. This form of hope, lacking any concrete grounding, is merely synonymous with faith.[90]

88. Jonas, *Imperative of Responsibility*, 143–44. Jonas specifically foresees "catastrophic climate changes caused by the heating up of the atmosphere" (190).

89. The claim is sometimes made that what Bloch meant by "concrete utopia" was "what can be approached by reflexion and action such that eventually it would become reality, contrary to what is purely utopian and therefore impossible" (Rainer Zimmermann, "Transforming Utopian into Metopian Systems: Bloch's *Principle of Hope* Revisited", in P. Thompson and Žižek, *Privatization of Hope*, 248). But this is a dubious distinction.

90. Thus Hudson writes that "modern European Utopian thought retained *surrogate religious aspirations*, while offering a purely worldly Utopia to which, at best, only reduced expectations could be appropriate" (*Reform of Utopia*, 18).

Secondly, Bloch is also frequently linked to ideas of a "desire for a better world" or a "dream of a better life". But this is so far subsumed under general theories of progress that it fails to illuminate what is distinctive about utopianism. The latter usually seeks, we will see, neither merely marginal improvements of the sort implied by the term "better" nor the perfection usually associated with religion. Instead, this book argues, utopianism is more usefully conceived as a form of both imagined and practical ideal sociability. Nor, finally, can we accept any longer the official "*obligatory* optimism", in Cioran's phrase, of "real-existing" socialism, which led Bloch to pronounce that "a Marxist does not have the right to be pessimistic".[91] Where pessimism reflects a realistic assessment of humanity's condition and prospects, it may facilitate change rather than retarding it.[92] This is what we today call "following the science". To argue otherwise is simply to accede to the now-debunked myth of progress, and indeed also to the idea that such myths are necessary tools in humanity's movement. It thus elides, or confuses, the category of hope as a fetishised, quasi-theological principle, with rational, empirical, and historical assessment. If, however, we can adapt the concept of the "concrete utopia" to encompass the possibility of other realisable ends, some indebted to Marx and others not, we might project a concept which is viable in the present. This requires a realistic, not a mystical, vision of utopia.

Three more recent studies also discuss or project "real utopias", albeit without the threat of imminent catastrophe facing writers on the subject today. The first, Erik Olin Wright's *Envisioning Real Utopias* (2010), focuses on several examples, including participatory city budgeting, Wikipedia, the Mondragon worker-owned co-operative, and universal basic income. These are described as "grounded in the real potentials of humanity, utopian destinations that have accessible waystations, utopian designs of institutions that can inform our practical tasks of navigating a world of imperfect conditions for social change." The ideals focus on enhancing well-being and happiness, and "living fulfilling and meaningful lives". There is no sustained discussion here of catastrophic environmental degradation, which was not at that time so imminent, or of the worth of the utopian tradition in discussing alternatives to capitalism. The

91. E. M. Cioran, *History and Utopia* (Quartet Books, 1996), 96; quoted in Peter Thompson, "Religion, Utopia, and the Metaphysics of Contingency", in Thompson and Žižek, *Privatization of Hope*, 91.

92. Cf. Hudson: "Pessimistic forms of utopianism are compatible with realistic views of human capacities and dispositions" (*Reform of Utopia*, 37).

examples discussed are seen as halfway houses between capitalism and socialism, extending nineteenth- and twentieth-century arguments aiming at abolishing capitalism, but on the grounds that it inevitably created poverty and inequality rather than its destruction of the planet. Wright agrees, however, that capitalism normally "contains a systematic bias towards turning increases in productivity into increased consumption rather than increased 'free time.'" He argues against the view that "it is arrogant for left-wing intellectuals to disparage the consumption preferences of ordinary people." Desires for goods are "heavily shaped by cultural messages and social diffused expectations". Wright recognises that "Consumerism as a cultural model for living a good life . . . hinders human flourishing." The possibility exists that "voluntary simplicity" might well appeal to substantial numbers of people. This is central to my argument here. But Wright also warns that any movement championing "lower consumption and much more leisure time" would be perceived to "precipitate a severe economic crisis", and would likely be resisted by pro-capitalist governments. This argument needs to be accommodated in any Green New Deal proposal.[93]

Secondly, Lucy Sargisson's *Fool's Gold? Utopianism in the Twenty-First Century?* (2012) reviews both literary and historical developments in the recent past, and stresses climate-change literature and the catastrophist visions it often portrays, while emphasising the plurality of utopian responses to such scenarios. While primarily concerned with fiction, Sargisson examines architecture as well as intentional communities in the present, particularly in reference to efforts to counter environmental destruction by organic food production, rewilding, and reforestation. She gives particular attention to the most recent trends which push back at the frontiers of what we conceive of as "utopian", including issues of sexuality and gender identity, and internet gaming. Since lived experiments are also regarded as "utopian", Sargisson works with a realistic or realised concept, though without deploying the concept of "heterotopia", while warning that "attempts to realize Utopia should be approached with great caution". So long as utopia is not confused with perfectionism, however, some at least of its vision can be plausibly described as realised or realisable, even if the concept itself resolutely remains "no place", and as such doomed to fail. Indeed, Sargisson acknowledges that "The realization of a perfect eutopia would necessarily suppress difference and this makes a dangerous starting point for a Utopian project in the twenty-first century", specifically instancing "a fatal combination of perfectionist utopianism and religious fundamentalism" as an impulse tending in this direction.[94]

93. Erik Olin Wright, *Envisioning Real Utopias* (Verso, 2010), 6, 66–69.

94. Lucy Sargisson, *Fool's Gold? Utopianism in the Twenty-First Century?* (Palgrave Macmillan, 2012), 98–115, 144, 39, 53.

Finally, Davina Cooper's *Everyday Utopias: The Conceptual Life of Promising Spaces* (2014) focuses on special areas and/or activities where particular social norms are irrelevant or in abeyance, such as nudist beaches and sexually promiscuous bathhouses, local exchange trading schemes, Summerhill School, and Speakers' Corner in London. Such spaces "perform regular daily life . . . in a radically different fashion" by "creating the change they wish to encounter, building and forging new ways of experiencing social and political life." In "building alternatives to dominant practices" they promote "radical" transformations of various kinds. Foucault and Bloch are acknowledged inspirations here, and this represents a useful extension of heterotopian theory. Cooper also suggests that the slow food and slow city movements, discussed below, might be similarly appraised.[95] This gives us a good sense of what the "promise" of "radically different" spaces and activities might consist in.

95. Davina Cooper, *Everyday Utopias: The Conceptual Life of Promising Spaces* (Duke University Press, 2014), 2, 10. See also Erin McKenna, *The Task of Utopia: A Pragmatist and Feminist Perspective* (Rowman and Littlefield, 2001).

PART II

Utopian Sociability in Fiction and Practice

4

The Varieties of Utopian Practice

THIS CHAPTER EXPLORES the proposition that in designated spaces utopian principles may be introduced temporarily, or for longer, with the aim of promoting more intensive sociability than that usually prevailing in the wider society. Such attempts assume many forms, from festivals and pilgrimages to long-term intentional communities. All demonstrate the possibility of viable alternative ways of living in which equality, solidarity, communal identity, and belongingness are prioritised as a kind of halfway house to utopia, or even approximating it. We commence with festivals, some of which are overtly defined by utopian principles.

Festivals as Utopian Spaces

Part of the folk memory or myth of the Golden Age, we have seen, was embodied in various European festivals. Originating in religious ceremonies and the seasonal and agricultural calendar, such celebrations even now sometimes possess a sense of sacred time and space. In more religiously oriented events, proximity to the divine and some anticipation of eternity are central, especially where life and death are at issue. Here, Mircea Eliade writes: "the sacred dimension of life is recovered, the participants experience the sanctity of human existence as a divine creation."[1] Others have a more mundane function, like promoting the fertility of the soil or its inhabitants, or demonstrating athletic prowess. They may also exhibit the possibility of spontaneous playfulness and a performative otherness in which we act out inverted roles in an upside-down world, the world we briefly feel we would rather live in but imagine we can never inhabit permanently. Some historians of utopia, notably Bronislaw Baczko, describe festivals as "a social elsewhere, a *heterotopia*"

1. Mircea Eliade, *The Sacred and the Profane: The Nature of Religion* (Harper and Row, 1961), 89.

and "a model of sociability".[2] This particularly applies to those which aim to construct an egalitarian space, to mock the pretentiousness of the powerful, even to invert social relations. If utopian qualities are present, they may invoke a divinely ordained or original social equality, or protest against a repressive work/sexual ethos and excessive discipline.

Yet festivals are also ephemeral, and defined precisely by their consciously temporary breach from normality. Every day cannot be Christmas, logically, as we have seen, because then what is special about Christmas is lost. *Sonntäglichkeit* cannot be everydayness. Sundayness implies no work, a special meal, a feeling of family and friends together. This might work for the long weekend, with a four-day working week. But utopia, by contrast, needs to last longer, and so must accept a diminished amount of whatever qualities define Sundayness. What are these? In the west, Christmas, in particular, is special because it notionally suspends competition, rivalry, antagonism, and dislike, and briefly extends a spirit of equanimity and toleration. As often riotous as reverent, its pagan origins never entirely abandoned, it is a temporary, often contrived, enhanced sociability and feeling of unity, togetherness, and belonging. A credit to our natural love of celebration, as children have become ever-more important it has been increasingly linked with their pleasures. We may imagine it now to be marred by excessive materialism, but fourth-century Roman commentators already noted that during the season of the festival of the Kalends, just after the Saturnalia, "The impulse to spend seizes everyone".[3] The season satirises our inability to extend our generosity more widely for the rest of the year, and yet tantalisingly reminds us of the value of so doing. To this degree, at least, its alterity is utopian.

Gift giving as a mode of sharing wealth has been associated with many other festivals throughout European history, including New Year's, All Souls' Day (which later merged into Halloween), and some saints' days, notably St Martin and St Nicholas, where apples and nuts predominated. Our chief interest here is in festivals which promote equality. The trend of temporarily inverting social ranks seemingly began with a Babylonian celebration called Sakeas (*c.*1700 BCE), and its Persian successor, the Sacaea (*c.*550 BCE). Here slaves temporarily dominated their masters, and a mock king, in some accounts a prisoner already condemned to death, was hanged as a human sacrifice after a period of revelry.[4] The Greek cult of the long-haired god of wine and ecstasy, Dionysus or Bacchus,

2. Bronislaw Baczko, *Utopian Lights: The Evolution of the Idea of Social Progress* (Paragon House, 1989), 184.

3. Clement A. Miles, *Christmas in Ritual and Tradition, Christian and Pagan* (T. Fisher Unwin, 1912), 168.

4. Andrew Lang, *Magic and Religion* (Longmans, Green, 1901), 118–20. See generally Photeine P. Bourboulis, *Ancient Festivals of the "Saturnalia" Type* (Spoudon, 1964).

for Barbara Ehrenreich "the first rock star", commenced around 1500 BCE.[5] Dionysian festivals, which continue today in less extreme forms, were traditionally associated with the "orgiastic", in the French sense of *jouissance*—that is, the momentary transgression of community norms across a range of behaviour, with one aim being the cementing of group bonds in the daring feeling of shared excess.[6] This licensed anarchy and regulated ecstasy seems to contradict the utopian principle of order, but in fact it is only utopia's boundary line. Such festivals were often accompanied by sacrifices, and, if not the literal, at least the symbolic eating, among the Bacchae, the Aztecs, and some Christians, of the body, and the drinking of blood.[7] In Rome they included a sexual element, and secret festivals at night involved both men and "women in the wildest frenzies of religious excitement" who "thought nothing a crime". When the Senate prohibited them in 186 BCE some seven thousand men were implicated.[8]

Such occasions were often linked with a fictitious or mythical original equality. Sometimes, at least as filtered through Euripides' *The Bacchae*, this exuberance also involved "a particularly virulent form of violent nondifferentiation", not only between men and women and social ranks, but even human and beast, with people cross-dressing and being attired as animals.[9] Bacchus' followers' adornment with animal skins and goat-horn headdresses has been linked to an invocation of earlier times and the subsequent portrayal of centaurs, satyrs, and other monstrous creatures.[10] Cross-dressing and wearing animal skins also featured in the Roman Kalends.[11] Emile Durkheim thought such occasions revitalised the cultural order by re-enacting its conception. To René Girard, these are moments "when the fear of falling into interminable violence is most intense and the community is therefore most closely drawn together".[12] Social

5. Barbara Ehrenreich, *Dancing in the Streets: A History of Collective Joy* (Grant Books, 2007), 41.

6. Michel Maffesoli, *The Shadow of Dionysus: A Contribution to the Sociology of the Orgy* (State University of New York Press, 1992), xvii, xviii, 141. But in the specific sexual sense it also implies a weakening of direct personal bonds and erotically driven individualism in favour of group identity through public coitus (5, 57).

7. Maffesoli, 119.

8. Otto Kiefer, *Sexual Life in Ancient Rome* (George Routledge and Sons, 1934), 118–22.

9. René Girard associates this with the "sacrificial crisis", chiefly the failure of sacrifice to act as a purge of social violence, which sometimes occurs through sheer repetition. Indeed, he sees "the destruction of differences" as "an essential part of the sacrificial crisis" (*Violence and the Sacred* (Johns Hopkins University Press, 1977), 39, 127–28).

10. Lafitau, *Customs*, 2: 24.

11. The Church forbade such excesses in 743 CE: C. Miles, *Christmas*, 171.

12. Sigmund Freud, *Group Psychology and the Analysis of the Ego* (1921; Bantam Books, 1960), 81; Girard, *Violence and the Sacred*, 120. To Girard, religion mainly functions to "humanize violence" by subjecting it to ritualised transcendence (134).

distinctions were swept away in revelry which to Girard aimed at a "gesture of harmony". This represented a temporary bursting of the boundaries of social norms as defined by gender, class, and even humanity, a kind of generalised rebellion against repressive social roles. The idea of ordered normality itself, indeed, seems threatened by such outbursts of irrationality. Yet it was also reinforced: we require such disorder in order to recall the advantages of order. This is how festivals serve as utopia's boundary, taunting us by implying through limiting excess that no more substantial or prolonged equality is possible. They are yet another instance of the utopian Tantalus, forever tempting us with a brief taste of the unreachable and then withdrawing it again. This much freedom you may have, they seem to say, but no more: the rest must remain in your imagination.

In the early western tradition the chief such event was the week-long Roman Saturnalia festival (17–23 December), which both licensed and contained a temporary excess of revelry and intoxication. This originated as early as 500 BCE in imitation of Greek festivals, like that of Hermes on Crete, where the master–slave relationship was reversed, slaves could even beat their masters, and sexual license was permitted.[13] In Rome it also commemorated the supposed reign of a pre-Roman monarch named Saturnus, when no slaves or private property existed, and whom a mock king supposedly represented. Freud described it as a legitimised "infringement on the prohibition" of everyday ego desires.[14] Long synonymous with excess, the feast was also associated with fertility, and sowing the winter seed. Visiting friends was customary, and gift giving was common, including wax candles, thought by some to be a substitute for human sacrifice, and representing the power of the returning sun after the solstice: this is one origin of Christmas.[15] Processions included people with blackened faces or dressed in animal skins. Persistently associated with nocturnal debauchery, the Saturnalia was finally banned in 186 BCE. Augustus Caesar (d. 14 CE) restored it to bolster his popularity. Soon sixty thousand people gathered in the Coliseum for games and mirth in the greatest of all Roman festivals.[16]

For us, the Saturnalia's key quality was its invocation of an original Golden Age of equality, and its association with what the Enlightenment called natural liberty. Lucian's well-known satire on the festival (*c.*165 CE) describes singing whilst naked, drinking and dice playing, and the "appointing of kings and

13. Hans Licht, *Sexual Life in Ancient Greece* (George Routledge and Sons, 1932), 131.

14. Freud, *Group Psychology*, 81.

15. Lang, *Magic and Religion*, 108–9; C. Miles, *Christmas*, 165–66.

16. A. W. Verrall, *Collected Literary Essays* (Cambridge University Press, 1913), 58.

feasting of slaves", though a poor man is told that "you will be presented with a sorry cloak, or a worn-out tunic; and a world of ceremony will go to the presentation."[17] Here the ruler of the Golden Age, Cronos, reminds participants that the activity's aim is that:

> men may remember what life was like in my days, when all things grew without sowing or ploughing of theirs—no ears of corn, but loaves complete and meat ready cooked—, when wine flowed in rivers, and there were fountains of milk and honey; all men were good and all men were gold. Such is the purpose of this my brief reign; therefore the merry noise on every side, the song and the games; therefore the slave and the free as one. When I was king, slavery was not.[18]

In Macrobius' description (*c.* 400 CE), Romans recalled this as a time which was "the happiest, both because of its material abundance, and also because the distinction between slavery and freedom did not yet exist, as is made plain by the fact that slaves are allowed almost complete license during the Saturnalia."[19] Writing in 1844, Mary Hennell associated it with ancient Crete, where the slaves who tilled the fields, "Once a year, at the feasts of Mercury . . . were waited on by their masters, to remind men of their primitive equality."[20] Roman slaves even wore their masters' clothes, and displayed the *pilleus*, or badge of freedom. The social world, in other words, was turned upside down, as an episodic or temporary utopia recalled both a lost past and, in Mikhail Bahktin's view, echoing Goethe, envisioned a future Golden Age, "the victory of all the people's material abundance, freedom, equality, brotherhood", a "new utopian order."[21] Equality and violence were also closely associated here, though festivals rarely induced disorder. Fighting was usually prohibited during Saturnalia, just to make sure, and wars could not be begun or criminals punished. In the eastern provinces, a mock king was also chosen by lot and played the part of a Lord of Misrule, for instance by carrying a flute girl on his back.[22]

Such celebrations thus in some sense merely mimicked, even satirised, the idea of the Golden Age, mocking in ritual and an immersion in hedonism what was originally a condition of virtue and peace based on an austere sense of

17. Lucian, *The Works of Lucian of Samosata*, 4 vols (Clarendon, 1905), 1: 23.

18. Lucian, *Works*, 4: 110–11.

19. Macrobius, *Saturnalia*, 3 vols (Harvard University Press, 1977), 1: 77.

20. [Mary Hennell], *An Outline of the Various Social Systems & Communities Which Have Been Founded on the Principle of Co-operation* (Longman, Brown, Green, and Longmans, 1844), 2.

21. Mikhail Bakhtin, *Rabelais and His World* (MIT Press, 1968), 256, 265.

22. E. O. James, *Seasonal Feasts and Festivals* (Thames and Hudson, 1961), 176.

duty. Cornucopia partly links these visions, but much else separates them. The well-to-do ignored their more utopian implications, with some, like Cicero entertaining Julius Caesar in 45 BCE, serving food graded according to the degree of friendship of the guests.[23] Saturnalia has accordingly often been disdained as a juvenile imitation of an adult Golden Age, an infantile fantasy which only reveals the impossibility of recreating any original condition, and perhaps even mocking its very existence. The analogy, indeed, of childhood with the Golden Age is a tempting one, and is treated further below.

After Rome's fall, aspects of Saturnalia persisted in various medieval festivals. Carnival at Shrovetide, before Lent, was commonly associated with legends of the "peasant utopia", the Land of Cockaigne (Cocaigne, in French, Schlaraffenland in German). This is not really a utopia, but a cornucopia of abundance and gluttony without work, a form of the Golden Age.[24] Its graphic depiction in Peter Bruegel the Elder's paintings "The Fight Between Carnival and Lent" (1559) and "Het Luilekkerland" (1567) illustrates the delights of debauchery and indulgence. (The first, a playful portrayal of celebration, hints at some key themes explored here.)[25] Typical depictions of Cockaigne show pigs with slogans written on their sides saying "eat me", and nature's abundance effortlessly pouring forth everywhere. The overlap with Carnival is clear. Peter Burke argues that "Cockaigne is a vision of life as one long Carnival, and Carnival a temporary Cockaigne, with the same emphasis on eating and on reversals".[26] Other types of festival in the medieval and early modern periods included *bals-masqués*, or masquerades, which offered an opportunity for freedom from social convention through the fantasy of anonymity, especially for women, for whom, some claim, a utopian space opened up in a manner hitherto unknown.[27] Carnivals could and can also be used as a form of personal revelation, an anarchic or Dionysian letting go which discloses inner secrets, or a collective and public confession or "revelation of

23. Strong, *Feast*, 26.

24. On Cockaigne's relation to utopia, see Herman Pleij's *Dreaming of Cockaigne: Medieval Fantasies of the Perfect Life* (Columbia University Press, 2001).

25. Most notably in the juxtaposition of and struggle between Carnival, and the celebration of excess, and Lent, or prudence, fasting, and restraint from the consumption of luxuries.

26. Peter Burke, *Popular Culture in Early Modern Europe* (Temple Smith, 1978), 190. See also, more generally, Christopher Kendrick's *Utopia, Carnival, and Commonwealth in Renaissance England* (University of Toronto Press, 2004), which argues that "More's text silently presents Utopia as the negation of Carnival, as its impossible logical end" (74), which is to say a regulated rather than topsy-turvy world.

27. Terry Castle, *Masquerade and Civilization: The Carnivalesque in Eighteenth-Century English Culture and Fiction* (Stanford University Press, 1986), 257–58.

truth", as in modern Trinidad.[28] In Britain the election of Lords of Misrule over the Christmas season extended even to the universities of Oxford and Cambridge, and the Inns of Court at London.[29]

The relation of festivals to utopia thus rests not on their celebration of excess, gluttony, and wild abandon, which clash with the Morean utopia's ideal of moderation and restraint, but on their pointed and often irreverent, anarchic egalitarianism. To Chris Humphrey, this "festive misrule" long remained one of their distinctive themes.[30] Sometimes abolishing social ranks was hinted at, even "a king without ministers and a kingdom without taxes".[31] Other forms of identity were also inverted. The lesser clergy, who organised the Feast of Fools, celebrated around 28 December–1 January, which lasted into the sixteenth century, wore bawdy and animal masks, mimicked church rituals while cursing their congregations, dressed as women, and satirised everyday hierarchies, often dancing (literally) perilously close to earlier Dionysian and Saturnalian revelries.[32] Sometimes priests elected an archbishop and bishop, and ate, drank, and played dice at the altar during the mock service, seemingly "giving vent to that which they felt during the whole year, that religion was a fable, and their duties the acts of a play."[33] Lords of Mis-rule and Abbots of Un-reason appeared to symbolically overturn everyday life.[34] Even the natural world could be inverted. In the "world turned upside down", relations between human and beast were reversed, the horse shoeing its master, the ox turned butcher, cutting him up. People dressed up as wild animals or monstrous half-animals, fools, clerics, devils, or supernatural beings, and men as women and vice-versa.[35] This was a feast of the imagination, amongst much else.

A clearer example of the temporary utopia-as-alterity/heterotopia idea we could not hope to find, except that its content is expressed as ecstasy, defining

28. Daniel Miller, *Stuff* (Polity, 2010), 18.

29. Theodor Gaster, *New Year: Its History, Customs and Superstitions* (Abelard-Schuman, 1955), 17.

30. Chris Humphrey, *The Politics of Carnival: Festive Misrule in Medieval England* (Manchester University Press, 2001).

31. Yves-Marie Bercé, *History of Peasant Revolts* (Cornell University Press, 1990), 325.

32. Burke, *Popular Culture*, 192.

33. George Zabriskie Gray, *The Children's Crusade* (Sampson Low, Son, and Marston, 1871), 18–19.

34. Max Harris, *Sacred Folly: A New History of the Feast of Fools* (Cornell University Press, 2011), 255–56.

35. Roger D. Abrahams and Richard Bauman, "Ranges of Festival Behavior", in *The Reversible World: Symbolic Inversion in Art and Society*, ed. Barbara Babock (Cornell University Press, 1978), 193–208; Burke, *Popular Culture*, 183–89. See Sandra Billington, *A Social History of the Fool* (Harvester, 1984). On linkages between monstrosity and the dystopian, see my *Dystopia*, 58–78.

the limits of order, rather than as order and structure as such. To Harvey Cox, the radical dimension of such festivals "exposed the arbitrary quality of social rank." The linkage to utopia is obvious: the festival acts out, as satire, the more equal world, mocking hierarchies while acknowledging that departures from them can be only temporary, and that utopia can never be achieved for long.[36] Once the everyday masks go back on, the festival proves by its temporality the permanence of normality. Every day cannot be Sunday. But Sundayness can be extended, though not indefinitely. Carnival sometimes occasioned popular uprisings, as a devotion to folly might well do.[37] Continuous rebellion, however, was sometimes explicitly disallowed. The chief figure of Carnival was a king or pope of fools who was ceremoniously killed after a brief reign of debauchery, thus symbolically reintroducing the reign of order and hierarchy, and hinting at the impossibility of rebelling against it, or returning to the Golden Age. So the irony of organised disorder was that it reinforced order as such, rather than rebellion.

Festivals thus represent an inversion or polarity at a number of levels, including oppositions of order/disorder, dominance/subjection, harmony/disharmony, male/female, human/animal, purity/filth, hidden/manifest, and public/private.[38] They were, and sometimes still are, a form of licensed misrule, or controlled disorder, often inverting and parodying both religious ceremony and sexual norms. Here many phenomena which were usually feared, including madness, sexual perversion, and bestiality and the monstrous, were actually laughed at, in a temporary triumph over fear of the other. The lure of the forbidden occasioned a double pleasure, both in violating norms and in the acts themselves. But such events were also paradoxically essential to maintaining social order. They offered a vent to release pressure, while hinting at the futility of rejecting existing hierarchies more permanently.[39]

36. Harvey Cox, *The Feast of Fools: A Theological Essay on Festivity and Fantasy* (Harvard University Press, 1969), 5, 82.

37. For some utopian aspects, see Kendrick's *Utopia, Carnival, and Commonwealth* and Humphrey's *Politics of Carnival*. More's friend Erasmus of course was the author of *In Praise of Folly*, also called *Moriae encomium*, which he wrote while living with More and dedicated it to him, More's counterbalance being the portrayal of the life of reason by comparison to the unreason of folly, or perhaps the cure for the illness. For rebellion, see the famous instance documented by Emmanuel Le Roy Ladurie in *Carnival: A People's Uprising at Romans, 1579–80* (Scholar Press, 1980). To Erasmus, "close friendship grows up only between equals" (*In Praise of Folly*, ed. Hoyt Hopewell Hudson (Princeton University Press, 1941), 26–27).

38. Stuart Clark, *Thinking with Demons: The Idea of Witchcraft in Early Modern Europe* (Oxford University Press, 1997), 31 (describing the work of M. Ingram).

39. S. Clark, 16–17.

Such symbolic misrule has been described as "ubiquitous" in early modern England, and probably was elsewhere too. It found its way into utopian literature in François Rabelais's portrayal of the famous "anti-monastery", the Abbey of Thélème, in his immensely popular *Gargantua and Pantagruel* (*c.*1532), where the influence of Thomas More is evident. Here all were free to enter or leave without being bound by vows of chastity, poverty, or obedience. Its motto, "Do What Thou Wilt", embodied this yearning for anarchic, libertine, "natural" freedom. Not long in his grave, More would have been aghast at any hint that this was serious.

Festivals clearly serve as prototypes of some forms of utopian experience. They exemplify our love of celebration, and, while flaunting our episodic capacity for togetherness, seemingly mock our ability to prolong the impulse, as we retreat into the bunker mentality of Januaryness or Mondayness. The "Christmas spirit", the ceremony of the family holiday, the extended rave, or pop festival, celebrations today like Burning Man, all give us later modern instances of temporary altruism, unity, solidarity, and brotherhood. Most such occasions do not entail enmity. Instead, the group licenses what is usually impermissible to the individual. So a crowd may be merely "*a device for indulging ourselves in a kind of temporary insanity by all going crazy together*."[40] We let our hair down, vent our frustrations, abandon the artificial conventionalities of being on our best behaviour. Little harm results beyond the occasional accident or hangover. We have tested the boundaries and know that no permanent breaching of them is possible.

The relatively benign festival-as-letting-off-steam has several other variants. Festivals can be dystopian, when licensed disorder focuses on enemies, and exclusivity bonds the in-group. The anti-festival caricatured in Orwell's Hate Week, mocking Nazism and Stalinism, is a case in point.[41] Some religious celebrations express hostility towards other religions. In India, for instance, Muslims may slaughter a cow in front of a Hindu temple, or Hindus make loud noises in front of mosques on holy days.[42] Antagonism may dominate the entire proceedings, and justify plunder, rapine, and slaughter in their righteous

40. Everett Dean Martin, *The Behavior of Crowds* (Harper and Bros, 1920), 37.

41. René Girard cites the example of a Gold Coast tribal "antifestival", where public conversations and many loud noises are forbidden for several weeks a year, in the evident aim of suppressing potential violence: *Violence and the Sacred*, 121–22.

42. Robert S. Robins and Jerrold M. Post, *Political Paranoia: The Psychopolitics of Hatred* (Yale University Press, 1997), 91.

proclamation of "holy war" or "crusade". Some crusades have involved mass religious enthusiasm bordering on the hysteria or "contagion" associated with an intense desire for salvation. In the Children's Crusade of 1212, for instance, thousands left without their parents' consent "in quest of the Holy Cross", mostly never to return.[43] Here, salvation was "almost made to depend on the recovery of the Holy Sepulchre", and was whipped up by the Pope and a thousand bishops and priests.[44] Sometimes it is claimed that liberating a "holy land" will free participants from sin, and even secure through martyrdom their salvation "to live with the saints".[45] The more "holy", indeed, individuals feel, the more they seem (paradoxically?) free of any restraint or fear of retribution; this is akin to a state of grace, holiness, or freedom from normal everyday moral restraints, seen also in a state of war—hence the connection to certain festivals.

Other types of group gathering have resulted from mass anxiety or latent hysteria rather than the organised release of stress, though here too religion may be involved. Here behaviour exemplifies a kind of collective panic attack. Ecstasy was both the outlet for and antidote to hysteria and often proved unpredictable, chaotic, and destructive. Mass dancing frenzies, beginning as early as the seventh century and recurring in Europe between the thirteenth and fifteenth centuries, are one such instance. Thousands of the poor, as it seems most were, engaged in uncontrollable ecstatic singing and dancing to the point of collapse, evidently as a mass response to trauma, and sometimes with visions of the heavens opening and the Saviour and Virgin Mary appearing.[46] Another form of mass panic associated with stress, tarantism, was prevalent in Italy and involved bizarre reactions to supposed spider or scorpion bites. Both indicate quasi-religious mass hysteria where fear becomes infectious by emotional contagion. The association of guilt with violence against outsiders is also sometimes evident here. The fourteenth-century German Brotherhood of Flagellants, or Brethren of the Cross, who whipped themselves bloody while leading onslaughts on Jewish moneylenders, demonstrates medieval links between masochism, sadism, and anti-Semitism. Later revivalist and Methodist assemblies in Britain and its American colonies during the Great Awakening of the eighteenth century witnessed ecstatic weeping, fainting, laughing, dancing, barking, jerking, and wailing.[47] At an 1801 camp meeting in Kentucky,

43. Gary Dickson, *The Children's Crusade: Medieval History, Modern Mythistory* (Palgrave Macmillan, 2007), 47, 133.

44. G. Gray, *Children's Crusade*, 8.

45. Carl Erdman, *The Origin of the Idea of Crusade* (Princeton University Press, 1977), xxii, 344–45.

46. Ehrenreich, *Dancing in the Streets*, 85.

47. Charles Platt, *The Psychology of Social Life* (George Allen and Unwin, 1922), 156–59.

twenty-five thousand people "danced, laughed, shrieked and fell over as the Spirit gripped them".[48] Here a profound sense of guilt and sin existed within congregations, though, crucially, such hysteria was not generally wedded to external aggression. These "extraordinary popular delusions", as an early chronicler, the nineteenth-century historian Charles Mackay, termed them, could also encompass such phenomena as financial speculation (the tulip mania), imperial enthusiasm, witchcraft, and much else.[49] The line between controlled and spontaneous mass hysteria was thus often perilously close.

The epoch of the French Revolution witnessed a richly symbolic secular revival of some earlier festivals. Most important was the Festival of Unity, whose explicitly utopian quality is undisputed. Here a Fountain of Regeneration was erected at the Bastille, where water was drunk from the breasts of an Egyptian goddess to symbolise mass baptism or rebirth into a state of freedom.[50] To Mona Ozouf, these rituals aimed to make sacred the new values of the revolution, which involved "a fantasy return to the equality of origins".[51] Everett Dean Martin suggests indeed that "*Every revolutionary crowd of every description is a pilgrimage set out to regain our lost Paradise.*"[52] If this is correct, we have a virtually unbroken linkage between Dionysian and Saturnalian "remembrances" of the ancient Golden Age, Carnival, and modern celebrations of revolutionary aspirations. However faint, however desperate, however vain and mocking, we have not forgotten our origins, or at least our imaginary beginning. Nor it seems has our desire for redemption dimmed much over the centuries, though the manner of its accomplishment has altered greatly.

A lesser variety of institutionalised festival worth mentioning is the latest form of celebration, the modern holiday, often seen as the chief attempt to escape everyday life by getting away from it all and letting our hair down. In the popular mind, at least by the late twentieth century, holidays and utopias were intimately interwoven, the image of rocking in a hammock beneath tropical skies being part of their shared content. In Europe the Club-Med-style holiday resort or Ibizan utopia is a peculiarly heterotopian space where suspending some of the normal rules of propriety at home is nearly obligatory, especially—for the under-thirties—where sex and alcohol are concerned. If everyday life is a prison, this is its antithesis, "liberation". The weekend, when

48. D. Thompson, *End of Time*, 117. The population of Kentucky at this time was about 220,000.

49. Charles Mackay, *Memoirs of Extraordinary Popular Delusions*, 3 vols (Richard Bentley, 1841).

50. Mona Ozouf, *Festivals and the French Revolution* (Harvard University Press, 1988), 84.

51. Ozouf, 114.

52. Martin, *Behavior of Crowds*, 207.

observed in the same spirit, can function similarly relative to the working week. Both juxtapose a ludic, pleasurable episode to the monotony of everyday treadmill life. Not without reason is utopia sometimes imagined to be a perpetual holiday.

Large numbers also participate in ceremonies marking that liminal transition at the end of each year when we exit the old and welcome the new, make resolutions to improve our lives, and renounce the melancholy of the last year's failures. The "oldest and most universal of festivals", New Year symbolises the passage of life, the hope for fertility in the new agricultural season, and our wish to be rid of the sin, error, and mistakes of the old, and to recommence with a clean slate. Once distinguished by four phases defined by rites of mortification, purgation, invigoration, and jubilation, the progress of civilisation has now mostly left us with only the last of these. So we wash and cleanse ourselves, champagne, a symbol of luxury, being the preferred liquid. We dress up to symbolise the new; the Romans wore white on 1 January. We make noise to scare away the evil spirits, who have often been presumed to be present on New Year's Day.[53] "Reborn", we ritually proclaim our morally reformed selves with New Year's resolutions to limit our excesses and draw a line under last year's failings. Symbolically we renew our vows to impose order on chaos.[54] The improvements generally last a couple of weeks and we then revert to our former incarnations, none the worse for the effort. Sometimes New Year's customs have retained an egalitarian element. In Westphalia, villagers once rode "hobby horses" adorned with white sheets and assembled outside the houses of the well-to-do making noise until they were entertained.[55] Birthday celebrations, too, mark our sense of liminality and transition, each one taking us further towards maturity and responsibility, at least until our dotage. All these traditions provide us with a sense of home, of belonging, and of attachment to familiar narratives.

Some more recently established festivals have also been explicitly anti-materialistic in orientation. Anti-consumerist celebrations imply a new, more conscious theory of time where we prohibit certain types of specifically later modern harmful activity. We have television- or internet- or phone-free days, for instance, with three hundred million people having participated in Screen Free Week, which originated in 1994;[56] non-shopping days such as Buy Nothing Day, which began in Canada in 1992, and Green Friday, an alternative to

53. Gaster, *New Year*, xi.

54. The classic account here is Mircea Eliade's *The Myth of the Eternal Return* (Princeton University Press, 1974).

55. Gaster, *New Year*, 95–96.

56. See Kalle Lasn, *Culture Jam: The Uncooling of America* (William Morrow, 1999).

Black Friday, the commercial variant on Thanksgiving; children-centred days; pet-centred days; no-driving days; or neighbour days where gift giving is encouraged. Eco-Christmas is now being widely discussed. In Melbourne, the Big Switch Off began in 2006 in an endeavour to reduce electric consumption.[57] Such events sometimes emphasise using local produce, display an explicitly anti-corporate mentality, and prioritise being over acquiring. In the Burning Man festival, which commenced in 1986, decommodification is explicitly promoted by the organisers.[58] At the Rainbow Family of Living Light gatherings, begun in 1992, mobile phones are banned entirely, and the aim is to create a "family dinner table" atmosphere.[59] But such festivals are much more marginal to mainstream society than was the Saturnalia to ancient Rome or Carnival to medieval European cities. Involving perhaps a few thousand people, they do not define the festive calendar of nations, or the common ethos of cultures. They rarely exhibit inversions of specific hierarchies; this is now regarded as too risqué. Modern culture is rarely so collectivist, and seldom succeeds in countering the greater isolation, alienation, competition, and loneliness of our cities. This limits, but does not prohibit, our capacity for utopian sociability. We will take up this problem in part III below.

Pilgrimage as a Utopian Activity

Pilgrimage provides another instance of episodic, even prolonged, utopian experience, normally defined by religion, though there are secular variants on this form of devotion. Many religious ceremonies involve the desire to be pardoned for wrongdoing, to free believers from sin; even, for Eliade, from the burden of time itself. Pilgrimages thus function as elaborate purification rituals.[60] They may bridge religion and utopia insofar as they create heterotopian spaces where suffering, atonement, a release from everyday life, participation in holiness and the possibility of glimpsing God, as the believer imagines, induce a morally and spiritually transformative experience. Belonging to the group of believers and the potentially saved is also a central element here. Pilgrimage can indeed serve as an allegory of life itself, with the promise of salvation as the reward.

The classic age of Christian pilgrimage dates from the tenth to the late fifteenth century. During this period, millions undertook often-perilous

57. Kim Humphery, *Excess: Anti-consumerism in the West* (Polity, 2010), 12.

58. Roxy Robinson, *Music Festivals and the Politics of Participation* (Ashgate, 2015), 115–16.

59. Michael I. Niman, *People of the Rainbow: A Nomadic Utopia* (University of Tennessee Press, 1997), 71.

60. Eliade, *Myth*, 86; Robert A. Scott, *Miracle Cures: Saints, Pilgrimage, and the Healing Powers of Belief* (University of California Press, 2010), 83.

devotional journeys to reach locations rendered holy by the presence of some saint or miracle, with some even reaching the proto-utopian Holy Land, a veritable portal to heaven. Their chief motive appears to have been the assumption that prayers would most likely be answered when offered in sacred places, and the more sacred the better. These wormholes into Eternity offered a sense of proximity to the holy, and enflamed spiritual intensity through the feeling of closeness to God. Early medieval miracles typically took place near shrines. Even if they were defined only by the saint's toenail clippings, these sacred spaces were very valuable indeed. They offered a respite from and hope of cure for a range of psychological and physical maladies. Faith was the key to their secret.

What interests students of utopia about pilgrimages, firstly, is the sense of the place to be reached, and the expectations of its extraordinariness. At one level all Christian pilgrimage aimed to reach the "Heavenly Jerusalem", the most perfect imaginary, and, in the Holy Land, real, space. Its shrines, and many along the way, opened the most direct window into God's own terrain, defined by the potency of the reliquary. Even today, some Christians believe they might ascend directly from Jerusalem to heaven on Judgment Day, and have acquired property there in anticipation thereof.[61] As in ancient Greece, special rules governed such sacred sites, which were regarded as inviolable and might offer asylum.[62] The sense of divinity, and especially of touching anything directly connected to God, was central here. The Holy Land was the supposed source for many relics of the holy dead, including pieces of the True Cross, Holy Blood, the tip of Lucifer's tail, the Holy Foreskin, and the ever-popular Milk of the Virgin.

In all of these and more there was a roaring trade by medieval times, and fabricating them was a booming business. Relics were thought to exude a kind of "holy radioactivity" by which, as early as the sixth century, it was believed objects grew heavier when placed in their proximity.[63] Saints' graves were the most impressive shrines: "The tomb itself manifested the bodily presence of the saint to the onlooker."[64] Their bones were a key source of relics. Rather than rising to heaven, saints were thus assumed to be at some level still present in their earthly tombs.[65] They could perform many invaluable tasks, including interrogating demons, or releasing the dead from Purgatory.[66] Catastrophes like shipwrecks on a forthcoming journey could be forestalled or avoided. To

61. The specific location is usually assumed to be the Valley of Jehoshaphat.

62. Matthew Dillon, *Pilgrims and Pilgrimage in Ancient Greece* (Routledge, 1997), 28.

63. Ronald C. Finucare, *Miracles and Pilgrims: Popular Beliefs in Medieval England* (St. Martin's Press, 1995), 26.

64. Diana Webb, *Medieval European Pilgrimage, c. 700–c.1500* (Palgrave, 2002), 166.

65. Peter Brown, *The Cult of the Saints: Its Rise and Function in Latin Christianity* (SCM, 1981), 4.

66. Cynthia Hahn, *Strange Beauty: Issues in the Making and Meaning of Reliquaries, 400–circa 1204* (Pennsylvania State Press, 2012), 16.

the afflicted or diseased, the lame, the childless, the crippled and blind, divine mercy could result from the saint's healing intercession. Sin being regarded as a cause of sickness, a full confession was obligatory before any cure could commence.[67] The sanctified space, its holiness magnified by myth and legend, came as close to providing a stairway to heaven or portal to the divine as this life permitted. Such miraculous spaces, where natural laws were apparently suspended, were in fact divine heterotopias, but as such very different from the socially defined heterotopias considered above. The opposite of such spaces are those defined by taboo, or prohibition, whose violation might have very grave consequences, including instant death. Such deaths have indeed occurred, so strong is the power of belief.[68]

All sacred locations were thus awe-inspiring. They were also very lucrative, both for the Church and for assorted hangers-on. Cities measured their affluence by the number of such sites they possessed, for holy tourism was a flourishing industry. Two hundred thousand pilgrims a year visited medieval Canterbury (population ten thousand), and in 1300 an equal number Rome, the "eternal city" (population *c.* one hundred thousand). Europe's total population was perhaps seventy million at this time, so it is not unreasonable to assume that as many as 1 per cent of the population were on the road seeking their mite of holiness at any one time. Countless numbers made the arduous journey to the Holy Land. Here were many shrines, and abundant piety. Later, Santiago de Compostela, Lourdes, Lisieux, and a thousand lesser and closer sites of pilgrimage beckoned.

To pilgrims, visiting such places might induce profound changes in this life. While en route they were often released from taxes, and their property was secure from lawsuits. The longer the journey the more arduous it was, with perils and irritations, from lice to pirates, at every turn. Each challenge helped build up the sense of anticipation. On reaching the "desired country", as John Bunyan's title page to *The Pilgrim's Progress* (1678) calls it, multitudinous ailments might dissolve through grace, blessings, indulgences, relics, or flasks of holy water. As with some festivals, the possibility of profound, even cathartic emotional release was always present. Some felt a joy, Donald Howard writes, "close to hysteria".[69] At one prophet's tomb in the Holy Land, men were reported howling like wolves, roaring like lions, and hissing like snakes.[70] Pulled by curiosity, pushed, sometimes, by the compulsion of doing penance, men and women sought relief in the holy. But it was not all hard work and grim piety.

67. R. Scott, *Miracle Cures*, 81.

68. See my *Dystopia*, 10.

69. Donald R. Howard, *Writers and Pilgrims: Medieval Pilgrimage Narratives and Their Posterity* (University of California Press, 1980), 43.

70. P. Brown, *Cult of the Saints*, 106.

Some seem to have gone along just for the ride, seeking adventure and a respite from everyday tedium. For many, indeed, the prospect of a change of life made the pilgrimage a kind of moveable feast, a carnival in motion. Thomas More himself complained that most pilgrims "cometh for no devotion at all, but only for good company to babble thitherward, and drinke dronke there, and then dance and reel homeward."[71] This implies that the lively sociability and group feeling pilgrimage offered was crucial to its popularity. So the distance between pilgrimage and festival was sometimes small. Pilgrimage, indeed, was a community in transit, whose bustle, variety, and air of expectation must have been riveting for those who lived on the land and had never left their hamlet.

Interesting for our purposes, secondly, is the clearly utopian nature of the group pilgrims joined. This was well defined: "the pilgrim's badge was a sign of Christian fellowship and the revered token of international brotherhood."[72] With the holy cross and distinctive badges sewn upon their costumes, and having received the blessing of the Church, pilgrims were protected by Saint Peter and the Pope and were above all law except ecclesiastical.[73] They wore similar clothes, usually a rough coat of grey and a broad, plumed hat. Many went barefoot, suffering being regarded as integral to feeling holy. They carried a staff with a piece of metal on its knob, on which was inscribed *Haec tute dirigat iter*, "May this guide thee safely on thy way." The sense of a shared goal of aspiring to holiness and of common experiences of suffering and of seeking and attaining sacred spaces thus bonded believers as fellow travellers.

These attributes are also shared by other religions. Islam's chief pilgrimage, for instance, is the Hajj, meaning aspiration, to another world, which each believer is enjoined to perform at least once. Hospitality is a key element in this journey, as it was for the tribes across whose lands many pilgrims journeyed. This drew on existing Islamic traditions. The great traveller Sir Richard Burton, who reached Mecca in disguise in 1853, noted that the desert Arabs, or Badawins, greeted each other with their foreheads pressed together, exchanged mutual inquiries, and could not turn their backs on someone.[74] Pilgrims within sanctuaries in Mecca endeavour to imitate what they think conditions in paradise are like, and cannot cut their hair or fingernails, or wear perfume or sewn clothing, paralleling the nakedness of Adam and Eve.[75]

71. In "Dyalogue on the Adoracion of Images", cited in Sidney Heath, *Pilgrim Life in the Middle Ages* (T. Fisher Unwin, 1911), 44.

72. Sidney Heath, *In the Steps of the Pilgrims* (Rich and Cowan, 1950), 12–13.

73. S. Heath, 19.

74. Captain Sir Richard F. Burton, *Personal Narrative of a Pilgrimage to Al-Madinah & Meccah*, 2 vols (1853; Tylston and Edwards, 1893), 2: 85.

75. Brannon Wheeler, *Mecca and Eden: Ritual, Relics, and Territory in Islam* (University of Chicago Press, 2006), 64–65.

A third relation between utopia and pilgrimage lies in the latter's suspension and reorganising of secular time, space, and order. To Victor Turner, pilgrimage is a "liminoid phenomenon" exhibiting the three stages of the rites of passage: separation, limen or margin, and reaggregation. It involves a deliberate flight "from the reiterated 'occasions of sin' which make up so much of the human experience of social structure." Its aim is at least partly, as in Saturnalia, the symbolic inversion of social roles. Renouncing or fleeing from past sins is assisted when the "pilgrim 'puts on Christ Jesus' as a paradigmatic mask, or persona, and thus for a while *becomes* the redemptive tradition, no longer a biopsychical unit with a specific history". And there is some longevity in this transformation: "It is true that the pilgrim returns to his former mundane existence, but it is commonly believed that he has made a spiritual step forward." Pilgrimage thus "provides a carefully structured, highly valued route to a liminal world where the ideal is felt to be real, where the tainted social persona may be cleansed and renewed."[76] Thus while we still return from sacred space to the profane, we may be morally transformed. To a degree, at least, we may carry something of utopia back with us, not just as a souvenir but as a feeling of genuine self-transformation.

Festivals and pilgrimages thus function to concentrate moral energies on higher aspirations, while also enabling participants to enjoy more intense social bonds, at least temporarily. We must now consider more long-lived efforts to preserve these bonds, particularly in relation to the problem of needs. Such groups as we have seen are usually defined by discussions of community, a central concept in modern sociology and social theory, and in some respects the chief generic focus of utopianism as such. Intentional communities are formed deliberately with a view to promoting a different way of life, and usually have an ideal sociability as part of their aspirations.

Intentional Communities

Much of the debate about realistic utopias focuses on small-scale intentional communities. Their "commitment to group cohesion and solidarity", Rosabeth Moss Kanter stresses, makes them stronger than most types of group.[77] In Timothy Miller's definition, they usually share seven features: a common sense of purpose and separation from the wider society, some type of self-denial and preference for group well-being over individualism, geographical proximity of their members, personal interaction, economic sharing, a real

76. V. Turner and Turner, *Image and Pilgrimage*, 1–39, here 7, 11, 15, 30.

77. Rosabeth Moss Kanter, *Commitment and Community: Communes and Utopias in Sociological Perspective* (Harvard University Press, 1972), 72–73.

existence, and securing a critical mass sufficiently large to make such experiments genuinely social.[78]

Our aim here in briefly introducing some of these groups is to isolate their specifically utopian qualities so we can later ask whether or how far these might be adapted today in light of our environmental imperative of sustainability. Monastic orders are not examined here, as being essentially irrelevant to large-scale urban life today, their influence on earlier utopianism notwithstanding. For most of history the bulk of humanity has lived on the land or in small-scale settlements. Isolated voluntary groups like monasteries have been common, and offer a stark contrast to life in great cities. Most have required stringent discipline and self-sacrifice and an arduous devotion to subsistence agriculture, which often amounts to voluntary poverty for their inhabitants. For this reason, we will see, a few examples should illustrate how limited the idea of retreating to the land will likely be in the coming decades. But this does not exclude learning other lessons from such groups.

Over the past few centuries, however, the largest and longest-lived intentional communities have been religious, though more modern communes also have secular, ecological, and countercultural origins. Some have been described in terms of "an attempt to cultivate friendship".[79] These are clearly utopian, according to the definitions of enhanced sociability and belongingness, and the modified Morean template adopted here. Others, however, display markedly dystopian characteristics, or a mixture of both qualities. Those dominated by charismatic leaders and an ethos of extreme devotion are sometimes termed cults or sects—we think of the Moonies, the followers of Sun Myung Moon, or Charles Manson's "Family", of the Shree Rajneesh sect, or Aum Shinrikyo. Some of these involve coercion and violence, like the People's Temple at Jonestown established by Jim Jones, who was fascinated by Orwell's *Nineteen Eighty-Four*, which resulted in mass murder and group suicide, or the Branch Davidians led by David Koresh, which met a similarly violent end.[80] Groups like ISIS might also qualify for treatment under this heading, for while its ideal caliphate is clearly utopian, the means adopted to achieve this are

78. Timothy Miller, *The 60s Communes: Hippies and Beyond* (Syracuse University Press, 1999), xxiii–xxiv. Miller expands on this definition in "A Matter of Definition: Just What Is an Intentional Community?" (*Communal Societies* 30 (2010): 1–15).

79. Philip Abrams et al., *Communes, Sociology and Society* (Cambridge University Press, 1976), 26.

80. On Jones and Orwell, see Philip G. Zimbardo's "Mind Control in Orwell's *Nineteen Eighty-Four:* Fictional Concepts Become Operational Realities in Jim Jones's Jungle Experiment", in *On Nineteen Eighty-Four*, edited by Abbott Gleason et al. (Princeton University Press, 2005, 127–54).

extremely harsh and coercive. Typically cults neither invite nor analyse any information which contradicts their narratives. Their mental world is hermetically sealed and self-sustaining, so that even failure validates it.

Other religious foundations have rejected such methods. One well-known and fascinating early modern example is that of the great Spanish explorer Vasco de Quiroga (1470–1565), who arrived in the New World in 1531. An admirer of Lucian's account of the Saturnalia, and of More's *Utopia*, he believed the latter's scheme was the only means of civilising the Mexican natives. These he proceeded successfully to organise into three highly regulated and quasi-monastic "Republicas des indios", comprising some thirty thousand people, which lasted for many decades and ceased their legal existence only in 1872.[81] An even more famous and substantial early modern communitarian experiment was the Jesuit "reductions" in Paraguay (1609–1767), where, in the ex-Jesuit Abbé Raynal's description, "magistrates chosen by the people" oversaw the needs of the native Guarani people. Here, consent was "the result of voluntary submission, and proceeds from inclination, founded on conviction, and where nothing is done but from choice, and full approbation". Without "avarice, ambition, luxury, a multitude of imaginary wants", all produce was deposited in public warehouses for distribution, and "the people enjoy the advantages of trade, and are not exposed to the contagion of vice and luxury". There was "no distinction of stations; and it is the only society on earth, where men enjoy that equality which is the second of all blessings", "for liberty is undoubtedly the first." By 1702 these groups comprised over 89,000, eventually reaching 200,000. This religious communism was an immense success; here "public justice has never been reduced to the cruel necessity of condemning a single malefactor to death, to disgrace, or to any punishment of a long duration; where the very names of a tax, or a law-suit, those two terrible scourges which every where else afflict mankind, are unknown".[82]

As in Paraguay, the success of most subsequent intentional communities has usually resulted less from the threat of violence than from internalising

81. Amongst More's ideas included here were rotation of families, communal dining, and a restriction on clothing to two pieces of coarse white garments: Fernando Gomez, *Good Places and Non-places in Colonial Mexico: The Figure of Vasco de Quiroga (1470–1565)* (University Press of America, 2001), 91. See also Geraldo Witeze Jr, "Vasco de Quiroga rewrites *Utopia*", in *Utopias in Latin America Past and Present*, ed. Juan Pro (Sussex Academic, 2018), 53–57; Toby Green, *Thomas More's Magician: A Novel Account of Utopia in Mexico* (Weidenfeld and Nicolson, 2004).

82. Guillaume Thomas Raynal, *A Philosophical and Political History of the Settlements and Trade of the Europeans in the East and West Indies*, 3 vols (J. and J. Robinson, 1811), 2: 94–96. For commentary, see Philip Caraman's *The Lost Paradise: An Account of the Jesuits in Paraguay 1607–1768* (Sidwick and Jackson, 1975).

group norms, often enforced by charismatic authority, to provide a basic consensus on outlook and goals. This is the inner psychological secret of most utopias: we do the right thing without any other external compulsion than conscience (which may be the internalised fear of God), in the shared belief that we should do so. Where deviance from norms is rare, occasions for disorder are few. The need for armies and police is accordingly limited. This ideal is very old: Rousseau praised Lycurgus for having "established such morality as practically made laws needless—for laws as a rule, being weaker than the passions, restrain men without altering them".[83] An analogy with ideas of an original Golden Age is thus evident.[84] Many utopians have assumed that small group dynamics would suffice to regulate behaviour, often using the model of the small-scale, tightly knit Protestant sect. The British founder of philosophical anarchism and former Sandemanian Baptist William Godwin wrote in 1793 that a "general inspection that is exercised by the members of a limited circle over the conduct of each other" would suffice to regulate small communities.[85] Some early socialists, many also former Protestant Dissenters, agreed that "The Eye of the Community, and the inward feeling produced, will soon either create a change of conduct, or make the individual retire from the Society".[86] Thus would voluntary conformity be substituted for coercion, with the threat of ostracism a sufficient deterrent to most misbehaviour.

This assumption also permeated later revolutionary movements, but with an important twist: the small-group context was usually missing. The utopia of many anarchists as well as Marx, Engels, and Lenin assumes that conduct is largely self-regulated, and external laws minimised if not eliminated. Here abolishing private property is often assumed to be the key to securing social harmony, if not quite, as critics like Niebuhr suggest, redeeming "society from sin".[87] Coercive political authority virtually disappears. Lenin followed Engels in suggesting that after the state "withered away" a sense of communal life, or *Gemeinwesen*, would emerge, implying the affective ties Tönnies described.[88] Ernst Bloch, we have seen, adopted a similar view. For the Hungarian Marxist György Lukács, the future society's ultimate objective was "the construction of a society in which freedom of morality will take the place of legal compulsion

83. Jean-Jacques Rousseau, *The Social Contract and Discourses* (E. P. Dent, 1973), 109.

84. To Mircea Eliade, "Marx's classless society and the consequent disappearance of historical tensions find their closest precedent in the myth of the Golden Age that many traditions put at the beginning and the end of history" (*Sacred and the Profane*, 206).

85. W. Godwin, *Enquiry Concerning Political Justice*, 2: 565.

86. Orbiston *Register* 2, no. 1 (17 January 1827): 10.

87. Niebuhr, *Faith and History*, 218.

88. Olson, *Millennialism, Utopianism, and Progress*, 275–76.

in the regulation of all behaviour."[89] Internalised morality, effectively government by conscience and social pressure, requires no police to enforce. A sense of autonomy seems dominant, for we will our own order.

These assumptions are all inherited from the psychology of small-group conformity. Such systems all tacitly rely on a variant of shame- and conscience-centred theory, for guilt, shame, and conscience necessarily play a vastly more important role where legal constraints are minimal. Amartya Sen claims that an ideal society requires being able to lead a "life without shame", conceived as a key element of "social flourishing".[90] But shame at breaking social norms is an important utopian concept: it is how the group retains control over its members. Guilt may be conceived of as violating the group's norms, or as Freud saw it, as the social anxiety which ensued. So does utopia imply giving a heightened role to guilt and shame? Perhaps, if the price is diminished coercion of the harder variety. Yet these psychological mechanisms are of course themselves coercive, if of the softer sort. They may also be ineffective: we glower at people who refuse to wear masks in a pandemic when requested to do so. But they may just glare back at us, or tell us that our lives are a reasonable price to pay for their freedom to infect us.

Most small-scale intentional communities do accept such psychological pressure as the price to be paid for their sense of security and well-being. They unite like-minded people associating voluntarily, where persuasion, patriarchy, and group pressure rather than violence settle disagreements. Their forms of government range from charismatic leaders through rule by elders to consensual democracies and anarchical ordering. They have rarely challenged mainstream culture and society. They serve as reminders of the utility of experimental social organisation, as well as the worth and practicability of values often superseded in mainstream society. They remain the friendly, accessible face of everyday utopia. Only when the fist is raised, or the fear of ostracism appears in a harsh and punitive manner, or threatens suffering by deprivation, do they cross over to the dark side. So where coercion or exploitation does not get the upper hand, such communities are by definition utopian ventures which consciously engender a greater sense of closeness and mutual support than is evident in the outside society.

Our interest in intentional communities here is twofold, both as examples of everyday utopias which achieve a high degree of belongingness, and in relation to our central theme here, as demonstrating sustainability by limiting

89. Georg Lukács, *Tactics and Ethics: Political Writings, 1919–1929* (Verso, 2014), 48.

90. As discussed in Tim Jackson's *Prosperity without Growth: Economics for a Finite Planet* (Earthscan, 2009, 146–47).

needs. The utopian quality of such communities lies clearly in their fostering a strong sense of mutual dependency and support, akin to an extensive friendship network or even a familial bond. Typically this is created and sustained by the group's commitment to a core set of values, whether religious or secular. Often these include equality, though patriarchy has been common, and contempt for greed, ostentation, and excessive egoism. Sustainability, we will see, has rarely been a source of contention for these communities. The more rural and agricultural they are, the less likely it is that luxury or personal flamboyance in dress and decoration present a problem. This is particularly true of the many communities founded in North America from the seventeenth century onwards, which are the focus of much of the secondary literature on the subject. We will introduce a few examples of these, before moving on to more secular experiments.

Christian Intentional Communities

In the last three centuries, most of the intentional communities founded on the land were in the United States, during the nineteenth century.[91] This movement originated partly in religious persecution in the Old World and also responded to specific economic crises, notably in the 1840s. Such communities are often conceived as a temporary phase in America's expansion, helping to ensure the pioneers' survival without their permanently committing themselves to more intensely communal institutions like collective property.[92] Until the frontier ended around 1890 this kind of mutual assistance was common. In this period most Americans still engaged in small-scale farming, not urban commerce or industry. Frugality and frontier life were synonymous, since subsistence took priority over luxury or ostentation. Farming was the main activity in most intentional communities, though some later went into

91. The literature here is vast. A good start can be made with: J. Noyes, *Strange Cults and Utopias;* Mark Holloway, *Heavens on Earth: Utopian Communities in America 1680–1880* (Dover, 1966); Yaacov Oved, *Two Hundred Years of American Communities* (Transaction, 1988); Donald Pitzer, ed., *America's Communal Utopias* (University of North Carolina Press, 1997); Robert P. Sutton, *Communal Utopias and the American Experience: Secular Communities, 1824–2000* (Praeger, 2004). A 1991 survey showed that there were more Hutterite colonies in North America (391) than all other types put together (*c.*335): Timothy Miller, *The Quest for Utopia in Twentieth-Century Communes* (Syracuse University Press, 1998); Robert S. Fogarty, *All Things New: American Communes and Utopian Movements* (University of Chicago Press, 1990). On current trends, see Eliezer Ben-Rafael, Yaacov Oved, and Menachem Topel's *The Communal Idea in the 21st Century* (Brill, 2013).

92. See Bestor, *Backwoods Utopias;* Pitzer, *America's Communal Utopias,* xv–xxi, 3–13.

manufacturing and prospered. Many were located in remote rural areas far from the distractions of great cities, and had little means of knowing what they were missing. But while a virtue was made of this necessity, pleas for austerity diminished as the wider society, which eventually swallowed up most of these experiments, grew in opulence. Their rustic charm, such as it was, is largely lost on us today. But the ability of Christian communities to resist the temptations of luxury, and much else, for the love of God and expectation of salvation, remains remarkable.

Before American independence, many such communities were of religious sectarians fleeing persecution in Europe, like the Quakers. A second wave took place between the 1820s and the 1860s, and a third towards the end of the century. Of the 140 or so communities founded between 1860 and 1914, it has been claimed that "All were inspired by the exhortation found in Revelation that the time had come to make 'all things new'"[93] "All things dull" might have been a motto for the more sin obsessed. For strict morals were the consequence of so much godliness, and all forms of pleasure and sensuality suffered as a result. At the Pietist Amana colony in Iowa, frivolity was severely frowned on. Cards, games, dancing, and musical instruments except flutes were banned. At meals, men and women were separated to "prevent silly conversation and trifling conduct".[94] Women could not wear jewellery, and had to dress drably and wear black caps. Additional rules included injunctions for men to "Fly from the society of women-kind as much as possible, as a very highly dangerous magnet and magical fire", for all to avoid "Dinners, weddings, feasts . . . at the best there is sin", and "Constantly practice abstinence and temperance, so that you may be as wakeful after eating as before."[95] There are many variants on these themes.

Amongst the best-known early groups defined by simplicity and hostility to luxury were the Shaking Quakers, or Shakers, a millenarian sect founded in the 1740s who eventually had up to six thousand members living in eighteen communities in New York and elsewhere.[96] In their "families" members called themselves "brother" and "sister", and having renounced sexual intercourse to avoid sin, maintained voluntary celibacy. They were also notable for welcoming formerly enslaved blacks and Jews. They too feared pleasure, and their resistance to things of the flesh extended to many areas. Eating was made

93. Fogarty, *All Things New*, 3.

94. Charles Nordhoff, *The Communistic Societies of the United States* (1875; Dover, 1966), 33. See Holloway, *Heavens on Earth*, 174.

95. Nordhoff, *Communistic Societies*, 51. See Robert P. Sutton, *Communal Utopias and the American Experience: Religious Communities, 1732–2000* (Praeger, 2003), 55.

96. Their formal title was the United Society of Believers in Christ's Second Appearing.

boring by cutting food into squares, while dining in silence, with the sexes separated.[97] Little meat, tea, and coffee were consumed, and musical instruments were prohibited.[98] After photography was invented, pictures of individual members were discouraged as promoting vanity, idolatry, and disunity.[99] At least until the 1880s their home-made clothing was all of one colour, and designed for economy, uniformity, and disguising bodily contours.[100] By 1805, Brethren had adopted "long, baggy trousers", while Sisters were clad "in long gowns, high collars, and neck handkerchiefs" whose "shapeless appearance was appropriate for a celibate society." Of the four-hundred-strong colony at Mount Lebanon, New York, then a century old, one observer said that "While here they have every comfort and every sane luxury; they have no desire to have anything so long as obtaining it means in any way suffering to others." Another said its members' "lives are a rebuke to that spirit of greed, selfishness, and love of luxury, which is the curse of modern civilisation."[101] Their villages were clean and orderly, their famous furniture minimalist yet elegant and stylish, according to the principle "simple is beautiful".[102] By the 1860s, however, at their peak, an observer noted that with the sect's prosperity, "the sense is continually aspiring after more, becoming more and more tasty about clothing, and articles of fancy, the use of high colors of paint, varnish &c., perhaps more than is virtuous or proper."[103] French perfume seems to have been particularly alluring. By the early twentieth century, having prospered greatly through hard work and inventiveness, the circular saw being amongst their products, the Shakers became more closely integrated into the outside society, even spending winters in a Florida colony. They began buying cars and

97. Chris Jennings, *Paradise Now: The Story of American Utopianism* (Random House, 2016), 75; Holloway, *Heavens on Earth*, 74.

98. Nordhoff, *Communistic Societies*, 205.

99. William Alfred Hinds, *American Communities and Co-operative Colonies* (1878; Porcupine, 1975), 47.

100. A general account of communitarian clothing is given in Seymour R. Kesten's *Utopian Episodes: Daily Life in Experimental Communities Dedicated to Changing the World* (Syracuse University Press, 1993), 66–70. Here a general tendency "to dress differently" is noted (66). Perhaps by way of compensation, many Shakers had visions of being richly adorned in gold and coloured silk (Lawrence Foster, *Religion and Sexuality: Three American Communal Experiments of the Nineteenth Century* (Oxford University Press, 1981), 66).

101. Frederick William Evans, *Autobiography of a Shaker* (United, 1888), 263, 265.

102. Sutton, *Religious Communities*, 26.

103. Priscilla J. Brewer, *Shaker Communities, Shaker Lives* (University Press of New England, 1986), 166. See Robley Edward Whitson, ed., *The Shakers* (Paulist, 1983), 20, and Evans, *Autobiography of a Shaker*, 265.

acquired electricity. "Even before 1900", one historian writes, "the Shakers no longer maintained the fiction of the simple life. They had become eager participants in nearly every aspect of modern life and thought".[104] It was a not uncommon progression.

Another early group, formed in 1785, was the Harmonists or Rappites, named after their leader, the German Pietist George Rapp (1757–1847), who had three main communities with about twelve hundred members by 1847. Rapp expected the millennium to arrive imminently. Likening his society to a garden, he carefully controlled contact with the outside world to keep the weeds out. Property was held in common, and after 1807 celibacy was obligatory. Needs were met as required, in illness or in health. One observer noted that "Every man has his station appointed him according to his ability, and every one has his wants supplied according to his wishes. He applies to the mill for his supply of flour; to the apothecary for medicine; to the store for cloaths, and so on for everything necessary for human subsistence."[105] The "Clothing materials which they receive are of the best quality, just as their food", their dress being a uniform blue jacket, the women in grey dresses with their hair tucked into a cap. It was reported that "among you no vicious habits are in vogue, no cursing, swearing and lying is heard, no excess is noticed; and there is no temptation to cheat because each individual needs no money and need fear no want". The colonists' success evidently lay in the fact that they "discountenance ostentation, luxury, pride and vanity". Visiting them in 1825, the economist Friedrich List thought that "they really have an advantage over the Americans who bind themselves too slavishly to fashion." Rapp himself told an admirer that "We all possess in fact the peace and quietness, which your feeling expresses. The tumult of the world is quite strange to us, the splendor of luxury is unknown, our houses, though far inferior to those in cities, are filled with peace and love." He summarised the colonists' successes as resulting from three principles:

> In general the rules and measures determined by us at the very start have protected us against all luxury and from financial indebtedness. Namely, (1) Denial of all unnecessary things, e.g., the use of tobacco was given up; (2) preference for clothing materials manufactured by ourselves; and (3) as much as possible to limit our food to the products of our own land. Self-sufficient thereby, almost independent, we also cannot sympathize with our

104. Stephen J. Stein, *The Shaker Experience in America* (Yale University Press, 1992), 303. See Sutton, *Religious Communities*, 28.

105. Karl Arndt, *George Rapp's Harmony Society 1785–1847* (University of Pennsylvania Press, 1965), 201.

> free Americans who are accustomed to fashion and luxury and sunk in debt, and who do not follow our example, when not only families but also cities and states have gone bankrupt.[106]

Also worthy of mention in this context are the Pietist sect founded in 1683 known as the Labadists, whose colony, dedicated to the "inward illumination of the spirit", seemingly required "tasteless meals, uncomfortable living quarters, and no heating even during the coldest winter."[107] The Dunkers of Ephrata, Pennsylvania, founded in 1732 by Conrad Beissel, were also virtually monastic, living individually in cells four paces long and two broad, and sporting a white Capuchin monk's habit. They addressed each other as "thou", and kept to the injunction "Neither be a Glutton or luxurious, lest thou mayst ruin thy Estate by debauching in Prosperity: and then suffer Want in Need."[108] Celibacy, a rigorous work regime, and a strict diet, where meat was rare and only milk and water were drunk, helped to keep sin at bay.[109]

The most successful North American religious communitarians are descended from the sixteenth-century heretical central European sect of Anabaptist Protestants. These too have been uniformly hostile towards vanity and luxury. Amongst the best known are the Hutterites, founded in the late 1520s by Jacob Hutter. Expelled from Moravia in 1622, they may have followed Thomas Müntzer in adopting community of goods. Arriving in North America in the 1870s, they had some 391 communities by 1991, mostly spread across Canada and the western United States, and 480 by 2004. Their current population, split into half a dozen or so subgroups, is about fifty thousand. Their success results from seven factors: ideological monopoly, socio-economic dependence, social control, regulating outside interaction, limiting colony size, selective change, and religious legitimation. Each of their communes, or Bruderhofs ("brother-places"), of about 3500 acres and ninety members, is conceived as a "small enclosed porch outside of heaven." Private automobiles are forbidden, though agricultural machinery is widely used. Television, radio, and other forms of mass media are prohibited, Hollywood being, in the eyes of one, "the sewer pipe of the world". But singing frequently in a mesmerising "loud, shrill, nasal voice" is an integral part of Hutterite life. Clothing is uniform, and is largely bought in bulk, sometimes second-hand, with each person providing a "wish-list" twice a year. Dining, in silence, is communal. Women are subordinate to men in all matters. Like the Anabaptist "neo-Hutterite"

106. Arndt, *George Rapp's Harmony Society*, 322, 324, 347, 440, 290, 120, 571.

107. Oved, *Two Hundred Years*, 20.

108. Robert S. Fogarty, *American Utopianism* (F. E. Peacock, 1972), 3, 7.

109. Sutton, *Religious Communities*, 5.

Bruderhof sect, founded in 1920, birth control is rarely practised, and families once averaged ten children, though five is now common.[110] Unlike other so-called Old Order groups, the Hutterites have no private property except in personal possessions. All work without pay and are provided for equally. Each Bruderhof buys and sells produce on the market, and barters with other colonies. Dedication to God and community by abandoning selfishness is the core Hutterite ideal. This is enforced by the *Ordnung*, a set of rules focusing on *Gelassenheit*, or yielding one's sense of self to the good of the community. This entails cultivating "plainness, simplicity, obedience, humility, lowliness and meekness". So from kindergarten onwards children are taught to share their toys and to learn co-operative behaviour. They learn German before English. Baptism at age eighteen to twenty-five marks their commitment to the community. They do permit alcohol, but in great moderation: forty-eight bottles of beer and some wine annually. Private space is extremely limited, yet while "a hundred eyes are watching . . . loneliness is unheard of."[111]

The best-known Anabaptist group are the Amish Mennonites, who emigrated to what is now Pennsylvania in the 1730s, and number some 250,000 today. (There are over two million Mennonites in total.) Amongst the Mennonites and Anabaptists individual differences in wealth are rare, though community of goods was never adopted. All reject military service and violence in general. They do not swear oaths, hold public office, or recognise the supreme sovereignty of the nation-state or the political system. They remain largely apart unto themselves, proud of their five-hundred-year tradition of isolation, alien from and to a degree hostile to outsiders, whom the Anabaptist Brethren call Worldlings. All esteem the community over the individual and value tradition over change, personal sacrifice over pleasure, and work over consumption. Religious rituals are central to their identity and cohesion. Old Order members also insist that "Friends are more important than status, fame, or wealth", and that "Newer, bigger, and faster are not necessarily better."[112]

In such groups the strong bonds of a common faith, piety, and devotion eliminate or diminish many obstacles which endanger more worldly communities. As a result "the overwhelming ethic" is "production in the service of the community, not conspicuous consumption for leisure and personal pleasure."

110. Benjamin David Zablocki, *The Joyful Community: An Account of the Bruderhof* (Penguin Books, 1971), 129; Robert P. Sutton, *Heartland Utopias* (Northern Illinois University Press, 2009), 119–26; Sutton, *Religious Communities*, 95.

111. Donald B. Kraybill and Carl F. Bowman, *On the Backroad to Heaven: Old Order Hutterites, Mennonites, Amish, and Brethren* (Johns Hopkins University Press, 2001), 183, 28. See Sutton, *Religious Communities*, 97.

112. Kraybill and Bowman, *Backroad to Heaven*, 19.

All affect plain dress, though ties, shoes, skirt length, and the use of colour have been controversial issues in recent decades. The Mennonites prohibit jewellery and photography, and discourage commercial entertainment like golf or bowling. The Amish forbid all forms of make-up as encouraging personal vanity.[113] Ornamentation and bright colours are generally proscribed. Decorating buildings is not allowed. Large mirrors, statues, photographs, and wall paintings are out. Weddings are kept simple, and no names can be attached to gifts.[114] Yet the Amish are by all accounts a very gregarious people who probably spend considerably more time visiting their neighbours than do most outsiders. Common worship and the need for mutual assistance in agriculture, of course, play central roles here. But pride, conceit, self-exaltation, luxury, too much ease, indiscipline, and an obsession with "looks" or ostentation are everywhere suppressed. The "fast pace of this world" is avoided through a deliberate slowness, and refraining from "time-saving" devices.[115]

Many Old Order Mennonites are hostile to technology. The Amish, too, are renowned for engaging only reluctantly with machinery. But as with the other sects descended from Anabaptism, their practices vary. Some refuse to ride bicycles, and use only horses for transport and field work. But many use modern tractors and other agricultural devices. Having banned telephones from their homes in 1908 as too worldly and likely to encourage gossip, they permit them on the roadside, and some now have mobile/cell phones. They have calculators but prohibit computers.[116] Gas grills and chainsaws are used in some groups, and modern medicine is widely accepted. Mennonites see "the car as a symbol of independence, individualism, speed, status, and mobility", and prohibit ownership of cars but not riding in them. The Hutterites own trucks and vans but not private cars. They use CB radios, which were proscribed by the Brethren in 1925, who unlike the Amish drive cars, but remove their radios. The Brethren allow computers, but only for work, not to play games. Old Order Mennonites often have electricity and telephones in their homes. Some Ontario Mennonites prohibit lightning rods as interfering with divine providence.[117] The net effect has been to slow the pace of modernisation, not to halt it, and to promote a mistrust of excessive speed rather than the worship of it. We will later consider to what degree the wider world can learn from these examples.

113. Kraybill and Bowman, x, 67, 77, 107.

114. John A. Hostetler, *Amish Society* (Johns Hopkins Press, 1963), 59–60.

115. Donald B. Kraybill, *The Riddle of Amish Culture* (Johns Hopkins University Press, 1989), 43–44.

116. Kraybill, 1.

117. Kraybill and Bowman, *Backroad to Heaven*, 13–15, 80, 94, 130, 132, 167–68, 221.

Christian utopianism thus often represents a kind of halfway house between millenarianism and the more secular and republican traditions associated here with More's paradigm. In Christian utopias, loving one's neighbour is contingent on loving God, and orderliness on fulfilling divine commandments, often under the paternal supervision of a charismatic leader. Both forms of love are often linked to renouncing excessive desire for sensual gratification or material possessions. While western society remained overtly Christian, until the early twentieth century, such variants on utopia were common. To experience intensified anxiety within a deeply Christian society has often meant envisioning utopia primarily through the prism of millenarianism. All religions, moreover, provide bonds whose strength is difficult if not impossible to attain in more secular communities for lengthy periods of time, except for dystopian ideologies like extremely xenophobic nationalism. Nearly all such religious communities have necessarily been limited in scope, size, and appeal. They are by definition not universalist, for they exclude non-believers, and usually promote salvation only for the few. They have a counterpart in both literary and religious–political utopias which aim at extirpating sin as such and attaining a state of grace. Let us now consider some secular alternatives to such aspirations.

Secular Intentional Communities

The chief difference between Christian and more secular, mostly socialist, utopian communities is that the latter usually assume that institutions rather than sin produce most social misbehaviour, which results mainly from poverty and inequality. Since ideology rarely achieves the same degree of intensity and unity as religion, or sustains it as long, and often pays a higher price for what consensus it does achieve, secular communities have often faced greater problems than the more religious. Without the powerful glue and the compulsion to bow to conscience and peer pressure that religion provides, group strength is weaker, and differences of opinion more frequent. At a Fourierist Phalanx he visited, for example, the Oneida community founder John Humphrey Noyes, whom we meet below, noted that "Their own passions torment them. They are cursed with suspicion and the evil eye. They quarrel about religion. They quarrel about their food. They dispute about carrying out their principles."[118] Basic organisational principles are more often contested. Many communities debated whether private family life did not hinder, in Noyes's view, "the universal good, which does not permit the building up of supposed self-good, and therefore forecloses all possibility of an individual family."[119] But

118. J. Noyes, *Strange Cults and Utopias*, 352.

119. J. Noyes, 146.

greater tolerance of personal difference has made proscriptions on personal decoration and idiosyncrasy, which mark many religious communities, much rarer. Some uniformity has been present, but it has rarely been as actively promoted as in the more religious settlements.

BRITISH SOCIALISM

The most important modern secular communities have emerged from socialism, the chief utopian response to industrialisation and capitalism. A tradition often linked to Thomas More, it is usually explicitly identified with a theory of sociability, and in Britain was founded by the Welsh philanthropist, factory reformer, and wealthy cotton spinner Robert Owen (1771–1858).[120] The name "socialism" emerged etymologically in the mid-1820s by shortening Owen's phrase the "social system" of united interests, which he contrasted to the "individual system" of selfishness, private property, and market competition. It was also conceived as a "change from the individual to the social system; from single families with separate interests, to communities of many families with one interest".[121]

Owen was an atheist who flirted with millenarian language and conceived that a superior social order with vastly improved "circumstances" would ameliorate human behaviour dramatically.[122] His scheme for introducing this system commenced with ameliorating workers' conditions at his mill at New Lanark in Scotland from 1800.[123] His ideas met with sympathy from Tory paternalists like the poet Robert Southey, who depicted Owen as a latter-day Thomas More.[124] When the Napoleonic War ended, economic crisis loomed, and these proposals were extended to encompass the relief of unemployment and poverty.[125] Hatched around 1816, Owen's "Plan" was a communist community of a few thousand inhabitants who combined manufacturing and

120. E.g., by Robert Southey, in *Sir Thomas More; or, Colloquies on the Progress and Prospects of Society* (2 vols, John Murray, 1829).

121. "Oration Containing a Declaration of Mental Independence" (1826), in Robert Owen, *Selected Works of Robert Owen*, 4 vols, ed. G. Claeys (Pickering and Chatto, 1993), 2: 46, 39.

122. On Owen and millenarianism, see W. H. Oliver's "Owen in 1817: The Millennialist Moment", in *Robert Owen: Prophet of the Poor*, edited by Sidney Pollard and John Salt (Macmillan, 1971, 166–87).

123. For details, see Ophélie Siméon's *Robert Owen's Experiment at New Lanark* (Palgrave Macmillan, 2017).

124. See the portrayal in Southey's *Sir Thomas More.*

125. On the communities founded as a result, see R. G. Garnett's *Co-operation and the Owenite Socialism Communities in Britain, 1825–45* (Manchester University Press, 1972), and,

agriculture to ensure self-sufficiency. Communities might trade with one another on an egalitarian basis of the labour and material costs of commodities, but no extensive production for export was anticipated, and wage-labour would end.

While not opposed to machinery in principle, Owen admired the simplicity of the North American Christian communities. Basically hostile towards luxury, his focus was on providing necessities for all. Many of the trends of the late eighteenth century thus needed to be reversed. Owen thought that:

> The acquisition of wealth, and the desire which it naturally creates for a continued increase, have introduced a fondness for essentially injurious luxuries among a numerous class of individuals who formerly never thought of them, and they have also generated a disposition which strongly impels its possessors to sacrifice the best feelings of human nature to this love of accumulation.

This growing lust for luxury increased the burden on the poor: "The rich wallow in an excess of luxuries injurious to themselves, solely by the labour of men who are debarred from acquiring for their own use a sufficiency even of the indispensable articles of life." The wealthy, for example, demanded "to purchase fine lace and muslins at one-fourth of the former prices; but, to produce them at this price, many thousands of our population have existed amidst disease and wretchedness".[126]

The answer could only be to restrain desires for more than a reasonable amount of labour could create, thus exchanging artificial needs for free time. Fashion was a key target here. In 1820 Owen thought that "fashions will exist but for a very short period, and then only among the most weak and silly part of the creation". In future, "All things will be estimated by their intrinsic worth, nothing will be esteemed merely for its cost or scarcity, and fashions of any kind will have no existence."[127] Owen viewed his first communal experiment at New Harmony, Indiana (1824–28), as an exercise in personal as well as collective frugality, reducing his meals to two per day, and rejecting changes of fashion in apparel. At New Lanark he had devised a kind of toga for the boys "somewhat resembling the Roman and Highland garb". Now he thought "a costume of the best form and material that can be devised" was preferable to

---

more broadly, J.F.C. Harrison's *Quest for the New Moral World: Robert Owen and the Owenites in Britain and America* (Charles Scribner's Sons, 1969).

126. Owen, *Selected Works*, 1: 113 ("Observations on the Effect of the Manufacturing System", 1815), 239, 244 ("Letter to the Earl of Liverpool on the Employment of Children in Manufactories", 1818).

127. "The Social System" (1826–27), in Owen, 2: 70.

"the waste of capital, materials and labour—the loss of health, the deterioration of intellect, and the immorality, which the manufacture and use of perpetually changing fancy dresses" induced.[128] Some members did sport distinctive attire, with pantaloons for both sexes and women adding a knee-length jacket.[129] What was intended as a gesture towards equality, however, became a cliquish distinction when the intellectuals, calling themselves the "literati", took to wearing it as a "badge of aristocracy".[130] In 1826 Owen reiterated that:

> With regard to dress, an object upon which so large a share of the industry of civilized states is now so uselessly and injuriously expended, the members of the community, having once ascertained the best materials and the form best adapted to the health of the wearer, will have no disposition to introduce afterwards any of the frivolous, fantastical and expensive varieties that may be current elsewhere. They will adopt the rational course of employing the time which the manufacturer of such useless decorations would consume, in the pleasures of social intercourse, and intellectual pursuit, and in healthful recreations.[131]

At New Harmony, a visiting German nobleman observed, Owen aimed to make needs equal.[132] But equality proved an elusive goal. In principle each member had a fixed sum of up to $180 per year to use in the community store for clothing and food by "free choice", to be measured against the value of their services.[133] But some discerned "partiality" in the system of distributing provisions and clothing, especially regarding the trustees' families, leaving poorer members deprived, in 1825, of winter clothing in particular.[134] Owen himself was accused of boarding at the tavern, where coffee and tea were available,

128. Robert Owen, *A New View of Society and Other Writings*, ed. G. Claeys (Penguin Books, 1991), 286 ("Report to the County of Lanark", 1820); Owen, *Report of the Proceedings at the Several Public Meetings Held in Dublin* (1823), 70.

129. Carol Kolmerten, *Women in Utopia: The Ideology of Gender in the American Owenite Communities* (Indiana University Press, 1990), 56.

130. Bestor, *Backwoods Utopias*, 179. See Bernhard, Duke of Saxe-Weimar Eisenach, *Travels through North America, during the Years 1825 and 1826*, 2 vols (Carey, Lea, and Carey, 1828), 2: 116; Sutton, *Secular Communities*, 10.

131. Owen, *Selected Works*, 2: 70, 315.

132. Arndt, *George Rapp's Harmony Society*, 335; George Browning Lockwood, *The New Harmony Communities* (Chronicle, 1902), 104.

133. Lockwood, *New Harmony Communities*, 108, 158.

134. Paul Brown, *Twelve Months in New Harmony* (C. H. Woodward, 1827), 76; A. E. Bestor, ed., *Education and Reform at New Harmony: Correspondence of William Maclure and Marie Duclos Fretageot 1820–1833* (1948; Augustus M. Kelley, 1973), 373; Thomas Pears and Sarah Pears, *New*

while others had to make do with cheap substitutes like rye coffee. Some criticised him for being too permissive, and complained about card playing, a profusion of musical instruments, dancing, and even a puppet show.[135] Amongst other factors, a lack of social discipline and multiple competing private aims brought about the rapid dissolution of the enterprise. Owen did not retreat from his views of fashion, however, reiterating in 1836 that it would exist for only a short period.[136] Later he conceded the need for the "merely ornamental", and occasionally he acknowledged that "beneficial luxuries" might exist in the future system.[137] This retreat from a more austere ideal, we will see, was common amongst socialists.

Other Owenite communities faced similar problems. Where meat was a luxury, as at the Ralahine colony in Ireland, fancy dress could scarcely be an issue.[138] In some later American socialist communities, too, the former New Harmonist Josiah Warren recounted, uniformity of dress was demanded "as one of the most necessary external signs of that equality of condition desirable among men".[139] A distinctive costume could be a badge of pride. At the Owenite community at Manea Fen, Cambridgeshire, in 1840, it was reported that:

> The men wear a green habit . . . presenting an appearance somewhat like the representation of Robin Hood and his foresters, or of the Swiss mountaineers. The dress of the females is much the same as the usual fashion, with trousers, and the hair worn in ringlets . . . They are quite the lions of the villages round about.[140]

As the wider socialist movement developed in the 1830s and 1840s the issues raised by these experiments were widely debated. Gradually, more puritanical attitudes to luxury and consumption were displaced by concessions to the value of a higher standard of living, and greater variety of attire and decoration. Owen's followers, of whom there were around fifty thousand in Britain in the early 1840s, thus oscillated between condemning luxury, grudgingly

---

Harmony: An Adventure in Happiness; Papers of Thomas and Sarah Pears (1933; Augustus M. Kelley, 1973), 24.

135. Bestor, *Backwoods Utopias*, 188–89.

136. *New Moral World* 2, no. 99 (17 September 1836): 369.

137. Owen, *Selected Works*, 1: xxxvii, 316; 2: 218.

138. E. T. Craig, *An Irish Commune: The Experiment at Ralahine, County Clare, 1831–1833* (1920; Irish Academic, 1983), 201.

139. Josiah Warren, *Practical Applications of the Elementary Principles of "True Civilization"* (1873), 74.

140. The *Working Bee*, 28 November 1840, quoted in W.H.G. Armytage, *Heaven's Below: Utopian Experiments in England, 1560–1960* (Routledge and Kegan Paul, 1961), 163.

conceding demand for it, and embracing a vision of unlimited plenty based upon a just system of production and exchange. Owenism's attitude towards needs has been broadly described as implying "neither luxury nor want."[141] But we need to differentiate between attitudes towards luxury. A broadly puritan view of needs and consumption was adopted by many early Owenites.[142] Some likened Owen's principles to those of Lycurgan Sparta, writing that any doubts about "the correctness and practicability of the New Views of Society, were entirely removed by reflecting upon the accordance of the whole with those principles which contributed to the establishment and prosperity of the Spartan government."[143] The purpose of socialism, many assumed, was that it "augments the productions useful to all by banishing luxury and idleness."[144] Condemning those "whose obvious interest it is to invent perpetual changes", one insisted that in a "rational society":

> Something like a standard of good taste would be arrived at, because nobody would have an interest in capricious changes. Elegance would be only another name for comfort; and appropriateness would take precedence of variety. The labour of dress-makers, hair-dressers, plumassiers, florists, jewellers, and lace-workers would thus be sparingly required, and a hideous proportion of our "staple manufactures," of "new wants," in silks, cottons, and hardwares, would be discovered to be useless; the share of time and energy bestowed on them being directed to more profitable and more worthy objects.[145]

Many Owenites were former Protestant Nonconformists who readily embraced moralistic arguments against luxury and fashion. Like their contemporaries, they recognised that "every class, down to what is called the lowest, strives to imitate the appearance and fashions of the class above it."[146] Many thought the present system of extreme inequality only generated unhappiness, even for the wealthy, amongst whom competition "in show and luxury" brought "jealousies, envyings, and many other evil passions, which disturb their peace." The solution was that "If every one had the value of his labour,

141. Harrison, *Quest*, 58. See, generally, Noel Thompson, "Owen and the Owenites: Consumer and Consumption in the New Moral World", in *Robert Owen and His Legacy*, eds Noel Thompson and Chris Williams (University of Wales Press, 2011), 113–28; Thompson, *The Market and Its Critics: Socialist Political Economy in Nineteenth Century Britain* (Routledge, 1988), 58–118.

142. See my *Machinery*, xxviii, 153–55, 190–96.

143. *New Moral World* 1, no. 43 (22 August 1835): 340.

144. *New Moral World* 3, no. 117 (21 January 1837): 102.

145. *New Moral World* 4, no. 183 (28 April 1838): 211.

146. *New Moral World* 5, no. 12 (12 January 1838): 178.

but of no more than his own, those irregularities in the condition of mankind would never arise, and all the evils attendant upon excessive luxury, and excessive poverty could be avoided."[147] Owen's main socialist rival in the 1820s, the Irish landowner William Thompson, wrote of "all those extra articles of luxury called for by excessive wealth", and expected that under socialism "*the peculiar vices of luxury and want would almost cease.*" In the new communities, any "motive for exertion arising from mere love of distinction, of excelling, of exciting envy, by means of individual accumulations of wealth, and thus attracting public sympathy", would disappear. Instead:

> Objects of dress, elegance, luxury, will be estimated according to their intrinsic value, their real utility; not forgetting in the estimate any one pleasurable quality, the lustre and softness of the silk, or the peculiar flavour of the exotic production. All factitious importance given to articles of wealth as mere sources of distinction, will be forgotten with those distinctions which equality of wealth annihilates.[148]

Another early Owenite and founder of the short-lived Orbiston community (1825–28) near New Lanark, Abram Combe, distinguished between "the old system", where "the children are trained to believe that labour is degrading, and that Pomp, and Sloth, and Luxury are the *best*, if not the *only* means of obtaining happiness", and the "New System". In the latter

> they will be trained to believe that useful Labour is a most honourable employment, without which the real dignity of our nature cannot be supported and that temperance and industry are the best, if not the only means of securing that health and independence, without which all other earthly possessions are worse than useless.[149]

Orbiston's newspaper, the *Register*, prophesied that "Grandeur, Rank, and artificial Riches, would not be desirable under the new system, because in themselves they give no rational title, in their possessors, to the approbation or respect of the Community". In the future, instead, "all the members would know this, those who assumed any superiority from the mere possession of those, would inevitably become objects of pity".[150] An early co-operative journal, the *Associate*, posited in 1829 that "Expensive luxuries (which have the

147. *The New Political Economy of the Honey Bee* (W. C. Featherstone, 1823), 15.

148. William Thompson, *An Inquiry into the Principles of the Distribution of Wealth* (Longman, Hurst, Rees, Orme, Brown, and Green, 1824), 207, 259, 468, 533.

149. Abram Combe, *The Sphere of Joint-Stock Companies* (G. Mudie, 1825), 43–44.

150. Orbiston *Register* 2 (14 March 1827): 29.

effect of enlarging cupidity and diminishing our sympathy with others) [would] cease to be created when the producers of them shall have to weigh the trouble of producing them against the pleasure of displaying them in their own persons." As late as 1842, a leading Owenite lecturer, George Alexander Fleming, proposed that if it was impossible to assuage every desire, a "dignified simplicity" might suffice.[151]

A similar tendency was evident at the Ham Common Concordium outside London inspired by the Transcendentalist and self-styled "Paradise planter" Amos Bronson Alcott, the father of Louisa May Alcott. Founded as Alcott House in 1838, this was initially overseen by the celibacy advocate and "Sacred Socialist" James Pierrepont Greaves, who warned of "the tendencies of comforts and luxuries to soften the Spirit, to weaken its self-command, and increase its sensibility to hardships and exposure".[152] It lasted ten years. Here, all rose between four and six and bathed in cold water. The dress was plain and utilitarian, being "a brown holland or cotton blouse, a neat check shirt without neck cloth or any other clumsy wrapper round the neck." The food was even plainer, since consuming "flesh, butter, cheese, eggs, mustard, vinegar, oil, spices, beer, wine, tea, coffee, and chocolate" was forbidden, with meals taken in silence. All slept on mattresses, it being considered "that lying on feathers is both enervating and unhealthy", as well as relying on animal exploitation. One member thought that "Every step towards simplicity is good, and has Divine sanction", and predicted the trend would be from cooked to raw food. By 1844 nothing was cooked at all, and "When the salt was conceded it was concealed in paper under the plate, lest the sight of it should deprave the weaker brethren."[153] A visitor, the "Pontifarch of the Communist Church" John Goodwyn Barmby, thought a balance might be fruit and vegetables in the summer and meat in the winter, with no alcohol unless it was home-brewed or received as a gift.[154]

These practices resulted from a careful assessment of moral principles. Another visitor commented that "Self-denial and asceticism were enjoined, as a means of rehabilitating the fallen nature of man; and the use of animal food was regarded with as much horror as by the votaries of Brahma", a "sanguinary diet" being associated with "a sanguinary code", though a wish to avoid sexual stimulants is also evident. The Owenite Samuel Bower thought that "The

151. *Associate*, no. 3 (March 1829): 15; *Union*, no. 3 (1 June 1842): 69.

152. James Pierrepont Greaves, *Letters and Extracts from the MS. Writings of James Pierrepont Greaves*, vol. 1 (Ham Common, 1843), 125.

153. Armytage, *Heavens Below*, 179, 181.

154. J.E.M. Latham, *Search for a New Eden: James Pierrepont Greaves (1777–1842): The Sacred Socialist and His Followers* (Associated University Presses, 1999), 140, 165, 169–70.

general reason for their abstinence from the accustomed dietary, and other modes and usages of the world, is, that the universe law may not be transgressed by unnecessary cruelty". He believed that since "The insatiable cravings of the appetitive organs can never be fully satisfied . . . the human being at length becomes a vast accumulation of wantful essences or instincts, which cannot be satiated." This process only obstructed "the expansion both of the intellectual faculties, the divine sympathies, and the establishment of Being affinities with the divine nature." When winter came, carrots and cold water proved insufficient, however, and many Concordians returned "to the outer world".[155] Asceticism was clearly in the ascendant, and George Jacob Holyoake thought the prevailing idea "was that happiness was wrong". Here too the ex-Owenite William Galpin, whose "life was all self-denial", stayed on his way from the Owenite community at Queenwood to be converted to White Quakerism in Dublin, growing long hair and a beard, going barefoot, and encouraging the residents to learn to see in the dark like animals, and to use sign language rather than speech. Some also thought nudity accompanied the natural life. A German resident was found digging in the garden one morning "mit nodings on", though he was later deemed too extreme even for this group.[156]

From the outset, however, other socialists rejected such austerity. The first prominent Owenite journalist and publisher, George Mudie, thought that so long as the system of rewarding labour was just, many "additional comforts and luxuries besides" could be enjoyed in the new system.[157] The sole proviso was that these should "be attainable without interfering with the arrangements for the production of *necessaries*, and as shall not require from them a greater exertion of labour than is consistent with the rational enjoyment of human life."[158] An influential pamphleteer with Owenite leanings, John Gray began in the 1820s with a marked prejudice against luxuries, arguing that a lace dress, for instance, was "useless. It can neither be eaten nor drank; and it forms no part of useful wearing apparel. It is made only to please the fancy and to be looked at. It will not compare, in point of real utility, with a penny loaf or a glass of cold water".[159] By the late 1820s, however, he shifted his emphasis to the limitations which capitalist competition placed on demand. Moving away from what he described as "speculative theories upon the perfectibility of

155. [Samuel Bower], *A Brief Account of the First Concordium* (1843), 2–3; Thomas Frost, *Forty Years' Recollections* (Sampson Low, Marston, Searle, and Rivington, 1880), 46–47.

156. Latham, *New Eden*, 162, 168, 182.

157. *Alarm Bell; or, Herald of the Spirit of Truth* (c.1842): 2.

158. *Economist* 1, no. 2 (3 February 1821): 27.

159. John Gray, *A Lecture on Human Happiness* (Sentry, 1826), 25.

man", he proclaimed instead the possibility of virtually unlimited production.[160] Now, he thought, it was evident that:

> in a properly constituted society, one in which money should be of a rational kind, the demand for that which is ornamental, pleasing, and luxurious, including the fine arts in all their approved branches, must ever increase as fast as the increased powers of production should enable a smaller proportion of the community to perform the more humble and laborious operations.[161]

The difference was that all produce would be shared equally, rather than luxuries accruing only to the few. Other socialists agreed. The Leeds printer John Francis Bray thought the new community system would "create and enjoy every necessary and luxury in the greatest abundance."[162] Still other Owenites despaired of settling the issue, and said the new communities would supply "in due time, the beneficial luxuries, if there be any luxuries which are permanently beneficial."[163]

We now begin thus to see proposals for a shift from private to public luxury. The Christian Owenite John Minter Morgan portrayed this in literary form in the ideal community described in his didactic novel *The Revolt of the Bees* (1826):

> All my wants are readily supplied; and I have access to libraries, museums, concerts, groves and gardens, superior to those which any private fortune, however ample, could command. I possess all the benefits of almost unbounded wealth, without any of its cares and anxieties. For, instead of a train of ignorant, servile, and rapacious followers, or a numerous retinue of disorderly or dissipated servants to control, I am surrounded by intelligent and affectionate friends, united to me and to each other by an interchange of kind offices and by mutual sympathy.

Morgan rejected the Spartan model as implying a "martial character", but thought it might be modified to form an "intelligent and benevolent" alternative.[164] He was certain, though, that luxury corroded sociability, quoting the evangelical leader William Wilberforce to the effect that "Prosperity and

160. Claeys, *Machinery*, 118–19.

161. Gregory Claeys, ed., *Owenite Socialism: Pamphlets and Correspondence*, 10 vols (Routledge, 2005), 8: 110 (Gray, *An Efficient Remedy for the Distress of Nations* (1842)). See J. Gray, *Lecture on Human Happiness* (1826), 56.

162. John Francis Bray, *Labour's Wrongs and Labour's Remedy* (David Green, 1839), 124.

163. *New Moral World* 1, no. 3 (15 November 1834): 18.

164. John Minter Morgan, *The Revolt of the Bees*, 3rd ed. (1826; 1839), 184, 109.

luxury gradually extinguishing sympathy, and puffing up with pride, harden and debase the soul."[165] And the "thing called *fashion*", he thought, was "better named *folly*".[166]

Many early socialists thus opted for what Noel Thompson terms "social opulence, private asceticism", and restraining "artificial" needs or "unnatural" wants which entailed excessive labour for the working classes.[167] The most ambitious Owenite community actually constructed, called Harmony, or Queenwood, in Hampshire (1839–45), was well provisioned precisely to indicate the opulent lifestyle workers could expect of socialism.[168] But its luxury was conspicuously public rather than private. The dining room was sumptuous, but dress was simple. No expense was spared in the internal construction and decoration of the central building, even to the use of mahogany, decoration with chandeliers, and devising a dumb-waiter system to deliver food to and remove dishes from the dining room to the kitchen, which, it was claimed, rivalled the best London hotels.[169] Sadly the food itself was dull, and to many overly vegetarian. When the community faltered, a few ascetic members embraced wholesale simplicity, meaning not only vegetarianism and water drinking but no shoes or stockings, and some thought they would have embraced nudism had the law permitted it.[170]

By the 1830s, however, the general trend within Owenism and even more in the co-operative movement was in the other direction, towards a less Spartan and puritan position with respect to needs. A Lancashire co-operator wrote in 1831 that when labourers began "to WORK FOR THEMSELVES, they will be supplied abundantly, not only with all the necessaries, but with all the luxuries, and all the elegancies of life". It was, echoed a Salford co-operator the same year, "not to lessen their comforts or to abridge their luxuries, but to increase our own, that the efforts of Co-operative Societies are directed". "In the perfect state of Co-operation", he or she continued, "could it be reached, the comforts, nay the luxuries of existence would be abundant: there could be

165. J. Morgan, 147.

166. John Minter Morgan, *Hampden in the Nineteenth Century*, 2 vols (Edward Moxon, 1834), 2: 7.

167. Noel Thompson, "Social Opulence, Private Asceticism: Ideas of Consumption in Early Socialist Thought", in *The Politics of Consumption*, eds Martin Daunton and Matthew Hilton (Berg, 2001), 51–68, and, more generally, Thompson, *Social Opulence and Private Restraint: The Consumer in British Socialist Thought since 1800* (Oxford University Press, 2015).

168. See Edward Royle, *Robert Owen and the Commencement of the Millennium: A Study of the Harmony Community* (Manchester University Press, 1998).

169. Garnett, *Co-operation*, 192.

170. Frost, *Forty Years' Recollections*, 47.

no cause for envy, insofar as the enjoyment of those things raise that passion." At the Third Co-operative Congress, held in 1832, it was even claimed that "by our united exertions, we shall enjoy tenfold the comforts and luxuries of life which are now enjoyed by those among us who are considered favourably circumstanced."[171] So, as the Coventry ribbon maker Charles Bray put it in 1840, once necessities were available to all, some "might be employed in the acquisition of comparatively useless luxuries and ornaments". Yet Bray retained a contrast between "artificial wants" and "real wants", writing that:

> The standard of utility would supplant that of caprice and fashion; and as useless articles of luxury and vanity would no longer be an indication of the extent of private property, or marks of superiority, being possessed by all if by any, they would no longer be desired, and distinction would be sought where alone it ought ever to be found, in useful and ennobling qualities.[172]

That year, another writer demanded a system "that shall meet the new, and increasing wants, of progressive civilized man: and one that will secure a full supply of all the wants of humanity, to every class, without curtailing the pleasures, the conveniences, the comforts, or the luxuries of any".[173]

The Christian "Model Town" planner James Silk Buckingham, writing in 1849, thought that "the highest degree of abundance in every necessary of life, and many luxuries" could be produced without entailing excessive labour.[174] At the technological utopian extreme of this spectrum, the German-American John Adolphus Etzler promised a "terrestrial paradise" of rapid travel, floating islands, and luxury without limit—even the eradication of snakes, mosquitoes, "and other troublesome vermin".[175]

Various socialists also made a direct connection between "community" and sociability. The Owenite Joseph Marriott wrote a play depicting community life as encouraging closer bonds between members and insisted that "From the favourable circumstances in which we are placed, we must have more real friends than the rest of mankind can possibly have in any other situation." Marriott's friends would also be abstemious to a degree, usually eating only plain foods; and while "any adult may take a certain quantity of ale and wine . . . not one of them is allowed to touch spirituous liquors."[176]

171. Claeys, *Owenite Socialism*, 3: 292, 338–39.

172. Claeys, 7: 280.

173. Claeys, 8: 2.

174. James Silk Buckingham, *National Evils and National Remedies* (Peter Jackson, 1849), 141.

175. John Adolphus Etzler, *The Collected Works of John Adolphus Etzler*, ed. Joel Nydahl (Scholars' Facsimiles and Reprints, 1977), 6 ("The Paradise within the Reach of All Men").

176. Joseph Marriott, *Community: A Drama* (A. Heywood, 1838), 6, 13.

FRENCH SOCIALISM

The early French socialists present a similar contrast between private and public luxury, while offering some startling departures from Owenite themes. Every historian of utopia pauses on encountering their leading theorist, the self-styled "Messiah of Reason" Charles Fourier (1772–1837), astonished at his engagement with the passions, so apparently at odds with utopia's obsession with rationality, and at his pure whimsicality. A passion for every occasion and an occasion for every passion might have been Fourier's motto. Once we recover from some of his excesses, including dreams of "anti-lions" and seas of lemonade, everyone growing to 2 metres in height and reaching the age of 144, and, cruel irony, the poles heated up to make the climate more comfortable, we discover a thinker of remarkable complexity and ingenuity. Even today he offers abundant insight into our central problem here, that of exchanging a mania for possessing things for enjoying life more fully and sustainably.

Though he thought agriculture the healthiest activity, Fourier was no ascetic. His ideal Phalanx of 1600 people would be a small town in itself, and would cater to many tastes. Occupants would happily reside in apartments of three or four rooms, "because in the new order the social relations of the series will be too active for anyone to spend much time in his apartment." "In Harmony", Fourier wrote:

> the rich will not spend their excess wealth on the construction of useless private castles; instead they will construct fine workshops and beautiful buildings for their favorite groups and series. Once this practice is widely followed, luxury will itself become a productive force. In Harmony luxury will be associated with useful work, with the sciences, the arts, and especially with cooking. Luxury will serve, along with many other vehicles, to make these functions attractive for both children and adults.[177]

Fourier resolutely opposed overzealous egalitarianism, and especially the Spartan image of utopia, "the virtuous republic sustained by cabbage and gruel", in Jonathan Beecher's apt phrase. He thought Rousseau's hostility to luxury, discussed below, was misplaced. Instead, he viewed satisfying the "luxurious passions" driven by the five senses as essential to human happiness.[178] "Civilisation", he believed, had been progressive up to the point when commerce came to predominate, when it became destructive for most. Yet its

177. Charles Fourier, *The Utopian Vision of Charles Fourier*, eds Jonathan Beecher and Richard Bienvenu (Beacon, 1971), 294.

178. Jonathan Beecher, *Charles Fourier: The Visionary and His World* (University of California Press, 1986), 247.

inventions and ingenuity could still benefit the majority. Instead of "general community of goods", the Phalanxes, or Phalanstères, would divide profits, giving a third to capital, five-twelfths to labour, and a quarter to talent, with a guaranteed minimum income for all, even in sickness and old age. If all were partners in the enterprise, no levelling equality would be required.[179] In keeping with Fourier's psychological theory, work would consist in up to eight activities daily, thus sating the variating, alternating, or "Butterfly" passion for variety. Five elaborate meals per day would sate the palate. "Courts of Love" would even guarantee minimal sexual gratification, for in the "new amorous world" avoiding such frustration was essential for social happiness—here Fourier was the forefather of Freud. To facilitate such pleasures, people would wear insignia indicating their particular aptitudes and desires.

In these details Fourier's radicalism is generally deemed deeper and more psychologically incisive, as well as vastly more sensual, than that of his socialist contemporaries.[180] Concerning pleasure and consumption his guiding premise was that "a general perfection in industry will be attained by the universal demands and refinement of the consumers, regarding food and clothing, furniture and amusements". Here, however, several principles were at work. "Let us refute", Fourier insisted,

> a strange sophism of the economists who maintain that the unlimited increase of manufactured products is an increase of wealth; the consequence of that would be, that if every person could be induced to use four times as many clothes as he does, the social world would attain to four times its present wealth in manufactured products.

There was "No truth whatever in this; their calculation is as false on this point as it is on the desirability of unlimited increase of population, or *food for cannon.*" So priorities were necessary. Gastronomy would be a key focus. Instead of having the largest amount of everything, there was to be the "greatest possible consumption of varieties of food" but the "smallest possible consumption of varieties of clothing and furniture", a position which we will see was anticipated by some eighteenth-century physiocrats. The Harmonians would pursue a policy "totally contrary to our ideas of commerce, which promote waste and the changes of fashion, under the pretext of maintaining the workman". In Harmony:

179. See Charles Fourier, *Design for Utopia: Selected Writings of Charles Fourier* (Schocken, 1971); Fourier, *The Theory of the Four Movements* [1808], eds Gareth Stedman Jones and Ian Paterson (Cambridge University Press, 1996).

180. See Charles Fourier, *Le nouveau monde amoureaux: Oeuvres complètes de Charles Fourier*, 12 vols (Éditions Anthropos, 1966–68), vol. 7.

> the workman, the agriculturist, and the consumer, are one and the same person; he has no interest in practising extortion upon himself, as in civilisation, where everyone strives to promote industrial disturbance occasioned by changes of fashion, and to manufacture poor goods or poor furniture, in order to double consumption, to enrich the merchants at the expense of the people and of real wealth.[181]

Harmonians would thus understand that "changes of fashion, defective quality, or imperfect workmanship, would cause a loss of five hundred francs per individual, for the poorest of the Harmonians possesses a wardrobe of clothes for all seasons, and is accustomed to using furniture, trappings, and appurtenances, for work or pleasure, of a fine quality." One consequence, Fourier hoped, was that clothing and furniture might possess "prodigious durability". So a new pair of shoes, which commonly wore out in six months, might last ten years. The prevailing principle would in any case be public rather than private luxury.[182]

Fourier nonetheless did not envision that the desire for social emulation, competition, and distinction would disappear. Instead it would be transferred to public enterprises, or work groups called Series, and become part of the community's *esprit de corps*. "In Harmony", he proclaimed, "luxury is corporative; everyone wants his favorite groups and series to shine."[183] At work, a

> like emulation will prevail among the Series of every description. It is sufficient for a rich man to make any of them shine, to incite all the neighbouring districts to vie with it in some shape, if not in luxury, at least in neatness, in perfection. This mania will seize all people of great means; it will cause luxury to be expended upon labour and workshops, so repulsive to-day by their poverty, coarseness, and filth.[184]

In the Phalanstères the "luxurious" passions, derived from the five senses, which generally required wealth to satisfy, were to be gratified to a minimal level, and particularly those of taste and touch. Such passions were natural, and not to be avoided or renounced. Nonetheless many workers currently employed in luxury trades would become redundant in the future. "Deprived of life's necessities, the civilized worker is tormented by a display of increased affluence which the savage does not see", thought Fourier. The answer was not to renounce refinement, especially of the senses, but to allow industry, science,

181. Charles Fourier, *Selections from the Works of Fourier*, introd. Charles Gide (Swan Sonnenschein, 1901), 196.

182. Fourier, 64, 194–97.

183. Fourier, *Utopian Vision*, 293–94.

184. Fourier, *Selections from the Works*, 194.

and art to serve all the Harmonians, thus producing "corporative" or "compound or collective forms of luxury". In "the societary state the common people will enjoy a minimum level of subsistence higher than that of our comfortable bourgeoisie".[185] "Let us impose on every new science", Fourier insisted, "the condition of raising the peoples to general riches, and not to exceptional riches, which magnify in the present day the wants and the privations of the mass."[186] "In civilization", he claimed, "we can only conceive of luxury in the simple mode; we have no conception of the compound or collective forms of luxury".[187]

Fourier stressed that this should include expanding the "social harmonies or pleasures of the soul, which at present are well-nigh unknown to the great". The new order would raise "sensual refinements to a perfection of which the civilised world is incapable of forming any conception." Excessive private luxury would be useless where so much gratifying public activity existed, and where psychological needs were recognised and happily sated. The "collective luxury" of the Phalanx would take precedence. To Fourier, thus, the pleasure to be derived from group membership and through work, both of which promoted unity and harmony, were primary to human happiness, and could act as a substitute for private luxurious pleasures.[188]

A lonely man who never married, Fourier understood the centrality of forming explicitly utopian groups to the new social ideal better than any other early socialist. It was, he wrote, only "by the analysis of groups that the study of the social man should have been initiated,—a thing entirely neglected". He even gave a name to the desire to associate in groups: "groupism". Each person "progresses in sociableness (*sociabilité*) only in so far as he succeeds in forming one, or two, or three, or four groups . . . Sociableness, then, depends upon the formation of groups, or passionate leagues". Here "friendship predominates in childhood, as love does in youth . . . ambition prevails in mature life, and . . . old age, isolated from the world, concentrates itself upon family affection".[189] Fourier listed friendship as amongst the four main social or "affective" passions—love, corporate association, and family affection being the others. Each of these constituted an "affective group", with "unityism", a term very close to what we have here termed belongingness, resulting from their collective bonds. In friendship, ties of self-interest were least developed, but

185. Fourier, *Utopian Vision*, 147, 243, 258.

186. Charles Fourier, *The Passions of the Human Soul*, 2 vols (Hippolyte Bailliere, 1851), 2: 290.

187. Fourier, *Utopian Vision*, 243.

188. Fourier, *Design for Utopia*, 75, 195.

189. Fourier, 155–56.

"civilizee" friendships were deficient in many other areas, too, such as being too subordinated to family, or restricted by age, or subverted by the falsity of commercialism. Fourier recognised that "The effects of collective friendship are very rare and of very short duration in civilization", for there were many divisions of interests, pleasures, and tastes amongst the 810 types of character. He also conceded that "the philosophers' sacred equality . . . is the essential attribute of the group of friendship." This meant that among the opulent class, "it is not possible to dissimulate the inequality of ranks and fortunes there. Without this oblivion of ranks, a group seeks in vain to adopt the tone of Friendship", for the "dominant tone" of friendship was "the confusion of ranks, playful gaiety", the very spirit, in other words, of the Saturnalia. Fourier recommended that at least one hour per evening be devoted to the task of promoting friendship, which was intertwined with enjoying "an extreme variety of pleasure". In fact, Fourier's entire system was geared towards promoting "harmony" through association. Socialism was, in his view, friendship as such.[190] One statement illustrates how this might operate:

> In private *fêtes* a shadow of transit is often attempted to be organized. The repast is divided into several little tables, where the guests sit together as they please without formality or stiff propriety. These tables may be visited alternately by every one connected with the majority of the company; too little friendship reigns however in civilization for these transits to be animated. Yet they offer a gleam of that kind of pleasure it is wished to organize. Care is taken to reproduce this same transit in the convivialities that follow the repast, and all sorts of games are offered to the company: cards and billiards for staid people, dances for the young folks, nonsense for the children; in the refreshments, liqueurs are presented to warm the blood of the grand-dads, ices to calm the fires of the dowagers, and lemonade for the boiling youth; in short, it is endeavored, in all the details of the *fete*, to organize a transit of delights, a kind of pleasure so rare in civilization, where some simple germs of it are created with great trouble; whereas in harmony the poorest of men can hope in the course of one day for more than one transit, and can vary from day to day the nature of his transits, obtain every day those of different kinds, either in composite, in papillon, or in cabalist, and obtain frequently unityist transits, combining three kinds of pleasure. This participation in the well-being of the rich is the spring that binds the harmonian people to its social order. If the poor were, as they are with us, reduced to die of hunger whilst the rich man swims in plenty, they would soon become the enemies of social order. But in harmony everything is

190. Fourier, *Passions*, 1: 263, 273–74, 284; 2: 90–91.

arranged in such a way that the well-being of the rich is communicated to the poor, and that a dose of well-being is spread gradatively over all classes; while in civilization the well-being of the great in no way secures the humble from misery: we find them poorest, on the contrary, in countries of large fortunes, like England, Spain, and Russia.[191]

The great theoretical question was how to promote these bonds, and ensure their functioning in work, where discord and emulation might also promote healthy competition, and at play. Fourier's leading American disciple, Albert Brisbane, reiterated that "Friendship tends to social equality and to the levelling of ranks."[192] Fourier understood that consuming luxury goods in private was pointless. Art, music, food and drink, and much else was best enjoyed in the company of others: the social context was crucial. Collective luxury and collective pleasure gave meaning to individual lives, not the isolated possession and consumption of luxury goods, which was the bourgeois norm. The point, as Dolores Hayden insists, with respect to the North American Fourierists, was that they "accepted the argument . . . that even the most elaborate private dwellings lack real luxury, because they are isolated from collective services. Facilities for collective child care, laundry, cooking, and dining can free all members of a community for equal participation in its activities."[193]

There was alas little opportunity to test many of these principles in the Fourierist communities founded in the 1840s, after the economic depression of 1837, though some manufactured furniture, shoes, and even books and musical instruments.[194] A few, notably the North American Phalanx, achieved an enviable diet which included chocolate, ham, beef, fruits, and pastry. But several families left the Trumbull Phalanx "because they had been in the habit of living on better food." By way of compensation, however, some achieved a rich social life, conscious in the desire that Fourier's principles encouraged amusement as such.[195]

The best-known phalanx, Brook Farm (1841–47), during its Fourierist phase, was founded by the Transcendentalist George Ripley, and aimed "to diminish the desire of excessive accumulation by making the acquisition of

191. Fourier, 2: 114.

192. Albert Brisbane, *Social Destiny of Man; or, Association and Reorganization of Industry* (C. F. Stollmeyer, 1840), 453.

193. Dolores Hayden, *Seven American Utopias: The Architecture of Communitarian Socialism, 1790–1975* (MIT Press, 1976), 353.

194. Carl J. Guarneri, *The Utopian Alternative: Fourierism in Nineteenth-Century America* (Cornell University Press, 1991), 183.

195. Guarneri, 183.

individual property subservient to upright and disinterested uses".[196] A Transcendentalist precept was "do not crave costly luxuries to make a show withal."[197] But when Fourier's ideas were adopted here in 1844 they were interpreted to mean, in Parke Godwin's formulation, that "Harmony can not exist without riches and luxury", a "tendency to luxury" being one of the passions, subdivided into health and wealth.[198] A key problem, work, was solved by having everyone do every task in rotation, though the intellectuals found farming difficult. Brook Farm had no dress code as such, though a general simplicity and similarity of clothing prevailed. Nothing made from cotton, wool, or leather was allowed because this was deemed to require exploiting slaves or animals.[199] There were vegetarians, who dined at their own table. Nonetheless, there "was almost insatiable desire for pleasure: music, dancing, card playing, charades, tableau vivants, dramatic readings, plays, costume parties, picnics, sledding, and skating".[200] One participant boasted that "The greatest harmony prevails among us; not a discordant note is heard; a spirit of friendship, of brotherly kindness, of charity, dwells with us and blesses us . . . There is a freedom from the frivolities of fashion, from arbitrary restrictions, and from the frenzy of competition; we meet our fellowmen in more hearty, sincere and genial relations".[201] But self-sufficiency in food was never achieved, and "civilizees" had to be hired to make up the forthcomings. After the community's main building was destroyed in a devastating fire, the farm was sold in 1849.

In the eyes of its best-known chronicler, Nathaniel Hawthorne, sadly, Brook Farm "as regarded society at large . . . stood in a position of new hostility, rather than new brotherhood."[202] But within the community a strong sense of comradeship prevailed. This is beautifully captured in one participant's recollection of the small talk and interaction of daily life in the community: "It is only to those loved most by us that we recite the trivial things, for we know that those trivialities link us closer than anything else, filling all the chinks in

196. John Thomas Codman, *Brook Farm: Historic and Personal Memoirs* (Arena, 1894), 12.

197. John Van der Zee Sears, *My Friends at Brook Farm* (Desmond Fitzgerald, 1912), 28.

198. Parke Godwin, *A Popular View of the Doctrines of Charles Fourier* (J. S. Redfield, 1844), 30, 45.

199. Sterling F. Delano, *Brook Farm: The Dark Side of Utopia* (Harvard University Press, 2004), 118.

200. Amee Rose, *Transcendentalism as a Social Movement* (Yale University Press, 1981), 131, quoted in Brian J. L. Berry, *America's Utopian Experiments* (University Press of New England, 1992), 101.

201. Codman, *Brook Farm*, 197.

202. Nathaniel Hawthorne, *The Blithedale Romance*, 2 vols (Chapman and Hall, 1852), 1: 44.

our friendship or love."[203] Brook Farm long stood as a model of self-sufficiency, thanks especially to its association with Ralph Waldo Emerson, whose essay "Society and Solitude" (1857) did much to popularise the idea of solitude, which he regarded as an "organic" necessity, and in many respects superior to "vulgar" society.[204]

Amongst the other early founders of French socialism, Henri de Saint-Simon (1760–1825) was famed for proclaiming that "The Golden Age of mankind does not lie behind us, but before; it lies in the perfection of the social order".[205] Like Fourier, and against Rousseau, Saint-Simon developed an idea of public luxury. "For a long time", he thought,

> luxury has been concentrated in the palaces of kings, in the dwellings of princes, in the mansions and the châteaux of a few powerful men. This concentration is very harmful to the general interests of society, because it tends to establish two distinct levels of civilization, two different classes of men, that of persons whose intelligence is developed by the habitual sight of works of art and that of men whose imaginative faculties receive no development, since the physical labors with which they are exclusively occupied do not stimulate their intelligence at all.
>
> Conditions today are favorable for making luxury national. Luxury will become useful and moral when the whole nation enjoys it. Our century has been vouchsafed the honor and the advantage of utilizing in a direct manner in political combinations the progress made by the exact sciences and by the fine arts since the brilliant epoch of their regeneration.[206]

To Saint-Simon such public luxury could be particularly exhibited and enjoyed in festivals and celebrations hosted by the government. This would end the current problem that "The nation holds as a fundamental principle that the poor should be generous to the rich, and that therefore the poorer classes . . . daily deprive themselves of necessities in order to increase the superfluous luxury of the rich."[207]

203. Codman, *Brook Farm*, 106.

204. Ralph Waldo Emerson, *The Conduct of Life and Society and Solitude* (Macmillan, 1892), 273, 277.

205. Henri de Saint-Simon, *The Political Thought of Saint-Simon*, ed. Ghita Ionescu (Oxford University Press, 1976), 98.

206. Quoted in Frank E. Manuel, *The New World of Henri Saint-Simon* (Harvard University Press, 1956), 315.

207. Henri de Saint-Simon, *The Organiser* (1819), in Frank E. Manuel and Fritzie P. Manuel, eds, *French Utopias: An Anthology of Ideal Societies* (Schocken Books, 1966), 277.

Saint-Simon's leading disciples took up the suggestion of creating a society rooted firmly in industry and mass production and also proposed gradually replacing political power by the administration of industry and commerce by experts, the "industriels". Their main statement, *The Doctrine of Saint-Simon: An Exposition* (1828–29), cast the new ideal in the form of a religion, as did Owen, some Fourierists, and the Icarians.[208] The Saint-Simonians' personal tastes soon turned to luxury, and they became well known in the early 1830s for hosting sumptuous dinners where politicians, artists, and intellectuals mingled.[209] The Saint-Simonian scheme influenced J. S. Mill and also Thomas Carlyle, whose *Past and Present* (1843), later denounced as "feudal socialism" by Marx, offered one of the first programmatic descriptions of a *dirigiste* statewide form of capitalism, in which industrial armies would be paid reasonable fixed wages in return for loyal service. This foreshadows in part, we will see, the writings of the most influential nineteenth-century literary utopian, Edward Bellamy.

Another form of French communitarian socialism in the 1840s was established by Étienne Cabet (1788–1856), and his followers, the Icarians or Cabetists. They developed a major political movement in France itself and also founded several colonies in the United States, one of which lasted fifty years. Cabet established the principle that in the community "Equality is *relative* and *proportional*. Each has an equal right in the benefits of the Community, *according to his needs*, and each has the equal duty of bearing the burdens, *according to his abilities*."[210] This slogan, later famously paraphrased by Marx in a utopian lapse in 1875,[211] was hung in the dining room at the Nauvoo community in Illinois. It was to be fulfilled in a style commensurate with bourgeois or even aristocratic expectations, in the belief that this increase would be immensely appealing to the working classes.

Cabet published a literary utopia, the *Voyage in Icaria* (1840), whose influences included Fénelon's *Télémachus*, Babeuf, Mercier, and above all Thomas

208. A selection of Saint-Simon's writings can be found in his *Selected Writings on Science, Industry and Social Organization*, edited by Keith Taylor (Holmes and Meier, 1975). The *Doctrine* has been reprinted, edited by Georg Iggers (Schocken Books, 1972).

209. Pamela Pilbeam, *Saint-Simonians in Nineteenth-Century France: From Free Love to Algeria* (Palgrave Macmillan, 2014), 18.

210. *History and Constitution of the Icarian Community* (1917; AMS, 1975), 255.

211. Namely, "From each according to his abilities, to each according to his needs!" which would be possible only "In a more advanced phase of communist society, when the enslaving subjugation of individuals to the division of labour, and thereby the antithesis between intellectual and physical labour, have disappeared" (Karl Marx, *The First International and After: Political Writings*, vol. 3, ed. David Fernbach (Penguin Books, 1974), 347).

More, who is classified with Socrates and Christ as a martyr to "la *Communaute de biens*" (community of property) who had seen "que la *Propriété* est la cause de tous les maux" (property is the cause of all evil).[212] Cabet described Icaria as labouring constantly to invent and extend every form of pleasure.[213] The Icarians enjoy gourmet food, elegant machine-made clothing, indoor plumbing and waste disposal, plentiful leisure, and sumptuous festivals to boot. Invisible machines even pipe music into hospitals, which like schools and public assemblies would be "palaces".[214] Cabet proclaimed that "The republic does not even forbid luxuries or superfluities, for you cannot call an amusement superfluous if it has no drawbacks." All such consumption was governed by three rules: firstly, "that all our amusements must be sanctioned by the law or by the people. The second is that we may not seek the pleasurable until we have provided ourselves with that which is necessary and useful. The third is that we do not allow any pleasures other than those that every Icarian can enjoy equally." As for attire, the community would aim to "harmonize" "variety with unity and equality".[215] "All individuals of the same social condition wear the same uniform": "Childhood and youth, puberty and adulthood, celibacy or marriage, widowhood or remarriage, the diverse professions and functions: all are indicated by dress." Women's clothing would be abundantly decorated, with feathers and jewels made of "alloys or of other gilded or non-gilded metals", or manufactured "precious stones", though the women themselves would "scorn and despise all conventional beauties and all feelings of childish vanity". But some limits to fashion would exist. A committee of experts would be formed to study clothing in all countries. A handbook would demonstrate permissible future styles, ordering that "feminine fashion 'should never change'"—women were clearly second-class citizens here—that there should be only "a certain number of determined forms of hats, toques, turbans and bonnets, and that the model of each of these forms be decided and arrived at by a commission of modistes and painters."[216]

For better or worse these theoretical problems were never put to the test of practice. In the actual community of Icaria, founded in 1849 at the former

212. Etienne Cabet, *Voyage en Icarie* (1840; Paris, 1848), 216, 479.

213. Etienne Cabet, *Travels in Icaria*, transl. Leslie J. Roberts, introd. Robert Sutton (Syracuse University Press, 2003), liii, 223. See Christopher H. Johnson, *Utopian Communism in France: Cabet and the Icarians, 1839–1851* (Cornell University Press, 1974).

214. Cabet, *Travels in Icaria*, 40; Robert P. Sutton, *Les Icariens: The Utopian Dream in Europe and America* (University of Illinois Press, 1994), 18–19.

215. *History and Constitution of the Icarian Community*, 255, 260.

216. Cabet, *Travels in Icaria*, 48–53, 223; Sylvester A. Piotrowski, *Etienne Cabet and the Voyage en Icarie* (Hyperion, 1975), 93.

Mormon colony of Nauvoo, Illinois, Cabet was insisting by 1853 that everything had to be "temperate, frugal, simple". Tobacco and whiskey were banned, and Cabet hoped there would be no more "objects of luxury, nor of vanity, nor of silk, nor lace, nor embroidery."[217] Dress distinctions he saw as "one of the principal causes of all the discussions amongst women, of all the divisions, and of all the difficulties for the administration", and a cause of "jealousies, quarrels and troubles". The sooner a standard wardrobe could be arranged, the better, Cabet concluded.[218] But his desire for a uniform was found impracticable owing to the amount of sewing it would require.[219] Later Icarian experiments were less *dirigiste* in these matters. At the last founded of the movement's seven colonies, the Icaria-Speranza community in California (1881–86), members were given accounts with local merchants to acquire their own clothing, and the constitution stipulated that "each member shall be at liberty to select whatever object of clothing that suits him."[220]

More influential than Cabet in France was Pierre-Joseph Proudhon (1809–65), who is at the more Spartan end of the utopian spectrum. Proudhon attracted a substantial following in mid- and later nineteenth-century France, and offered a more decentralised, mutualist alternative to state socialism.[221] More than Fourier or Saint-Simon, he praised the virtues of austerity and poverty, maintaining "that the happiest man is he who best knows how to be poor", and even that:

> It is not good for man to be in easy circumstances; on the contrary he must always feel the stimulus of want. Easy circumstances would be more than corruption, they would be slavery; and it is important that man should be able, on occasions, to rise above want and to go without even the needful.[222]

Proudhon lauded Lycurgus for having "understood perfectly that the luxury, the love of enjoyments, and the inequality of fortunes, which property engenders, are the bane of society". But he rejected Lycurgus' solution to the

217. Sutton, *Secular Communities*, 62; Sutton, *Les Icariens*, 83.

218. Quoted in Diana M. Garno, *Citoyennes and Icaria* (University Press of America, 2005), 93. But as Garno goes on to note, Cabet's "association of women's fashion with vanity and luxury" ignored the fact that women's labour on producing and maintaining clothing was enormously important to the community.

219. Sutton, *Les Icariens*, 73.

220. Albert Shaw, *Icaria: A Chapter in the History of Communism* (Knickerbocker, 1884), 212.

221. On Proudhon's ideas, see Alan Ritter's *The Political Thought of Pierre-Joseph Proudhon* (Princeton University Press, 1969), and on the movement, K. Steven Vincent's *Pierre-Joseph Proudhon and the Rise of French Republican Socialism* (Oxford University Press, 1984).

222. Henri de Lubac, *The Un-Marxian Socialist: A Study of Proudhon* (Sheed and Ward, 1948), 48.

problem, contending instead that "How much wiser he would have been if, in accordance with his military discipline, he had organized industry and taught the people to procure by their own labor the things which he tried in vain to deprive them of."[223]

Amongst the German writers of note in the burgeoning communist movement, the tailor Wilhelm Weitling (1808–71) was neo-Spartan in his desire to reduce the amount of labour involved in changing fashions.[224] The philosopher Moses Hess (1812–75), who converted the young Friedrich Engels to communism in 1842, took up Babeuf's ideas in Paris in the 1830s. But he eventually decided that Babeuf's "equality of poverty" involved "killing every kind of desire". Rather than dividing produce equally, Hess thought, "it is according to human nature that the means for life or activity should be different, so that everyone will always receive those materials which he needs at any given time for his life activity." So restricting needs as such was impossible.[225]

Several later nineteenth-century communitarian experiments also contributed to the debate over luxury and consumption. A few communities were evidently undisturbed by issues of personal style. Members of the Christian Catholic Church community at Zion City, Illinois, in the 1880s, for instance, reportedly "wore the latest styles of clothing when not in church garb".[226] At the anarchist colony at Modern Times, Long Island, variety in women's clothing was noted by a visitor, Moncure Conway, though this was seemingly matched by monotony in male attire.[227] Many more settlements demonstrate restraint in this area. A Transcendentalist community, Fruitlands, co-founded by a former Ham Common Concordium resident, which lasted only seven months in 1843–44, banned both animal products and animal labour. The novelist Louisa May Alcott reported that here, at her family's table, "No milk, butter, cheese, tea, or meat, appeared. Even salt was considered a useless

223. Pierre-Joseph Proudhon, *What is Property?* (Humboldt, *c.*1890), 325–26.

224. Carl Wittke, *The Utopian Communist: A Biography of Wilhelm Weitling* (Louisiana State University Press, 1950), 63.

225. Moses Hess, *The Holy History of Mankind and Other Writings*, ed. Shlomo Avineri (Cambridge University Press, 2004), 108, 122. See also Shlomo Avineri, *Moses Hess: Prophet of Communism and Zionism* (New York University Press, 1985).

226. Philip L. Cook, *Zion City, Illinois: Twentieth Century Utopia* (Syracuse University Press, 1996), 150.

227. Moncure Conway, *Autobiography, Memories and Experiences*, 2 vols (Cassell, 1904), 1: 235–36.

luxury and spice entirely forbidden by these lovers of Spartan simplicity". The community aimed "to supersede the labor of cattle by the spade and the pruning-knife." As for clothing, "A new dress was invented, since cotton, silk, and wool were forbidden as the product of slave-labor, worm-slaughter, and sheep-robbery. Tunics and trowsers of brown linen were the only wear". Wide-brimmed linen hats were also worn by both sexes. But, Alcott noted, "self-denial was the fashion, and it was surprising how many things one can do without."[228] Here most accepted the need "to simplify clothing as well as food", though some questioned other radical proposals being mooted, such as abolishing ships and railroads. The most extreme solution to the clothing problem was of course to have none at all. So here the former Owenite Samuel Bower took to nudity, though he was forced to practise it at night only, and then in a white shift, leading to rumours of ghosts.[229]

One of the most remarkable of these experiments was John Humphrey Noyes's community at Oneida, New York (1848–81). A millenarian whose doctrine of "Christian Perfectionism" claimed that freedom from sin was possible in this life, Noyes (1811–86) believed himself absolved from following many moral norms. In 1834, in the early days of his theological odyssey, he announced "that men can be perfect—and that he himself was perfect".[230] He began his first community at Putney, Vermont, in 1844. In 1846 he proclaimed that "All individual proprietorship of either persons or things is surrendered and absolute community of interests takes the place of the laws and fashions which preside over property and family relations in the world." His followers described Noyes as "the father and overseer whom the Holy Ghost has set over the family thus constituted." His seven disciples (at this point) agreed that "as such we submit ourselves in all things spiritual and temporal, appealing from his decisions only to the spirit of God, and that without disputing." In 1847 Noyes declared that "*the Kingdom of God Has Come*". By 1852 the community, now some 208 strong, proclaimed a "new state of society", which included the "*Abandonment of the Entire Fashion of the World*—especially marriage and

228. Louisa May Alcott, *Transcendental Wild Oats and Excerpts from the Fruitlands Diary* (Harvard Common Press, 1975), 27, 37, 49.

229. Richard Francis, *Fruitlands: The Alcott Family and Their Search for Utopia* (Yale University Press, 2010), 89. Bower ate only raw food like beans and grains, and refused to eat potatoes because they grew downwards rather than upwards towards the sky. He regarded killing any living thing as interfering with the natural order of things and would not cut vegetables because it interfered with their life. He later became a Shaker, then left them too.

230. Quoted in Roger Wunderlich, *Low Living and High Thinking at Modern Times, New York* (Syracuse University Press, 1992), 56.

involuntary propagation"—and promoted the "*Cultivation of Free Love, Dwelling together in Associations or Complex Families*."[231]

Thereafter Oneida became well known, or infamous, for practising "complex" or group marriage in an "enlarged family" where all were conceived as married to everyone else. Noyes deduced from his theory of "Bible Communism" the idea that celibacy was a "perversion". Sexual relations were encouraged, but regulated by "male continence", or coitus interruptus, a technique he thought older men could teach young girls, and older women young men. A rotation scheme for sexual intercourse commenced, in which Noyes himself seems to have played a particularly active part, all the women under forty having signed a statement acknowledging him as "God's true representative". From 1869 the practice began of planned parenthood based upon a proto-eugenic ideal called "stirpiculture", where younger people mated with their supposed spiritual superiors. Of the 58 babies born according to this plan, 9 were Noyes's. Children were raised in common, and men and women were more equal than in the outside society.[232] Paradoxically the scheme led to closer bonding by couples, and a weakening of community ties. But eventually a rebellion occurred against Noyes's insistence on initiating all the teenage girls and choosing people's mates. Oneida was attacked by outsiders as a "Utopia of obscenity".[233] Noyes stepped down as president of the community in 1877, and monogamous marriage was reintroduced in 1879.

Dress was an important issue at Oneida. Initially all drew equally on the central supplies of the community, with a clothing committee allocating resources. According to Bible Communism, God possessed all property, and adherents were urged to imitate the "Primitive Church" and anticipate a time "when fashion follows nature in dress and vocation".[234] Noyes early on suggested that "the uniform of vital society" should be similar for both sexes.[235] His choice of something like "bloomers" was intended to "crucify the dress spirit", as well as, ultimately, to make woman "what she ought to be, a *female*

231. Constance Noyes Robertson, ed., *Oneida Community: An Autobiography, 1851–1876* (Syracuse University Press, 1970), 10–11, 267.

232. Wunderlich, *Low Living*, 59. See also George Wallingford Noyes, ed., *Free Love in Utopia: John Humphrey Noyes and the Origin of the Oneida Community* (University of Illinois Press, 1971).

233. Robert S. Fogarty, ed., *Special Love/Special Sex: An Oneida Community Diary* (Syracuse University Press, 1994), 87; Sutton, *Religious Communities*, 82 (quote).

234. John Humphrey Noyes, "Bible Communism", in *John Humphrey Noyes: The Putney Community*, ed. George Wallingford Noyes (Oneida, 1931), 121.

235. Harriet M. Worden, *Old Mansion House Memories* (Kenwood, 1950), 10.

man".[236] In June 1848, after some discussion, the women took to wearing simple calf-length gowns or frocks with pantaloons, after the children's fashion of the day.[237] A year later, at a time when false hair was fashionable, and with a view to "discouraging feminine vanity", the women cut their hair short. Some claimed that the women found long hair "incompatible with true simplicity in feminine attire", and that several "women declared it distasteful and burdensome. The idea of wearing their hair short occurred to them". Others asserted that the community generally found the time spent caring for long hair "a degradation and a nuisance." But "no one could deny that short hair made the women appear younger; and all rejoiced in this new freedom." These restrictions on fashion made travel outside the community difficult for them, however. So "Few women were sent out, because in the entire feminine wardrobe of the Community, there remained but one dress and hat made according to the fashions of the day; one woman at a time wore this costume on their sallies into the world."[238] Any "woman who wore too-elaborate dresses, or any but the simplest jewelry, was sure to be criticized for surrendering to the 'dress spirit,' Noyes's name for feminine vanity."[239] In 1865 he claimed that:

> I can truly say that I am proud of our *women*, young and old. I see that in will and principle, and to a good extent in practice and feeling, they have conquered the fashions of the world in themselves, and are substantially free from bondage to the spirit of dress and ornament, and special love. The long discord I had with them about these things has passed away.[240]

Yet there was no enforced rule in this area. In 1864 it was reported that:

> We are not careful about equality in this thing. There is a clothing committee which has power to veto any great extravagance, but generally persons

236. *First Annual Report of the Oneida Association*, 41, quoted in Lawrence Foster, *Women, Family, and Utopia: Communal Experiments of the Shakers, the Oneida Community, and the Mormons* (Syracuse University Press, 1991), 92. Bloomers were loose-fitting wide pantaloons, and were widely associated at the time with reform movements. On dress reform in utopian fiction in this period, see Darby Lewes's *Dream Revisionaries: Gender and Genre in Women's Utopian Fiction, 1870–1920* (University of Alabama Press, 1995, 48–53).

237. Pierrepont Noyes, *My Father's House: An Oneida Boyhood* (Rinehart, 1937), 127; Robertson, *Oneida Community*, 298; Maren Lockwood Carden, *Oneida: Utopian Community to Modern Corporation* (Johns Hopkins Press, 1969), 44.

238. Robert Allerton Parker, *A Yankee Saint: John Humphrey Noyes and the Oneida Community* (G. Putnam's Sons, 1935), 170–71, 204.

239. Spencer Klaw, *Without Sin: The Life and Death of the Oneida Community* (Penguin Books, 1993), 137.

240. Robertson, *Oneida Community*, 301.

> have what dress they wish for, and some more than others. There is no sum appropriated, and no rule of equal distribution—and no prescribed uniform.[241]

There seems to have been considerable pride in the results. In 1871 the community paper reported that:

> People sneer at our dress, and talk slightingly of our looks; but, alas! they know not what they say. We have no "heavers," nor "plumpers," nor "false calves," nor "rouge"; and perhaps we don't look as well as your city belle who is puffed and padded and painted; but we are *genuine* from head to foot.[242]

In other areas the community suffered shortages early on. Sometimes butter was available only once or twice a week at "good dinners", and greens otherwise, though Noyes was seemingly less deprived than the others. Initially "Luxuries were no problem", one of Noyes's sons recalled, "since vanities and all forms of dissipation were abhorrent to men and women striving for salvation."[243] Later, however, prosperity solved any problems which resulted. As the business began to thrive, an early historian of the community noted, "it was almost inevitable that personal wants should expand, that the Community should fall into habits of luxury, even of extravagance."[244] In the 1860s the desire for private possessions, evidently originating in some boys' request for their own pocket watches, provoked an investigation, and evidently opened up the floodgates of desire.[245] Soon the community possessed "a Turkish bath, a photographic studio, a 'chemical laboratory,' elaborate properties for theatrical performances, and musical instruments for the orchestra", and even a two-storey summer house. Hired labour did much of the harder kitchen and gardening work, while the expanding family moved into newly built private houses nearby. Noyes still discouraged personal indulgence generally, such as excessive decoration of private rooms in Mansion House, the large communal building. Music, art, and theatre were promoted, however. A croquet lawn was laid out in 1862 and soon became a centre of activity, and by the 1870s, Ellen Wayland-Smith notes, the community had become "full-fledged members of the emerging American bourgeoisie." Somewhat ironically, the community's great success in highly fashionable silkware and then silverware after 1865

241. Robertson, 79.

242. Robertson, 308.

243. P. Noyes, *My Father's House*, 16.

244. R. Parker, *Yankee Saint*, 232.

245. Ellen Wayland-Smith. *Oneida: From Free Love Utopia to the Well-Set Table* (Picador, 2016), 177.

became a leading source of its prosperity.[246] But here, clearly, a principle of private frugality and public luxury was in practice.[247]

By the early 1870s a well-thought-out system of distribution existed. In 1874 the community's paper explained how it worked:

> A Community family, though in many respects like a common one, is so large as to make it necessary to divide the care; and so arise various departments, clothing, washing, furniture, bedding, and so on, each having a competent head appointed to the office. If you need a new window curtain, stand-spread, an easy chair, a foot-stool, a different bedstead, a looking-glass, or a larger bureau, just apply to Mrs. S., who has charge of the furniture, and she will be sure to do her best to accommodate you. When the carpet in your room grows threadbare and shabby, let Miss K. know of it and she will provide you with a better one. If anyone of the 200 beds needs a tick mended, more husks supplied, a new mattress, a wider counterpaine, larger or smaller pillows, in short if in any way it needs changing or repairing, she who needs it has only to speak to Mrs. N. and as if by magic the work is done. If one of your dresses is new and you would like to have it carefully washed, or you need a piece of cloth bleached, mention it to Mrs. K., and you will find yourself as well served as though you had not left it to others. Mrs. C. will furnish goods for your own clothing, and Mrs. V. will provide for your children.[248]

In 1875, however, the system of distribution changed, and an annual allowance was granted to each member. This seems to have instigated a veritable flood of consumption, with a local paper reporting: "No more bloomer costumes in public; no more severed hair; but in their places elegant dresses and waving tresses will be the rule . . . Fashionable hats, costly switches, silk dresses, etc., etc., almost daily go from our village to the Community." As late as the 1920s a woman who married a descendant of the community met criticism for dressing too fashionably.[249] Yet in the same period Oneida's manufactures, and especially silverware, were being advertised "in an ever more unabashed manner to American consumers' aspirations to class distinction and chic". A 1928 ad depicts "two cool turbaned flappers at a restaurant [who] idly apply lipstick while their waiter serves tea. The caption warns, 'Look at your silver—your guests do.' 'Women of position today are unwilling to use make-shift silver . . .

246. Wayland-Smith, 103; Klaw, *Without Sin*, 104.

247. Wayland-Smith, *Oneida*, 201; Carden, *Oneida*, 43–44; Fogarty, *Special Love*, 21.

248. Robertson, *Oneida Community*, 95–96.

249. Klaw, *Without Sin*, 268.

dessert forks for salad, or teaspoons for oranges,'" and another ad insists that "fashion in table service today requires that each course, each dish, have its appropriate silverware."[250]

One other nineteenth-century American trend is worth mentioning, involving a plea for simplicity which was individualist rather than communitarian. Henry David Thoreau's *Walden* (1854) lamented that society was overburdened with etiquette and politeness, and proclaimed that nature was a surer companion. Disdainful of fashion worship, Thoreau was the most influential nineteenth-century advocate of the simple life, and of the claim that material accumulation impeded spiritual aspiration. His central ideal, "voluntary poverty", proceeded from the assumption that "Most of the luxuries, and many of the so-called comforts of life, are not only not indispensable, but positive hindrances to the elevation of mankind". Luxury was destroying humanity, he thought, and "the only cure for it as for them is in a rigid economy, a stern and more than Spartan simplicity of life and elevation of purpose". Thoreau thought that "When he has obtained those things which are necessary to life, there is another alternative than to obtain the superfluities; and that is, to adventure on life now, his vacation from humbler toil having commenced", which adventure for Thoreau involved immersion in nature.[251] It is a moot point as to whether this vision is "utopian", since it does not project an ideal society as such.[252] But its enormous influence on subsequent ideas of voluntary simplicity, we will see, is undoubted.

Twentieth-Century Communitarianism

By the early 1900s much of the enthusiasm for communitarian settlement had dwindled. In North America, as the frontier vanished, co-operation against the elements was less necessary. Some new communities were founded, and others persisted.[253] But as the lure of city life became ever stronger, urban affluence made socialist farming colonies less attractive. An early history of the movement noted that the "general experience of Communities and Phalanxes" tended to be that once the initial sacrificial stages had passed, when "privations and hardships are endured with great cheerfulness", there was "likely to be more discontent and grumbling after they have grown prosperous and are

250. Wayland-Smith, *Oneida*, 200.

251. Henry David Thoreau. *Walden; or, Life in the Woods* (Walter Scott, 1886), 15, 90, 14.

252. Joshua Kotin's *Utopias of One* (Princeton University Press, 2018) explores this argument.

253. The best overall survey for this period is Timothy Miller's *The Encyclopedic Guide to American Intentional Communities* (2nd ed., Richard W. Couper, 2015).

supplied with all the necessaries of life and some luxuries than when they were struggling for the bare means of subsistence."[254]

At various points in the twentieth century, however, intentional communities once again became fashionable. The counterculture, in particular, instigated a wave of new communities, mostly in the United States, many of which aimed at self-sufficiency and embraced frugality. Some of at least 3500 communes founded in the 1960s—like The Farm in Tennessee, which began to unravel in the 1980s—have remade themselves as ecovillages.[255] Here and elsewhere some New Age groups, such as the ecovillage at Findhorn, Scotland, have survived for decades. There are about 400 communes in the United Kingdom today.[256]

The most impressive such movement took place in Palestine, where kibbutz communes were first established in the 1920s. They reached nearly 180,000 members in some 255 communities by the late 1980s. Here, internal cohesion was provided by strong ideological, nationalist, religious, and explicitly utopian bonds. Initially few luxuries were available in these largely isolated rural settlements.[257] Hostility to bourgeois "decadence" was a prominent theme early on in the movement, with asceticism fed by three separate value systems: socialist, nationalist, and personal.[258] Initially a "rise in the standard of living was also regarded as a potential threat to interpersonal relationships. And finally, the craving for luxury was said to threaten the self-sufficiency of the kibbutz".[259] Thus, at least into the 1960s, make-up was "avoided as 'bourgeois'", and distinction in apparel was discouraged.[260] Here, at least for a short time, "Utopia . . . was realized. But not for long."[261] By the 1990s the movement had begun to decline, undermined by the competing vision of a materialist utopia. Its ideology of frugality, equality, and self-sacrifice seemed dated, and younger

254. Hinds, *American Communities*, 283.

255. Bill Metcalf, *Shared Visions, Shared Lives: Communal Living around the Globe* (Findhorn, 1996), 80, 83. A recent account is Timothy Miller's *Communes in America, 1975–2000* (Syracuse University Press, 2019).

256. For the early twentieth century, see Chris Coates's *Utopia Britannica: British Utopian Experiments, 1325–1945* (D&D, 2001).

257. Henry Near, *The Kibbutz Movement: A History*, 2 vols (Oxford University Press, 1992), 1: 184.

258. Talmon, *Family and Community*, 216.

259. Talmon, 218.

260. Melford E. Spiro, *Kibbutz: Venture in Utopia* (Harvard University Press, 1956), 165. See Spiro, *Children of the Kibbutz* (Harvard University Press, 1958), 327.

261. Henry Near, "Utopian and Post-utopian Thought: The Kibbutz as Model", *Communal Societies* 5 (1985): 46.

members were lured by the attractions of city life and the higher wages and standard of living there. Austerity to some was inevitable at first, but avoidable as the wider society grew wealthier.[262] Some historians insist nonetheless that the model proves that "it is possible to create a viable egalitarian society."[263]

Notable experiments were also made in Australia and New Zealand from the nineteenth century onwards.[264] These mostly mirror the patterns evident elsewhere, being largely rural, self-sufficient, and motivated by left, environmentalist, and occasionally religious commitments. One of the best-known early communitarians in Australia was William Lane, who formed a colony there and then moved to Paraguay. Initially sympathetic to Edward Bellamy's arguments, Lane established a Bellamy Society in 1887, but then conceived the idea of founding "a colony of say 1,000 people, amply supplied with capital, and imbued with sound social ideas, including the ideas of Simplicity and Art, as opposed to Bellamy's luxury idea". Like many at the time he seems to have been struck by the difficulty of crafting any kind of socialism in the midst of capitalist society. "We must go where we shall be cast inwards," he reportedly said, "where we shall be able to form new habits, uninfluenced by old social surroundings, where none but good men will go with us." Success must result, thought Lane, who asked:

> what do we expect? Not mansions, but cottages; not idle luxury, but work-won plenty; for each a home and marriage—honest, lifelong marriage—with sturdy children growing all around to care for us when we are old. We expect that the earth will yield, and that the flocks will increase, and that the axes will fell and the hammers weld, under our own hands, as under the hands of all others who toil. We expect that the song and the dance will come back to us, and that with our human instincts satisfied, we shall joy in living. If this is a wild dream, an impossible hope, what hope is there for humanity?

Neither New Australia nor its successor colony in Paraguay, Cosme, survived long. The clothing was simple, and there was enough to eat, if your taste ran to steamed iguana. The example nonetheless provides us with a clear contrast of how competing ideals of socialism based in urban luxury or rural simplicity were understood in the period, and how plans for reconstructing society oscillated between these extremes.[265]

262. Talmon, *Family and Community*, 214–15.

263. Ran Abramitzky, *The Mystery of the Kibbutz: Egalitarian Principles in a Capitalist World* (Princeton University Press, 2018), 6.

264. For Australia generally, see Bill Metcalf's *From Utopian Dreaming to Communal Reality: Cooperative Lifestyles in Australia* (University of New South Wales Press, 1995).

265. Gavin Souter, *A Peculiar People: The Australians in Paraguay* (Sydney University Press, 1981), 21, 16; Anne Whitehead, *Paradise Mislaid: In Search of the Australian Tribe of Paraguay* (University of Queensland Press, 1997), 16.

By the late nineteenth century, Australia also hosted numerous religious communities, including some Moravians and a kibbutz. The 1960s counterculture also spawned various religious and secular groups. Many found it very difficult to achieve the "naive vision of 'the simple life of self-sufficiency'", as a resident of the Moora Moora commune near Melbourne put it. Buying solar panels requires cash, which requires selling something for an export market, which requires knowledge of how to do so, and so on.[266] Unusually for the period, both Australia and New Zealand also had substantial late twentieth- and early twenty-first-century communitarian movements, many of which have now passed their thirtieth anniversary, and at least one, its sixtieth.[267] Like communities elsewhere, however, "success" should not be equated with longevity. Thousands of people have experimented with different lifestyles, discovered much about themselves, others, and the world, and returned to mainstream society the wiser. Much of what was first viewed as "alternative" in such locations later became the norm in the wider society.

So far we have considered only attempts to return to the land. How do more urban communities vary from this pattern of development? Here, all the temptations of modern life are usually present, and a higher degree of individuality is almost inevitable. Urban communities may nonetheless foster a strong ethos of equality, and even a degree of simplicity. After 1968, a *Wohngemeinschaft* movement of informal apartment or house sharing included many hundreds in Germany in particular. A cohousing initiative in Denmark has met with considerable success. Similar communes have appeared in many other countries, as have ecovillages. Suburbs also represent a utopian (or anti-utopian) phenomenon of their own, as do nostalgic recreations of an ersatz small-town life like Disney's Celebration community, with its promise of "a return to a more sociable and civic-minded way of life".[268]

Another type of communalism involves promoting solidarity, co-operation, and equality in the workplace rather than communal living. Many forms of workers' co-operatives and other collective self-help organisations can be termed "utopian" in proportion to the degree of communal property they promote, the equality they manifest, and the sense of well-being and belonging

266. Metcalf, *From Utopian Dreaming*, 165.

267. Sargisson and Sargent, *Living in Utopia*, xiv, 169.

268. From a Celebration brochure, quoted in Robert H. Kargon and Arthur P. Molella, *Invented Edens: Techno-cities of the Twentieth Century* (MIT Press, 2008), 146. See further Roger Silverstone, ed., *Visions of Suburbia* (Routledge, 1997); Robert Fishman, *Bourgeois Utopias: The Rise and Fall of Suburbia* (Basic Books, 1987).

their members actually achieve.[269] There are many different forms of co-operation, and some efforts, like Mondragon in Spain, have become world-renowned. As a practical means of establishing unity of purpose, social justice, and equality, co-operation has an excellent track record. In some countries the movement has given rise to political parties and has developed substantial international networks. The utopian component in such efforts varies enormously, from being negligible or non-existent in "divi" schemes of consumer co-operation, where profits are divided amongst members, to promoting a powerful sense of solidarity in worker-owned and managed enterprises.

On the Possibility of Everyday Utopia

The communitarian and co-operative experiments briefly introduced here demonstrate the viability of the idea of the realistic or everyday utopia. Even utopianism's most strident critics, like Kolakowski, concede that, in small-scale experiments, "Undeniably . . . people are able to create conditions in which aggressiveness, hostility, and selfishness, if not eradicated, are really minimized."[270] This is utopia: it is not merely fantastic speculation about ideal pasts or futures, or alternative spaces, or confined to a psychological "principle of hope", or a futile quest for secular salvation. It includes the possibility of practising greater sociability in real time and space, and with it the prospect of a vast improvement in participants' lives. With varying degrees of success, such experiments realise enhanced sociability and belongingness on a small-group scale. Individual behaviour is modified sufficiently to define an alternative space where mutual support and solidarity are stronger than in the outside society. Sometimes this experience is extended over a considerable time, and into a wider space. A crucial question, we have seen, is just how long and how widely such sociability can be sustained.

But opportunities are plentiful for the experiment. Such moments of episodic behavioural modification can occur in a hundred different contexts, including the sublime, idealised garden;[271] the rural or pastoral idyll; the fervent

269. An early history of such efforts is Henrik F. Infield's *Co-operative Communities at Work* (Kegan Paul, Trench, Trubner, 1947).

270. Kolakowski, *Modernity on Endless Trial*, 144. He concludes that "the idea of human fraternity is disastrous as a political program but is indispensable as a guiding sign" (144). Kolakowski, who became increasingly religious, accordingly aligns himself Kant's theory of "radical evil" to describe the chief barrier to any successful utopia.

271. But the proximity of the garden ideal and the small-scale rural utopia presents some interesting problems; here Henry Thoreau's *Walden* is germane. In the ancient world the enclosed, lush garden was also a prototype for ideas of "paradise", the term in Persian meaning

loss of self in religious ceremonies, festivals, or carnivals;[272] the merging of the individual in the crowd in musical or sporting events; or, for longer periods, the sense of community and equality instilled by common struggle during wartime, and nostalgically recalled for its feeling of common sacrifice. These utopic moments and/or spaces share a heightened sense of communal belonging or identity, or *esprit de corps*, and the dissolving of the ego in the collective, often in an ecstatic, cathartic manner. Here we joyfully fuse with others, rising above and beyond ourselves to share in the strength, wisdom, and longevity of the mass or the whole, indulging our deeply felt desire to be accepted by the group. A new-found sense of equality is often their defining characteristic. In utopian literature this was the core ideal of the Morean paradigm of utopia, whose origins we must now examine.

---

indeed exactly this. See Annette Lucia Giesecke, *The Epic City: Urbanism, Utopia, and the Garden in Ancient Greece and Rome* (Harvard University Press, 2007).

272. A particularly important example here; one work which weds some of these themes is Christopher Kendrick's *Utopia*.

5

Luxury, Sociability, and Progress in Literary Projections of Utopia

FROM THOMAS MORE TO THE EIGHTEENTH CENTURY

OVER THE PAST five hundred years utopian fiction has often functioned as a reflection of social conscience, warning the moderns of the dangers of extreme inequality of property, vanity, and greed, and judging societies by the standards of an ideal where these were absent or greatly restrained. Utopia thus often presents itself as a discourse on virtue, sociability, moderation, and restraint. Prior to Thomas More (1478–1535) some of these themes had been explored in the ever-popular imaginary voyage, a genre dominated by Sir John Mandeville's bestselling *Travels*, which looked backward to classical authors in its discovery of lands inhabited by monstrous beings, and still hinted at a paradise in the east.[1] With More, a radically new approach to the ideal society emerged, and the term "utopia" entered our vocabulary.

Thomas More

Every tradition needs a starting point, and More's *Utopia*, whose title is also *The Best State of a Commonwealth*, offers a well-defined template for thinking about utopianism, regardless of how much we may bicker about its particulars.[2] *Utopia* was originally published in Latin, of which some twenty editions

1. As late as 1735 Linnaeus was still classifying supposedly slant-headed Canadian tribes under the heading of "monstrous"): Robert F. Berkhofer, *The White Man's Indian: Images of the American Indian from Columbus to the Present* (Vintage Books, 1979), 40.

2. Russell Ames offers fourteen possibilities for interpreting the text in *Citizen Thomas More and His Utopia* (Princeton University Press, 1949, 5):

appeared.[3] It is rooted in the Renaissance humanist revival of studying the classics and promoting an ideal of active public service which Quentin Skinner terms "civic self-government". In *Utopia* this involves above all a willingness to share much in common, to labour for the communal good, and to live virtuously.[4] Though it falls short of modern expectations in many areas, *Utopia* would come to stand for a humanitarianism of ever-extending toleration, and antipathy towards violence, cruelty, and oppression. While it is not set in the future, More's great work also does not merely nostalgically hearken back to a Golden Age. It hints tantalisingly at the possibility of attaining a more just society. It addresses head-on, if satirically, the problems of luxury and inequality and the impossibility of reconciling utopia and plutocracy, and makes any utopia contingent on resolving these issues. In this imagined other place, the best qualities of humanity shine forth.

With More, the concept of utopia thus emerges as a humanist critique of market society. Its central theme is dispossession. Book one details how thousands of peasants in Britain were being evicted by greedy landlords enclosing common land for sheep farming. Uprooted, and deprived of their ancestral rights, they became citizens of nowhere. So displacement and, by implication, belongingness are associated with utopia from the outset. A contemporary account tells us that "there noblemen and gentlemen: yea, and certain Abbots . . . leave no ground for tillage, they enclose all into pasture: they throw down houses: they pluck down towns and leave nothing standing but only the church to be made a sheep-cote."[5] As a character in the book, More describes

---

1. A fantastic escape from unpleasant reality. 2. A blueprint for a better society which More thought men might soon establish. 3. A better society which might exist in some far-off time. 4. A better society which More desired but did not believe possible. 5. A reconstruction of medieval social virtues. 6. A revival of primitive Christian communism. 7. A speculative portrait of rumored American societies, like that of the Incas. 8. A strictly rational philosophic construction, minus Christianity, for the purposes of moral instruction. 9. A pleasant fable written by a humanist for the amusement of himself and his scholarly friends. 10. A fruit of classical studies, following Plato's *Republic*. 11. An early plan for British imperialism. 12. A Christian humanist account of a scholar's paradise, where philosophers are kings and the church is purified. 13. A society constructed as the direct opposite to England for the purpose of disguising social criticism. 14. A description of a desirable and possible organization of city republics.

3. On the history of the Latin editions, see Cave's *Thomas More's Utopia*, 14–31.

4. Quentin Skinner, "Thomas More's *Utopia* and the Virtue of True Nobility", in *Visions of Politics*, vol. 2, *Renaissance Virtues* (Cambridge University Press, 2002), 213–31.

5. Thomas Becon, *The Jewel of Joy*, cited in Maurice Beresford, *The Lost Villages of England* (Lutterworth, 1954), 84.

how sheep have "started eating men" by the village, and thieves are hung by the hundreds. The sailor Raphael Hythloday, who recounts to More his experience in the distant island Utopia, declaims against the "conspiracy of the rich, whose objective is to increase their own wealth while the government they control claims to be a commonwealth concerned with the common welfare."[6] In More's view, the avarice of the wealthy directly caused the misery of the poor. So the "senseless luxury of his age", wrote the German socialist Karl Kautsky, induced More to give his Utopians "frugal manners", and launched what Edward Surtz and J. H. Hexter term a virulent attack on "the whole aristocratic way of life, even its tastes and pastimes."[7] From this time on, generically, utopia would often stand for exalting the human over the monetary, sociability over greed, humility over pride, and the public interest over that of the wealthy few.

The key question invariably addressed to More, however, is whether he meant to suggest that Utopia's institutions might be adapted in Europe, or anywhere else, on a similar scale. Did he imply that utopia was realisable, or is its literary representation just another way of reminding us what we might be capable of if we were better human beings than we are ever likely to become? More says Hythloday's account of Utopia contains "things he noticed that it would be sensible for us to imitate", even that, with reservations about overweening pride, which alone prevents people from adopting these institutions, "I might wholeheartedly wish that the whole world would imitate the structure of the Utopian society".[8] But More's text is also presented as an imaginary or fantastic voyage. So even without the one-eyed giants and other monsters which populate the genre, readers would expect fanciful exaggeration if not downright lies.[9]

Yet More is not such an author, even if "Hythloday" means "talker of nonsense". *Utopia* is much more than this. Critics of communism emphasise the book's absurdities, with its many puns, starting with u-topia, no place, and eutopia, the good place. It is admittedly a satire, even an inversion of the extreme degeneracy of the English aristocracy, "a kind of hair shirt for the nobility of England", rather than a plan whose particulars, especially with respect to

6. Thomas More, *Utopia*, ed. David Wootton (1516; Hackett, 1999), 66, 157. Subsequent references to the text are to this edition.

7. Karl Kautsky, *Thomas More and His Utopia* (1888; A. and C. Black, 1927), 210; Thomas More, *The Complete Works of Thomas More*, vol. 4, eds Edward Surtz and J. H. Hexter (Yale University Press, 1965), lii.

8. More, *Utopia*, 59, 159.

9. See Percy G. Adams, *Travelers and Travel Liars: 1600–1800* (University of California Press, 1962).

religion and property, More expected could be adopted.[10] The description of large-scale communal property by Hythloday refers to New World accounts with which More was familiar. Hythloday mentions Amerigo Vespucci by way of verifying the apparent plausibility of his own narrative.[11] There are also hints of Spartan and monastic practices. More had lived for some four years with Carthusians in the Charterhouse in London. He followed their regimen, rising at 2 a.m. daily for several hours of prayer and devotion.[12] He wore a hair shirt to punish his flesh and more than once regretted not becoming a monk, or a Franciscan friar. To Hexter, he "seems to have been seeking some way to combine the austerity of the cloister with a life in the world, and his quest was life-long."[13]

We can reasonably presume that More thought that such a life was possible on this scale, in effect utopia in microcosm. There are significant differences between the monastic and the Utopian lifestyles, however. The Carthusians sought seclusion and isolation, with even dining being solitary except on Sundays and feast days. But the Utopians are eminently sociable. More knew that the monastic life of contemplation would never interest the majority. These are held together rather, by a kind of patriotism, or love of the common good and its territory and people. Yet a higher ideal was viable for a few. This is an important step in formulating a realistic theory of utopia. So More made space in *Utopia* for this ethos of monastic rule, by portraying one group who specially devote themselves out of piety to public works like tending to the sick and repairing roads. More prioritises, but does not fetishise, hope, insofar as belief in a future life is a religious dogma for the Utopians. The more ascetic sect amongst them, the Buthrescas, "reject all the pleasures of this life as being wicked", and are celibate vegetarians, like the Carthusians. One of their roles is to "censure the moral failings of the people".[14] But their devotion and abstinence are not shared by the majority, who enjoy many pleasures, provided they are natural, innocent, honest, and involve no harm to others. Few in More's time thought all the world could live the ascetic life or would want to.

More's is thus a delicate balancing act. He appears to play devil's advocate with his own key themes. His character in the text challenges Hythloday's

10. Nisbet, *Idea of Progress*, 111, following Hexter.

11. See my *Searching for Utopia* (Thames and Hudson, 2011), chap. 5. An account which promotes this argument is Lorainne Stobbart's *Utopia: Fact or Fiction? The Evidence from the Americas* (Alan Sutton, 1992).

12. The Carthusians were distinctive for their silence, and rejection of meat eating. They were devoted to learning, but wept a great deal, perhaps as a consequence.

13. J. H. Hexter, *More's Utopia: The Biography of an Idea* (Princeton University Press, 1952), 88.

14. More, *Utopia*, 150.

account of common goods using Aristotle's arguments against Plato, contending that "there can never be prosperity where all things are held in common. For how can there be a plentiful supply of goods if everyone gives up working? The prospect of turning a profit will not act as an incentive, and everyone will work half-heartedly, trusting they will be fed by the labor of others."[15] He also takes up the objection that "you can't hope to make everything perfect unless you only have perfect people to deal with—which isn't likely to be the case for a few years yet".[16] Hence the complaint that "More does not invent ideal institutions for mankind, but an ideal mankind for their institutions".[17] The moral of the story thus seems to be that Europeans have forgotten how to be good Christians, but may never remember either. *Utopia* observes that "Christ approved of the communal way of life practiced by his first followers, and . . . among the most authentic Christians common ownership is still practiced."[18] More's great friend Erasmus of Rotterdam thought his "purpose was to show whence spring the evils of states", and that *Utopia* portrayed a "holy commonwealth that Christians ought to imitate."[19] He also reported that, when young, More had defended Plato's communism and even community of wives, a doctrine not advanced in *Utopia*, where the patriarchal family is the root of social authority.[20] But in his *Dialogue of Comfort against Tribulation*, written some twenty years after *Utopia*, More insisted that

> if all the money that is in this country were tomorrow next brought together out of every man's hand, and laid all upon one heap, and then divided out unto every man alike, it would be on the morrow after worse than it was the day before. For I suppose when it were all egally [equally] thus divided among all, the best should be left little better then than almost a beggar is now.[21]

Some take this as More's final word on the subject. But this does not devalue monastic ideals, only attempts to extend them too widely. The monastery might be one attainable utopia, but it was not for the many. The question of scale, indeed, was and remains central to the entire subject. Erasmus thought

15. More, 87.

16. More, 84.

17. Gerard Dudok, *Sir Thomas More and His Utopia* (A. H. Kruyt, 1923), 174.

18. More, *Utopia*, 145.

19. Quoted in E. E. Reynolds, *The Field Is Won: The Life and Death of Saint Thomas More* (Burns and Oates, 1968), 101; and Hexter, *More's Utopia*, 47.

20. H. W. Donner, *Introduction to Utopia* (Sidgwick and Jackson, 1945), 29.

21. Saint Thomas More, *A Dialogue of Comfort against Tribulation* (Yale University Press, 1977), 183.

community of goods "was only possible when the Church was small, and then not among all Christians: as soon as the Gospel spread widely, it became quite impossible. The best way towards agreement is that property should be in the hands of lawful owners, but that out of charity we should share one with another."[22] Universalising an apostolic community is what Thomas Müntzer thought he was doing, if we can believe a confession after torture.[23] C. S. Lewis writes that "It is certain that in the *Confutation* (1532) More had come to include communism among the 'horrible heresies' of the Anabaptists and in the *Dialogue of Comfort* he defends private riches."[24] So these excesses, for More the product of Protestantism, were to be condemned; this was not how to get people to live like real Christians.

Perhaps the key point, then, was to make them better citizens. *Utopia* may suggest a kind of halfway house between apostolic Christianity and the model republic, limited by the feasibility of modifying behaviour suitably. More did not view Utopia as the heavenly city, or analogous to the Anabaptists' coming paradise on earth. Surtz reminds us that Utopians, though well behaved, "are not saints".[25] But they are indisputably virtuous citizens. In the Manuels' rendition:

> the kingdom of Utopia was not a second paradise . . .
>
> . . . More's vision of the good society in the *Utopia* did not require a conception of a perfect man. Christian humanism was not as absolutist as religious millenarianism, though it was doubtless a utopian project in its own right. More did not anticipate the reign of the Holy Ghost on earth and his Utopian is not a 'new man' in the Abbot Joachim's or Thomas Müntzer's sense.[26]

22. Quoted in Donner, *Introduction to Utopia*, 72. Others in this circle also wrote against community of goods, like Juan de Vives in *De Communione Rerum* (1538).

23. David Weil Baker, *Divulging Utopia: Radical Humanism in Sixteenth-Century England* (University of Massachusetts Press, 1999), 12.

24. C. S. Lewis, *English Literature of the Sixteenth Century* (Clarendon, 1954), 169. Kautsky is more precise about what More condemned in his *Confutation of Tyndall's Answer* (1532): "all those that say the baptising of children is void and that they say that there ought to be no rulers at all in Christendom neither spiritual nor temporal, and that no man should have anything proper of his own, but that all lands and all goods ought by God's law to be all men's in common, and that all women ought to be common to all men, as well the next of kin as the farthest stranger and every man husband to every woman and every woman wife unto every man and then finally that our blessed saviour Christ was but only man and not God at all" (Kautsky, *Thomas More*, 189). See Thomas More, *The Complete Works of Thomas More*, vol. 8, eds Louis A. Schuster, Richard C. Marius, James P. Lusardi, and Richard J. Schoeck (Yale University Press, 1973), pt. 2, 664.

25. Saint Thomas More, *Utopia*, ed. Edward Surtz (Yale University Press, 1964), xxix.

26. Manuel and Manuel, *Utopian Thought*, 123, 135.

So how does More conceive of Christian association? To Germain Marc'Hadour, Utopian solidarity is not defined by brotherly love.[27] More knew of discussions about friendship by Plato, Aristotle, Cicero, Aquinas, and others.[28] At least inside Utopia, he projected people as linked by a "natural fellowship" based upon "mutual good will".[29] This corresponds to what has here been termed *enhanced sociability*, "enhanced" because such good will is generally in short supply in modern societies and therefore requires a constant effort to muster and sustain. Most Utopians are easy-going, and united by their commitment to the commonwealth and their shared beliefs, dress, and way of life. They are much more sociable than the English of More's day. All are not necessarily "friends" as such all of the time; they cannot be, in a total population (larger than Britain's at the time) of perhaps six and a half million, formed into groups of about sixty thousand.[30] But their shared ideals of fairness, justice, and equity help to avoid multitudinous quarrels. (Private discussions of public affairs are also forbidden on pain of death.) They are thus prone to co-operation rather than conflict, and trust rather than suspicion. Their common faith in the immortality of the soul, disbelief in which brings about ostracism but not more severe punishment, permits tolerance of religious diversity, their contempt for superstition notwithstanding. Their belief that virtue will be rewarded in the next life constrains them to limit their mutual pleasures to "those that are good and honest" and conducive to virtue, and to disdain luxury. They recognise that pleasure is not a constant immersion in sensual gratification.[31] They universally condemn pride, the root of so much dissent and competition. They also reject hunting and needlessly slaughtering animals, or inflicting senseless suffering on them. So already with More the diminution of violence is a central principle. Their sociability goes beyond that of neighbours, or commercial

27. Germain Marc'Hadour, "Utopia and Martyrdom", in Olin, *Interpreting Thomas More's Utopia*, 75.

28. On this interpretation of More, see David Wootton's "Friendship Portrayed: A New Account of Utopia" (*History Workshop Journal* 45 (1998): 28–47), where a debt to Erasmus in particular is noted, and Wootton's introduction to his edition of *Utopia* (Hackett, 1999), esp. p. 8. For the classical background, see, e.g., A. Price's *Love and Friendship*. On community of goods, see Peter Garnsey's *Thinking about Property: From Antiquity to the Age of Revolution* (Cambridge University Press, 2007).

29. More, *Utopia*, 135.

30. Utopia has 54 cities composed of 6000 households each, perhaps of ten people, with a rural population about the same. England and Wales combined had a population of less than three million at this time.

31. More, *Utopia*, 116.

contacts, or even citizens. But it does not extend to the universal love commended by Christianity.

Equality is the basis of More's Utopian sociability, which reflects the Aristotelian maxim that all things are common among friends. Utopian friendship encompasses a very large group. Yet equality is not absolute. Social ranks exist, based around a hierarchy of age, the eldest ruling each household, wives waiting on husbands, children on parents. There is slavery. There is imperialism, whose chief rationale is that the law of nature allows the Utopians "to wage war against any society that does not use the land it owns but leaves it waste and uncultivated, while at the same time preventing others from using and owning it", a justification for conquest we encounter as late as the twentieth century and which practically invites the extermination of hunter-gatherer peoples.[32] And, curiously, the Utopians do not seem to want to export their way of life to others. But vanity is minimal. No competition over external symbols of inequality, particularly luxuries, exists. The Utopians are well-off, "prosperous burghers after their six hours' manual labor was over", in Russell Ames's phrasing. The dominant motif of Utopian life is simplicity.[33] "They define virtue", we are told, "as living according to nature", a view More's marginal annotator described as "like the Stoics".[34] They are wary of importing luxuries, but have no taste for them either. They mock ambassadors adorned with precious stones, which they regard as fit only to amuse children. Like the Mayans and Spartans, they are generally contemptuous of gold and silver as personal decoration, except one type of gem. But they nonetheless store up gold to pay their mercenaries, the tough herdsmen known as Zapoletes, whom they are happy to see die in their cause.[35]

If, as Pierre Versins suggests, "clothes make the utopian",[36] More's text champions minimal personal adornment and plain, utilitarian, inconspicuous, but comfortable clothing with little distinction between persons. Utopians class among the "counterfeit pleasures" "the conviction many people have that

32. More, *Utopia*, 103. For its later uses, see my *Imperial Sceptics: British Critics of Empire, 1850–1920* (Cambridge University Press, 2010, 13–18).

33. Ames, *Citizen Thomas More*, 153.

34. More, *Utopia*, 116.

35. It is surprising how often this is overlooked; most commentators merely note that "The Utopians do not value gold or silver", while not acknowledging how they use them (Peter Ackroyd, *The Life of Thomas More* (Random House, 1998), 168). H. W. Donner too insists that the Utopians put gold "to the most hateful and contemptible uses", notably in making chamber pots, ignoring this dimension (*Introduction to Utopia*, 32). Karl Kautsky is much more accurate: "Gold and silver constitute the war chest of the nation" (*Thomas More*, 212).

36. Quoted in Daniel Roche, *The Culture of Clothing: Dress and Fashion in the "Ancien Régime"* (Cambridge University Press, 1996), 420.

the better the clothes they wear the better people they are."[37] So they despise silk. But their clothes are "not unattractive" and are made to last at least two years. This simplicity is a key to their society's success. They are satisfied with little and do not compete with one another. There is no idleness, so no resentment of the greater leisure of a few. Houses are exchanged every ten years, so no jealousy can arise here either. Without vanity, no great variety of goods is required or produced. Natural hierarchy is balanced with equality. When dining, for instance, More tells us, the older men receive the choicest morsels, then equal portions are distributed among the rest, thus heeding the principles of both seniority and equality. Contentment seems the norm where money cannot purchase distinction. Personal strife is rare, but some crime exists, and divorce still occurs. In Hexter's view, "The Utopians are consequently free of all the anxiety and all the resultant hostile emulation with which men in a pecuniary society are oppressed and with which they oppress one another."[38]

This common life is not, thus, necessarily being satirised by More in his comment near the end of *Utopia* that "All nobility, magnificence, splendor, and majesty is utterly destroyed" by such arrangements. This can also be read as a disingenuous satire on European luxury, grandeur, pride, greed, and folly.[39] A society based on money getting is juxtaposed to one based on virtue, and virtue triumphs, at least in the imagination. So More may hint that it is improbable, but not impossible, that such a state might be recreated in fallen Europe. All of these "realist" components—Sparta, the New World, the monastery, the actual practice of friendship—played a role in *Utopia,* and were married to various philosophical debates concerning the optimal possibilities of human behaviour. All had existed, or were believed to have existed, or did exist, somewhere, if nowhere quite in this combination.

Readers can reasonably surmise, then, that *something like* the well-ordered society described in *Utopia* lay, in More's own mind, within the bounds of possibility.[40] But the critical question then is: on what scale? Whether or not he thought that similar institutions and practices had been discovered in the New World remains a moot point. So is the hope that this type of friendship might be extended to a nation. Why did he project Utopia on a scale larger than Britain? More probably did not think that private property could ever be abolished in Utopian fashion on the scale of a city-state, as opposed to a

37. More, *Utopia,* 118.

38. Hexter. *More's Utopia,* 60.

39. More, *Utopia,* 159; Q. Skinner, "Thomas More's *Utopia*", 240–41.

40. Modern anthropologists and psychologists have identified a spectrum of primitive societies from the least to the most aggressive. See the discussion in Erich Fromm's *The Anatomy of Human Destructiveness* (Fawcett Crest, 1973, 196–203).

monastery.[41] Utopia thereafter came, more than anything else, to imply a condition of institutionally supported, enhanced sociability and friendship, resting upon a broadly egalitarian foundation, often including communism.[42] How widely it could be practised was left unanswered.

Utopian Fiction after More

How did these themes develop after More? Already in the sixteenth century he had many imitators, like the Italians Anton Francesco Doni and Ortensio Lando, who, as Antonio Donato shows, translated *Utopia* and used the genre to criticise contemporary vice. Many Italian utopias combined authoritarian rule with an appeal to personal simplicity, with Doni in particular stressing the progressive degeneration of mankind since the Golden Age. In Doni's ideal city, all dine in inns on the same simple dishes, live in identical modest houses, and wear similar clothing colour-coded by age. Deformed children are destroyed, and marriage abolished by establishing community of women, with the goal of eradicating aristocracy. No crime would exist, it is assumed, once property has been abolished.[43] In Lodovico Zuccolo's early seventeenth-century *Porto; or, The Republic of Evandria* no one may wear gold or use it to make luxury goods. Here, we are told,

> the population lives in a state of utmost contentment, they are the most restrained, modest, and ready to [attend to] the needs of the country's people ever encountered in new and old republics. Their attitude is brought about by their strict, continuous public and private education, and their unremitting diligence that are very effective in keeping luxury away. In Evandria, there is not even a single beggar and there are not excessively wealthy people.

41. Q. Skinner, "Thomas More's *Utopia*", 244.

42. This generalisation is not meant to encompass every form of utopia or community. A challenging instance might be the mid-nineteenth-century individualist settlement of Modern Times, which was based on the ideas of Josiah Warren and encouraged a maximum of privacy, toleration, and non-interference in the behaviour of others. See Wunderlich, *Low Living*. Later anarchist communes also enter into this debate. But the proposition that any stronger form of sociability (say a republican ideal of civic duty) is inherently at odds with an ideal of *increasing*, rather than static, individuation (a growing cultivation of individuality and the sense of personal uniqueness) cannot be taken up here.

43. Paul F. Grendler, *Critics of the Italian World: Anton Francesco Doni, Nicolò Franco & Ortensio Lando* (University of Wisconsin Press, 1969), 165–66, 173–74; on Doni, see Antonio Donato's *Italian Renaissance Utopias: Doni, Patrizi, and Zuccolo* (Palgrave Macmillan, 2019). See, generally, Eliav-Feldon's *Realistic Utopias* and Houston's *Renaissance Utopia*.

Here too "The magistrates create amongst the rich, a constant desire to outdo each other by sponsoring public works in order to prevent them from hording treasures to the detriment of the country's freedom."[44]

Two other early seventeenth-century utopias exemplify the role the genre played in the next century or so in promoting virtue and equality. The Dominican monk Tommaso Campanella's (1568–1639) *The City of the Sun* (1623) describes a great city some 7 miles in circumference located near the equator in the southern hemisphere. Its inhabitants, the Solarians, are sober, generous, and chaste and love knowledge. This is a more egalitarian vision of *Utopia*'s proposals. Like Plato, Campanella extends communism to wives, children, houses, and food. Private property is condemned as promoting self-love. After its abolition, "there remains only love for the state", and the inhabitants "burn with so great a love for their fatherland" as scarcely seems imaginable.[45] Friendships of many types are encouraged. Those based on exchanging favours are unnecessary, because all possess equal things, and

> friendships develop in time of war, in sickness, and in the pursuit of knowledge wherein they help and encourage one another. Youths address each other as brothers; they address those who are their seniors by fifteen years or more as fathers and are, in turn, addressed by them as sons.[46]

Friendship thus rests on equality and frugality as well as uniformity and a sense of mutual endeavour. Sexual intercourse, however, is strictly regulated by the prince in charge of Love, assisted by many lesser magistrates. Women who use make-up face the death penalty, and pride is regarded as "the most execrable vice". Magistrates ensure the birth only of the most fit, with their time of conception being determined by astrologers. Children are "bred for the preservation of the species, and not for individual pleasure", and are reared in common.[47] The workday is a mere four hours, and diet and exercise are regulated to promote the well-being of the population, who often live to the age of a hundred. There is some division of labour according to sex, women doing the clothes making, and men all woodwork and arms manufacture. Social equality is also secured through universal free education and mandatory agricultural labour, and all have everything they require. A familial model of hierarchy exists. There are no servants or slaves. Men and women dress

44. Donato, *Italian Renaissance Utopias*, 217.

45. Bacon and Campanella, *Two Classic Utopias*, 50–51.

46. Tommaso Campanella, *The City of the Sun*, transl. Daniel J. Donno (University of California Press, 1981), 40–41.

47. Bacon and Campanella, *Two Classic Utopias*, 60–61.

similarly in white, and all go barefoot. At meals the young wait on their elders and also, with some unwillingness, on each other. Magistrates get larger portions, though they distribute some of these to the more studious boys at their table. People learn many trades, then practise what they excel at. The government is theocratic, and Campanella hints that Spain is destined to impose God's plan on the world through this monarchy—more utopian imperialism. There are hints, too, of the promise of scientific and technological inventions, like ships which move without wind or sails. Only one book exists, however, called "Wisdom", which contains the nation's entire knowledge. War with the four "impious" neighbouring kingdoms on the island is frequent, though fortuitously all are invariably defeated.

A second key early modern utopia crucial to our own problematic, but of a very different cast of mind, is Francis Bacon's *New Atlantis* (1626), which was written by 1617 and had reached eleven editions by 1676. Here, the grand theme is scientific and technological advancement, with their end being "the happiness of mankind". Bacon (1561–1626) was a leading statesman, serving as Lord Chancellor of England, and made a seminal contribution to knowledge through promoting science. *New Atlantis* represents a nodal point in the history of utopian ideas insofar as it makes humanity's mastery of nature a central theme. The narrative invokes Plato's description of Atlantis. English mariners discover an island called Bensalem in the South Sea of the Pacific Ocean whose Christian inhabitants are vastly more virtuous and accomplished than their European counterparts. Chaste, honest and righteous, they are singularly lacking in corruption, especially in politics. They are devoted to scientific knowledge. Amongst their leading institutions is an establishment named Salomon's House, the "noblest foundation . . . that ever was upon the earth", where "natural philosophy", as science was then termed, is freely explored with a view to improving society.[48]

Bacon's *Instauratio Magna* (1620) had suggested that state-supported experimental laboratories might advance scientific knowledge. *New Atlantis* describes how investigators use "knowledge of the causes, and secret motions of things" to accomplish the "enlarging of the bounds of human empire, to the effecting of all things possible". This fulfilled the promise of mankind's domination of nature first suggested in Genesis, where God grants dominion over his creation to Adam and Eve.[49] New Atlantis sends out explorers every twelve years to secure knowledge from every part of the earth. In deep caves and high towers it creates artificial metals. Lakes, gardens, orchards, baths, parks, and enclosures exhibit all manner of cultivation and experiment, the prolongation

48. Bacon and Campanella, 20.

49. Bacon and Campanella, 31.

of life being a special interest. The population lives fairly luxuriously. Instruments of war are also produced, and both air and submarine travel exist. This is indeed a scientists' paradise: those who devise valuable inventions are rewarded liberally and have statues of themselves erected, of brass and even gold. Some of their discoveries are hidden from the state. There is even a house of "deceits of the senses" where impostures and illusions are contrived, a veritable "fake news" factory. We learn relatively little of the wider society, though different religions are tolerated. There are Jews, but they are described as being "far different" from those elsewhere, having accepted Christianity. The happy inhabitants of Bensalem prohibit polygamy, and reject Thomas More's proposal that engaged persons view one another naked before marriage, instead allowing their friends to do so.

By the seventeenth century, Hans Blumenberg argues, at least one strand of the idea of progress was derived from a revolt against classical learning and in favour of modern science and the idea of the mastery of nature—the Baconian ideal.[50] Bacon's aspirations were parodied in Jonathan Swift's *Gulliver's Travels* (1726), where an institute on the floating island of Laputa undertakes such tasks as extracting sunshine from cucumbers. Bacon's vision of boundless scientific progress and faith in technological innovation came nonetheless to dominate modern thought. But it is at odds with most seventeenth-century utopias, which continued to trumpet the virtues of frugality. Utopia's ambiguity about progress and simplicity was thus well established by this time. Indebted to Luther, More, and Campanella, Johannus Valentinus Andreae's *Christianopolis* (1619) took as one contemporary model Geneva, where moral censors made weekly investigations into citizens' morals, and gambling, luxury, and other vices were successfully suppressed.[51] Christianopolis aims to avoid luxury and intoxication as barriers to Christian love towards all. Make-up and high heels are proscribed. Money was to be abolished, and equality and a "contempt for riches" fostered by the government.[52] Samuel Gott's *Nova Solyma: The Ideal City; or, Jerusalem Regained* (1648–49), too, contended that money should be used only "in the just expend of it for simple necessaries", and condemned luxury in both food and dress.[53]

50. Blumenberg, *Legitimacy*, 106.

51. Here in 1567 a poor widow was fined and spent four days in prison for repeatedly wearing an expensive silk handkerchief, and sumptuary laws were in place even in the mid-eighteenth century (E. William Monter, *Calvin's Geneva* (John Wiley and Sons, 1967), 216).

52. Johann Valentine Andreae, *Christianopolis*, ed. Edward H. Thompson (Kluwer Academic, 1999), 245.

53. Samuel Gott, *Nova Solyma: The Ideal City; or, Jerusalem Regained*, 2 vols (1648; John Murray 1902), 2: 132–33.

As we move into the modern era, the utopian literary depiction of these themes undergoes several alterations. Literary satires on folly and the *jeux d'esprit* of wildly imaginative projections remained part of what the concept implied. But a new seriousness crept in, driven to an impressive degree by advances in science and technology which gave substance to some utopian dreams, and to the idea of secular perfectibility. A new realism emerged by the French Revolution, then was extended much further a century later. As the appeal of millenarianism declined, that of utopia accelerated. Thereafter, utopianism became a definitive mode of modernity as such, and most of the moderns became, willy-nilly, utopians in the sense of aspiring to an indefinite improvement of social and economic conditions. Capitalism was now seen as the great barrier to greater equality. So the Bolshevik Revolution of 1917 became the quintessential utopian moment of all times. Now the hint of satire in More's description of communism in *Utopia*, of improbability, of uncertainty as to whether humanity is capable of this much virtue and sacrifice for the common good, and this much collective identity, was replaced by a brash self-confidence in the grand superiority of Lenin's updated Morean system. The satirical ambiguity presented in the pun upon *utopia*, nowhere, and *eutopia*, the good place, disappears, replaced largely by the wedding of *eutopia* to *euchronia*, the good time just over the horizon, a dazzling sun about to burst upon a hitherto-unenlightened world.

The Eighteenth Century

Many of these transitions were foreshadowed in the eighteenth century. Utopianism in this period still lacked the confidence of later modernity, for the full force of the technological revolution was not yet evident. Indeed, despite Baconian assumptions about scientific improvement, most utopian writers before 1700 commended restraining wants to bolster social order. In a time of rapid social change, curiosity about the distant past increased, intermingled with a sense of loss. Primitivism not only remained appealing, it enjoyed a revival. Great interest was aroused in the 1760s by the forged poems of "Ossian", which purported to illustrate ancient Scottish Gaelic culture. More influential still was the South Sea island shipwreck narrative. A great icon of the age was the frontispiece to one of the bestsellers of the day, another imperialist fantasy, Daniel Defoe's *Robinson Crusoe* (1719). This portrayed Crusoe clad in goatskin, a true man of nature enjoying full natural liberty and master of all he surveyed—namely, his own island, which was a kind of miniature landed estate. This book has enthralled millions of readers. At one level the narrative is a fantasy of the enforced simplification of life, an appealing challenge to many younger readers. Defoe's portrait drew on the true story of Scottish mariner Alexander Selkirk, who was shipwrecked for four years on an island some 360 miles off the coast of Chile. Its moral, the well-known contemporary journalist Richard Steele observed, was that "he is happiest who confines his Wants to natural Necessities; and he that goes further in his Desires, increases his Wants in Proportion to his Acquisitions".[54]

Viewed from afar, as fantasy, the idea of returning to nature now took on a new reality. It does not matter for our purposes that Defoe's text is not really a utopia, as it does not concern a society, and can easily be read as an introspective psychological study, a confessional reckoning with one's conscience, a voyage within, a whimsical flight from the complexity of modernity, as well as an imperial fantasy. Though reduced to a "meer state of nature", Crusoe is also not really a "man of nature" either, since he rescues so much useful equipment from his ship.[55] But as a psychological trope the text came to epitomise the island retreat from civilisation and the return to natural origins and, as a kind of Stoic alter ego for the moderns, a simpler, autonomous, autarkic self. It spawned a host of imitators across Europe, the Robinsonades, for whom the central theme was escaping the trammels of civilisation to a life of virtuous simplicity, especially in the South Seas.[56] It was one of Rousseau's favourite works, and the first book the subject of his famous treatise on education, *Emile*, who ends his life on a desert island, reads, to learn self-sufficiency.[57]

The trope was highly contagious. In an age of extensive navigation, the rediscovery of simplicity seemed to throw into high relief the falsehoods and corruption of European society. The shipwreck narrative seemed indeed to symbolise civilisation's own degradation. In an early prominent example, on his first voyage (1768–71) to what became Australia, Britain's greatest explorer, Captain James Cook, found that the natives "live in a Tranquillity which is not

54. Diana Souhami, *Selkirk's Island* (Weidenfeld and Nicolson, 2001), 177. The island was named Robinson Crusoe Island in 1966. It was said of Selkirk that "he thought his wants sufficiently supplied, fashion having no longer any empire over him" (John Howell, *The Life and Adventures of Alexander Selkirk* (Oliver and Boyd, 1829), 31).

55. Daniel Defoe, *Robinson Crusoe* (1719; Penguin Books, 1965), 130.

56. A good account is Walter de la Mare's *Desert Islands and Robinson Crusoe* (Faber and Faber, 1930). See also Ian Kinane, *Theorising Literary Islands: The Island Trope in Contemporary Robinsonade Narratives* (Rowman and Littlefield, 2017); Artur Blaim, *Failed Dynamics: The English Robinsonade of the Eighteenth Century* (Lublin, 1987).

57. Jean-Jacques Rousseau, *Emile; or, On Education* (Dartmouth College Press, 2010), 114–15.

disturbed by the Inequality of Condition". They were, he reported, "far more happier than we Europeans, being wholly unacquainted not only with the Superfluous, but with the necessary Conveniences so much sought after in Europe; they are happy in not knowing the use of them." Indeed "they seem'd to set no Value upon anything we gave them, nor would they ever part with anything of their own for any one Article we could offer them."[58]

Better than Robinson Crusoe was Crusoe with Eve attached. At the same time, slightly further north, French explorers were inventing Tahiti as the paradigmatic erotic South Sea tropical paradise. Nakedness and relaxed sexual morals, themes in some seventeenth-century utopias, like Henry Neville's *The Isle of Pines* (1668), where "at liberty to do our wills", public lovemaking results, were central to this narrative.[59] Amongst the first to cement the association of exotic and erotic was the French voyager Louis de Bougainville, who had witnessed the Jesuits' expulsion from Paraguay and visited Polynesia in 1767.[60] Here, wondering if he had stumbled upon the Garden of Eden, he found the women—whom he described as "white"—"very hospitable, very gracious, and quick to caress".[61] "These people breathe only rest and sensual pleasures. Venus is the goddess they worship", he recorded in his journal. He named Tahiti "La Nouvelle Cythère" after the Greek isle mentioned by Fénelon, whom we will shortly meet, where the goddess of love Aphrodite supposedly arose from the sea.[62] Its inhabitants, he thought, followed the Laws of Nature, and still retained "the openness of the Golden Age". This was "perhaps the happiest society which the world knows", and he added, "as long as I live I will celebrate the happy island of Cythera: it is the true Utopia".[63] Cook too noted

58. James Cook, *Captain Cook's Journal during His First Voyage round the World* (Elliot Stock, 1893), 323.

59. Gregory Claeys, ed., *Restoration and Augustan British Utopias* (Syracuse University Press, 2000), 121.

60. He was very critical of the experiment, describing the Jesuits as (in Diderot's words) "those cruel sons of Sparta in black robes [who] treated their Indian slaves quite as badly as the ancient Spartans treated their helots" (Denis Diderot, *Rameau's Nephew and Other Works* (Anchor Books, 1956), 191.)

61. Others wrote that the women were amazed by white skin, which they associated with higher status: Anne Salmond, *Aphrodite's Island: The European Discovery of Tahiti* (University of California Press, 2010), 103.

62. Louis-Antoine de Bougainville, *The Pacific Journal of Louis-Antoine de Bougainville, 1767–1768* (Hakluyt Society, 2002), 63, referring to François Fénelon, *Les aventures de Télémaque*, 2 vols (1699; Librairie Hachette, 1920), 1: 238, 253, 257.

63. *News from New Cythera: A Report of Bougainville's Voyage, 1766–1769* (University of Minnesota Press, 1970), 23, 44–45; Bougainville, *Pacific Journal*, 257. See Charlotte Haldane, *Tempest*

"more than one half of the better sort of the inhabitants" had the "resolution of injoying free liberty in Love, without being Troubled or disturbed by its consequences", but killing all children born as a result.[64] His companions remarked that unmarried women engaged in public sexual intercourse without guilt or jealousy. (Adultery carried the death penalty, though men could lend their wives to others.) Others noted that they had only a few, elected chiefs. Many, however, also condemned the idleness of indigenous peoples, seeing no link between being obsessed with work and lusting for conquest and possessions, and failing to acknowledge that they themselves often regarded paradise as plenty without work. Few today see this approach as anything other than another variant on European imperialism: gold, sex, the paradise of perpetual warm breezes and natural abundance—so much to appropriate, now compounded by male fantasy.[65] Hundreds died of venereal disease as a result of Bougainville's visit.

Amongst leading Enlightenment figures, the French philosopher Denis Diderot did most to popularise Bougainville's image of the Tahitians, which was widely discussed, Rousseau's ideas on the primitive coincidentally being "all the rage" in Parisian salons in the 1760s.[66] Diderot thought "The Tahitian is close to the origin of the world, while the European is close to its old age", and reflected that European mores seemed mere shackles for those to whom "the most profound feeling is a love of liberty." He has one old Tahitian tell Bougainville:

> You are welcome to drive yourselves as hard as you please in pursuit of what you call the comforts of life, but allow sensible people to stop when they see they have nothing to gain but imaginary benefits from the continuation of their painful labors. If you persuade us to go beyond the bounds of strict necessity, when shall we come to the end of our labor? When shall we have

over Tahiti (Constable, 1963), 2. The same was said of New Zealand by Joseph Banks, who noted of "this Island of Sensuality" that in "Otaheite . . . Love is the Chief Occupation, the favourite, nay almost the Sole Luxury of the inhabitants" (*The Endeavour Journal of Joseph Banks, 1768–1771*, 2 vols (Public Library of New South Wales, 1962), 2: 330).

64. J. Cook, *Captain Cook's Journal*, 95.

65. On the development of these images, see Jonathan Lamb's "Fantasies of Paradise", in *The Enlightenment World*, edited by Martin Fitzpatrick, Peter Jones, Christa Knellwolf, and Iain McCalman (Routledge, 2004, 521–35).

66. Bougainville, *Pacific Journal*, lvii, 74, 257, 263, 283. See Denis Diderot's *Supplement to the Voyage to Bougainville* (1771), reprinted in *Political Writings*, edited by John Hope Mason and Robert Wokler (Cambridge University Press, 1992, 31–76). Diderot distinguished between "good" and "bad" luxury, where in the former state opulence was widespread, and thus was by no means opposed to luxury as such (130).

> time for enjoyment? We have reduced our daily and yearly labors to the least possible amount, because to us nothing seemed more desirable than leisure. Go and bestir yourselves in your own country; there you may torment yourselves as much as you like; but leave us in peace, and do not fill our heads with a hankering after your false needs and imaginary virtues [*ne nous entête ni de tes besoins factices, ni de tes vertus chimeriques*].[67]

Diderot also put forward a theory, much loved by his contemporaries, of the primitive as representing the psychological alter ego of the moderns, writing:

> Shall I outline for you the historical origin of nearly all our unhappiness? It is simply this: Once upon a time there was a natural man; then an artificial man was built up inside him. Since then a civil war has been raging continuously within his breast. Sometimes the natural man proves stronger; at other times he is laid low by the artificial, moral man. But whichever gains the upper hand, the poor freak is racked and torn, tortured, stretched on the wheel, continually suffering, continually wretched, whether because he is out of his senses with some misplaced passion for glory or because imaginary shame curbs and bows him down. But in spite of all this, there are occasions when man recovers his original simplicity under the pressure of extreme necessity.

He warned, in turn, that the choice of modern rulers was:

> If you want to become a tyrant, civilize him; poison him as best you can with a system of morality that is contrary to nature. Devise all sorts of hobbles for him, contrive a thousand obstacles for him to trip over, saddle him with phantoms which terrify him, stir up an eternal conflict inside him, and arrange things so that the natural man will always have the artificial, moral man's foot upon his neck. Do you want men to be happy and free? Then keep your nose out of his affairs—then he will be drawn toward enlightenment and depravity, depending on all sorts of unforeseeable circumstances.[68]

The moral was that "that men become more wicked and unhappy the more civilized they become . . . Without going through the list of all the countries in the world, I can only assure you that you won't find the human condition perfectly happy anywhere but in Tahiti."[69]

67. Denis Diderot, *Rameau's Nephew and Other Works*, 194, 197; Diderot, *Supplément au Voyage de Bougainville* (Librarie E. Droz, 1935), 123.

68. Diderot, *Rameau's Nephew*, 234–35.

69. Diderot, 237.

Diderot concluded that Tahitian morals proved that the happiest societies aimed to satisfy basic needs, not superfluous luxuries. This did not imply that the moderns could regain their "original simplicity", for returning to poverty might mean that "our conventional virtues" would give way to unscrupulousness and immodesty.[70] Diderot accepted that luxury was entwined with modernity; the problem was rather, as Anthony Strugnell emphasises, the social inequality which usually accompanied it. Thus he distinguished between *mauvais luxe* and *bon luxe*, the former signifying the ostentation of the rich accompanied by the misery of the poor who try to emulate them, the latter a wealthy society where such extremes were avoided by a sound system of taxation and administration. But he offered no plan as to how the latter might be achieved.[71]

If women featured as the objects of desire in these masculine fantasies, it is a moot point as to how this was mirrored, or not, in female utopias. Here, the most extreme objections to men are met by simply eliminating them, as in Mary Bradley Lane's much later pioneering feminist tract *Mizora* (1880), where women reproduce without men (the "dark-skinned races" are also disposed of on eugenic grounds), and in Charlotte Perkins Gilman's *Herland* (1915). In the Enlightenment, withdrawal from men was a more common option. Ideals of separatist women's communities, "feminiotopias" or "microsocieties" based on what Mary Astell called the "Noble Vertuous and Disinteress'd Friendship" of women, are usually dated from the late seventeenth century.[72] To Alessa Johns, this period certainly produced what German scholars call "'*Freundschaftsutopien*,' Utopias of intimacy rather than a fully invented society."[73] These projected an extrapolation of specifically female forms of friendship onto a wider scale, or over a longer period. Other female utopias, Johns stresses, tend to be process- rather than end-state oriented, and gradualist and pragmatic, with a common focus on education.[74] Clearly by eliminating patriarchal oppression such communities take a measured step towards utopia. But does gender define their distinctive sociability?

70. Denis Diderot, *Selected Writings* (Macmillan, 1966), 149; Anthony Strugnell, *Diderot's Politics* (Martinus Nijhoff, 1973), 37. His idea of the Tahitians was also indebted to images of North American indigenous peoples, especially the Hurons: Todorov, *On Human Diversity*, 276.

71. Strugnell. *Diderot's Politics*, 187, 221–22.

72. Mary Astell, *A Serious Proposal to the Ladies* (1696), cited in Nicole Pohl and Brenda Tooley, eds, *Gender and Utopia in the Eighteenth Century* (Ashgate, 2007), 11.

73. Alessa Johns, *Women's Utopias of the Eighteenth Century* (University of Illinois Press, 2003), 142. On female friendship, see further Rebecca D'Monté and Nicole Pohl's *Female Communities, 1600–1800* (Macmillan, 2000), esp. 23n22.

74. Alessa Johns, "Feminism and Utopianism", in *The Cambridge Companion to Utopian Literature*, ed. G. Claeys (Cambridge University Press, 2010), 178.

And can a specific form of female association be universalised, implying not only the overcoming of male oppression but a superior form of human interaction, perhaps defined by the absence of domination as such? To Janice Raymond, "gyn/affection" describes a female bond common to nunneries and convents, where, hierarchies notwithstanding, greater trust and confidence may exist between women than is common in the outside society.[75] To Carol Pearson, women's utopias are distinctive insofar as they rest on ideas of shared power rather than power over people, an ideal discussed by Carol Gilligan in terms of a "network of connection".[76] To Nicole Pohl, too, writers like Sarah Scott, the author of the country-house "female utopia" (in Jane Spencer's term) *Millenium Hall* (1762), indicate that female spaces could be conceived in terms of utopian openness, intimacy, and the absence of what Scott calls the "boundaries and barriers raised by those two watchful and suspicious enemies, Meum and Tuum", with "all property laid in one undistinguished common".[77] Utopian spaces could also be feminised by removing them from male entanglements, or (a common plot) converting men to their values. The convent may have often been a prototype here, but later spaces were less inhibiting, and could open up erotic as well as platonic emotional or romantic potential. To Pohl, *Millenium Hall* thus provides a particularly good example of "an emancipatory and liberating space for women".[78]

The juxtaposition of male spaces defined by power, intrigue, and corruption with friendship-centred female spaces is particularly important for utopian theory. Today we concede that no utopian project which promotes either misogyny or misandry, or the primacy of any form of gender identity, can be worthwhile, any more than one proposing the domination of any other group over others, or enforced servitude. Groups whose identity is explicitly based on oppressing others, and whose virtues are described in terms of denouncing the vices of others, excepting of course oppressive minorities, cannot be made to fit into universalist utopianism. Separatism, forming groups of like-minded people, is another issue and is justifiable so long as no harm to others results. Female separatism too, as Johns points out, could have a cosmopolitan dimension in which antagonism to nationalism was pointedly present.[79] Given the

75. See Janice Raymond, *A Passion for Friends: Towards a Philosophy of Female Affections* (Women's Press, 1991).

76. See the discussion in Angelika Bammer's *Partial Visions: Feminism and Utopianism in the 1970s* (Routledge, 1991, 25–26).

77. Sarah Scott, *Millenium Hall* (1762; Virago Books, 1986), xi, 41. Class barriers remain in this otherwise idyllic portrayal.

78. Nicole Pohl, *Women, Space and Utopia, 1600–1800* (Ashgate, 2003), 74.

79. Johns, *Women's Utopias*, 160.

later emergence of a self-conscious tradition of feminist utopian writing, and the influence here and in science fiction of authors like Ursula Le Guin, this trend needs to be underscored. To many such writers, not only are women's friendships with women different, so are their utopian projections of them.[80] Nan Bowman Albinski notes the relative absence in the Victorian period of "any feeling for eutopian communities of women".[81] But various twentieth-century writers have developed ecofeminist arguments, which also form a part of some environmentalist strategies.

Luxury, Simplicity, and Utopian Satire

Following the English Revolution and a puritanical cultural interlude, the restoration of the monarchy in 1660 under Charles II commenced an age of excess. The new commercial wealth which began to be more widely dispersed by the century's end set off a veritable explosion in the emulation of courtly and aristocratic styles. A much more indulgent attitude towards vice soon prevailed, and satirising luxury thus became a leading theme in the literary utopias of this period.[82] *Memoirs concerning the Life and Manners of Captain Mackheath* (1728) comments on the epoch that:

> there arose among us a general and uncommon Desire of Money, and after this an extraordinary Appetite for Power; the two great Fundamentals of every Evil. Avarice immediately overthrew all Probity, and Trust, and mutual Confidence; . . . After this extraordinary Change of Property, Virtue seemed to become Vice, and Vice, Virtue; and all Men inclined to think, that if they had Wealth, they had a Right to every thing; . . . and this Poison having thus mixed with the Blood and Spirits of the People, they became weak and enervate: the Desires of Mankind after Wealth being insatiable, were not to be diminished either by Want or Abundance. After this followed Rapine, Injustice, a general Dissolution of Morals; and in each Man

80. This implies an essentialist premise that women are more companionate and/or less aggressive than men, or that patriarchal dominance compels them to different forms of association, which cannot be explored here.

81. Nan Bowman Albinski, *Women's Utopias in British and American Fiction* (Routledge, 1988), 37.

82. Surveys of British utopianism in this period include Lois Whitney's *Primitivism and the Idea of Progress in English Popular Literature of the Eighteenth Century* (Johns Hopkins Press, 1934), Christine Rees's *Utopian Imagination and Eighteenth Century Fiction* (Longman, 1996), and Artur Blaim's *Gazing in Useless Wonder: English Utopian Fictions, 1516–1800* (Peter Lang, 2013).

> was found a Desire after the Goods of his Neighbour, and the Rich oppressed the Poor without Modesty or Moderation.[83]

Similarly, *Memoirs of the Court of Lilliput* (1727) laments that "wherever Luxury and Idleness presides, there will be room for Pride, for Vanity, and Lust; and that led me to a reflection how much an elevated Station is an Enemy to Virtue; and how greatly we deceive ourselves in believing that Riches are the Source of Happiness".[84]

Another satire from 1744 contrasts the dissolute manners of Europe to those of Madagascar, and notes that "It is our own luxurious Effeminacy, that has stripped us out of our natural Simplicity, and cloathed us with the Rags of Dissimulation." By contrast were the "Happy People, unto whom the Desire of Gold hath not yet arrived", for modern times "may be truly called the Age of Gold,/For it, both Honour, Love, and Friends are sold."[85]

The literary utopia thus came frequently to symbolise resistance to, or at least contempt for, increasing inequality and luxury. It doubtless echoed nostalgia for a world being rapidly lost, a sense of guilt at the abundant selfishness of the age, as well as the appeal of new primitive worlds now being discovered and conquered. To focus on Britain briefly, four models of virtuous restraint dominate eighteenth- and nineteenth-century debates: the idea of an arcadian state of nature, often without private property, where luxury does not yet exist; the primitive Christian community, often with uniform dress and consumption and prohibitions on frivolity and luxury; the classical republican ideal, where property and often trade are limited; and a Tory or Country Party ideal, where corruption is associated with the growing predominance of a Whiggish commercial interest, and contrasted with a virtuous landed interest and patriot-king.[86] Utopian texts thus echoed the wider debate of the period between opponents of commercial development, notably Jean-Jacques Rousseau, and its defenders, like David Hume and Adam Smith, whose views are considered below.

Both the "hard primitivism" of Spartan and puritan utopias and the "soft primitivism" of arcadian plenty are represented in Enlightenment utopias.[87]

83. *Memoirs concerning the Life and Manners of Captain Mackheath* (A. Moore, 1728), 11–12 (style modernised).

84. *Memoirs of the Court of Lilliput*, 2nd ed. (J. Roberts, 1727), 11.

85. "A Paradox: Proving the Inhabitants of the Island, called *Madagascar*, or *St. Lawrence* (in Things temporal) to be the happiest People in the World", *Harleian Miscellany* 1 (1744): 257–58.

86. See Gregory Claeys, ed., *Modern British Utopias*, 8 vols (Pickering and Chatto, 1997), 1: xxviii–xxxii, which gives examples of each type.

87. See, generally, C. Rees, *Utopian Imagination*; Blaim, *Gazing in Useless Wonder*.

The Golden Age or Cockaigne-like trope of discovering an unknown land, often in the South Seas or Australasia, where pursuing gain and riches is unknown, is adopted in, for instance, *The Island of Content* (1709). "Nature", we are told here, "is here so lavish of her Plenty, that we abound in Variety of Dainties, without human Labour". No meat is eaten "because we look upon it sinful to destroy one of God's Creatures for the Preservation of another".[88] No need exists for "Mercers and Drapers, to dun and plague our Quality". All have

> the Liberty, without the least Expence, of chusing such Apparel as shall best humour their own Fancy; for which Reason our very Women here are wholly innocent of Pride, not at all regarding superficial Ornaments, endeavouring only to excel each other in Vertue, Modesty, Eloquence, Musick, and such like Female Graces, that are Ornaments to the Mind, as well as to the Body.[89]

Similarly, in *The Adventures and Surprizing Delivrances of James Dubourdieu* (1719), a group called the "children of love" have been chosen by God for their innocence of sin. Amongst them:

> indeed there was no occasion for magistrates, when there was no ground for contention; there being no property among them, but a perpetual and uninterrupted course of a perfect love of one another. What the earth produced was a sufficient stock, plentifully to provide for their subsistence; and their cultivation of these products was so far from being laborious to them, that it was only their exercise and diversion.[90]

Many Enlightenment literary utopias envisioned these scenarios occurring on a small or monastic scale. Some such projections were correspondingly austere. Two works evidently by the same author express this theme. *An Essay concerning Adepts* (1698) asks, "why could not all Superfluous expences be regulated, and all the occasions of them cut off[?]", including the "unnecessary Arts, unnecessary Ornaments in Clothing and Furniture, and unnecessary Eating and Drinking". "To live Luxuriously", we are informed, "is but a Custom: If it was broke off, no Body would miss it, and evidently it would be of infinite advantage to the Society that it were so." Meals might be made common by law, thought the author, and limited to an hour's duration.[91] Similarly,

88. Gregory Claeys, ed., *Utopias of the British Enlightenment* (Cambridge University Press, 1994), 5.

89. Claeys, 8.

90. Claeys, *Modern British Utopias*, 1: 84.

91. Claeys, *Restoration and Augustan*, 227.

Annus Sophiae Jubilaeus: The Sophick Constitution; or, The Evil Customs of the World Reformed (1700) contends for equality and universal charity, and for imitating Lycurgan Sparta, with common living in "colleges". Luxury is the great enemy. All current disorders come "from superinduced necessaries, from a custom of being used to them, and desiring them more inordinately than those things which are actually necessary." Uniform attire for all of the same age is recommended, as well as plain but wholesome food. Alchemy would ensure that "every Body knows how to make easily infinite quantities of Gold and Silver", such that "Money must needs grow vile and good for nothing; nobody will slave for wages, and all will be Rich alike". Thence iron money alone would exist. In addition, "Painting, Ingraving, Weaving of Ribbon, or Lace, or the like frivolous Arts, are blotted out of the Catalogue of our Manufactures."[92] Indeed, "All those things therefore which are superfluous, and cause unnecessary Expences should be forbidden by express Laws. Even excess in Cloths and Apparel should be punishable". Streets would be identical, with no lanes or alleys, and each house cleaning its respective section thereof. Four parishes would make up a "little Town . . . like a Kingdom or a little World, which may subsist of itself,—and need not know what there is beyond its limits."[93] Equality is here proclaimed to be the law of God, which would have endured had Adam not sinned. "Adepts" are thus described as approving

> best of simplicity of Life. An Adept will not have rich Cloths, Furniture and Equipage, nor will accept of Titles, or Honour; those Things, and Philosophy are inconsistent: He looks upon all Men as being equal and Brethren, and upon himself as no better than others; he would not have others respect him, more than he would respect them: As for Titles and Dignities, he looks upon them, as Things very dangerous, and not easily reconcileable with Christian Humility; he would have all Men go plain, even plainer than the Quakers.[94]

"Soft primitivist" texts in this period usually condemn the degeneracy of modern Europe. John Kirkby's popular novel *The Capacity and Extent of the Human Understanding; Exemplified in the Extraordinary Case of Automathes* (1745), for example, juxtaposes "following nature" to the "life of luxury". Here, a shipwrecked couple discovers that while

> they had not now the same Advantages of Society, which their former Life afforded; but this was balanced to them, when they considered, that they

92. Claeys, 228 224, 217.

93. Claeys, 215.

94. Claeys, *Modern British Utopias*, 1: 14.

> had so much less of Vanity and Impertinence. In fine, they began to look upon this Change as so far from being an Evil, that they blessed God for using such a Means to bring them to a true Knowledge of themselves.

As a result, they now had

> a more true and solid Felicity than what their former Condition had ever afforded them . . . They now looked upon themselves to be as sufficiently supplied with all the real Necessities of Life; as ever, though not in so splendid a Manner. Their homely Fare went down with as good a Relish, as when they were entertained with more costly Dishes; and their Sleep was as sweet upon their Beds of Moss, as what they formerly enjoyed upon those of Down: The Reason was, because they now eat and slept only to satisfy Nature, and not Luxury.[95]

Some satires in this period cut in the other direction, and mocked the whimsicality of the primitivists. Sometimes read literally, Edmund Burke's well-contrived send-up of Bolingbroke's deism, *A Vindication of Natural Society; or, A View of the Miseries and Evils Arising to Mankind from Every Species of Artificial Society* (1756), is a case in point. Here, "a state of nature", where it was "an invariable law, that a man's acquisitions are in proportion to his labors", is contrasted to "a state of artificial society", where it was "a law as constant and as invariable, that those who labor most enjoy the fewest things; and that those who labor not at all have the greatest number of enjoyments." The appeal of a state "founded in natural appetites and instincts, and not in any positive institution", by contrast to "political society", was obvious:

> Here there are no wants which nature gives, and in this state men can be sensible of no other wants, which are not to be supplied by a very moderate degree of labor; therefore there is no slavery. Neither is there any luxury, because no single man can supply the materials of it. Life is simple, and therefore it is happy.[96]

In those utopias which assumed that the function of satire was reforming manners, institutional regulation and the enforced restraint of consumption are often prescribed. In *A Description of New Athens in Terra Australis Incognita* (1720), one of the first English-language texts set in this location, wages and prices are fixed at a level "sufficient to maintain them, their Families and Dependants", and forced agreements for lower wages are prohibited. Oppressing

95. Claeys, 2: 68–69.

96. Claeys, 3: 36. William Godwin quoted the text as approving of natural society in *Enquiry Concerning Political Justice* (1: 10).

the poor has been abolished, along with coaches, so that all except the sick and lame are obliged to walk. A Christian ideal of equality and brotherhood prevails, without sectarian division.[97]

Some utopias projected the love of wealth being converted into a zeal for public service. A widely read work, *The Adventures of Sig. Gaudentio di Lucca* (1737), recounts the discovery of a land hidden in the African deserts where grain, gold, fruits, and inventions permit even "the Magnificence of Life". Its laws aim "to keep up the equality of Brotherhood and Dignity, as exact as they can", and to this end patriarchal rulers distribute property for the benefit of the public.[98] Here "they are all Masters, and all Servants, every one has his employment; generally speaking, the younger sort wait on the elders, changing their Offices as it is thought proper by their superiors, as in a well regulated Community."[99] Public employments are described as being "rather an honorary Trouble than an Advantage, but for the real Good of the whole". As a result:

> They place their great Ambition in the Grandeur of their Country, looking on those as narrow and mercenary Spirits, who can prefer a part to the whole; they pride themselves over other Nations on that Account, each Man having a proportionable share in the publick Grandeur, the Love of Glory and Praise seems to be their greatest Passion. Besides, their wise Governours have such ways of stirring up their emulation by publick Honours, Harangues, and Panegyricks in their Assemblies, with a thousand other Arts of Shew and Pageantry, and this for the most minute Arts, that were it not for that fraternal Love ingrafted in them from their Infancy, they would be in danger of raising their emulation to too great a height. Those who give Indications of greater Wisdom and Prudence in their Conduct than others, are marked out for Governours, and gradually raised according to their Merit.[100]

More often, however, it was degeneracy which utopian authors satirised. In *Memoirs . . . concerning Captain Mackheath* (1728) the middle ranks are portrayed as enfeebled by luxury through aspiring to keep up with their betters, with the lower in turn imitating them.[101] *Private Letters from an American in England to His Friends in America* (1769) describes Britain as beyond redemption, having suffered a "decay of virtue for near a century past". Every twenty

97. Reprinted in Claeys, *Utopias*, 27–54.

98. Claeys, *Modern British Utopias*, 1: 363.

99. Claeys, 1: 366.

100. Claeys, 1: 367.

101. *Memoirs concerning the Life and Manners of Captain Mackheath* (1728), 13–14.

houses is a French hairdresser, and even servants learn Italian and fencing. Finally Britain's government passes to the American colonies.[102] (Colonial Americans fondly contrasted their own frugality and virtue with the excessive luxury, corruption, and degeneration of Britain.)[103] Like other satirists of the age, utopian writers agreed that national decline was chiefly occasioned by the lower orders emulating the upper. In one of the most illuminating utopias of the period, *The Travels of Hildebrand Bowman* (1778), the narrator visits countries at different levels of economic development. He recalls that under "Queen Tudorina" (Elizabeth I) the people were brave and virtuous, wine was drunk only in moderation, and lawyers and physicians returned their fees when they failed to protect their clients. In "Luxo-volupto", however—contemporary Britain—"trade and manufactures having brought immense wealth into the country, luxury followed fast on their steps". Now, inspired by the examples of imperial conquerors, clearly pointing to the vast wealth of India being used to corrupt British politics, "Wealth is become the only object which all men aim at to support that luxury, and all crimes of course are perpetrated to attain it."[104] So the dire warning is issued that "your nation is following exactly the steps of all rich and powerful kingdoms; luxury has got in among you, and will soon destroy you", as evidenced particularly in the manners and mores of women.[105]

In such texts suppressing luxury was often linked with closer social bonds: simplicity and fellowship go hand in hand. A key problem, however, was that the further luxury advanced, the more the prospect of realising any more primitive utopia seemed to recede. So how could such a transition to a new and more virtuous society be envisioned? A consideration of several influential French authors helps us to answer this question.

The Transformation Problem

From the 1600s to the present France has been more closely identified with luxury than any other nation. Not so paradoxically, perhaps, it has also championed the cause of primitivism and simplicity with great vigour. It thus plays a central role in the modern luxury debate. Here, unsurprisingly, the utopian literary form was commonly used in the seventeenth and eighteenth centuries

102. Claeys, *Modern British Utopias*, 3: 367.

103. David E. Shi, *The Simple Life: Plain Living and High Thinking in American Culture* (University of Georgia Press, 2007), 50–73.

104. Claeys, *Modern British Utopias*, 4: 82–83.

105. Claeys, 4: 50.

to describe alternative societies of purity and virtue. In Denis Veiras's *The History of the Severambians* (1675), the founding legislator of the utopia enjoins its people not to "employ themselves in vain and unprofitable Arts; which minister only to Luxury and Vanity, cherish Pride, and, by engendering Envy and Discord, lead Mens Minds astray from the Love of Virtue."[106] Such prospects seemed, however, to belong to the realm of fiction and fantasy rather than political theory. When the historian Montesquieu compared Veiras's utopia to Sparta under Lycurgus some decades later, there was no hint that the model might be emulated.[107] In his *Terre Australe Connue* (1676), the unfrocked Franciscan monk Gabriel de Foigny described a land of hermaphrodites who "do not know the meaning of *thine* and *mine*, all is held in common with such complete sincerity that it surpasses even the intimacy of man and woman among Europeans."[108] But no one sought to apply his principles, which the utopian genre perhaps indeed pushed further into the realm of improbability by mixing them with what we now call science fiction. So did anyone seriously consider imagining an actual return from the present to a more virtuous state, by renouncing luxury or severely mitigating its worst effects? The answer is *yes*.

Far and away the most influential effort to do so was Archbishop Fénelon's *The Adventures of Telemachus* (1699). Under Louis XIV, François de Salignac de la Mothe-Fénelon (1651–1715) became tutor to the Duke of Burgundy, second in line to the throne. An opponent of absolutism much influenced by Plato and Stoicism, he wrote *Telemachus* as a crash course in the possibility of reconciling Christian virtue, public liberty, and national harmony, and encouraging greater frugality from the monarch down.[109] Little known today, *Telemachus* became the bestselling book of the following century, in part precisely because it proposed what Istvan Hont calls "the surgical correction of developed luxury".[110] As such its relevance now is much greater than might be readily apparent.

106. Denis Veiras, *The History of the Severambians*, eds John Christian Laursen and Cyrus Masoori (State University of New York Press, 2006), 234.

107. Montesquieu, *The Spirit of the Laws* (1748; Hafner, 1949), 34–35.

108. Gabriel de Foigny, *The Southern Land Known*, ed. David Fausett (1676; Syracuse University Press, 1993), 48.

109. See Andrew Mansfield, *Ideas of Monarchical Reform: Fénelon, Jacobitism and the Political Works of the Chevalier Ramsay* (Manchester University Press, 2015), 83–105. On the progress of Stoicism in this period, see Christopher Brooke's *Philosophic Pride: Stoicism and Political Thought from Lipsius to Rousseau* (Princeton University Press, 2012).

110. Istvan Hont, "The Early Modern Debate on Commerce and Luxury", in *The Cambridge History of Eighteenth-Century Political Thought*, eds Mark Goldie and Robert Wokler (Cambridge University Press, 2006), 383–87. See further Michael Sonenscher, *Before the Deluge: Public*

In *Telemachus* Lycurgan Sparta is a clear precedent, with the saint-king Louis IX (1214–70), who lived "without ostentation or luxury", being one later model.[111] The plot centres on the lawgiver Mentor's reform of the corrupt state of Salentum, with the aim of "reducing everything to a noble simplicity and frugality". Mentor divides the society into seven classes, giving each an appropriate costume distinguished by colour, and regulating the furniture and ornaments of their houses, the use of gold and silver therein being prohibited, and all furniture being plain and long-lasting. Those engaged in luxury trades are returned to the countryside as cultivators, and an agrarian law restricts property holdings. All arts "subservient to pomp and luxury" are banished. Even the diet of the upper ranks is rendered modest by renouncing "high sauces". Only "useful" commodities are manufactured and traded. Music which is "soft and effeminate . . . that tended to corrupt the manners of youth" is confined to "festivals in temples, there to celebrate the praises of the gods and heroes". Sculptors and painters are restricted to the same themes, and wine drinking limited to sacrifices and high festivals. Commerce excludes luxurious or superfluous goods. Wants are thus reduced "to the real exigencies of nature".[112] Self-love is mitigated in part by a devotion to the "pure love of order", the "source of all political virtues" (in Hont's phrasing). But while those inhabiting the new order would be "obedient without being slaves" (again Hont), absolute power would be required to introduce it.[113] The moral is simple: luxury corrupts manners, and by contagious imitation leads even "those in low life" to "affect to pass for people of fashion". Thus "all live above their rank and income, some from vanity and ostentation, and to display their wealth; others from a false shame, and to hide their poverty." Even "those who are poor will affect to appear wealthy; and spend as if they really were so." But "it is the pride and luxury of certain individuals that involve so many of their fellow-creatures in all the horrors of indigence." Because some wallowed in

---

Debt, Inequality, and the Intellectual Origins of the French Revolution (Princeton University Press, 2007); Sonenscher, *Jean-Jacques Rousseau: The Division of Labour, the Politics of the Imagination and the Concept of Federal Government* (Brill, 2020), 26–50.

111. François Fénelon, *Letters*, ed. John McEwen (Harvill, 1964), 140–41. At least one modern account, however, denies that Fénelon wished to "return to the lost world of the ancients", because he championed neither poverty nor austerity: see Ryan Patrick Hanley, *The Political Philosophy of Fénelon* (Oxford University Press, 2020), 18, 50.

112. François de Salignac de la Mothe-Fénelon, *The Adventures of Telemachus*, ed. O. M. Brack (University of Georgia Press, 1997), 148–49, 235, 149–50.

113. Istvan Hont, "The Early Modern Debate on Commerce and Luxury", in *The Cambridge History of Eighteenth-Century Political Thought*, eds Mark Goldie and Robert Wokler (Cambridge University Press, 2006), 387.

luxurious idleness, others suffered wretched poverty. Only "by changing the taste, manners, and constitution of a whole nation" could these processes be reversed.[114]

How seriously was such advice meant to be taken? Around 1694–95, Fénelon warned Louis XIV that his predecessors had "reduced France to destitution in order to maintain a state of prodigal and incurable luxury at Court."[115] His warning went unheeded. Opposed by Mandeville and Montesquieu, amongst others, Fénelon was a key source for Louis-Sébastien Mercier, whose central position in the history of literary utopianism in this period is explored below.[116] What Fénelon embodied was the recurrent association of greater austerity with personal integrity. To Daniel Roche, thus, "The polemic about harmful or useful luxury permeates the utopias" of this period. For their authors, the simplicity of external appearance corresponded in some sense with simplicity of the soul, and an authentic presentation of the self, as well as personal modesty and social restraint.[117] In this period, Frank and Fritzie Manuel conclude, "the hard utopias predominate over the visions of the soft and luxurious blessed isles in the manner of Diderot", with the "stark *Manifesto of the Equals*" being "the final programmatic statement of the more common Spartan ideal of the age."[118] These results would continue, we will see, into the nineteenth century.

114. Fénelon, *Adventures of Telemachus*, 269, 152, 269.

115. Fénelon, *Letters*, 300.

116. [Louis-Sébastien Mercier], *Memoirs of the Year Two Thousand Five Hundred*, 2 vols (G. Robinson, 1782), 2: 17–18.

117. D. Roche, *Culture of Clothing*, 428n115.

118. Manuel and Manuel, *French Utopias*, 8.

6

The Triumph of Unsocial Sociability?

LUXURY IN THE EIGHTEENTH CENTURY

WE TURN NOW to consider how the problems of luxury and simplicity which we have so far examined in relationship to utopianism were conceived in mainstream thinking, and the non-literary expression of utopian ideas. We have seen that throughout history inequality has been regarded as the greatest threat to *Gemeinschaft*-style communities. Luxury in turn has long been regarded as a key cause and effect of inequality, and as such subversive of both sociability and utopia. But the diffusion of luxury is practically synonymous with modernity as such. Its pursuit has stimulated and delighted our senses and hastened our pace of life, while fuelling our greed and inhumanity. It drove the imperialism of the sixteenth century and following, when, to the Abbé Raynal, "a kind of luxury, unknown to the ancients, . . . infected Europe with a multitude of new tastes", such that "those nations which can furnish the rest with the means of gratifying them, must become the most considerable".[1] It fostered the slave trade and resulted in the killing of millions of Africans and indigenous Americans in the quest for spices, gold, silver, sugar, and cotton. Then came a profusion of suffering in the new industrial age, when amidst great accomplishments, discoveries, and innovations, a new wave of exploitation and oppression commenced. This process seemingly rendered utopia more remote than ever, leaving only the possibility of what Immanuel Kant famously called "unsocial sociability", a "propensity to enter into society, bound together with a mutual opposition which constantly threatens to break up the society."[2] To appreciate

1. Raynal, *Philosophical and Political History*, 3: 67.

2. Kant, *On History*, 15. Kant rejected the idea that the Golden Age was anything other than uncomfortable, arguing instead that history was "a gradual development from the worse to the

how impressive the attempt was to vindicate luxury and to portray it as harmless and even beneficial, we must first recall how it was traditionally regulated in Europe.

Regulating Luxury: Sumptuary Laws

Sumptuary laws restricting personal consumption and display were common in late medieval and early modern Europe. Various motives inspired them, including a desire to impede expensive foreign luxuries, to curtail the vices associated with consuming them, and to prevent the confusion of ranks—hampering our ability to spot the affluent in the street at a glance—largely with a view to retaining distinctions between the nobility and the rising nouveau riche, the merchant classes. As growing affluence intensified emulation and demands for social recognition, the newly enriched too sought in turn to restrict the display of those beneath them. For as we move into the seventeenth and eighteenth centuries the lower orders, too, sought equality in display. Desire was democratised, and a revolution in appearance commenced.[3] To Alan Hunt, fashion thus became a major battleground in the class struggles of the early modern period.[4]

Attempts to suppress these urges were obviously driven by the wish to sustain the existing system of ranks and the exhibition of status it required. But other motives were also at work, including ensuring male domination over women. The body as such has been a "site of regulation" in many societies. The amount of flesh women could display in public has been widely limited, notably by face, hair, and arm coverings. Hairstyles have often been restricted where they are held to represent "deviant" ideals. So have make-up, perfume, jewellery, and decoration of other kinds, as well as the materials used in clothing and the cost of attire.

In Europe, efforts to regulate luxury began in the ancient world, and were common until the eighteenth century. As we saw, the Greeks and Romans were well aware that luxury endangered both private and civic virtue.[5] But

---

better", with inequality a major cause of this progress: see "Conjectural Beginning of Human History", in Kant, 68.

3. This is a central theme in an excellent general introduction to the subject, Gilles Lipovetsky's *The Empire of Fashion: Dressing Modern Democracy* (Princeton University Press, 1994).

4. Alan Hunt, *Governance of the Consuming Passions: A History of Sumptuary Law* (Macmillan, 1996), 169.

5. See generally John Sekora, *Luxury: The Concept in Western Thought, Eden to Smollett* (Johns Hopkins University Press, 1977); Christopher Berry, *The Idea of Luxury: A Conceptual and Historical Investigation* (Cambridge University Press, 1994).

they were often unable to resist succumbing to temptation when it presented itself, or as rank or status dictated. Conspicuous consumption assumed forms which are thankfully mostly forbidden to us today. That the Roman emperor Elogabulus devoured a plate of nightingale's tongues was remembered nearly two thousand years later.[6] That Cleopatra dissolved in vinegar a pearl worth £15 million today is still retold.[7] Conspicuous consumption, as it would come to be called, was not a modern invention.

Ancient attempts to allay extravagance focused mostly on clothing and entertainment. In 594 BCE the Greek legislator Solon limited to three the number of garments women could wear and restricted elaborate funeral processions. In 215 BCE, the first major Roman law of this type, the *lex Opia*, prohibited women from owning more than half an ounce of gold, from wearing coloured robes or silk, and from riding in carriages inside the city except when travelling to religious festivals. Later laws focused on banquets, restricting the number of guests, and the cost, menu, and value of silverware used, and prohibiting foreign wines.[8] Peasants were limited to one colour of clothing, officers to two, and commanders to three. Guards were sent to markets to intercept forbidden foodstuffs, like dried figs and Atlantic oysters.[9] The wearing of their masters' clothing by slaves was confined to the Saturnalia.

In Rome the role of greed in political corruption was well known, which may be why banquets were targeted, though gluttony and drunkenness were early on recognised as accompanying such excesses.[10] The great Roman moralist Cicero thought luxury caused avarice, and held "no vice more detestable . . . more especially in great men".[11] How "base and unworthy a thing it is", he lamented, "to dissolve in luxury, softness, and effeminacy; and how brave and becoming it is, on the other hand, for a man to lead a life of frugality and temperance, of strictness and sobriety". "Hannibal himself", he warned, who had invaded Italy and nearly vanquished Rome, had been "conquered . . . by pleasure."[12] The decline and fall of the Roman Empire itself would long be associated with the debilitating effects of luxury upon its upper classes. The gendered nature of this debate is also evident from now on, though it is clear

6. Emile de Laveleye, *Luxury* (Swan Sonnenschein, 1891), 15. See James Burgh, *Political Disquisitions*, 3 vols (E. and C. Dilly, 1774–75), 3: 83.

7. McNeil and Riello. *Luxury*, 2.

8. C. Berry, *Idea of Luxury*, 76–82.

9. McNeil and Riello, *Luxury*, 21.

10. Zanda, *Fighting Hydra-like Luxury*, 71.

11. Cicero, *Offices*, 108.

12. Cicero, 47; Cicero, *Agrarian Speeches* (Oxford University Press, 2018), 87.

that while men's desire for luxury may assume different forms from women's, its intensity is as great.

Christianity drove home the message of the dangers of vanity and greed and condemned luxury as sinful. Vanity was often associated with women, the result both of Christian misogyny and a misunderstanding of male egotism, which, given patriarchalism, has not focused as much on external representations of the self as other manifestations of power. Medieval and early modern restrictions focused on the type and value of apparel permitted to each class, and specifically limited the wearing of gold and silver, ermine and other furs, and jewels. Rules regulating dress according to social class were introduced by Charlemagne as early as 808. Public ostentation was also often at issue. In fourteenth-century Nürnberg, the number of guests at weddings was set at twelve, six of each sex, and giving gifts and eating there were prohibited. Gambling, drinking, dancing, and profanity were also suppressed. Christian writers stressed the role of pride in the desire for luxury. Sensual gratification as such was dangerous, but so was desiring to please others by means of finery and cosmetics, which was how the Church father Tertullian defined *luxuria*. Since demonstrating one's importance was often associated with desirability, such propensities were often linked to lust. Misogyny intensified from the perceived impediment of sexuality to achieving salvation.[13] Then there were the financial costs. Some worried that foreign trade carried too much bullion away and impoverished a nation's coffers, a concern later associated with mercantilism. Others complained that luxury undermined national power by inhibiting population growth.[14] Straightforward killjoys clearly resented other people enjoying themselves, or were jealous, and denounced all pleasure as an insult to God or a detraction from devotion to him.

Clothing was everywhere a key target in such regulations. In early modern Nürnberg, women were prohibited from wearing silk, while men over fifty could not sport red buckram.[15] Here the complaint was heard in 1657 that "the different classes are barely to be known apart" because of "extravagance in dress and new styles to such shameful and wanton extremes."[16] Modes of dining continued to be a focus. An English law of 1336 limited dinner or supper for all classes to two courses, "and each mess of two sorts of victuals at the utmost, be it Flesh or Fish", excepting feast days, when three courses were

13. C. Berry, *Idea of Luxury*, 90.

14. See, generally, Joseph J. Spengler's *French Predecessors of Malthus* (Duke University Press, 1942), esp. 77–169.

15. See Kent Roberts Greenfield, *Sumptuary Law in Nürnberg: A Study in Paternal Government* (Johns Hopkins Press, 1918).

16. Hunt, *Governance*, 121.

allowed.[17] Here too, when wages rose rapidly after the Black Death, the Statute Concerning Diet and Apparel (1363) defined what dress each class might wear and fixed meals at two courses, again excepting on the forty or so feast days a year. Another law, which lasted twenty-six years, required most people older than six to wear woollen caps on Sundays, to bolster the woollen industry. Further laws followed in 1509 and 1533, the last being in 1604. Elizabeth I also regulated the style of men's beards, and decreed that:

> no great ruff should be worn, nor any white color, in doublet or hosen, nor any facing of velvet in gowns, but by such as were of the bench. That no gentlemen should walk in the streets in their cloaks, but in gowns. That no curled or long hair be worn, nor any gown but such as be made of sad color.[18]

In 1566, four inspectors stood for eight hours daily at the gates of London to intercept those wearing prohibited forms of hose.

The chief aim of such measures was to bolster the differentiation of ranks, particularly by restricting how far the middling gentry could ape the nobility, and to limit costly imports. But gluttony, especially feasting and drunkenness, was equally widely condemned. As late as 1621 a Member of Parliament proposed an act to establish a "settled fashion" in clothes and hoped that "in time", "newfangleness of apparel" would "die of itself".[19] But such efforts seemingly had little effect. The desire to flaunt was inextinguishable. Already in the fourteenth century, Henry Knighton complained that "the lesser people were so puffed up in those days in their dress and their belongings, and they flourished and prospered so in various ways, that one might scarcely distinguish one from another for the splendour of their dress and adornments".[20] In 1600 William Vaughan noted that many servants "would bestow all the money they had in the world on sumptuous garments".[21] Once ignited, it seemed, the passion for display, and for equality, could not be suppressed.

17. Hunt, 1.

18. Elizabeth B. Hurlock, "Sumptuary Law", in *Dress, Adornment, and the Social Order*, eds Mary Ellen Roach and Joann Bubolz Eicher (John Wiley and Sons, 1965), 298. Sad or "sadd" colours, meaning serious, could include blue or green, and did not mean black.

19. Quoted in Keith Thomas, "The Utopian Impulse in Seventeenth-Century England", in *Between Dream and Nature: Essays on Utopia and Dystopia*, eds Dominic Baker-Smith and C. C. Barfoot (Rodopi, 1987), 37.

20. Quoted in Joanne Sear and Ken Sneath, *The Origins of the Consumer Revolution in England* (Routledge, 2020), 107.

21. Quoted in Neil McKendrick, John Brewer, and J. H. Plumb, *The Birth of a Consumer Society: The Commercialization of Eighteenth-Century England* (Hutchinson, 1983), 38.

Elsewhere a similar story is evident. In France the bourgeoisie were early on subject to quite specific restrictions, the wearing of gold and precious stones, squirrel and ermine being strictly prohibited in the thirteenth century. Here, the case against luxury was framed around five arguments: its sinfulness; the growing competition of the bourgeoisie with the aristocracy; the latter's tendency to dissipation and vain display; the loss of precious metals, especially gold leaf, in the everyday wear and tear of luxury goods; and the export of bullion abroad to pay for imported goods.[22] Under Charles V (1338–80) excessive ornaments on shoes were banned as conducive to "worldly Vanity and extravagant Presumption."[23] Charles VI (1368–1422) restricted dinners to soup and two dishes. A 1485 edict forbade all except the nobility to wear silk, or cloth containing gold or silver, it being noted that, besides being "displeasing to God", the "public welfare of our kingdom is seriously damaged by the great expenditure and outlay that a number of our kingdom make on clothes that are too ostentatious and sumptuous, and unsuitable to their estates".[24] Charles IX (1550–74) confined silk wearing to princesses and duchesses, and regulated ornamentation by rank, laws which were repealed only after the 1789 revolution, in, tellingly, one of the first acts of the new regime.[25] In 1560 "the greater part of the evils of France" was blamed on "the pomp and superfluity of dress of both men and women". Specially lamented was "the quantity of money which goes out of the kingdom for perfumes, scented gloves, embroideries and the like, of which the cost is great and the enjoyment short." In 1571 it was even demanded that to "bring the poor people back to their ancient humility" labourers "and other village people" should "not wear any colored clothes, but only grey undyed", and that no one should dress in foreign clothes. In 1576 Jean Bodin, following Plato, while disdaining the so-called Golden Age as merely one of iron, condemned luxury as "the source of all the vices and calamities of a commonwealth", and urged such heavy taxation on luxury articles that "only the rich and the well-fed can use them." In 1614 the Estates General complained that:

22. Charles Woolsey Cole, *French Mercantilist Doctrines before Colbert* (R. R. Smith, 1931), 12–13.

23. Jean-François Melon, in Henry C. Clark, ed., *Commerce, Culture, & Liberty: Readings on Capitalism before Adam Smith* (Liberty Fund, 2003), 260.

24. Cole, *French Mercantilist Doctrines*, 13–14.

25. When Louis XVI called the Estates General in 1789 the prohibitions on the Third Estate (half the deputies), and an injunction that all wear suits of black cloth to designate that they were not gentlemen, caused great offence (Aileen Ribeiro, *Fashion in the French Revolution* (B. T. Batsford, 1988), 45–46).

> One of the greatest disorders existing in this kingdom is the corruption and destruction of good morals produced by the liberty that almost all classes of persons without distinction have unwisely taken in clothing themselves too superbly, in dining too elaborately and too heavily, and in furnishing and fitting out [their establishments] too sumptuously.

A year later, in 1615, the economist Antoine de Montchrétien lamented that "The shopkeeper is dressing like the gentleman, the humble man like the noble. It is no longer a question of being, but merely of appearing. If this luxurious display continues it will ruin the army, make the cities insolent, the men effeminate and the women immoral".[26] We are already close, as we will see, to Mandeville's and Rousseau's starting points a century later. Further ordnances were passed in the 1660s and 1670s, particularly regarding the wearing of gold and silver fabric. They were widely ignored.[27]

A similar narrative existed elsewhere. Foreign clothes were banned in many times and places, "French fashions" being prohibited in Venice in 1509. Prostitutes were sometimes, as in Florence in 1337, the only group allowed to dress in finery above their station, but had to wear identifiable costumes marking their trade.[28] At Basel in 1637 a veritable social crisis seems to have been afoot when a statute noted that in "these troublous times" it was necessary "to plant and propagate all modesty as far as possible, and to so direct things that all destructive superfluity may be checked." To this end men and women were enjoined to avoid openly wearing gold chains, necklaces, or bracelets, and other clothes ornamented with pearls, such as ruffs, shirts, handkerchiefs, napkins, headdresses, pendant buttons, and neckcloths.[29] In Italy artisans who produced such forbidden items were also punished.[30]

Some forty Italian cities, in fact, passed far and away the largest number of sumptuary laws from the twelfth through the eighteenth centuries, at least 208, beginning at Genoa in 1157. The majority aimed at regulating women's clothes, expenditure on which was evidently inflationary, producing a growing demand for increased dowries, which was thought in turn to reduce the marriage and birth rate.[31] These cities were increasingly defined by a rising, opulent

26. Cole, *French Mercantilist Doctrines*, 15, 17–18, 55, 165, 127.

27. Jennifer M. Jones, *Sexing* La Mode*: Gender, Fashion and Commercial Culture in Old Regime France* (Berg, 2004), 31.

28. Hunt, *Governance*, 73, 245; Hurlock, "Sumptuary Law", 296.

29. John Martin Vincent, *Costume and Conduct in the Laws of Basel, Bern, and Zurich, 1370–1800* (Johns Hopkins Press, 1935), 56, 141.

30. McNeil and Riello, *Luxury*, 53–54.

31. Zanda, *Fighting Hydra-like Luxury*, 73, 92–94.

merchant class competing with the aristocracy; Marxists will recognise the spectre of a bourgeois revolution. The confusion of ranks was a widespread concern, and women's clothing was a key target, sometimes indeed with "misogynist venom".[32] In 1784 the Parisian bookseller Antoine-Prosper Lottin linked the new culture of appearances to a general trend towards making life a perpetual spectacle or masked ball "where everyone wants to lose the marks of their estate and tries to disguise themselves with the mask of a higher station."[33] The emergence of fashion was also bound up with what Gilles Lipovetsky, whose starting point is the French sociologist Gabriel Tarde on imitation, discussed below, calls a "relative devaluing of the past", and an investment of prestige in the new.[34] So marked were these developments that it has been suggested that "modern consumer society, with its insatiable consumption setting the pace for the production of more objects and changes in style, had its first stirrings, if not its birth, in the habits of spending that possessed the Italians in the Renaissance."[35]

Reactions against this process included campaigns for reforming manners and for reaffirming the Roman virtue of *frugalitas*. Notable here were the efforts of the sin-obsessed millenarian Dominican friar Girolamo Savonarola (1452–98), who attempted to stem the excesses of Florence's rulers, the Medici banking family, as well as its dissolute and fun-loving population, by cracking down on adultery, sodomy, immodest dress, gambling, swearing, drunkenness, and lewd dancing. Denouncing banking as usury, and condemning "the unjust taxes which are grinding down the poor", Savonarola warned that the rich would soon "suffer great affliction".[36] He urged his fellow monks to imitate the austere ways of their forefathers, to wear plain clothes, fast regularly, take only communal meals, and abjure elaborate decoration in their cells. Thousands of Florentines heeded his call, and at the Carnival of 1497, the year Columbus discovered the "terrestrial paradise", a "bonfire of the vanities" took place in a fusion of class antagonism and religious enthusiasm. This veritable mountain of objects some 60 feet (20 metres) high, arranged in seven tiers according to the deadly sins, included "shameful pictures and sculptures, gambling devices, musical instruments and musical books, masks, costly foreign

32. Catherine Kovesi Killerby, *Sumptuary Law in Italy, 1200–1500* (Clarendon, 2002), 111.

33. Quoted in Jennifer Jones, *Sexing* La Mode, 148.

34. Lipovetsky, *Empire of Fashion*, 18.

35. R. Goldthwaite, "The Economy of Renaissance Italy: The Preconditions for Luxury Consumption", *I Tatti Studies: Essays in the Renaissance* 2 (1987): 16, quoted in Killerby, *Sumptuary Law in Italy*, 1.

36. Paul Strathern, *Death in Florence: The Medici, Savonarola and the Battle for the Soul of the Renaissance City* (Jonathan Cape, 2011), 103.

draperies overpainted with immodest scenes" and more. Effusive declarations of repentance occurred. Some believed angels had descended to live amongst men and expected "Florence to become a new Jerusalem out of which would come the laws and splendour of the good life . . . the renovation of the Church, the conversion of the infidels, and the consolation of the righteous"; indeed, a new "paradise on earth".[37] Morals tightened for a while. The historian Guicciardini noted that there had never "been as much goodness and religion in Florence as there was in his time".[38] Then the Pope interceded against him. Savonarola was tortured, hung, and burnt.[39] Florentines soon reverted to their accustomed profligacy.

Outside Europe similar trends are evident. Though clothing was most widely addressed, playing games, drinking, and other public as well as private amusements were restricted at one time or another in many places. In Mexico a royal decree of 1510 banned wearing silk, which had become common amongst the native population.[40] In the American colonies, a Massachusetts law of 1639 prohibited drinking the health of one's companions as "a mere useless ceremony", and one in 1634 proscribed "new and immodest fashions," "great, superfluous, and unnecessary expenses", lace, slashed sleeves, and wearing "gold and silver trinkets".[41] In Connecticut a law forbade any person from spending "their time idly, or unprofitably."[42] Amongst the African Kaffirs, the lower classes were forbidden to decorate themselves like their social betters, while in ancient Peru gold and silver could not be used without government permission.[43] In Tokugawa Japan (1603–1868) even the types of toys children might possess were restricted.[44]

In seventeenth-century Europe laws were still being passed to try to regulate the dress of an ever-increasing number of ranks, orders, and classes, including

37. Strathern, *Death in Florence*, 261; Donald Weinstein, *Savonarola: The Rise and Fall of a Renaissance Prophet* (Yale University Press, 2011), 218, 1; Weinstein, "Millenarianism in a Civic Setting: The Savonarola Movement in Florence", in Thrupp, *Millennial Dreams in Action*, 199.

38. Francesco Guicciardini, *The History of Florence* (1510; Harper Torchbooks, 1970), 146.

39. Machiavelli later dignified him as one of those "unarmed prophets" whose "learning, prudence, and courage" had nonetheless failed to guarantee success: Niccolò Machiavelli, *The Prince and the Discourses* (Modern Library, 1950), 22, 229.

40. Green, *Thomas More's Magician*, 126.

41. Sanford, *Quest for Paradise*, 106–7.

42. Hunt, *Governance*, 39.

43. Mary Ellen Roach and Joanne Bubolz Eicher, eds, *Dress, Adornment, and the Social Order* (John Wiley and Sons, 1965), 297.

44. See Francis Elizabeth Baldwin, *Sumptuary Legislation and Personal Regulation in England* (1923; General Books, 2010); Hunt, *Governance*.

municipal officials, members of various guilds and societies, and their wives and daughters to boot (only female clothing was addressed), whose costumes were meant to indicate their position in the social hierarchy.[45] Such warnings were still present as late as Wilhelmine Germany (1890–1918), when the bourgeoisie in particular began to question the more philistine aspects of consumerism, while still deploring the "sick aping" and "delusion of grandeur" of the rising "new" middle classes.[46]

By now, however, such efforts were doomed. Most such laws fell into abeyance and were probably poorly enforced at best. Indeed, to Alan Hunt, sumptuary regulations only kindled longing for whatever was prohibited. Luxury represents some desires which do not abate, notably for status, and "competition of a social and a sexual kind, in which the social elements are more obvious and manifest and the sexual elements more indirect, concealed, and unavowed".[47] Luxury is also a very fluid concept. When a sense of rarity and status differentiation seems exhausted in one area, it moves to another: paintings, buildings, décor, travel (the "grand tour", the Orient Express, the private yacht or jet), clothing, jewellery, fine art, antiques, private zoos and rare breeds of animals—whatever can symbolise rank, taste, importance, money. For every great luxury, too, there were little luxuries to brighten the lives of even the poorest, who were driven to emulate those above them, and to revel in the small joys they sometimes achieved in doing so, even if only in fantasy. Even the picture of a princess clad in fur and gems cut out from a newspaper could bring a vicarious sense of inhabiting her world when tacked to the wall of a thatched hut in a hamlet. Cults around such forms of celebrity have always been common; think recently of Diana, Princess of Wales. The "sumptuary project" thus seems to have been flawed from the outset. We aspire to higher status, and its symbols. Where markets offer them, we desire what we can afford, and more. There is an infinite number of forms of luxury, and limiting one, like plugging up a leaky dam, seems only to push pressure elsewhere.[48]

Moralists nonetheless continued to warn of the consequences of unrestrained desire throughout the eighteenth century, when the floodgates of modernity opened for the first time. New desires meant new frustrations, and a constant lust for the unattainable. To critics, this vast new augmentation of

45. For the example of Tallinn, see Astrid Pajur's "The Fabric of a Corporate Society: Sumptuary Laws, Social Order and Propriety in Early Modern Tallinn", in *A Taste for Luxury in Early Modern Europe*, edited by Johanna Ilmakunnas and Jon Stobart (Bloomsbury, 2017, 21–38).

46. Warren Breckman, "Disciplining Consumption: The Debate about Luxury in Wilhelmine Germany, 1890–1914", *Journal of Social History* 24 (1990–91): 485–505.

47. J. C. Flügel, *The Psychology of Clothes* (Hogarth, 1930), 138.

48. Hunt, *Governance*, 102, 220.

wealth and the torrent of ambitions it unleashed heralded a decadent age. A new, more expansive sense of self was on the horizon, defined by unlimited wants. As early as 1690 the English physician and economist Nicholas Barbon declared that:

> The Wants of the Mind are infinite, Man naturally Aspires, and as his Mind is elevated, his Senses grow more refined, and more capable of Delight; his Desires are inlarged, and his Wants increase with his Wishes, which is for everything that is rare, can gratifie his Senses, adorn his Body, and promote the Ease, Pleasure, and Pomp of Life.[49]

In 1700 the pamphleteer Charles Davenant complained that:

> Trade, without doubt, is in its nature a pernicious thing; it brings in that Wealth which introduces Luxury; it gives a rise to Fraud and Avarice, and extinguishes Virtue and Simplicity of Manners; it depraves a People, and makes way for the Corruption which never fails to end in Slavery, Foreign or Domestick.[50]

What Joyce Appleby describes as the idea of "man as a consuming animal with boundless appetites" now appeared.[51] A link to economic growth was also established. John Houghton claimed in 1696 that "those who are guilty of Prodigality, Pride, Vanity, and Luxury, do cause more Wealth to the Kingdom, than Loss to their own Estates."[52]

Yet from the outset there was ambiguity about this prospect. The inexhaustible new desires seemingly undermined people's capacity for prudence, self-restraint, and self-mastery. In 1767 the political economist Sir James Steuart noted that while formerly men were "forced to labour because they were slaves to others, men are now forced to labour because they are slaves to their own wants." Yet Steuart also urged a grudging acceptance of this process, a compromise with desire. Where "excessive gratification" of natural wants occurred, he thought, this rendered luxuries liable to "vicious excess", or an "abuse of enjoyment". Mere enjoyment or "sensuality" without "vanity, pride, ostentation", however, was harmless.[53] In practice, differentiating between these

49. Nicholas Barbon, *A Discourse of Trade* (1690; Johns Hopkins Press, 1905), 14.

50. Charles Davenant, *The Political and Commercial Works*, 5 vols (1771), 2: 275.

51. Joyce Oldham Appleby, *Liberalism and Republicanism in the Historical Imagination* (Harvard University Press, 1992), 48.

52. Quoted in Joyce Oldham Appleby, *Economic Thought and Ideology in Seventeenth-Century England* (Princeton University Press, 1978), 170.

53. Sir James Steuart, *An Inquiry into the Principles of Political Economy*, 4 vols (1767; Pickering and Chatto, 1998), 1: 324, 327, probably following Hume. Steuart divided luxury into four

expressions of pleasure, or between subjective hedonistic enjoyment and the pleasures of display or showing off, was very difficult. And some thought vicious excess was in any case clearly in the ascendant.

By mid-century this process drew forth a new generation of critics, many of whom thought the general trend was towards degeneration of manners and away from simplicity. Shaftesbury followed Marcus Aurelius in asserting that "The senseless part of mankind admire gaudiness: the better sort and those who are good judges admire simplicity".[54] Oliver Goldsmith's famous poem "The Deserted Village" (1770) evoked a vanishing way of life and a more virtuous past "ill exchanged" for the tawdry baubles called luxury.[55] A widely discussed account of this period was John Brown's *Estimate of the Manners and Principles of the Times* (1757–58). To Brown, commerce had initially helped to satisfy needs, which had "increased beyond all Belief within these twenty Years", but had now changed "its Nature and Effects. It brings in Superfluity and vast Wealth; begets Avarice, gross Luxury, or effeminate Refinement among the higher Ranks, together with general Loss of Principle". "Luxury, in this last Period", he warned,

> being exhausted in it's Course; and turned, for want of new Objects of Indulgence, into Debility and Languor, would expire or sleep, were it not awakened by another Passion, which again calls it into Action. Nothing is so natural to effeminate Minds, as *Vanity*. This rouzes the luxurious and debilitated Soul; and the Arts of pleasurable Enjoyment are now pushed to their highest Degree, by the Spirit of delicate Emulation.
>
> Thus the whole Attention of the Mind is centred on *Brilliancy* and *Indulgence*: Money, tho' despised as an *End*, is greedily sought as a *Means*: And *Self*, under a different Appearance from the trading Spirit, takes equal Possession of the Soul.

He warned further that:

> where Luxury and Effeminacy form the ruling Character of a People, the Excess of Trade and Wealth naturally tends to weaken or destroy the Principle of

---

types: the moral, affecting the mind; the physical, affecting the body; the domestic (affecting income); and the political, in harming the state.

54. Anthony, Earl of Shaftesbury, *The Life, Unpublished Letters, and Philosophical Regimen* (Swan Sonnenschein, 1900), 179. He also wrote of "that simplicity of manners and innocence of behaviour which has been often known among mere savages, ere they were corrupted by our commerce, and, by sad example, instructed in all kinds of treachery and inhumanity" (quoted in Whitney, *Primitivism*, 35–36).

55. See Peter Laslett, *The World We Have Lost* (Methuen, 1965).

> *Honour*, by fixing the Desire of Applause, and the Fear of Shame, on improper and ridiculous Objects. Instead of the Good of others, or the Happiness of the Public, the Object of Pursuit naturally sinks into some unmanly and trifling Circumstance: The Vanity of Dress, Entertainments, Equipage, Furniture, of course takes Possession of the Heart.[56]

The price paid for such excess now seemed substantial, and we begin to see our own ancestors at work and play. More emulation meant labouring harder, for longer hours, and getting less rest, and then consuming more stimulants. The pace of life increased, and a sense of rarely catching up became prevalent. Desires outstripped the means of satisfying them. Ever-greater extravagance and profligacy now became an obsession which ruined even the wealthy. At the table, at the ball, at the races, in the street, in one's transport and the number of one's retainers, but especially in clothing, the desire for recognition, for standing out, demanded redoubled effort. Those who had not, must needs borrow to keep up. These excesses doubtless encouraged mass intoxication, as in Britain's early eighteenth-century gin craze, when the poor took to strong drink with a vengeance. ("Drunk for a penny, dead drunk for tuppence" was a slogan of the times.) But in fact drunkenness continued to be marked throughout the Industrial Revolution, when miniature heterotopias, "Temples of Oblivion", in W. J. Dawson's neat phrase, increasingly became temporary retreats from the pain of life.[57]

A sense of loss and regret continued to mount in the latter decades of the century. By the 1750s and 1760s many journals echoed the warning that "Amongst the many reigning vices of the present age none have risen to a greater height than that fashionable one of luxury, and few require a more immediate suppression, as it not only enervates the people, and debauches their morals, but also destroys their substance". In the face of such depravity and the threat of ruin it heralded, some advocated reviving Elizabethan sumptuary laws. A writer in 1773 warned that while

> brute creation . . . guided by instinct and natural desire, has happiness . . . man, anxious to be unhappy, industrious to multiply woe, and ingenious in contriving new plagues, new torments, to embitter life, and sour every present enjoyment, has inverted the order of things, has created wishes that have no connexion with his natural wants, and wants that have no connexion with his happiness.[58]

56. John Brown, *An Estimate of the Manners and Principles of the Times* (1757–58; Liberty Fund, 2019), 303, 319–20, 325.

57. Wolfgang Schivelbusch, *Tastes of Paradise: A Social History of Spices, Stimulants, and Intoxicants* (Vintage, 1992), 4, 148; W. J. Dawson, *The Quest of the Simple Life* (Hodder and Stoughton, 1903), 91.

58. Quoted in Whitney, *Primitivism*, 46–47.

A backlash was inevitable. In the 1790s a wave of nostalgia for the loss of rural simplicity and virtue appeared, in the writings of novelists like Robert Bage, William Godwin, Mary Hays, Thomas Holcroft, and Charlotte Smith. Such sentiments would be resurrected with marked regularity, every twenty years or so, in nearly every subsequent generation, down to the 1960s.

Mandeville's Paradox

Against such criticisms a counter-trend developed, which proved much more agreeable to the later moderns. Consumerism could not have flourished if luxury had not been exonerated, for its role in the rapid expansion of what Scottish Enlightenment authors called commercial society was obviously significant.[59] As Istvan Hont stresses, eighteenth-century theorists like David Hume and Adam Smith feared luxury only when it was "unregulated".[60] If it was "well-ordered", and brought more "polished manners", or what we today usually term "civilisation", this could justify the much greater inequality of property which accompanied commerce. If, however, the new system threatened social disorder, as unfulfilled desires exploded into resentful protest, it was clearly much riskier. And, certainly, we know petty criminality did increase as the "very dregs of the people", in the magistrate Henry Fielding's words, "aspiring to a Degree beyond that which belongs to them", took to stealing what they could not afford to buy.[61]

To those who thought the benefits of luxury trade and consumption clearly outweighed their disadvantages, it was necessary to dismantle the old moralistic discourse on luxury. This trend was instigated chiefly by Bernard Mandeville's *The Fable of the Bees* (1723–28), the text which most clearly foreshadowed the idea of "conspicuous consumption" popularised by Thorstein Veblen, which is considered in part III.[62] Mandeville transformed luxury from a vice

59. See Christopher J. Berry, *The Idea of Commercial Society in the Scottish Enlightenment* (Edinburgh University Press, 2013).

60. Hont, "Early Modern Debate", 379–418, here 380; see further Hont, *Jealousy of Trade: International Competition and the Nation-State in Historical Perspective* (Harvard University Press, 2005), 159–84.

61. Hunt, *Governance*, 83. See Peter Stearns, *Consumerism in World History: The Global Transformation of Desire* (Routledge, 2001), 54–55.

62. See Gordon Vichert, "The Theory of Conspicuous Consumption in the Eighteenth Century", in *The Varied Pattern: Studies in the Eighteenth Century*, eds Peter Hughes and David Williams (A. M. Hakkert, 1971), 251–67. On Mandeville, see Hector Monro's *The Ambivalence of Bernard Mandeville* (Clarendon, 1975), Thomas Horne's *The Social Thought of Bernard Mandeville* (Columbia University Press, 1978), M. M. Goldsmith's *Private Vices, Public Benefits: Bernard Mandeville's Social and Political Thought* (Cambridge University Press, 1985), and E. J. Hundert's

into a virtue, or at least a social good, albeit one resulting from unintended consequences. The chief reference point here was the state of nature, humanity's original primitive condition, which for these writers had now shed most of its classical and Christian idealised associations and was merely "barbaric" or "savage" compared with modern "refined" or "polished" civilisation. Mandeville thought the expansion of self-centred desires was a great impetus to humanity's advancement towards civility. Savage society was precarious, fraught with peril, and "a State of Simplicity, in which Man can have so few Desires, and no Appetites roving beyond the immediate Call of untaught Nature". Subsequently, by contrast, "The restless Industry of Man to supply his Wants, and his constant Endeavours to meliorate his Condition upon Earth, have produced and brought to Perfection many useful Arts and Sciences".[63] So now pursuing luxury became a great motor-force of civilisation.

Mandeville defined luxury as anything "not immediately necessary to make Man subsist as he is a living Creature." He noted that "fine Feathers make fine Birds, and People, where they are not known, are generally honoured according to their clothes, and other Accoutrements they have about them". This was linked to a distinction between self-love, or the desire for self-preservation, and self-liking, the propensity to "over-value ourselves". To Mandeville, "Man himself in a savage State, feeding on Nuts and Acorns, and destitute of all outward Ornaments, would have infinitely less Temptation, as well as Opportunity, of shewing this Liking of himself, than he has when civiliz'd". Thus "It is this that makes us so fond of the Approbation, Liking and Assent of others; because they strengthen and confirm us in the good Opinion we have of ourselves". This was a much more invidious tendency. Self-liking made us "seek for Opportunities, by Gestures, Looks, and Sounds, to display the Value it has for itself, superiour to what it has for others". It was indeed "so necessary to the Well-being of those that have been used to indulge it; that they can taste no Pleasure without it". It was "the Mother of Hopes, and the End as well as the Foundation of our best Wishes".[64]

Crucial here was the argument that luxury represented merely naturally expanding wants, whose economic benefits to a growing state were obvious. Luxury was anything desired beyond what mere subsistence demanded and was thus refined need. There was no "correct" ideal of consumption as such, and all the passions might be enrolled in its pursuit. This was a moral

The Enlightenment's Fable: Bernard Mandeville and the Discovery of Society (Cambridge University Press, 1994).

63. Bernard Mandeville, *The Fable of the Bees*, 2 vols (1723–28; Clarendon, 1924), 2: 132, 128.

64. Mandeville, 2: 285, 130, 136; 1: 107, 127.

revolution. Seemingly a vice, or "vicious", in the phrasing of the time, avarice was in fact "very necessary to Society". Emulation and opulence combined to make the less well-off habitually envious of the rich. Apparel was a key example. Pride induced "every Body, who is conscious of his little Merit, if he is in any ways able, to wear Clothes above his Rank, especially in large and populous Cities", where they are "esteem'd by a vast Majority, not as what they are, but what they appear to be".[65] Outward impressions were prized above all in the quest for the good opinion of others: this was indeed the dawn of the age of appearance, where image is all. So too even prodigality, "a noble sin", was an "agreeable good-natur'd Vice that makes the Chimney a smoke, and all the Tradesmen smile".[66] All excess, even in fornication and drunkenness, Britain being in the midst of the gin craze, created demand and enriched others. Prodigality, not frugality, made the world go round. What a vast "number of People", Mandeville reflected, "how many different Trades, and what a variety of Skill and Tools must be employed to have the most ordinary *Yorkshire* Cloth?"[67] No Stoic contempt for its display was thus justified. The frugality of the rich man sitting on his chest of coins inhibited the distribution of wealth, while the prodigality of the merrymaker stimulated it. The poor today "lived better than the rich before", and that was a great accomplishment. Yet Mandeville admitted that sumptuary laws might be useful in "an indigent Country, after great Calamities of War, Pestilence, or Famine, when Work has stood still, and the Labour of the Poor has been interrupted."[68] The present day, we will later see, might count as just such an exception.

In his time Mandeville's celebration of "private vices" was regarded as the apotheosis of immoralism. But his viewpoint resonated in an exuberant age excited by its own inventiveness, enamoured of the novelty of its pleasures, and increasingly prone to vindicate the force of the passions in human life. His influence was considerable, not least in France, where Jean-François Melon, Voltaire, and Montesquieu were amongst his disciples.[69] Enjoyment without guilt became fashionable, and eventually a life defined by hedonism became the dominant modern ideal. In Britain this process was assisted by the Scottish Enlightenment's leading philosopher, David Hume. Originally entitled "Of Luxury", his famous essay "Of Refinement in the Arts" (1742) contended that luxuries were merely "great refinement in the gratification of the senses",

65. Mandeville, 1: 101, 128.

66. Mandeville, 1: 103.

67. Mandeville, 1: 169.

68. Mandeville, 1: 251.

69. Spengler, *French Predecessors of Malthus*, 111.

a definition evidently indebted to Mandeville. They were harmful only "when they are pursued at the expense of some virtue, as liberality or charity", which left "no ability for such acts of duty and generosity as are required by his situation and fortune".[70] The "encrease and consumption of all the commodities" served "to the ornament and pleasure of life". Thus "Refinement on the pleasures and conveniences of life has no natural tendency to beget venality and corruption":

> Riches are valuable at all times, and to all men because they always purchase pleasures, such as men are accustomed to, and desire: Nor can anything restrain or regulate the love of money, but a sense of honour and virtue; which, if it be not nearly equal at all times, will naturally abound most in ages of knowledge and refinement.[71]

In 1751, Hume explained what an uphill battle the new idea entailed:

> Luxury, or a refinement on the pleasures and conveniencies of life, had long been supposed the source of every corruption in government, and the immediate cause of faction, sedition, civil wars, and the total loss of liberty. It was, therefore, universally regarded as a vice, and was an object of declamation to all satirists, and severe moralists. Those, who prove, or attempt to prove, that such refinements rather tend to the increase of industry, civility, and arts regulate anew our *moral* as well as *political* sentiments, and represent, as laudable or innocent, what had formerly been regarded as pernicious and blameable.[72]

Hume admitted that the impact of luxury was relative, and that "A degree of luxury may be ruinous and pernicious in a native of Switzerland, which only fosters the arts, and encourages industry in a Frenchman or Englishman."[73] Nonetheless, this was a turning point in the self-confidence and self-image of the moderns. The defence of "innocent luxury", in Hume's phrase, gave birth to the modern worship of and obsession with luxury. Refinements in gratification brought increased pleasure, and the pursuit of them spurred labour. This induced greater trade and production, which stimulated further desires for luxuries and the pleasure they brought. "Sloth and idleness" by contrast were "more hurtful both to private persons and to the public. When

70. David Hume, *Essays Moral, Political and Literary*, 2 vols (1758; Longman, Green, 1882), 1: 285–86.

71. Hume, 1: 299, 302, 305.

72. David Hume, *Enquiries concerning Human Understanding and concerning the Principles of Morals* (Clarendon, 1975), 181.

73. Hume, 337.

sloth reigns, a mean uncultivated way of life prevails amongst individuals, without society, without enjoyment."[74] Thus emerged the modern cycle of desire–work–consume–desire.

The new civility resulting from this process—at home, at least, since slavery and imperial conquest are not mentioned—had many advantages. Polite manners could neutralise humanity's tendency to be "commonly proud and selfish, and apt to assume the preference above others", for "a polite man learns to behave with deference towards his companions, and to yield the superiority to them in all the common incidents of society".[75] With Hobbes in mind, Hume thought that in its primitive or "natural" state humanity was prone to violence. By contrast, the moderns enjoyed an enviable degree of sociability. As he explained in a letter:

> By the expressions of politeness, I mean these outward deferences & ceremonies, which custom has invented, to supply the defect of real politeness or kindness, that is unavoidable towards strangers & indifferent persons even in men of the best dispositions of the world. These ceremonies ought to be so contriv'd, as that, tho they do not deceive, nor pass for sincere, yet still they please by their appearance, & lead the mind by its own consent & knowledge, into an agreeable delusion . . . The French err in the contrary Extreme, in making their civilities too remote from the truth, which is a fault.[76]

So the moderns could claim to be promoting a form of sociability which was distinctly superior. In civilised monarchies like France, where Hume spent several years, sociability might proliferate rapidly as commerce expanded. Extremes notwithstanding, this could provide much of that social order, later called "civil society", traditionally associated with the state. In the most optimistic gloss on this process, Hume famously claimed that:

> the more these refined arts advance, the more sociable men become: nor is it possible, that, when enriched with science, and possessed of a fund of conversation, they should be contented to remain in solitude, or live with their fellow citizens in that distant manner, which is peculiar to ignorant and barbarous nations. They flock into cities; love to receive and communicate knowledge; to show their wit and their breeding; their taste in conversation or living, in clothes or furniture. Curiosity allures the wise; vanity the foolish; and pleasure both. Particular clubs and societies are everywhere formed; both sexes meet in an easy and sociable manner; and the tempers

74. Hume, *Essays*, 1: 309.

75. Hume, 1: 192.

76. David Hume, *The Letters of David Hume*, 2 vols, ed. J.Y.T. Greig (Clarendon, 1932), 1: 20.

> of men, as well as their behaviour, refine apace. So that, beside the improvements which they receive from knowledge and the liberal arts, it is impossible but they must feel an increase of humanity, from the very habit of conversing together, and contributing to each other's pleasure and entertainment. Thus *industry, knowledge, and humanity* are linked together, by an indissoluble chain, and are found, from experience as well as reason, to be peculiar to the more polished, and, what are commonly denominated, the more luxurious ages.[77]

This may have been a sociability still tinged with commerce and self-interest, and thus "unsociable". But Hume's claim comes as close to a utopian account of sociability as any in the tradition usually termed "liberal". Agrarian simplicity was not virtuous, merely coarse and primitive. Commerce and civic and private virtue could be happily reconciled without frugality, because polished and urbane manners brought not merely greater sociability but greater humanity.

This ideal of sociability was wedded to the most powerful and successful of the modern visions, the "stark utopia", in Karl Polanyi's phrase, of unlimited opulence and universal plenty associated with the liberal model of the so-called "free", self-regulating market, which, supposedly without coercion, generates incessantly expanding production and consumption.[78] Drawing on the natural law tradition, the Scottish writers posited that an original period lacking private property had probably existed. This had some potentially utopian dimensions, for here, as Adam Smith conceded, "the whole produce of labour belongs to the labourer", who thus was relatively independent.[79] But it did

77. Hume, *Essays*, 1: 301–2.

78. Karl Polanyi, *Origins of Our Time: The Great Transformation* (Victor Gollancz, 1945), 13. Polanyi's central thesis is that the cataclysmic politics of the twentieth century originated "in the utopian endeavour of economic liberalism to set up a self-regulating market system" (29). To term this vision "utopian", however, given its central reliance upon private property, we would need the term to encompass a set of essentially unrealistic expectations about the improvement of human behaviour in modern commercial society, and the possibility of the production of unlimited wealth without engendering that social collapse which characterised the great example of Roman decline and fall. See, generally, Franco Venturi's *Utopia and Reform in the Enlightenment* (Cambridge University Press, 1971). Adam Smith thought it as "absurd" to expect that free trade would ever be established as any Utopia or Oceana (*An Inquiry into the Nature and Causes of the Wealth of Nations*, 2 vols (1776; Clarendon, 1869), 2: 44). But Hume's theory of greater humanity growing out of commercial sociability fits the utopian definition.

79. A. Smith, *Wealth of Nations*, 2: 44. Marx mocked this by saying that "the political economist postulates the original unity of capital and labour as the unity of the capitalist and the worker; this is the original state of paradise" (Marx and Engels, *Collected Works*, 3: 312).

not follow that returning to such a condition was possible or desirable. Land was farmed much more efficiently in the later stages of society than the earlier. This justified it having become private property as humanity progressed through the four stages—hunting and gathering, pastoral, agricultural, and commercial—which composed the Scots' main "savagery to civilisation" narrative.[80] Inequality obviously grew at the same time. But, historians like John Millar insisted, everyone eventually benefited, because in refined and luxurious nations people were "excited, with mutual emulation, to surpass one another in the elegance and refinement of their living".[81] This provided an incentive to labour lacking in earlier societies, and an energising drive to self-improvement. Mandeville had swept the field.

Smith described the movement through these stages in terms of a natural propensity defined by "The uniform, constant, and uninterrupted effort of every man to better his condition." This was "the principle from which public and national, as well as private opulence is originally derived", and was "frequently powerful enough to maintain the natural progress of things towards improvement".[82] In commercial society, the dangers of excessive private interest and selfishness were inhibited by a natural sympathy for others, coupled with other "social passions" like generosity, and a restraining sense, at least among the higher orders, of self-command, civic duty, and a willingness to sacrifice self- and class interest to the common good. Yet Smith, whose political sympathies were republican, above all "in his love of all rational liberty", as Sir James Mackintosh put it, was no vulgar apologist for the new system.[83] He feared that merchants and manufacturers had an interest "always in some respects different from, and even opposite to, that of the public", that they had "generally an interest to deceive and even to oppress the public", and had "upon many occasions, both deceived and oppressed it."[84] This indicated one major cause of potential instability in the existing system. The rich were always conspiring to undermine the "free" market for their own ends, while claiming that the same "free" market allowed them to do whatever they wanted. It is a familiar story.

80. On the origins of this account, see Istvan Hont's "The Language of Sociability and Commerce: Samuel Pufendorf and the Theoretical Foundations of the 'Four-Stages Theory'", in *The Languages of Political Theory in Early Modern Europe*, edited by Anthony Pagden (Cambridge University Press, 1987, 253–76).

81. John Millar, *The Origin of the Distinction of Ranks*, 4th ed. (1771; William Blackwood, 1806), 233.

82. A. Smith, *Wealth of Nations*, 1: 346.

83. John Rae, *Life of Adam Smith* (Macmillan, 1895), 124.

84. A. Smith, *Wealth of Nations*, 1: 264–65.

There were other problems with the new social model. Smith agreed with Mandeville that a key reason for demonstrating wealth was to attract the attention and good opinion of others. "It is", he thought,

> the vanity, not the ease, or the pleasure, which interests us. But vanity is always founded upon the belief of our being the object of attention and approbation. The rich man glories in his riches, because he feels that they naturally draw upon him the attention of the world . . . At the thought of this his heart seems to swell and dilate itself within him, and he is fonder of his wealth, upon this account, than for all the other advantages it procures him.[85]

This produced a potentially extreme subversion of the moral restraint by which alone, Smith believed, the system could avoid social crisis. He warned that the

> disposition to admire, and almost to worship, the rich and the powerful, and to despise, or, at least, to neglect, persons of poor and mean condition, though necessary both to establish and to maintain the distinction of ranks and the order of society, is, at the same time, the great and most universal cause of the corruption of our moral sentiments.

It was thus "from our disposition to admire, and consequently to imitate, the rich and the great, that they are enabled to set, or to lead, what is called the fashion". By contrast, a lamentably "small party" admired wisdom and virtue, almost to the disdain and contempt of the majority. The "great mob of mankind" worshipped wealth and power.[86] This appears much more pessimistic than Hume's account. For greed, pride, and a love of adulation now seemed more likely to outweigh the humanising effects of greater sociability, and any propensity to prioritise civic virtue or the sacrifice of private interest to the common good. Mandeville might have swept the field. But he might have also heralded the end of civility, and the commencement of a new epoch of overweening egotism.

Smith thus insisted that Stoic self-command and both personal and civic virtue required recognising the social duties of the well-to-do. The wise and virtuous man should be "at all times willing that his own private interest should be sacrificed to the public interest of his own particular order or society". Wealth was also better spent putting others to work than on trinkets or "frivolous ornaments of dress and furniture".[87] Yet the propensity of the wealthy to vain display threatened to undermine any such warnings, and to divert energy from productive to unproductive expenditure. This was not an unambiguous,

85. Adam Smith, *The Theory of Moral Sentiments* (1759; Henry Bohn, 1853), 71.

86. A. Smith, 84–85, 87.

87. A. Smith, 198; Smith, *Wealth of Nations*, 1: 184.

or blindly optimistic, vision of the future of commercial society: it was riddled with doubts.

Others at the time were still more alarmed by these trends. Smith's friend Adam Ferguson worried that growing wealth might

> serve to corrupt democratical states, by introducing a species of monarchical subordination, without that sense of high birth and hereditary honours which render the boundaries of rank fixed and determinate, and which teach men to act in their stations with force and propriety. It may prove the occasion of political corruption, even in monarchical governments, by drawing respect towards mere wealth; by casting a shade on the lustre of personal qualities, or family-distinctions; and by infecting all orders of men, with equal venality, servility, and cowardice.[88]

Ferguson also led Smith to concede that the process by which some of the new wealth was created in urban workshops might threaten the well-being of the workforce. In particular, growing specialisation meant that repetitive work so fundamentally stunted the workers as to virtually erode their civic identity and even their very humanity. Compared to "the sanguine affection which every Greek bore to his country" and "the devoted patriotism of an early Roman", in commercial society "the bands of affection are broken", and "man is sometimes found a detached and a solitary being: he has found an object which sets him in competition with his fellow-creatures, and he deals with them as he does with his cattle and his soil, for the sake of the profits they bring."[89] Ferguson insisted that "when wealth is accumulated only in the hands of the miser, and runs to waste from those of the prodigal; when heirs of family find themselves straitened and poor, in the midst, of affluence", a despotic government would result. His ideal character was largely military in nature, and he admired Sparta as the only nation which had "made virtue an object of state."[90] He worried that:

> having separated the arts of the clothier and the tanner, we are the better supplied with shoes and with cloth. But to separate the arts which form the citizen and the statesman, the arts of policy and war, is an attempt to dismember the human character, and to destroy those very arts we mean to improve. By this separation, we in effect deprive a free people of what is necessary to their safety.[91]

88. A. Ferguson, *History of Civil Society*, 255. See, generally, Ian McDaniel's *Adam Ferguson in the Scottish Enlightenment* (Harvard University Press, 2013).

89. A. Ferguson, *History of Civil Society*, 19.

90. A. Ferguson, 262, 161.

91. A. Ferguson, 230.

Such reservations, especially regarding luxury, were also wrapped up with the standard republican interpretation of the fall of Rome, as expressed by the seventeenth-century writer Algernon Sidney—namely, that "luxury and pride . . . destroyed discipline and virtue; [and] then ruin necessarily followed".[92] Then it was that virtue itself was undermined:

> when luxury was brought into fashion, and they came to be honour'd who liv'd magnificently, tho' they had in themselves no qualities to distinguish them from the basest of slaves, the most virtuous men were exposed to scorn if they were poor: and that poverty which had been the mother and nurse of their virtue, grew insupportable.

The antidote, he thought, was that

> in well-govern'd states, where a value is put upon virtue, and no one honoured unless for such qualities as are beneficial to the public, men are from the tenderest years brought up in a belief, that nothing in this world deserves to be sought after, but such honours as are acquired by virtuous actions: by this means virtue itself becomes popular, as in Sparta, Rome, and other places, where riches (which with the vanity that follows them, and the honours men give to them, are the root of all evil) were either totally banished, or little regarded.[93]

In Britain, these warnings continued in the early eighteenth century in the writings of the leading disciples of Machiavelli, John Trenchard, and Thomas Gordon. They described how in Rome:

> Pleasure succeeded in the room of temperance, idleness took place of the love of business, and private regards extinguished that love of liberty, that zeal and warmth, which their ancestors had shewn for the interest of the publick; luxury and pride became fashionable; all ranks and orders of men tried to outvie one another in expense and pomp; and when, by so doing, they had spent their private patrimonies, they endeavoured to make reprisals upon the publick; and, having before sold every thing else, at last sold their country.[94]

Other republicans, too, like Andrew Fletcher of Saltoun, bemoaned how the "luxury of all other ranks and orders of men makes every one hasten to grow

92. Algernon Sidney, *Court Maxims* (Cambridge University Press, 1996), 16.

93. Algernon Sidney, *Discourses Concerning Government*, 2 vols (G. Hamilton and J. Balfour, 1750), 1: 364–65.

94. John Trenchard and Thomas Gordon, *Cato's Letters; or, Essays on Liberty, Civil and Religious, and Other Important Subjects*, 2 vols (1720; Liberty Fund, 1995), 1: 130.

rich; and consequently leads them to betray all kind of trust reposed in them".[95] Where Rome had gone, might Britain follow?

Eighteenth-century debates about luxury could not but occasionally revert to Sparta as a reference point, though its attractions were declining. Scotsmen seem to have had a peculiarly soft spot for hard Sparta in this period, in part doubtless because they felt themselves to display one of the last vestiges of the warrior virtues in Europe. A key example is what is often regarded as the first text in political economy proper, Sir James Steuart's *Inquiry into the Principles of Political Oeconomy* (1767). This lauded Lycurgan Sparta's "most perfect plan of political oeconomy" as

> a system, uniform and consistent in all its parts. *There,* no superfluity was necessary, because there was no occasion for industry, to give bread to any body. *There,* no superfluity was permitted, because the moment the limits of the absolutely necessary are transgressed, the degrees of excess are quite indeterminate, and become purely relative. The same thing which appears superfluity to a peasant, appears necessary to a citizen; and the utmost luxury of this class frequently does not come up to what is thought the mere necessary for one in a higher rank. Lycurgus stopt at the only determined frontier, the pure physical necessary. All beyond this was considered as abusive.[96]

But did such hostility to luxury suit modern times? The answer was *no*. Steuart argued that its modern critics "instead of crying down luxury and superfluous consumption, ought rather to be contriving methods for rendering them more universal." Equality, he insisted, would not prevent luxury, unless "every one confined to an absolute physical necessary, and either deprived of the faculty of contriving, or of the power of acquiring any thing beyond it."[97] Luxury, rather, could be the means of diffusing wealth, by creating a desire for goods and thus preventing money from being hoarded. If a way could be found to export luxuries, thus stimulating manufactures while also endeavouring to "encourage oeconomy, frugality, and a simplicity of manners, discourage the consumption of every thing that can be 'exported', and excite a taste for superfluity in neighbouring nations", it might be possible to wed "all the advantages of antient simplicity to the wealth and power which attend upon the luxury of modern states".[98] By such means, he thought,

95. "An Account of a Conversation concerning the Right Regulation of Governments", in Andrew Fletcher, *Select Political Writings and Speeches*, ed. David Daiches (Academic Press, 1979), 100.

96. Steuart, *Inquiry*, 1: 275.

97. Steuart, 2: 46–47.

98. Steuart, 1: 278–79.

> elegance of taste, and the polite arts, may be carried to the highest pitch. The whole of the inhabitants may be employed in working and consuming; all may be made to live in plenty and in ease, by the means of a swift circulation, which will provide a reasonable equality of wealth among all the inhabitants. Luxury can never be the cause of inequality, though it may be the effect of it. Hoarding and parsimony form great fortunes, luxury dissipates them and restores equality.[99]

Following Hume, Steuart asserted that:

> *excess* in consumption is vicious, in proportion as it affects our *moral, physical, domestic,* or *political* interests; that is to say, our *mind,* our *body,* our *private fortune,* or the *state.* When the consumption we make does no harm in any of these respects, it may be called moderate and free from vice.[100]

Throughout this period, indeed until the early twentieth century, the fall of Rome remained the great example of the fatal consequences of neglecting such advice. Its most famous chronicler, Edward Gibbon, regarded the second century CE as the happiest in history, and reminded generations of readers that in Rome "The vigour of the soldiers, instead of being confirmed by the severe discipline of camps, melted away in the luxury of cities".[101] From now onwards, warnings about decadence as a cause of internal decline accompanied the rise and flourishing of every major empire, and notably the progress of Victorian conquest. By the end of the nineteenth century, most European societies had sybaritic and Spartan extremes. But prior to the French Revolution one man did more to sustain the memory of Sparta than anyone else: Jean-Jacques Rousseau.

Rousseau and Utopia

The debate about humanity's original state was central to seventeenth-century natural law writers like Thomas Hobbes, Samuel Pufendorf, and Hugo Grotius. To Pufendorf, like Hobbes, this condition, while superficially "attractive

99. Steuart, 1: 343.

100. Steuart, 2: 187. For commentary, see S. R. Sen's *The Economics of Sir James Steuart* (G. Bell and Sons, 1957, 35–37).

101. Edward Gibbon, *The Decline and Fall of the Roman Empire*, 3 vols (Modern Library, 1932), 1: 119. By 219 CE, Gibbon commented, "Rome was at length humbled beneath the effeminate luxury of Oriental despotism" (1: 219). But he accepted the Mandevillian argument that, for the moderns, "luxury, though it may proceed from vice or folly, seems to be the only means that can correct the unequal distribution of property" (1: 48).

in promising liberty and freedom from all subjection", was also "attended with a multitude of disadvantages", including "war, fear, poverty, nastiness, solitude, barbarity, ignorance, savagery".[102] But this did not mean that civilised life was without hazard. While not hostile to commerce, Pufendorf echoed Stoic reservations in warning that "Life is far better spent with modest and sober trappings, and with a manly body hardened by patience to endure adversities, than with excessive resources that burden rather than sustain us, or in tender luxury that exasperates even mild discomforts."[103] He thought it legitimate that "Idleness has to be banished; and the citizens recalled to habits of economy by sumptuary laws which prohibit excessive expenditures, especially those by which the citizens' wealth is transferred abroad".[104]

Conceived as a contest between luxury and simplicity, the eighteenth-century debate about utopia centres on one writer more than any other: Thomas More's "last and most significant heir", in Judith Shklar's acute reckoning, Jean-Jacques Rousseau (1712–78). Sometimes called "the first philosopher of the Greens", Rousseau admired Fénelon and Defoe.[105] His bold *Discourse on the Origins of Inequality* (1755) provoked a torrent of debate in positing that an ideal past stage, "nascent society", had indeed existed.[106] This was not, however, the original and most primitive state of mankind, which, contra Hobbes, he believed was not a condition of strife but one of sufficient inconvenience and unsociability not to be ideal. Humanity's happiest epoch was not that of the "noble savage", a phrase long associated with Rousseau, but which he never used.[107] It was, rather, the earliest state of society, when small villages and towns existed but before private property, the use of metals, and specialisation of functions generated great social inequality. This stage at points resembled

102. Samuel Pufendorf, *On the Duty of Man and Citizen according to Natural Law*, ed. James Tully (1673; Cambridge University Press, 1991), 117.

103. Samuel Pufendorf, *On the Natural State of Man* (1678; Edwin Mellen, 1990), 117.

104. Samuel Pufendorf, *Duty of Man*, 154. See, generally, Stephen Buckle's *Natural Law and the Theory of Property: Grotius to Hume* (Clarendon, 1991).

105. Judith Shklar, *Political Thought and Political Thinkers* (University of Chicago Press, 1998), 179; Maurice Cranston, *The Noble Savage: Jean-Jacques Rousseau, 1754–1762* (University of Chicago Press, 1991), ix.

106. The context for understanding this debate is provided in Hont's *Jealousy of Trade* and in Michael Sonenscher's *Sans-Culottes: An Eighteenth Century Emblem in the French Revolution* (Princeton University Press, 2008). On Rousseau and utopia, see also James F. Jones Jr's *La Nouvelle Héloïse: Rousseau and Utopia* (Librarie Droz, 1977).

107. See Hoxie Neale Fairchild, *The Noble Savage: A Study in Romantic Naturalism* (Cornell University Press, 1928); Whitney, *Primitivism*; Stelio Cro, *The Noble Savage: An Allegory of Freedom* (Wilfrid Laurier University Press, 1990).

classical Sparta, which Rousseau championed against his enemy Voltaire's Athenian model and his blanket vindication of luxury and private consumption, as expressed in his poem *Le mondain* (The man of the world) (1736). This satire on Fénelon, the Garden of Eden, and Christian and Stoic virtue generally was probably influenced by Mandeville and certainly by Jean-François Melon, another of Rousseau's targets. Sometimes seen as wedded to deism, it is usually regarded as a turning point in the debate about luxury in France.[108]

Nonetheless, Rousseau's supposed praise for the primitive state as a "lost paradise", and belief that all social institutions should aim at "the restoration of the liberty and equality of the state of nature", in Gierke's analysis, "spread like wildfire".[109] Rousseau soon stood for a kind of primitive alterity to the increasingly frenzied pace of European urban life, a critical psychological other, and mode of conscience, which as we saw above is one essential function of the concept of utopia. So James Boswell commented to Samuel Johnson: "You are tempted to join Rousseau in preferring the savage state. I am so too at times. When jaded with business, or when tormented with the passions of civilized life, I could fly to the woods".[110] The "savage" was already a mental trope of psychological retreat, and a state of mind. Only much later would there arise a counter-narrative of the "savage mind", in which a far less appealing portrait of primitive life was drawn, corresponding to older images of indigenous peoples as sodomites and cannibals. Both images are regarded usually today as mere prejudices externalised upon the other, and more indicative of the "European mind" than anything else. Even today we often retreat to nature to calm ourselves down and clear out the clutter of everyday life, without ever intending to abandon the bustle of normality.

108. See Voltaire, *Oeuvres complètes* (Voltaire Foundation, 2003), 16: 269–314, here 286, 279: "L'apologie du luxe est un article du deisme". Ernst Cassirer claimed that Voltaire later "expressly retracts his glorification of pleasure" in light of the Lisbon earthquake of 1755: *The Philosophy of the Enlightenment* (Princeton University Press, 1951), 147. In Voltaire's *Candide* (1759), a short utopian section on El Dorado similarly portrays affluence for the many, where "land had been cultivated as much for beauty as from necessity, for everywhere the useful was joined to the agreeable", and where children "covered with tattered garments of gold brocade" play quoits with gold, emeralds and rubies. When we find them being served six hundred humming birds in a single dish—and this for free, however, we begin to suspect a satire. And then we meet a man aged 172 . . . (*Candide; or, Optimism* (Penguin Books, 2005), 43–44).

109. Otto Gierke, *Natural Law and the Theory of Society, 1500 to 1800*, 2 vols (Cambridge University Press, 1934), 1: 101.

110. Quoted in Arthur O. Lovejoy, *Essays in the History of Ideas* (Johns Hopkins University Press, 1948), 38.

The battle lines were now drawn for one of the greatest conflicts in the history of ideas. But much more than a reading of the past was at stake here. At one level this was also a contest for the soul of the moderns. It is important to note the psychology associated with Rousseau's ideal state. Society proper commenced, he thought, when we began to see ourselves, as Mandeville had described, principally through the eyes of others. Our natural desire for self-protection, or *amour de soi-même*, was now transformed into *amour-propre*, a self-love formed on our vain desire for others' admiration.[111] To Rousseau this was a kind of original sin of psychological corruption, which bifurcated the self into warring factions. "What good is it to seek our happiness in the opinion of another", he asked in exasperation, "if we can find it within ourselves?"[112] This became the chief source of artificial inequality, as we come to seek, through pride, to surpass others, or at least to live vicariously in that image of ourselves which we think others have of us, which mischievously accentuates our best attributes and overlooks those faults we endeavour to conceal. Yet of course this process also fuels our desire for sociability. Rousseau was well aware of Mandeville's paradox. Sociability involves a desire for mutual recognition, but this entails recognising false, partial, inauthentic, pretentious, and merely external selves, our decorative stage masks, rather than any "real" or true self. It thus encourages dissimulation, fraudulence, insincerity, and artificiality. The moderns were slaves in their dependency on the opinion of others. And they were hollow, for what lay within was nothing but the rotted remnant of original authenticity. Romanticism would often identify the latter with feelings or sentiments, though doubts would creep in as to whether the "true" self ever existed, or whether, instead, the personality was very largely a social construct.

Parallel to the psychological development came the origins of social inequality. In some of the Enlightenment's most famous passages, Rousseau offered his secularised version of humanity's fall from the Golden Age and paradise. Degeneration commenced with private property: "The first man who, having enclosed a piece of ground, bethought himself of saying 'This is mine', and found people simple enough to believe him, was the real founder of civil society."[113] With the division of labour, agriculture, and the discovery of

111. Maurice Cranston, *Jean-Jacques: The Early Life and Work of Jean-Jacques Rousseau, 1712–1754* (Norton, 1982), 301.

112. Jean-Jacques Rousseau, *Discourses on the Sciences and Arts and Polemics* (University Press of New England, 1992), 21. See Edward Hundert, "Mandeville, Rousseau, and the Political Economy of Fantasy", in *Luxury in the Eighteenth Century*, eds Maxine Berg and Elizabeth Eger (Palgrave, 2003), 28–40.

113. Rousseau, *Social Contract and Discourses*, 84.

metals, new wants arose. These produced mutual dependency. Humanity now surrendered their independence:

> so long as they undertook only what a single person could accomplish, and confined themselves to such arts as did not require the joint labour of several hands, they lived free, healthy, honest, and happy lives, in so far as their nature allowed, and they continued to enjoy the pleasures of mutual and independent intercourse. But from the moment one man began to stand in need of the help of another; from the moment it appeared advantageous to any one man to have enough provisions for two, equality disappeared, property was introduced, work became indispensable, and vast forests became smiling fields, which man had to water with the sweat of his brow, and where slavery and misery were soon seen to germinate and grow up with the crops.
>
> Metallurgy and agriculture were the two arts which produced this great revolution. The poets tell us it was gold and silver, but, for the philosophers, it was iron and corn, which first civilized men, and ruined humanity.[114]

This critique, and his powerful defence of popular sovereignty *The Social Contract* (1762), made Rousseau the thinker who had the greatest intellectual influence on the French Revolution. What he stood for, above all, was equality. It is not just an end in itself, but also a means of achieving a more "real", "authentic", and substantial form of sociability. This also required, as *Emile* emphasised, not having "vain and useless" toys as a child, generally avoiding luxury, and keeping "as close as possible to nature".[115] Vanity was the greatest enemy. It was, Rousseau thought, "not so much the luxury of softness that ruins as the luxury of vanity. This luxury, which does not turn to anyone's good, is the true scourge of society".[116] But in *Emile* he nonetheless conceded to his model wife, Sophy, an indulgent coquettishness which included a love of clothing, if not expensive and extravagant dress.[117]

Rousseau recognised, indeed, that competitive emulation could not be eliminated. Instead it needed to be channelled into rewarding public virtue. He suggested how this might be done in 1772 in designing a new constitution for Poland. This proposed dividing the nation into thirty-three small states,

114. Rousseau, 92.

115. Jean-Jacques Rousseau, *Emile* (1762; J. M. Dent, 1911), 36, 311.

116. Jean-Jacques Rousseau, *The Plan for Perpetual Peace, On the Government of Poland, and Other Writings on History and Politics* (University Press of New England, 2002), 18.

117. Rousseau, *Emile* (1911 ed.), 335.

thus preserving a sense of local civic identity akin to Sparta's. He outlined a system of competitions, exercises, and uniforms for a citizens' militia, all designed to promote civic pride. Here

> by honours and public rewards, all the patriotic virtues should be glorified, that citizens should constantly be kept occupied with the fatherland, that it should be made their principal business, that it should be kept continuously before their eyes. In this way, I confess, they would have less time and opportunity to grow rich; but they would also have less desire and need to do so. Their hearts would learn to know other pleasures than those of wealth. This is the art of ennobling souls and of turning them into an instrument more powerful than gold.[118]

Rousseau also insisted that in frequent open-air festivals

> the various ranks of society will be carefully distinguished . . . the people should be together with their leaders on pleasurable occasions, that they should know them, that they should be accustomed to seeing them, and that they should often share their pleasures. Provided that subordination is always preserved, and that distinctions of rank are not lost sight of, this is the way to make them love their leaders, and to combine respect with affection.[119]

Leaders, however, were not to be rewarded with private wealth. Rousseau regarded "one of the most important functions of government" as being "to prevent extreme inequality of fortunes". His *Discourse on Political Economy* (1755) commended sumptuary laws to stem the concentration of wealth, and the pursuit of frivolities and "the arts that minister to luxury", so that the "purely superfluous arts" were not favoured "at the expense of useful and difficult trades".[120] Heavy taxes, he thought, "should be laid on servants in livery, on equipages, rich furniture, fine clothes, on spacious courts and gardens, on public entertainments of all kinds, on useless professions, such as dancers, singers, players, and in a word, on all that multiplicity of objects of luxury, amusement, and idleness, which strike the eyes of all".[121] But attitudes and mores were more important than laws. Elsewhere, Rousseau said that the Greeks' "profound contempt for Asian ostentation was the best sumptuary law they could have had", a disdain which "was even more palpable among the

118. Jean-Jacques Rousseau, *Political Writings*, ed. Frederick Watkins (Nelson, 1953), 169.

119. Rousseau, 172–73.

120. Rousseau, *Social Contract and Discourses*, 147.

121. Rousseau, 166.

Romans". He lamented the fact that "When these Peoples began to degenerate, when vanity and love of pleasure succeeded love of fatherland and virtue, then vice and softness made their way everywhere, and the only problem was Luxury and the money to provide for it."[122]

An outlook defined by simplicity was thus least corrosive of humanity. In *Julie; or, The New Heloise* (1761), Rousseau commended a household ideal of moderation, defined not by "the inconvenient and vain display of luxury, but plenty, the true comforts of life, and the necessities of needy neighbors". Here, "plenty and elegance" and "pleasure and sensuality without refinement or indolence" did not require "wealth and luxury".[123] Simplicity would prevail in clothing and manners.[124] Inequality undermined that "ardent love of country" which had raised the Spartans "above the level of humanity."[125] Rousseau advised the Poles to remember that "immense disparities of fortune . . . constitute a great obstacle to the reforms needed to make love of country the dominant passion."[126] Sumptuary laws would not suffice to stifle the love of luxury. It was only "from the depth of the heart itself that you must uproot it by impressing men with healthier and nobler tastes . . . Simplicity of manners and adornment is the fruit not so much of law as of education."[127] Those who encouraged material luxury would find that "luxury of spirit" was "inseparable from it", and would "create a scheming, ardent, avid, ambitious, servile and knavish people."[128] The key here was to "pay little attention to foreign countries, give little heed to commerce; but multiply as far as possible your domestic production and consumption of foodstuffs." As luxury and indigence disappeared, so would "the frivolous tastes created by opulence" and "the vices associated with poverty."[129] He reminded the Corsicans, too, that self-sufficiency was rooted in agriculture, "the only means of maintaining the external independence of the state." They were recommended to have no large towns, or luxuries like carriages, and to enact sumptuary laws to this end, applying them more severely to leaders and more leniently to the poor.[130]

122. Jean-Jacques Rousseau, *Social Contract* (1762; University Press of New England, 1994), 45–46.

123. Jean-Jacques Rousseau, *Julie; or, The New Heloise* (University Press of New England, 1997), 305, 363, 435.

124. See D. Roche, *Culture of Clothing*, 417–20.

125. Rousseau, *Political Writings*, 164.

126. Rousseau, 174.

127. Rousseau, 175–76.

128. Rousseau, 224.

129. Rousseau, 231.

130. Rousseau, 283.

Rousseau exposed to a greater degree than anyone else the seemingly insuperable problem as to how to reform societies already afflicted by the contagion of luxury. But while he flirted with both, he ultimately rejected both the idea of the Golden Age and that of returning to anything like the Spartan model. However tantalising, utopia could never be reached again by the most developed of the moderns. But some might come closer than others. His hesitancy was sufficient to satisfy both sides of the debate. At one point he suggested that the moral standards of Sparta and the early Roman Republic were "not beyond our reach".[131] Elsewhere he hinted that while a few Corsican and Swiss shepherds might retain such manners, "a vicious people can never return to virtue". Years later he added that "One can never go back to the time of innocence and equality, once one has departed from it".[132] For the rest, hope was difficult to muster, though a reformed Poland presented one possibility of approaching as close to Sparta as modernity permitted. Such intermediate stages might halt further degeneration. Once inequality was rife, the law could not "despoil any private citizen of any part of his property", it could only "prevent him from acquiring more".[133] But this principle too would soon be challenged.

After Rousseau

This debate raged throughout the later eighteenth century.[134] Some, like Étienne Bonnot de Condillac, writing in the 1770s, agreed that the simple life where all were employed and enjoyed a decent subsistence with little inequality was much the best. Luxury could be enjoyed only by the few; the peasants who came seeking it in the city found only unemployment and misery, and the land was depopulated to boot. As France led Europe in producing luxury goods, however, it also had ample defenders of their

131. "Voila ce que firent admirablement les lois a Lacedemone et les moeurs chez les premiers Romains, d'ou je conclus que la chose n'est point impossible": Jean-Jacques Rousseau, *Oeuvres complètes*, 5 vols (Gallimard, 1959–95), 3: 503.

132. Cranston, *Jean-Jacques*, 243. To Judith Shklar, Rousseau is closest to the Spartan model in his "Discourse on Political Economy": *Men and Citizens: A Study of Rousseau's Social Theory* (Cambridge University Press, 1969), 16. In *Emile* he commented that "Men say the golden age is a fable; it always will be for those whose feelings and taste are depraved. People do not really regret the golden age, for they do nothing to restore it. What is needed for its restoration? One thing only, and that is an impossibility; we must love the golden age" (438).

133. Rousseau, *Political Writings*, 322.

134. See, generally, Henry C. Clark's *Compass of Society: Commerce and Absolutism in Old-Regime France* (Rowman and Littlefield, 2010).

consumption. In 1770 the historian Raynal defended a pro-commercial position in contending that:

> Whatever may be said in praise of the Spartans, the Egyptians, and other distinct nations, who have owed their superior strength, grandeur, and permanency, to the state of separation in which they kept themselves; mankind has received no benefit from these singular institutions. On the contrary, the spirit of commerce is useful to all nations, as it promotes a mutual communication of their productions and knowledge.[135]

Besides Voltaire, such views were upheld by Jean-François Melon, who followed Mandeville in asserting that "Societies are removed from the Condition of Savages, only in Proportion to the Greater, and more general Conveniencies they procure to themselves." Trade stimulated demand and helped employ the poor, while sumptuary laws were a "Restraint upon Liberty" which tended to "reduce Workmen into dangerous Idleness, and take away a new Motive to Labour". "What Matter is it to a State", he thought, "if, through a foolish Vanity, a particular person ruineth himself, by vying with his Neighbour, in Equipage? It is a punishment he well deserveth."[136]

Others, like the philosopher Jean-François de Saint-Lambert, aspired to a middle ground between "the prejudices of Sparta and those of Sybaris."[137] The Baron d'Holbach (1723–89) reasoned that luxury resulted from the natural progress of desire, so that each person "every day invents a thousand new wants and discovers a thousand new modes of satisfying them". So Rousseau's pleas on behalf of simplicity were useless. Yet Holbach too warned of the risks of "commerce born of avidity", and the tendency of luxury to fuel greed, war, and exploitation.[138] Stronger opposition still came from the philosopher Claude Helvétius (1715–71), who proposed separating luxury from its cause, inequality. "Mankind, slothful by nature, cannot be drawn from his repose but by a powerful motive"—namely, superfluities called luxuries—he reasoned. Luxury thus helped prevent "a stagnation which is detrimental to society." To be avoided, however, was "that destructive luxury which produces intemperance, and

135. Raynal, *Philosophical and Political History*, 1: 122.

136. Jean-François Melon, *A Political Essay upon Commerce* (1734; 1738), 37, 188–89.

137. Quoted in Mark Hulliung, *The Autocritique of Enlightenment: Rousseau and the Philosophes* (Harvard University Press, 1994), 141.

138. [Baron d'Holbach], *System of Nature* (1770; B. D. Cousins, 1841), 2; Spengler, *French Predecessors of Malthus*, 143, 252; quoted in Richard Whatmore, *Republicanism and the French Revolution: An Intellectual History of Jean-Baptiste Say's Political Economy* (Oxford University Press, 2000), 27.

above all, that avidity of wealth, the corruptor of the manners of a nation, and forerunner of its ruin":[139]

> Luxury is not therefore injurious as luxury, but as the effect of a great disparity in the wealth of individuals. Accordingly, luxury is never carried to excess, when there is not too great inequality in the distribution of riches; it increases in proportion, as these are confined to fewer individuals, and arrives at its utmost height when a nation divides itself into two classes, one abounding in superfluities, the other wanting in necessaries.[140]

In countries where the wealthy got rich by exploiting the masses, they became degenerate, debased, and corrupt. The interest of "the order of great men" was "to be unjust with impunity; it is to stifle in the hearts of men every sentiment of equity." Here "the excessive power of the great" led to a "consequent contempt in which they hold their fellow-citizens."[141] As a result, Helvétius claimed, the very "nations most celebrated for their luxury and *police,* should be the very countries where the majority of the inhabitants are more unhappy than the savage nations, which are held in such contempt by the civilized."[142] But he also conceded that "luxury, when it is arrived to a certain pitch, renders it impossible to restore an equality between the fortunes of individuals". Accordingly "the epocha of the greatest luxury of a nation is generally the epocha preceding its fall and debasement."[143]

Helvétius thus proposed a scheme of "national luxury" to promote both equality and opulence, which included an equal division of labour. "This distribution", he insisted, "does not permit the citizens to live in the pomp and intemperance of a nabob, but in a certain state of ease and luxury, when compared with the citizens of another country", a situation he thought the English peasantry had reached in the 1760s. It was not the poorest and most frugal nations which were most free and virtuous, but those "whose riches are most equally divided".[144] Corruption of manners, meaning "nothing more than the division between public and private interest", happened when "all the riches and power of a state are collected into a few hands." Thus, he thought, "it is necessary that all the individuals be equally employed, and forced to concur equally in the general good, and that the labour be equally divided among

139. Claude Helvétius, *A Treatise on Man*, 2 vols (Albion, 1810), 2: 83–84. Published posthumously, this work was probably written in the 1760s.

140. Claude Helvétius, *De l'Esprit; or, Essays on the Mind* (1758; 1810), 14–16.

141. Helvétius, *Treatise on Man*, 2: 89.

142. Helvétius, *Essays on the Mind*, 17–18.

143. Helvétius, 16–17, 20.

144. Helvétius, *Treatise on Man*, 2: 85–86.

them".[145] So the problem was not luxury as such, it was inequality. And draconian, if not Lycurgan, steps were required to counter it.

Such observations led Isaiah Berlin to class Helvétius amongst the proto-totalitarian "enemies of freedom".[146] Yet Helvétius also thought that "Luxury . . . is in most countries the immediate and necessary effect of despotism", where "the riches of the whole nation" were "swallowed up by a small number of families." Once established, "Arbitrary power . . . only serves to hasten the unequal partition of the riches of a nation". It was, "therefore, despotism, that the enemies of luxury should oppose".[147] Resolutely opposed to "unlimited power", Helvétius was keen to promote "a reciprocal dependence between all the orders of citizens". There were, he thought, "a multitude of ways of preventing a too speedy accumulation of wealth in a small number of hands, and of checking the too rapid progress of luxury".[148] Progressive taxation could limit landholdings, while sumptuary laws might suppress superfluous consumption, since "the rich then having no longer the free use of their money, it will appear to them less desirable, and they will make fewer efforts to obtain it." "Large consumptions", not "the wants of the people", should always be the focus of taxation.[149] Yet Helvétius acknowledged that "In the present condition of Europe, all inquiry of this kind may appear superfluous. Whatever may be said, the French, English, and Dutch, will never be induced to throw their gold into the sea." Despair lingers behind his reflection that "The Lacedaemonians . . . without commerce and without money, were nearly as happy as a people could be." They had learned that "the temperate man is, at the end of the year, at least as happy as the glutton".[150] Their wants gratified, they repaired to the amphitheatre for sport and conversation, and found this the most agreeable of "all occupations proper to fill up *the interval between a satisfied and a rising want*". Divided into thirty republics, mutually supporting one another, and having "adopted the laws and manners of the Spartans", France too might avoid both "the invasion of foreigners and the tyranny of their countrymen."[151]

The appeal if impracticability of a Spartan solution was echoed by other French writers. Well acquainted with the leading English republican authors, Gabriel Bonnot de Mably (1709–85) endorsed the Spartan ideal enthusiastically,

145. Helvétius, 2: 88–89.

146. Isaiah Berlin, *Freedom and Its Betrayal: Six Enemies of Human Liberty* (Chatto and Windus, 2002), 11–26.

147. Helvétius, *Treatise on Man*, 2: 108, 106.

148. Helvétius, 2: 90, 110.

149. Helvétius, 2: 111, 139.

150. Helvétius, 2: 11, 115, 114.

151. Helvétius, 2: 114, 301.

dismissed luxury as incompatible with morality and martial courage, condemned foreign trade in anything but necessaries, and endorsed abolishing private property.[152] Yet Mably thought that while "luxury destroyed equality in Sparta and Rome", it had had a more positive effect in modern France by helping redistribute wealth.[153] Returning to Spartan mores, he concluded, was only an "agreeable dream" and a "chimera", given the moderns' "corruption" and the "degradation of our customs and manners". Spartan ideals did not suit larger states, which required a mixed system of government.[154]

Agrarian utopian ideas were also associated with Étienne-Gabriel Morelly (1717–78) and Nicolas-Edme Restif de la Bretonne (1734–1806). Property in land was their target; inequality of wealth generated by commerce and industry was not yet the central issue, as it would be after 1820. Morelly's prose-poem the *Basiliade* (1753) foresaw the perfectibility of society in ending private property. His *Code de la Nature* (1755) hoped that in future children would "be trained to behave good-naturedly toward their fellows, to seek their friendship, and never to lie", and insisted that commerce should not be allowed to introduce private property into the republic, the greatest of all vices being "avarice . . . the desire to have".[155] Bretonne's schemes concentrated on limiting luxury, defined as "that which gives rise to occupations which bring the worker a profit for sterile labor". In Paris, he thought, luxury "reaches its murderous hand even to the lowest class", and he warned that "such occupations, confined within the State, are deadly and destructive, that a country which employs a great number of hands in the building of vainglorious edifices must become impoverished". Accordingly, he advised that "luxury and its manufactures should be tolerated by the government only insofar as they keep the population from procuring them abroad". The "perpetuation of the inequality of riches" was "always antithetical to sound morals and to the true happiness of the human race". Thus, he concluded, "Only the abolition of every disastrous luxury will

152. See Rachel Hammersley, *The English Republican Tradition and Eighteenth-Century France* (Manchester University Press, 2010), 86–98.

153. Johnson Kent Wright, *A Classical Republican in Eighteenth Century France: The Political Thought of Mably* (Stanford University Press, 1997), 33; and, generally, Peter M. Jones's "The 'Agrarian Law': Schemes for Land Redistribution during the French Revolution" (*Past and Present* 133 (1991): 96–133).

154. J. Wright, *Classical Republican*, 103.

155. Étienne-Gabriel Morelly. *Naufrage des isles flottantes; ou, Basiliade du célèbre Pilpai* (1753), in Manuel and Manuel, *French Utopias*, 112; Morelly, *Code de la nature*, ed. Gilbert Chinard (1755; 1950), 86, 121.

put an end to vice."[156] Bretonne aimed at the greatest possible social equality, writing that:

> Through our sense of equality, our community feeling, moral behavior is uniform and public. We practice virtue in a body. We reject vice in a body. Laziness, uselessness, sumptuary excess, or luxury—all this becomes impossible among us. No man can gorge himself in a public assemblage of his fellow-citizens. Each one takes only what he needs.[157]

This he thought might be accomplished by "diminishing all immense fortunes", helping the peasants become proprietors, imposing heavy taxes on luxury, encouraging popular entertainments and festivals for young and old alike, and abolishing prostitution by rendering all sexual relations honourable.[158] His utopia also portrayed a "complex system of prizes consisting of differently colored clothing, badges, ribbons, and crowns", indicating how status and achievement could be symbolically acknowledged without disturbing social equality.[159]

These themes were also taken up by the French political economists known as the physiocrats, who worried that the luxury trades diminished the amount of food available to the poor. They included Anne Robert Jacques Turgot (1727–81), the Marquis de Mirabeau, and François Quesnay (1694–1774), of whom Fénelon is regarded as a precursor. Their key concern was averting the transfer of wealth from agriculture to manufactures and imported foreign luxuries, especially for personal ornamentation, or conspicuous consumption in decoration (*faste de decoration*) to the detriment of native produce. Neither manufactures nor trade were in their view productive of wealth. They strongly condemned the propensity of the great landed magnates to dissipate their fortunes on idle, vain fripperies and amusements in the capital. Foreign trade, in particular, Quesnay regarded as at best a "necessary evil".[160] He feared that "an opulent nation which indulges in excessive luxury in the way of ornamentation can very quickly be overwhelmed by its sumptuousness".[161] Another physiocrat, Nicolas Baudeau, defined luxury as "that subversion of the natural and essential order of national expenditure which increases the total of

156. Nicolas-Edme Restif de la Bretonne, *Les Nuits de Paris; or, The Nocturnal Spectator* (1739; Random House, 1964), 108–9.

157. In Manuel and Manuel, *French Utopias*, 173.

158. Restif de la Bretonne, *The Corrupted Ones* (1776; Neville Spearman, 1967), 200–201.

159. Mark Poster, *The Utopian Thought of Restif de la Bretonne* (New York University Press, 1971), 79.

160. Max Beer, *An Inquiry into Physiocracy* (George Allen and Unwin, 1939), 62.

161. François Quesnay, *Quesnay's Tableau Économique* (Macmillan, 1972), ij.

unproductive expenditure to the detriment of that which is used in production".[162] The moral of the story was that "no encouragement at all" should be "given to luxury in the way of ornamentation; for this is maintained only to the detriment of luxury in the way of subsistence, which sustains the market for raw produce, its proper price, and the reproduction of the nation's revenue".[163] Luxury in subsistence (*faste de consummation*), then, was to be promoted, but not in ornamentation. Desiring better food encouraged food production, and indeed might rightly be regarded, Ronald Meek writes, as a "patriotic type of expenditure".[164]

Adam Smith too shared a marked bias against "unproductive labour", which made the *Wealth of Nations* appear anti-aristocratic to some. In early nineteenth-century Britain such arguments would find some support, notably in William Spence's *Britain Independent of Commerce* (1807), which defended luxury insofar as it allowed the poor to benefit, in nations whose wealth resulted from their own national produce, but condemned the procuring of "certain luxuries, which we could do very well without . . . in exchange for which we give much more valuable necessaries". (He suggested that Britons drink home-made wine and abstain from tea and silk.)[165] This provoked varying responses from Robert Torrens, James Mill, Simon Gray, and William Cobbett. Another view was taken by William Paley, who warned that luxury limited marriage, particularly when the middle classes began aping their betters. Its use thus should be restricted to the elite, such that "the condition most favourable to population is that of a laborious, frugal people, ministering to the demands of an opulent, luxurious nation".[166]

The American Revolution intensified these debates greatly. The New World had less experience of luxury, and greater pride in the puritan virtues. Travellers like the Frenchman Brissot de Warville, who commented that from "inequality result envy, the taste for luxury, ostentation, an avidity for gain, the habit of mean and guilty measures to acquire wealth", congratulated the Americans on their freedom from such vices, while noting their creeping presence

162. Ronald Meek, *The Economics of Physiocracy* (George Allen and Unwin, 1962), 316.

163. Quesnay, *Tableau Économique*, appendix A.

164. Meek, *Economics of Physiocracy*, 317. See, generally, John Shovlin's *The Political Economy of Virtue: Luxury, Patriotism, and the Origins of the French Revolution* (Cornell University Press, 2006), 13–48, and Sonenscher's *Before the Deluge*.

165. William Spence, *Tracts on Political Economy* (Longman, Hurst, Rees, Orme, and Brown, 1822), 6.

166. William Paley, *The Principles of Moral and Political Philosophy* (1786; Baldwyn, 1819), 465.

in cities like Philadelphia.[167] In the colonies, both millenarianism and republicanism were appealing. The practicalities of settling a new continent also encouraged greater co-operation, sometimes with the direct inspiration of Thomas More, whose *Utopia* was carried by Sir Humphrey Gilbert, a late sixteenth-century explorer, with an eye to implementing its principles.[168] The sharing of goods had occurred at Jamestown in 1607. Labour was initially communal at the Plymouth Pilgrim colony (1620–91), and landed property was held in common for seven years before being divided. Some, like the preacher Robert Cushman, argued for a more permanent communism, blaming Satan for bringing "this particularizing first into the world," which led some to wish to "live better than thy neighbour".[169] Proposals for agrarian laws also emerged occasionally; the revolutionary general Charles Lee urged limiting landholdings to 5000 acres.[170] Some colonies, like those which became Pennsylvania and Massachusetts, had overtly utopian agendas from the outset—William Penn termed his a "holy experiment".[171] Godliness sometimes got the upper hand entirely. The so-called Great Awakening of the 1740s witnessed a substantial period of intense revivalism when apocalyptic and millenarian themes were common. In subsequent decades, American preachers happily adapted puritan ideas of the New Israel to wed millenarian themes to a civil theology or political religion which has been called "civil millennialism". Here, the colonists' struggle simultaneously heralded Christ's return and proclaimed "America as a new seat of liberty".[172] American exceptionalism was already rampant before the revolution. Already in 1765 John Adams stated that "I always consider the settlement of America as the opening of a grand scheme and design in Providence for the illumination and emancipation of the slavish part of mankind all over the earth."[173]

These fantasies gained momentum in the years before 1789. The Revolution of 1776 was widely interpreted in millenarian terms, even if "the boundaries between millennialism, civic republicanism, and the secular utopianism of the Enlightenment were often vague", and "the distinction between secular and

167. J. P. Brissot de Warville, *New Travels in the United States of America, Performed in 1788* (1792; Woodstock Books, 2000), 174.

168. Kumar, *Utopianism*, 68.

169. Arthur Lord, *Plymouth and the Pilgrims* (Houghton Mifflin, 1920), 170–71.

170. Charles Lee, *Memoirs of the Life of the Late Charles Lee* (J. S. Jordan, 1792), 75.

171. Quoted in K. Thomas, "Utopian Impulse", 30.

172. Nathan O. Hatch, *The Sacred Cause of Liberty: Republican Thought and the Millennium in Revolutionary New England* (Yale University Press, 1977), 23–24.

173. Quoted in Hannah Arendt, *Between Past and Future* (Faber and Faber, 1961), 176.

religious utopianism is difficult to make".[174] The historian Adam Zamoyski comments that some thought the new nation heralded "an entirely new kind of human polity, and "seemed to be on the point of bringing about the chiliastic dream of a utopian state on earth, to make up for the paradise which the children of the Enlightenment no longer believed in."[175] British republicans expected that the progress of reason, claims of universal natural rights, and the "light of philosophy" would vanquish tyranny and superstition. Republican ideals of "liberty, equality, and fraternity" now encapsulated the whole of human aspiration, and indeed, thought some, extended even to universal love. The Unitarian scientist Joseph Priestley, who called More's Utopia "visionary even to a proverb" and doubted it "would bear to be reduced to practice", looked forward to a millennial state of improvement rather than back to "the savage uniformity of Sparta", seeing this in particular in terms of the progress of liberty.[176] The events of the French Revolution, he thought,

> unparalleled in all history, make a totally new, a most wonderful, and important, aera in the history of mankind. It is . . . a change from darkness to light, from superstition to sound knowledge, and from a most debasing servitude to a state of the most exalted freedom. It is a liberating of all the powers of man from that variety of fetters, by which they have hitherto been held.[177]

Other republicans praised the equality of the new nation. Priestley's friend the Unitarian reverend Richard Price warned that in "the refined states of civilization property is engrossed, and the natural equality of men subverted", particularly where "great towns propagate contagion and licentiousness; luxury and vice; and, together with them, disease, poverty, venality, and oppression."

174. Ruth H. Bloch, *Visionary Republic: Millennial Themes in American Thought, 1756–1800* (Cambridge University Press, 1985), xvi, 92. The main difference is that "secular utopian prophecy was not based on the authority of Scripture, nor did it assume the active role of a providential God. Its key terms were not those of revealed Christianity but more exclusively those of the radical republican Enlightenment: liberty, reason, and the rights of man" (187).

175. Quoted in Landes, *Heaven on Earth*, 253.

176. Joseph Priestley, *Lectures on History and General Policy* (1788; Thomas Tegg, 1826), 39; Priestley, *Political Writings* (Cambridge University Press, 1993), 109. See Jack Fruchtman Jr, *The Apocalyptic Politics of Richard Price and Joseph Priestley: A Study in Late Eighteenth Century English Republican Millennialism* (American Philosophical Society, 1983); Clarke Garrett, *Respectable Folly: Millenarians and the French Revolution in France and England* (Johns Hopkins University Press, 1975).

177. Joseph Priestley in Gregory Claeys, ed., *Political Writings of the 1790s*, 8 vols (Pickering and Chatto, 1995), 2: 381.

The time might come when "all liberty, virtue, and happiness must be lost, and complete ruin follow." Britain, he thought, was "far advanced into that last and worst state of society, in which false refinement and luxury multiply wants, and debauch, enslave, and depopulate." But it might yet evolve into a millennial state where war ended and harmony reigned.[178] Referring to Plato, More, and Wallace, Price insisted that "it is out of doubt that there is an equality in society which is essential to liberty and which every state that would continue virtuous and happy ought as far as possible to maintain". It followed that:

> The happiest state of man is the middle state between the savage and the refined or between the wild and the luxurious state. Such is the state of society in Connecticut and some others of the American provinces where the inhabitants consist, if I am rightly informed, of an independent and hardy yeomanry, all nearly on a level, trained to arms, instructed in their rights, clothed in home-spun, of simple manners, strangers to luxury, drawing plenty from the ground, and that plenty gathered easily by the hand of industry and giving rise to early marriages, a numerous progeny, length of days, and a rapid increase . . . [179]

In Britain, more radical schemes for limiting accumulation included William Ogilvie's *Essay on the Right of Property in Land* (1781), which proposed limiting manufactures and introducing sumptuary laws and a "progressive agrarian" giving every citizen at most 40 acres.[180] The leading British radical of the revolutionary epoch, Thomas Paine, advocated a form of agrarian law through progressive taxation, though it would have left many large estates intact.[181] Paine considered that "One thing is called a luxury at one time, and something else at another; but the real luxury does not consist in the article, but in the means of procuring it". He cared not "how affluent some may be, provided none be miserable in consequence". Thus, while such extremes of property as constituted "a prohibitable luxury" might be limited, Paine thought it "would be impolitic to set bounds to property acquired by industry, and therefore it is right to place the prohibition beyond the probable acquisition to which industry can extend; but there ought to be a limit to property,

178. Richard Price, *Observations on Reversionary Payments*, 7th ed., 2 vols (1812), 2: 145–46.

179. Richard Price, *Richard Price and the Ethical Foundations of the American Revolution*, ed. Bernard Peach (1785; Duke University Press, 1979), 208–9.

180. William Ogilvie, *An Essay on the Right of Property in Land* (J. Walter, 1781), 41, 92–120.

181. Gregory Claeys, *Thomas Paine: Social and Political Thought* (Unwin Hyman, 1989), 80–81.

or the accumulation of it, by bequest."[182] These ideas were also framed within assumptions some regard as millenarian, though hyperbole is easily misconstrued. An ex-Methodist lay preacher raised as a Quaker, Paine proclaimed that "we have it in our power to begin the world over again. A situation, similar to the present, hath not happened since the days of Noah until now". The "present generation", he thought, "will appear to the future as the Adam of a new world".[183] This was perhaps just talking the issue up, understandably in the circumstances, and ought not to be taken too literally. But the language of religion and politics was very close in these years. Both millenarianism and revolution promised cleansing, or a freeing from sin, often performed in a ritualised manner—for example, in executing Louis XVI. Both promised moral renovation, and an improved human type, without the taint of original sin.[184] Here utopia, revolution, and the millennium overlapped, and shared several common sources. All now promised novelty, the expectation for which came to define the spirit of the age.

A Consuming Passion: Novelty and the Desire for Things

The luxury debate took place against a backdrop of ever-expanding commerce and growing optimism about its consequences. For about 150 years, until 1914, the moderns came to see themselves as ever more exceptional, as cyclical theories of the growth, decline, and inevitable collapse of even the greatest civilisations were replaced by ideas of indefinite improvement driven by science and technology, which spilt over into a great deal of applause about "national character". In the most advanced societies, life expectancy nearly doubled in this period. Materialist conceptions of history, notably as provided by the Scottish conjectural historians, were adopted by mainstream social theory and described progress in terms of a rising standard of living for the majority, as well as greater civility, morality, and social and political liberty. Ideas of God's benevolent Providence still remained in the background. To Ernest Lee Tuveson,

182. R. B. Rose, *The Enragés: Socialists of the French Revolution?* (Melbourne University Press, 1965), 84; Thomas Paine, *Rights of Man*, ed. G. Claeys (1791–92; Hackett Books, 1992), 207, 214–15; Paine, *Agrarian Justice* (J. Adlard, 1797), 11.

183. Thomas Paine, *Complete Writings*, ed. Philip S. Foner, 2 vols (Citadel, 1945), 1: 45, 449.

184. Millenarianism has also played an interesting role in non-western movements of rebellion, often against imperial conquest. Notable instances here are the Boxer Rebellion in China, Maori resistance to British rule in New Zealand, and cargo cults in various parts of the South Pacific. See Michael Adas, *Prophets of Rebellion: Millenarian Protest Movements against the European Colonial Order* (University of North Carolina Press, 1979); Vittorio Lantenari, *The Religions of the Oppressed: A Study of Modern Messianic Cults* (Macgibbon and Kee, 1963).

some strands of millenarianism fused in the seventeenth and eighteenth centuries with Baconian New Philosophy, resulting in the defining ideal of modernity, "the secular millennialism that we know as Progress".[185] Those Tuveson calls "millennialists" rather than millenarians believed in the gradual "triumph of the true Christian spirit rather than a visible descent from heaven", resulting in an undemanding utopia of unfolding moral and material progress. A few sceptics still prophesied that modernity threatened degeneration, and perhaps a reversion to barbarism.[186] But even after the debacle of two world wars few heeded them, until faith in the inevitability of an indefinitely improving future finally faltered in the early twenty-first century.

The idea of indefinite betterment both fuelled and corresponded with a growing worship of novelty as such. We seldom appreciate how momentous this transformation was. Since time immemorial "old" had meant "good", as in "the good old days". The customary and ancestral were venerated as having passed the test of time. The old names, the old stories, the old gods, all resonated with a sense of permanence and certainty, their mere invocation magically wedding us seamlessly to past glories and stabilising our sense of self around the very words, the names giving power by a process called word magic to generation after generation from the beginning.[187] A sense of degeneration from origins was widespread, as in the myth of the Golden Age. *Nihil mihi antiquius est*—"nothing is dearer to me than antiquity", said Cicero. "In China and Siberia you tell the passer-by, to please him, that he *looks aged*, and your interlocutor is deferentially addressed as *elder brother*", relates Gabriel Tarde.[188] In the west, the Greeks often remained, as one study puts it, "in the grip of the past", and obsessed by imitating and restoring the old.[189] But they also acknowledged and reckoned with novelty and valued innovation. While never escaping the rhythmic cycles of nature in the countryside, many lived in expanding cities. Athenians in particular were by the fifth century BCE more receptive to the foreign, if only because they were seafarers. Their creativity across a huge range of activities, from sculpture to poetry to founding colonies, was immense,

185. Tuveson, *Millennium and Utopia*, 34. These accounts of the secularisation of Christianity have been challenged, notably in Hans Blumenberg's *Legitimacy* (3–124). For a summary of this debate, see Malcolm Bull's *Apocalypse Theory and the End of the World* (Basil Blackwell, 1995, 1–20).

186. Tuveson, *Redeemer Nation*, 33–34. See, generally, Malcolm Jack's *Corruption & Progress: The Eighteenth-Century Debate* (AMS, 1989).

187. Stuart Chase, *The Tyranny of Words* (Methuen, 1938), 37, following Malinowski.

188. Gabriel Tarde, *The Laws of Imitation* (Henry Holt, 1903), 246.

189. B. A. van Groningen, *In the Grip of the Past: Essay on an Aspect of Greek Thought* (Brill, 1953).

and would cast a long shadow over western culture. They recognised "newness" while never becoming infatuated with it, acknowledging, Armand D'Angour writes, that "Innovation means change, and change means loss", with all this implied for memory, forgetting, nostalgia, cultural amnesia, and destruction.[190]

So the Greeks never adopted a world view which valued change for the sake of change, where custom and tradition are accordingly reviled, and what is old is derided as decrepit and outmoded. For this a mentality rooted in ideas of scientific and technological innovation, and a consumption-oriented market, was necessary. It was progress which overthrew the despotism of tradition, and finally even the long-dominant position of the ancients themselves in western intellectual life. Cutting these anchors to the past, the moderns leapt with wild abandon into the unfolding present and looming future. Increasingly, from about the seventeenth century onwards, the desire for novelty came to dominate everyday life. It was soon, and remains, highly addicting. Novelty stimulates and provides a constant sense of anticipation, which makes us nervous, anxious, and impatient. It is also highly democratic, requiring neither age nor wisdom, strength nor fortune, to provide instant gratification. Novelty personalises our possession of objects and invites the extension of our own ego into things, and our unique fusion with them.

These feelings were uppermost in the disruptive and disturbing world of the seventeenth and eighteenth centuries, when new discoveries, inventions, and commodities began to grip the imagination of millions. "New" now increasingly meant "better". Fashion, novelty, change, and youth conspired to overthrow the tyranny of custom, tradition, and ancestry. External appearance, and feigning superiority, was first and foremost in the lead, as vanity, envy, and a desire for equality drove a revolution in emulation. Clothing was in the vanguard. Already by the late sixteenth century the French essayist Michel de Montaigne was complaining that "we change so suddenly and promptly that the inventiveness of all the tailors in the world could never furnish us enough novelties."[191]

In the new world of appearance the best-clothed set the trend. But apparel also symbolised other, more deep-seated changes, most notably for greater social equality. Centuries of rigid distinctions of rank were now called into question. The Saturnalia seemingly sprang to life again, but on an epochal

190. Armand D'Angour, *The Greeks and the New: Novelty in Ancient Greek Imagination and Experience* (Cambridge University Press, 2011), 29.

191. "Of Ancient Customs", in Michel de Montaigne, *The Complete Works of Montaigne*, ed. Donald M. Frame (Hamish Hamilton, 1958), 216.

scale. To Gilles Lipovetsky, the "democratisation of fashion" was an "expression of the freedom of human subjects" which inevitably affected ideas and institutions. Here, "in the history of fashion, modern cultural meanings and values, in particular those that elevate newness and the expression of human individuality to positions of dignity, have played a preponderant role."[192] As a desire for greater freedom and a resentment of traditional fetters, especially of class, grew, clothes permitted the appearance of freedom and equality. To Don Slater, this is a process of "continuous self-creation through the accessibility of things which are themselves presented as new, modish, faddish or fashionable, always improved and improving."[193] More than ever before, clothing symbolised freedom, and the possibility of what Bauman terms "the offer of plentiful new starts and resurrections (chances of being 'born again')".[194] The moderns began their passionate quest for remaking themselves, and for what Richard Sennett calls "purified identity", which increasingly involved a kind of continuous liminal passage to the new mentality of the ever-better future.[195]

From individuals to nations, a universal passion for possessing the new rapidly defined the spirit of this impatient and ambitious age. Things and territory alike became new objects of desire. Nations could augment and remake themselves by becoming empires. In seventeenth-century France, some linked greed for new possessions and territorial conquest with ambition, luxury, and state expansion, all animated by "the desire for what they do not possess".[196] In Britain, self-transformation began in earnest in the eighteenth century as the pursuit of fashion became all-consuming. For women there were cosmetics, hair shampoos, hair dye, toothpaste, false breasts, and false hips. In 1783 "false fronts" came into fashion to imitate the pregnancy of the Duchess of Devonshire. For men, false calves were in style for a time. But a thousand other modes of decoration, and variations in colour, texture, fabric, and style, could set off the gentleman or dandy.[197] Some objects and symbols of fashion came into being and passed away so rapidly that, for devotees, the "mode", from the French word for style, needed to be followed almost daily. Now it was pointless to make goods to last, for they might rapidly be outmoded.

The mentality which accompanied these transformations was exhilarating and defines that special sense of arrogant pride in uniqueness often shared by

192. Lipovetsky, *Empire of Fashion*, 22, 34, 5.

193. Don Slater, *Consumer Culture and Modernity* (Polity, 1997), 10.

194. Zygmunt Bauman, *Consuming Life* (Polity, 2007), 49.

195. Richard Sennett, *The Uses of Disorder: Personal Identity and City Life* (Faber and Faber, 1996), 13.

196. Quoted in H. Clark, *Compass of Society*, 21.

197. Quoted in McKendrick, Brewer, and Plumb, *Consumer Society*, 82.

the later moderns. A large part of the suffocating legacy of the past could now be dispensed with. The ideas of the Golden Age and the Garden of Eden implied humanity's degeneration from a god-like condition to something far less worthy, even base, without hope of return in this life. This was a permanent chip on humanity's shoulder, depressingly inherited for centuries. Removing it was liberating. Now, rather than being embarrassed at their inferiority, the moderns could claim their superiority not only over the ancients but equally every other preceding epoch. Each generation could claim to be cleverer than its ancestors. Youth culture follows naturally on this. So does a disdain for antiquity, the elderly, and old things in general, even a conscious antipathy and contempt towards the past, and an obligatory amnesia for anything but feel-good nostalgia, or what is politically or ideologically useful, like constitutional "originalism". Now we can freely mock the ignorance of our forefathers, if we will. "It is from the folly not the wisdom of our ancestors that we have so much to learn", wrote the philosopher Jeremy Bentham, against Burke.[198]

This trend could not but affect political thinking. Hannah Arendt dates a desire for absolute novelty to the revolutionaries of the Enlightenment, who "saw things never seen before, thought thoughts never thought before".[199] Reinhart Koselleck notes that in the late eighteenth century, "the divide between previous experience and coming expectation opened up, and the difference between past and present increased, so that lived time was experienced as a rupture, as a period of transition in which the new and the unexpected continually happened."[200] "The habit of taking on faith one's priests and one's ancestors", thought Tarde, was "superseded by the habit of repeating the words of contemporary innovators", a process he termed "substituting the spirit of investigation for credulity", even though it was "merely a welcoming of foreign and persuasive ideas following upon a blind acceptance of traditional and authoritative affirmations."[201] Now technological advance, conquest, a sense of "enlightenment", growing affluence, and the resentment of privilege combined to form a heady mixture of heightened expectation. This was truly a rebirth of the species. Where the gods had failed, a promethean humanity might succeed.

Science, reason, and rationality were the watchwords of this process. Many leading Enlightenment figures promoted a sense of liberation from the trammels of superstitious theology. There was not yet the worry that mere propaganda,

198. "The Book of Fallacies", in Jeremy Bentham, *The Works of Jeremy Bentham*, 11 vols (Simpkin, Marshall, 1843), 2: 401.

199. Hannah Arendt, *The Human Condition* (University of Chicago Press, 1958), 249.

200. Reinhart Koselleck, *Futures Past: On the Semantics of Historical Time*, transl. Keith Tribe (Columbia University Press, 2004), 246.

201. Tarde, *Laws of Imitation*, 245.

new lies and new nonsense, might replace it, or that the old religions might yet hold sway over millions made anxious by the deeply unsettling crises of modernity. The diffusion of ideas across the masses by suggestibility, contagion, and susceptibility to symbolic manipulation, along with the possibility of a collective unconscious strata of thought, were as yet ill understood. The progress of science suggested the dissipation of gullibility, at least as long as deference to scientists existed. Europe's scientific and technological superiority heralded domination over the rest of the world, and eventually machines and the efficiency they personified defined the modern, as humanity struggled to keep pace with mechanical devices and measured themselves by the effort, a battle they were bound to lose. A new political order, soon called democracy, wedded to a new idea of freedom, would denote liberation from the age-old shackles of despotism, absolutism, and the oppression of the masses. "Revolution" captured many of these meanings and more, and eventually "modernity" too.

The Progress of Novelty

If pre-modern attempts to restrain luxury were often stymied, they became utterly fatuous as markets expanded and emulation stimulated demand. Growing opulence quickly undermined efforts to maintain a fixed society of ranks as defined by appearance. Luxury certainly marked sixteenth-century Italian Renaissance society. In northern Europe, a taste for a wide variety of curiosities, including Flemish tapestry, Chinese scarves and ceramics, textiles, interior furnishings, and much else, began in this period, and accelerated in the second half of the seventeenth century, as commerce expanded rapidly in the Dutch Republic after it achieved independence from Spain in 1581.[202] In the late seventeenth century, real income, exports, and domestic spending all rose, and hitherto unfelt wants for all things Indian became manifest, especially tea, cotton prints, and muslins.[203] By the mid-eighteenth century, many hitherto rare and curious objects foreign to the everyday life of most made their presence felt: dolls, clocks, wigs, books, hats, pens. Tastes arose for tea, coffee, sugar, silk, calico, pepper, spices, tobacco, and chocolate. In houses, ceramics replaced pewter, and metal, wooden eating utensils, while cushions, carpets, wallpaper, and prints became standard interior décor.[204]

202. Linda Levy Peck, *Consuming Splendor: Society and Culture in Seventeenth-Century England* (Cambridge University Press, 2005), 113.

203. Appleby, *Economic Thought and Ideology*, 73–98, 163, 167–68.

204. Roy Porter, *English Society in the Eighteenth Century* (Allen Lane, 1982), 236–37.

Imperial expansion played a key part here. Britain's eastern empire began with the quest for nutmeg and mace in what is now Indonesia in 1603.[205] As in ancient Rome, India was a key source of the new and most exotic commodities, as piratical adventurers seized much of its fabled wealth and then, as "nabobs", corrupted politics at home on their return. India's conquest became synonymous with greed, and with enrichment through downright plunder. The word "loot" was one of India's first gifts to the English language. The chief looter was the first prototype of the modern buccaneering corporation, the East India Company. Founded in 1600, "the Honourable Company" saw off its French, Portuguese, and Dutch rivals and by 1770 was carrying an astonishing half of Britain's trade. Eventually it would carry half the world's. By 1803 it possessed an army twice the size of Britain's. One of its founders, Robert Clive, became fabulously wealthy after conquering Bengal. On arriving, he discovered that:

> The sudden, and among many, the unwarrantable acquisition of riches, had introduced luxury in every shape, and in its most pernicious excess . . . infecting almost every member of each department. Every inferior seemed to have grasped at wealth, that he might be enabled to assume that spirit of profusion, which was now the only distinction between him and his superior.[206]

Some 40 per cent of British Members of Parliament (MPs) had shares in the Company, which by 1693 was spending £1200 a year bribing ministers and MPs—"the world's first corporate lobbying scandal".[207]

This wealth flowed back to Britain with the reckless force of a gold rush. As the pioneering work of Neil McKendrick, John Brewer, and J. H. Plumb demonstrates, the age was defined by an intensified desire for new food and drink, clothing, furniture, personal decoration, and much else, fuelled by growing opulence, increasing inequality and the egalitarianism it provoked, and a passion for display.[208] Concentrated populations, especially in London, the largest city in Europe, enhanced the psychological effects of emulation by multiplying the possibilities and effects of display a hundredfold. The anonymity of the city

205. John Keay, *The Honourable Company: A History of the East India Company* (HarperCollins, 1991), 3.

206. Quoted in James Mill, *The History of British India*, 6 vols (Baldwin, Craddock, and Joy, 1820), 3: 353.

207. William Dalrymple, *The Anarchy: The Relentless Rise of the East India Company* (Bloomsbury, 2019), 23. This is perhaps several million pounds today.

208. McKendrick, Brewer, and Plumb, *Consumer Society*.

rendered mere pretence and appearance much more likely to be rewarding. As Montesquieu observed in 1748:

> in proportion to the populousness of towns, the inhabitants are filled with notions of vanity and actuated by an ambition of distinguishing themselves by trifles. If they are very numerous, and most of them are strangers to one another, their vanity redoubles, because there are greater hopes of success. As luxury inspires these hopes, each man assumes the marks of a superior condition.[209]

The French political economist Jean-Baptiste Say noted that:

> in a great metropolis . . . the demand for luxuries is more urgent than elsewhere, and the dictates of fashion, however absurd, more implicitly obeyed than the eternal laws of nature . . . a man will, perhaps, be content to lose his dinner, so he may appear in the evening circle in embroidered ruffles.

This was exacerbated, he thought, where the court resided at the capital, for here "material products seem to be more wantonly consumed".[210]

The great cities were nothing if not exciting. So much could be seen so quickly, and so much passed off for what it was not. Appearance was everything: every occasion seemed an invitation to make believe, and to act out a role. Being a modern meant pretending to be someone else, to make a stage of all the world. Acting itself, disdained in the ancient world and Christianity, became a prized skill, and professional actors social heroes who embodied a new culture of deception. Novel impressions leapt out at every corner and had never been more plentiful, diverse, or overwhelming. For those inured to the slow pace of rural life it was enough to keep the head constantly spinning. So Britons lurched like drunken sailors forever clutching out at the next glittering experience. The most famous study of its type, Simmel's "The Metropolis and Mental Life" (1902–3), tells us that "the psychological basis of the metropolitan type of individuality consists in the *intensification of nervous stimulation* which results from the swift and uninterrupted change of outer and inner stimuli."[211] The power of suggestion seemed magnified in proportion to the volume of population and the plenitude of impressions. Larger towns in "civilised" countries were more "subject to the domination

209. Montesquieu, *Spirit of the Laws*, 95.

210. Jean-Baptiste Say, *A Treatise on Political Economy* (1802; Claxton, Remsen and Haffelfinger, 1880), 322, 408.

211. Simmel, *Sociology*, 409–10.

of fashion", thought Tarde.[212] To Norbert Elias, keen competition sharpened the observation of behaviour and intensified the drive to impress by appearance.[213]

No longer confined to the very few wealthy upper classes, this competition now expanded rapidly and explosively, and for some became all-consuming. Courtesans and mistresses often set the trend in conspicuous consumption. With a million inhabitants in the late 1700s, London had about fifty thousand prostitutes—one in ten women, perhaps one in seven, leaving children and the elderly out—and clearly a much larger proportion of poor women, for whom this stage of modernity was a tragedy of epic proportions.[214] Paris too was a magnet for provincials defined by "limitless grandeur . . . monstrous riches, and scandalous luxury", in the words of Louis-Sébastien Mercier, whom we will soon meet as a critic of these excesses.[215] Here a "culture of appearances" commenced in the seventeenth century which made fashion accessible outside the aristocracy for the first time.[216] Some thought it "the genius of the French people to be fashionable". And indeed the dazzling court model and aspirations of dynastic glory centred on Versailles often set the style for the rest of Europe.[217]

In Britain, London, with 10 per cent of the nation's population by 1800, soon acquired "a downright tyrannical control over social development, with modern economy an ally of truly inexhaustible resources."[218] This was clearly a major factor in the pattern of consumption and display, which spread out from a court-centred metropolis. The society was increasingly cohesive, owing to the proliferation of newspapers. With advertising fuelling demand, London fashions seduced the nation. The "girls in the country" were seized by "the most longing desires" to acquire them, while "the plough boys and the cowherds . . . desert their dirt and drudgery and swarm up to London to wear fine clothes".[219] Britain was (and remains) notoriously a deeply snobbish and class-ridden society where prejudice against the "lower orders" and the poor was extreme. This

212. Gabriel Tarde, *Penal Philosophy* (William Heinemann, 1912), 369.

213. Norbert Elias, *The Court Society* (Pantheon Books, 1983), 104–6.

214. The term was used more widely than today to include "immoral" women, e.g. those living with a man while unmarried.

215. Louis-Sébastien Mercier, *The Picture of Paris: Before & after the Revolution* (1788; George Routledge and Sons, 1929), 5.

216. D. Roche, *Culture of Clothing*, 32.

217. Jean Pottier de la Hestroye, quoted in H. Clark, *Compass of Society*, 21, 47.

218. René König, *The Restless Image: A Sociology of Fashion* (George Allen and Unwin, 1973), 146.

219. Quoted in McKendrick, Brewer, and Plumb, *Consumer Society*, 95.

peaceful rebellion of appearance clearly aimed to subvert the pretentiousness of the ruling elite. Dressing like the well-to-do deprived them of a vital marker of their superiority. But it was also deeply conservative insofar as it implied joining them, by gaining at least the appearance of "respectability"—the term dates from 1785.[220] Achieving "gentility" could be accomplished by aping the manners of the gentry, or keeping up with the "Sir Joneses", as Peter Stearns puts it.[221] Unlike France, where a large class of peasants tended to hoard any surplus it acquired, Britain had no such rank. Its class structure was much more fluid, as was noted in 1767:

> In England the several ranks of men slide into each other almost imperceptibly and a spirit of equality runs through every part of their constitution. Hence arises a strong emulation in all the several stations and conditions to vie with each other, and the perpetual restless ambition in each of the inferior ranks to raise themselves to the level of those immediately above them. In such a state as this fashion must have uncontrolled sway. And a fashionable luxury must spread through like a contagion.[222]

Such a structure, indeed, to Harold Perkin, "might have been designed for the spontaneous and rapid generation of demand for cheap consumer goods."[223] In the race of status competition fine apparel was particularly prized by servants, who amongst the poor were in closest proximity to luxury, but whose voices are rarely part of the historical record. They were commonly given their masters and mistresses' clothing by bequest, or even bought new attire. The "appearance of female domestics will perhaps astonish a foreign visitor more than anything in London", wrote one observer, for they were "clad in gowns well adjusted to their shapes", made even of silk and satin, and wore "hats adorned with ribbons".[224] Having well-dressed servants was itself a status symbol. Servants "were given every possible encouragement by their employers to dress in the height of fashion", writes Jean Hecht, and this led "in the course of the period to a general increase in the extravagance of the outfits".[225] And so, thought Thomas Day, "The laborious and simple youth

220. Woodruff D. Smith, *Consumption and the Making of Respectability, 1600–1800* (Routledge, 2002), 189.

221. Stearns, *Consumerism in World History*, 26.

222. Quoted in McKendrick, Brewer, and Plumb, *Consumer Society*, 11.

223. Harold Perkin, *The Origins of Modern English Society, 1780–1880* (Routledge and Kegan Paul, 1969), 91.

224. Quoted in McKendrick, Brewer, and Plumb, *Consumer Society*, 57–58.

225. J. Jean Hecht, *The Domestic Servant Class in Eighteenth-Century England* (Routledge and Kegan Paul, 1956), 211, 121.

that we are continually sending you out of the country, are, by a few months residence in your houses, transformed into the most worthless and contemptible characters: idle, luxurious, delicate, and abandoned, as their betters".[226]

This process induced a flood of complaints, and commentators frequently mocked the poor for their aspirations. In the 1720s Daniel Defoe noted that "the same flourishing pride has dictated new methods of living to the people; and while the poorest citizens live like the rich, the rich like the gentry, the gentry like the nobility, and the nobility striving to outshine one another, no wonder all the sumptuary trades increase".[227] By mid-century even day labourers, observed Adam Smith, were "ashamed to appear in public without a linen shirt".[228] In 1763 one critic thought that if this trend continued, "in a few years we shall probably have no common people at all".[229] "Time was", complained another observer in 1777:

> when those articles of indulgence, which now every mechanic aims to be in possession of, were enjoyed only by the Lord or Baron of a district [but now] men began to feel new wants [and] sighed for indulgences they never dreamed of before . . . [The] wish to be thought opulent . . . led them into luxury of dress. The homespun garb then gave way to more costly attire, and respectable plainness was soon transformed into laughable frippery . . . [and] every succeeding year gave way to fresh wants and new expences.[230]

A visitor to mid-eighteenth-century Nottingham lamented that "even a common Washer woman thinks she has not had a proper Breakfast without Tea and hot buttered White Bread!"[231] The exclamation says it all. A pound of tea had cost six months of a labourer's wage in 1650.[232] Now it was an essential which by 1790 could be had for twelve hours' labour. Some thought that universal ruin loomed. By 1800 the process was so far advanced that:

> The rapid increase of vanity and extravagance, in this island is a subject pregnant with mischief and alarm. Commercial monopoly and Eastern opulence . . . have already fostered dissipation and immorality into monsters of colossal magnitude, that every moment threatens (sic) the humbler

226. Hecht, 216.

227. Quoted in Perkin, *Modern English Society*, 92.

228. A. Smith, *Wealth of Nations*, 2: 466.

229. Lorna Weatherill, *Consumer Behaviour and Material Culture in Britain, 1660–1760* (Routledge, 1988), 194–95.

230. Quoted in McKendrick, Brewer, and Plumb, *Consumer Society*, 50–51.

231. Quoted in McKendrick, Brewer, and Plumb, 57–58.

232. Sear and Sneath, *Consumer Revolution in England*, 85–86.

> classes of independence, with ruin. That frugality which once characterized the middling and lower classes of society amongst us is no more: the little tradesman and mechanic of the present day, fatally though impotently, ape the luxuries and fashionable vices of their superiors.[233]

London's size and wealth made it possible to display the new wares to the greatest advantage. For the first time, large shop windows, which were even illuminated at night, offered a mesmerising spectacle to onlookers. Oxford Street was one of the first such emporia. Here, observed a German visitor, Sophie von la Roche, in 1786,

> It is almost impossible to express how well everything is organised in London. Every article is made more attractive to the eye than in Paris or in any other town . . . Behind great glass windows absolutely everything one can think of is neatly, attractively displayed, and in such abundance of choice as almost to make one greedy . . . [234]

What one bought could then be displayed at home for the admiration of family and friends, or paraded publicly, as new locations for sociability were constantly proliferating. Coffee houses, taverns, spas, clubs, inns, ballrooms, restaurants, and public houses provided abundant opportunities to flaunt and admire, to set the trend and to follow it.[235] To see and be seen in one of the new pleasure gardens of London, like Vauxhall or Ranelagh, where people watching and entertainment were all the rage, was for many the height of delight. A growing uniformity of attire facilitated the ready mixing of social ranks; by 1700 at Sadler's Wells, those taking the waters included "Modish sparks and fashionable ladies, good wives and their children, mingled with low women and sempstresses in tawdry finery; with lawyers' clerks, and pert shopmen; with sharpers, bullies, and decoys."[236]

By 1790 Britain was the first large-scale consumer society, the wealthiest nation in the world, and also one of the most unequal. One visitor that year observed that "England surpasses all the other nations of Europe in . . . luxury . . . and the luxury is increasing daily!" It seemed that "All classes . . . enjoy the accumulation of riches, luxury and pleasure."[237] By 1819, remarked the Swiss

233. Quoted in Perkin, *Modern English Society*, 92–93.

234. Sophie von la Roche, *Sophie in London in 1786* (Jonathan Cape, 1933), 87.

235. Werner Sombart, *Luxury and Capitalism* (University of Michigan Press, 1967), 51, 61, 107, 171.

236. Warwick Wroth, *The London Pleasure Gardens of the Eighteenth Century* (Macmillan, 1979), 16.

237. Quoted in McKendrick, Brewer, and Plumb, *Consumer Society*, 10.

political economist Jean Charles Sismondi, the "upper English aristocracy" had "actually achieved a measure of wealth and luxury which surpasses everything one could see in all other nations." Thirty years later still he regarded Britain as an "astonishing country" which demonstrated "a great experiment for the instruction of the rest of the world."[238]

The Fate of Imitation

And so indeed it would prove to be. For, more than any other nation, it was Britain's bourgeois revolution that first established the paradigm of consumerism which would define later modernity. As a work entitled *London Unmask'd* (1784) noted, a "passion for novelty" was driven by "the importation of exotic fashions, exotic wonders, and exotic manners". A "fascination with all things new" gripped multitudes, and people began to define themselves in terms of what Julie Park calls a "mimetic object", mirroring the qualities of things.[239] Endless surprise and delight, Maxine Berg stresses, accompanied this continuous exploration of new and hitherto undiscovered worlds.[240] Such variety was spell-binding to many, for the possibilities of self-expansion were infinitely greater than ever before. New objects could enlarge, elongate, embellish, and enrich a life, and by a multitude of extensions create novel identities of perpetually unfolding fantasies of a hitherto unknown type. The old fantasy of shapeshifting was now realised. A veritable machine of self-invention and self-mutation had been set in motion. Even across the centuries, the sense of wonder and intoxication is almost palpable.

One of the great discoveries of the age was the power of imitation driving this process of self-expansion. To later nineteenth-century commentators the revolution in appearance and style which so distinguished the period revealed the central role of emulation in society. To Gabriel Tarde and Georg Simmel, imitation came to the fore as a basic principle, "the primal soul of social life".[241] To Tarde, indeed, whose *Laws of Imitation* (1890) popularised the concept, society *was* imitation, a constant replication and multiplication of those facets of others' behaviour we find attractive, the repetition of which provides us with a reassuring sense of certainty, structure, and meaning, a

238. J.-C.-.L. Simonde de Sismondi, *New Principles of Political Economy* (1819; Transaction, 1991), 8; Sismondi, *Political Economy and the Philosophy of Government* (Chapman, 1847), 115.

239. Julie Park, *The Self and It: Novel Objects in Eighteenth-Century England* (Stanford University Press, 2010), 10–11, 79, 122.

240. Maxine Berg, *Luxury & Pleasure in Eighteenth-Century Britain* (Oxford University Press, 2005).

241. Tarde, *Laws of Imitation*, 189.

rudder to assist our forward progress. Imitation was the psychological glue holding society together. Watching others provides constant guidance and avoids bewilderment for the perplexed. Though Durkheim objected that it provided no evidence of solidarity, and thus overwhelmingly suggested an individualist model of society, Tarde thought it a key basis of friendship.[242]

But to Tarde this process had altered dramatically in modernity. While an impulse to emulate already existed, fashion allowed imitation to be freed from its original, customary and generational focus. In keeping with the "law of the imitation of the superior" most played follow-the-leader; the involuntary, passive, and irrational implications have worried some of Tarde's subsequent interpreters. But a few innovated and modified.[243] Imitation played "a role in societies analogous to that of heredity in organic life" in providing both repetition and variation, continuity and innovation.[244] Simmel saw fashion as focused on emulating the clothing of one's social superiors, driven by a desire for prestige and status.[245] In democracies, in particular, this process also gradually produced a convergence of style, with the elite eventually copying the dress and speech of the majority in order to fit in without their greater wealth fostering resentment. The spectacle of a billionaire in jeans is a familiar one to us today. For such is the force of modern egalitarianism that now even the rich have to appear industrious; as Huxley noted already in 1931, "The leisured class is vanishing; even the rich work. In other words, even the rich behave, to some extent, like members of the proletariat."[246] How this necessity emerged, through revolution, we must now consider.

242. Bruno Karsenti, "Imitation: Returning to the Tarde-Durkheim Debate", in *The Social after Gabriel Tarde*, ed. Matei Candea (Routledge, 2010), 44. Durkheim thought instead that imitation closely resembled contagion, or the transmission of a common feeling, and Le Bon's notion of the crowd was one of Tarde's sources. See Durkheim's *Suicide*, translated by John A. Spaulding and George Simpson (Routledge and Kegan Paul, 1952), which treats both the general principle and its application to suicide (123–42).

243. "A kind of sleepwalking", Robert Leroux terms it: "Tarde and Durkheimian Sociology", in *The Anthem Companion to Gabriel Tarde*, ed. Leroux (Anthem, 2018), 120.

244. Tarde, *Laws of Imitation*, 11.

245. "Fashion", in Georg Simmel, *On Individuality and Social Forms* (University of Chicago Press, 1971), 294–323.

246. Aldous Huxley, *Aldous Huxley's Hearst Essays* (Garland, 1994), 18.

PART III

Luxury and Sociability in Later Eighteenth- and Nineteenth-Century Utopianism

7

The Later Eighteenth Century and the French Revolution

Spartans, Neo-Harringtonians, and Utopian Republicans

In the later eighteenth century, utopian literary responses to growing affluence, luxury, and inequality were both satirical and programmatic. In Britain, literary and practical utopianism was mostly developed by two groups of republican writers. The first, in the tradition of Walter Moyle and Marchamont Nedham, were followers of James Harrington, or neo-Harringtonians, who promoted an agrarian law restricting the amount of landed property anyone could hold, chiefly to preserve popular liberty, which again seemed under threat from newly concentrated wealth and ever-expanding monarchy. The second advocated community of goods and can be termed utopian republicans because of their alignment with Thomas More. Both often exhibited more or less overtly Spartan sympathies.[1] Both used fiction, pamphlets, and books to promote their views. The controversy they entered into would come to a boil with the French Revolution, and then flow into socialism early in the new century.

In Britain, the main neo-Harringtonian republican utopian literary text of this period was written by a friend of Adam Ferguson, James Burgh (1714–75). This projected a New World colony where an "equal division of the land cut off the means of luxury with its temptations, checked pride and ambition, and established the habits of industry and diligence among them". *An Account of the First Settlement, Laws, Form of Government, and Police, of the Cessares, a*

1. On this spectrum of ideas, see Gregory Claeys and Christine Lattek's "Radicalism, Republicanism, and Revolutionism: From the Principles of '89 to Modern Terrorism", in *The Cambridge History of Nineteenth-Century Political Thought*, edited by Gareth Stedman Jones and Claeys (Cambridge University Press, 2011, 200–254).

People of South America (1764) praises Sparta's "extraordinary instance of zeal for the good of others". The new colony agrees "that every one should have an equal share, that so we might check every proud, ambitious, and destructive passion, and banish riches as well as poverty from us." Gold and silver mining are prohibited, and the prices of food and labour regulated. Clothing and ornamentation are also restricted:

> The senate is enjoined to establish sumptuary laws, and carefully to guard against the first introduction of all sorts of luxury: and to prohibit all those arts and trades, which minister only to idleness and pride, and the unnecessary refinements and embellishments of life, which are the certain forerunners of the ruin of every state. And though it is very commendable to be neat and cleanly in our apparel, yet nothing is more contrary to a wise and rational conduct, than to lay out too much thought and expence upon it; and a frequent change of fashions shews a vain and trifling mind. The senate have therefore regulated every one's dress according to their age and sex: it is plain, decent and becoming, but no diamonds or jewels, no gold or silver lace, or other finery are allowed of, lest pride and vanity, the love of shew and pomp should steal in among us by imperceptible degrees.[2]

Thus, here there is

> no gold or silver to boast of, no sideboards of plate to make a show of, no grand houses, no sumptuous furniture, no fine or gaudy apparel, nor any foreign trade and commerce to introduce among us those expensive fashions and needless superfluities, which come by degrees to be considered as the real necessaries of life.

The advantages of avoiding such temptations were obvious. Once established, Burgh warned, the chains of luxury were "the hardest to break of any in the world." Now, in Britain, "An emulous endeavour to outvie each other in all the elegant accommodations of life, seems to be not only the ruling principle of a few, but the main ambition of a vast majority, the characteristic and almost universal passion of the age." The inevitable result would be progression "from virtuous industry to wealth; from wealth to luxury; from luxury to an impatience of discipline, and corruption of manners: till by a total degeneracy and loss of virtue, being grown ripe for destruction, it falls a prey at last to some hardy oppressor."[3] "To maintain luxury", Burgh thought, "the great ones must oppress the meanest". Luxury was "hurtful to manners, and dangerous

2. Claeys, *Utopias*, 79–80, 117, 123.

3. Claeys, 125.

to states". It threatened to "break the martial spirit of a people", and to excite wants and desires which "being artificial, are without all *bounds* and *limits*", and which then invited bribery and corruption.[4] Within a few decades these complaints would also form part of a rising common evangelical refrain. In 1785 one minister proclaimed that "luxurious has now extended its baneful influence and spread its destructive poison through the whole body of the people", adding that "When the mass of blood is corrupt, there is no remedy but amputation."[5]

Such themes were also taken up later in this period in *Bruce's Voyage to Naples* (1798), which describes an underground world whose inhabitants "never kill any living creature for food, nor is any vice held in more detestation among them than luxury of any sort." While on earth's surface the "insolence of these creatures of fortune" is remarkable; here there is "nothing but friendship, hospitality, and a brotherly affection to all their fellow-creatures", for like the "wise Lycurgus" gold and silver are prohibited and equality of fortune prevails.[6] Neo-Harringtonian arguments were also supported in Catherine Macaulay's *Loose Remarks on Certain Positions to be found in Mr. Hobb's Philosophical Rudiments of Government and Society: With a Short Sketch of A Democratical Form of Government* (1767). This emphasises "fixing the Agrarian on the proper balance", with property equally divided among heirs, and shows a bias towards popular ownership, warning that otherwise Britain might face Rome's fate.[7] But David Hume regarded Harrington's "agrarian" as "impracticable", since some might disguise ownership by having titles in other people's names. So he set no limits to private property in his essay entitled "Idea of a Perfect Commonwealth" (1752).[8]

On the programmatic side, a rigorous plan of utopian republicanism was presented in Robert Wallace's *Various Prospects of Mankind, Nature and Providence* (1758). Invoking Plato, More, Harrington, and Hume, Wallace sketched a society where all men are trained to farm, no one is "distinguished by their houses, cloaths, nor food, but all of them should enjoy every thing in the same manner", and care for the ill and aged is guaranteed—all, citing More, with a

4. Burgh, *Political Disquisitions*, 3: 30, 59–60.

5. John Thornton, cited in Ford K. Brown, *Fathers of the Victorians: The Age of Wilberforce* (Cambridge University Press, 1961), 75.

6. Claeys, *Utopias*, 257, 267, 280. This text, sometimes attributed to James Bruce, is adapted from *A Voyage to the World in the Centre of the Earth* (1755).

7. Catherine Macaulay, *Loose Remarks on Certain Positions to be found in Mr. Hobb's Philosophical Rudiments of Government and Society: With a Short Sketch of a Democratical Form of Government* (T. Davies, 1767), 33.

8. Claeys, *Utopias*, 58.

nod to Sparta, without private property. Wallace conceded that every system of equality faced four obstacles: emulation or ambition, the love of ease or sensual pleasure, the love of liberty, and the desire for scarce goods by more than could acquire them. All these he thought could be rendered compatible with a "Utopian constitution". Distinctions could be founded on "real virtue or excellency" rather than "false and unjust conceptions of wealth and dignity".[9] Equality would extinguish envy. Idleness would excite contempt, and love of country would inspire industry. Authority would be sufficiently divided to avoid tyranny. "As by removing property, we destroy theft and robbery; so by maintaining an equality, we prevent hardships, banish discord, and restore the golden age."[10] But Wallace also revealed the classic conundrum for modern utopian reformers, noting of Sparta that

> It would be impossible at present to persuade any civilized nation whatsoever to agree to such a distribution. The rich and powerful part of the people have too many advantages above others ever to part with them, and put themselves on a level with their inferiors. In truth, no such generosity nor self-denial can be expected, nor ought to be demanded . . . If any such equal government is possible at present, according to the ordinary course of affairs, it must be erected in some wild and uncultivated country, where there are few inhabitants.[11]

Outside of Britain a similar ambiguity is evident. In the Norwegian baron Ludwig Holberg's underground world (1742) we encounter a country named Miho, whose founder "made sumptuary laws, which forbid all luxury on the severest penalties; and accordingly this society, for its great continence and parsimony, may be justly called another Sparta." But we soon discover that "New Sparta" suffers from many problems. The consequence of these reforms is that "all luxury being prescribed, and the rich baulking their genius, and giving into no indulgences, the common people of course must lead an indolent, idle, and beggarly life, for want of matter to make a proper gain of." The moral is thus, pace Mandeville, that profligacy produces employment, and abstemiousness, poverty.[12] Sparta is thus satirised here, not promoted.

One of the most impressive French literary utopias of this period to focus centrally on luxury was written by the influential economist Jean-Baptiste Say

9. Robert Wallace, *Various Prospects of Mankind, Nature and Providence* (A. Millar, 1758), 44, 76–77.

10. Wallace, *Various Prospects of Mankind*, 100–101.

11. Wallace, *Various Prospects of Mankind*, 57–59.

12. [Ludwig Holberg], *Journey to the World Under Ground* (1755), in *Popular Romances*, ed. Henry Weber (John Murray, 1812), 157.

(1767–1832). *Olbie* (1800) aimed to answer the question, "What are the means of establishing moral behaviour among a people?" and thus necessarily tackled the problem of luxury.[13] Say is best known for his defence of Smithian principles of free trade, and the natural propensity of markets to balance supply and demand, which is termed "Say's law". But he had many reservations about the threat of commercial expansion to republican manners, luxury in his view being "deadly to small and large states", "the country where there is the least luxury" being "the richest and the happiest." In 1795 he wrote that "luxurious display comes only with the sacrifice of what is useful, and is always the neighbor of poverty". By contrast "true luxury, luxury as best understood, consists in the abundance of necessary goods, the good quality of what is consumed, and the convenience and cleanness of whatever we touch."[14]

"True luxury" could thus be reconciled with sound manners and simplicity. It was, indeed, Say wrote elsewhere, "in the middle state between luxury and indigence that enlightenment and virtues are found." This ideal would allow a people to live

> in ease rather than in opulence, where everyone knew how to read . . . where there are no unproductive individuals who are a burden on society, and no poor men who cannot, with labour and good conduct, win an easy subsistence, and lead the kind of life which the English call *comfortable*,

meaning "something commodious but without luxury".[15] Say also addressed the effects of luxury on sociability. The Olbiens erect a temple to friendship, because "money produces loosely-attached servants rather than faithful friends and capable citizens". The new regime aims to ensure that "the majority of families comprising it lives in genuine ease" and "extreme opulence is as rare as extreme poverty." Thus

> To lessen its power more and more, the leaders among the Olbiens professed a rather imposing contempt for ostentation. Simplicity of tastes and manners was in Olbie a motive for preference and an object of consideration. The chiefs of the state adopted a general system of simplicity in their clothing, pleasures, and social relationship. Neither their domestics nor the soldiers of their guard ever showed a sign of deference [*une déférence*

13. Jean-Baptiste Say, *Olbie; ou, Essai sur les moyens de réformer les moeurs d'une nation* (1802; 1830), 44, 3–4, transl. Roy Arthur Swanson, in *Utopian Studies* 12 (2001): 79–107. For commentary, see Richard Whatmore's *Republicanism* (126–35).

14. J.-B. Say, *An Economist in Troubled Times*, ed. and transl. R. R. Palmer (Princeton University Press, 1947), 20, 33; Whatmore, *Republicanism*, 127.

15. Quoted in Whatmore, *Republicanism*, 121, 124;

> *stupide*] to the flunkeys of luxury. The people by degrees formed the same habit and soon there was no longer seen a crowd of imbeciles stupefied at the sight of a diamond suiting or some other gewgaw of this kind. People were no longer admired in proportion to the consumption they effected. What happened? They consumed nothing beyond what was truly necessary to their utility or their pleasure. Luxury, attacked at its base, which is opinion, gave way to a more generally diffused ease; and, as always happens, happiness increased at the same time as morals were reformed.[16]

If a contempt for luxurious excess could be cultivated, modern republicanism might be reconciled with commerce, and the fate of Rome and many other empires avoided. As Say later put it:

> The luxury of ostentation affords a much less substantial and solid gratification, than the luxury of comfort . . . Besides, the latter is less costly, that is to say, involves the necessity of a smaller consumption; whereas the former is insatiable; it spreads from one to another, from the mere proneness to imitation; and the extent to which it may reach, is as absolutely unlimited.[17]

But Say also offered a cautionary note, observing in reference to Helvétius's claim that "Sparta was without luxury, without abundance, and . . . Sparta was happy", that "I do not say that men ought to imitate the institutions of Lycurgus; I merely assert that men are what one makes them."[18]

The Utopian Turn towards the Future

The appeal of Sparta, then, was far from dissipated. But the overwhelming attractions of novelty could not but eventually infect utopia as well. The process of commercial, scientific, and technological development inevitably turned expectation towards the future, and, with faith in Christianity waning, towards the idea of indefinite improvement in this life. In utopian thought, two substantial shifts resulted in the late eighteenth century. Firstly, the ideal polity was increasingly envisioned as plausible at the level of large-scale,

16. Say, *Olbie*, 92, 91, 96.

17. Say, *Treatise on Political Economy*, 397.

18. Say, *Olbie*, 106. See Helvétius, *Treatise on Man*, 2: 113. In his *Treatise on Political Economy* (1803), Say defined luxury as "the use of superfluities", but then confessed, "I am at a loss to draw the line between superfluities and necessaries", and noted that "ostentatious indulgence, the excess and refinement of sensuality" were "equally unjustifiable" (105–6). For context, see Sonenscher's *Before the Deluge* (334–40).

populous nations, or even as a world-state. Secondly, it was conceived as no longer something to be discovered after a long voyage, or by leaping imaginatively through time or space, taking us sufficiently great distances from the taint of corrupt old Europe to permit the cognitive displacement needed to posit such momentous changes. Far from being an unattainable, imaginary good place, or eutopia, the ever-impatient moderns would demand that utopia be realised in the secular here and now, and preferably, if this was the only life there was, as soon as possible. Utopia now became a potential future just over the horizon, to be created voluntarily through human agency, aided perhaps by a benevolent Providence, but inscribed at any event in the historical process.

These shifts were linked to the irresistible pull of scientific and technological progress, the extension of luxury down the social ranks, and the concomitant spirit of social and political equality which emerged from the British revolutions of 1649 and 1688. The explosive impact of consumerism upon parallel developments in thought and politics has been little considered. But there is a reasonable case here for cause and effect: the expansion of desire introduced by commercial society, in particular, forced utopia to adopt a future-oriented *topos*. Utopias now *had* to be set in the future because ideal pasts, though still appealing to some, were always more primitive, when what people wanted was novelty, innovation, and more stuff. So utopia now became fused with the modern theory of progress as an infinite process of improvement in health and prosperity.[19] In what Koselleck terms the "temporalization of Utopia", "the metamorphosis of utopia into the philosophy of history", the "imagined perfection of the formerly spatial counterworld" now made a secularised millenarianism central to modern consciousness.[20] The parallel is nonetheless not an entirely exact one. Millenarianism implies moral renewal, cleansing, and a massive settling of scores with respect to past sins, with the prospect of salvation at the grand finale. Its secular equivalent is more gradual, never reaches an end point, and does not promise any literal redemption. Its forward-looking vision of "betterment" is distinguished from "progress" generally by its egalitarianism and promise of intensified sociability, neither of which is central to what we usually call capitalism. But a growing richness of experience and delight in novelty did not and does not translate into moral improvement as such. Progress might conceivably be material without being moral. Indeed, an obsession with the material might

19. See Baczko, *Utopian Lights*.

20. Reinhart Koselleck, *The Practice of Conceptual History*, transl. Todd Samuel Presner et al. (Stanford University Press, 2002), 85.

well detract from the possibility of leading a more moral life. And so it has often been.

On the literary front, four texts, three obscure and one very famous, set the trend for projecting utopia into the future.[21] The first is Michel de Pure's unfinished *Épigone, histoire du siècle future* (1659), which weaves romance, adventure, and an account of a world ruled by women in an elaborate parody.[22] Secondly, Samuel Madden's *Memoirs of the Twentieth Century* (1733) is set in 1997. This, however, points as much towards dystopian degeneration, contrasting the extreme poverty of the many who "want even the common Necessaries of Life", to "the Luxury of the Nobility and Gentry [which] is increas'd beyond all Bounds, as if they were not only insensible of, but even rejoyc'd in the publick Calamities of their Fellow-Subjects." Addicted to gluttony, gambling, and vice, the society seems destined to collapse. But a wise government intervenes to tax luxury heavily, particularly foreign imports, thus bolstering employment in agriculture, and constructs public granaries to avoid famine. The book flopped: some 10 copies were sold, and 890 of the 1000 printed had to be pulped.[23]

Thirdly, *The Reign of George VI 1900–1925* (1763), is preoccupied with great power relations, particularly Russia's growing pre-eminence and ambitions, but singularly ignores the likely outcome of trends in commerce and manufacturing.

Far better known, fourthly, is the 1771 text Koselleck terms the "first futuristic novel", Louis-Sébastien Mercier's famous *L'an 2440, rêve s'il en fut jamais* (The year 2440, a dream if ever there was one; confusingly translated into English as *Memoirs of the Year Two Thousand Five Hundred*).[24] One of the

21. On some antecedents, see I. F. Clarke's *The Pattern of Expectation 1644–2001* (Jonathan Cape, 1979, 15–34).

22. A modern text is edited by Lise Leibacher-Ouvrard and Daniel Maher (Presses de l'Université Laval, 2005).

23. Samuel Madden, *Memoirs of the Twentieth Century: Being Original Letters of State under George the Sixth* (Osborn, Longman, Davis and Batley, 1733), 85; Margolis, *Brief History of Tomorrow*, 27–28.

24. Koselleck, *Practice of Conceptual History*, 85, 88. Just why Britain required sixty years more to reach the ideal future is unclear. On the later development of the future-oriented fantasy, see Clarke's *Pattern of Expectation*. A modern French edition is *L'an 2440: Reve s'il en fut jamais*, with introduction and notes by Christophe Cave and Christine Marcandier-Colard (La Decouverte, 1999).

great bestsellers of the epoch, it had reached eleven editions by 1793, with translations into English, Dutch, and German, and had sold at least 63,000 copies by 1814. It spawned numerous imitations, and proved so controversial that in Spain it was burnt at the order of the Inquisition by the king himself.[25] Mercier (1740–1814) began the novel as an essay on "The Savage" to illustrate, as Bury puts it, that "the true standard of morality is the heart of primitive man, and to prove that the best thing we could do is to return to the forest". But he then abandoned the project to argue the opposite principle.[26] In *L'an 2440* elements of both approaches are intermingled. Chiefly, however, Mercier developed his friend Turgot's idea of intellectual progress, outlined in "A Philosophical Review of the Successive Advances of the Human Mind" (1750). This contended for the infinite perfectibility of humanity, and in turn spurred Condorcet, then Saint-Simon and Comte, in a similar direction.[27]

To Robert Darnton, Mercier's great success resulted from his "vision of the world where writers and readers made Rousseau's dream come true and where life was at last an open book."[28] Luxury, as destructive of virtue, was Mercier's principal target, and he later acknowledged Rousseau as a source of his antagonism to the "insolent luxury" of Versailles.[29] In the Paris of 2440 the "horrid inequality" of "extreme opulence and excessive misery" has disappeared. External commerce has ceased, and its linkage with imperialism been acknowledged and extinguished. "Just sumptuary laws have suppressed that barbarous luxury" which blighted the *Ancien Régime*. Since there are "no monks, nor priests, nor numerous domestics, nor useless valets, nor workmen employed in childish luxuries, a few hours of labour are sufficient for the public wants".[30] Offensive weapons are banned. Women are demure and wear no make-up. Coffee, tea, and other "poisons" are prohibited. Everyone dresses in a "simple

25. Robert Darnton, *The Forbidden Best-Sellers of Pre-revolutionary France* (HarperCollins, 1996), 136; Everett C. Wilkie Jr's "Mercier's L'An 2440: Its Publishing History during the Author's Lifetime" (*Harvard Library Bulletin* 32 (1984): 5–35) reports sales of 63,000 by 1814. It also spawned a number of future-oriented imitations, such as, in Germany, *Das Jahr 1850* (1777) (W. W. Pusey, *Louis-Sébastien Mercier in Germany* (Columbia University Press, 1939), 40). Herder was one fan.

26. J. B. Bury, *The Idea of Progress* (Macmillan, 1921), 194.

27. Repr. in Anne Robert Jacques Turgot, *Turgot on Progress, Sociology and Economics*, ed. Ronald L. Meek (Cambridge University Press, 1963), 41–60. Turgot distinguished between "extravagant luxury" and "temperate luxury" (103).

28. Darnton, *Forbidden Best-Sellers*, 336.

29. Louis-Sébastien Mercier, *De J. J. Rousseau considéré comme l'un des premier auteurs de la Révolution*, 2 vols (Buisson, 1791), 1: 197.

30. [Mercier], *Year Two Thousand Five Hundred*, 1: 5, 29, 183.

modest manner", though those who have saved someone's life or performed other acts of public utility wear an embroidered hat. Only "useful and necessary luxury", unmixed with "pride and ostentation", which instead "promotes industry . . . creates new commodities [and] adds to our conveniences" exists.[31] Cosmopolitanism is the norm: "We regard all men as our friends and brethren. The Indian and the Chinese are our countrymen, when they once set foot on our land." The pleasures of conversation and open sincerity have been restored, and the imprudence and hypocrisy of past interchange are gone. The "most affable people in the world" regard a "happy mediocrity" as the ideal of "sovereign wealth".[32] Since "Foreign traffic was the real father of that destructive luxury, which produced in its turn, that horrid inequality of fortunes, which caused all the wealth of the nation to pass into a few hands", the new regime commences by "destroying those great companies that absorbed all the fortunes of individuals, annihilated the generous boldness of a nation, and gave as deadly a blow to morality as to the state."[33] Thus "We cultivate an interior commerce only, of which we find the good effects; founded principally on agriculture, it distributes the most necessary aliments; it satisfies the wants of man, but not his pride." So "All that promotes ease and convenience, that directly tends to assist nature, is cultivated with the greatest care. All that belongs to pomp, to ostentation and vanity, to a puerile desire of an exclusive possession of what is merely the work of fancy, is severely prohibited."[34] Mercier's condemnation of the moderns is savage:

> You thought yourselves highly ingenious in the refinements of luxury, but your pursuits were merely after superfluities, after the shadow of greatness; you were not even voluptuous. Your futile and miserable inventions were confined to a day. You were nothing more than children fond of glaring objects, incapable of satisfying your real wants. Ignorant of the art of happiness, you fatigued yourselves, far from the object of your pursuits, and mistook, at every step, the image for the reality.[35]

The moral of the story is clear: the worst effects of trade and commerce could be reversed, at least in the imagination. Reforming great corporations was the starting point. Simplicity of manners was the end. Yet by 1795 Mercier began to doubt his earlier proposals, writing of the distinction between

31. [Mercier], 1: 32; 2: 238.

32. [Mercier], 1: 208; 2: 2.

33. [Mercier], 2: 186, 188.

34. [Mercier], 2: 186, 189.

35. [Mercier], 2: 190.

"laudable and pernicious luxury" that "I should at present be almost equally afraid either to abolish luxury or to give it a still greater extension". He now accepted that luxury might well stimulate agricultural demand, defining it as "the perpetual spur which incites man to labour", such that "without this attraction, the hands of the cultivator would grow languid".[36] "Inequality", too, he proclaimed "a thing so essential to the welfare of society, that did it not exist, it would be necessary to create it politically." He still wished to "condemn the luxury which engrosses vast enclosures for the chace; but I cherish that luxury which creates amusements, Aelean games, and theatrical entertainments; those entertainments which, by softening the manners of the people, enlarge their understanding". Mercier now also seemed to accept that luxuries progressed hand in hand:

> Man is not rendered happy by your moral precepts, but by furniture, clothes, utensils, commodious houses, wholesome and well-prepared food: and without the luxury of enamelled gold-boxes, diamonds, pictures, bronzes, and statues, we should not have a multitude of agreeable and useful articles which are reckoned essential to our comforts.[37]

This ambiguity would haunt much thinking on the subject during the upheavals following the revolution of 1789.

The French Revolution

The central event of this epoch, the overthrow of Louis XVI in 1789 and subsequent founding of the French Republic, awakened powerful desires for liberty, equality, and fraternity whose influence continues to the present day. The American colonists' revolt took place in a relatively equal and largely agricultural country. The French revolutionaries confronted a corrupt and highly unequal state where commerce was well advanced, and the love of luxury, regarded as essential to monarchies, deeply ingrained. The crisis which produced the revolution was in part driven by the increasing disjuncture between the splendours of Versailles and the growing degradation of the poor, as well as the unfairness of taxation, and the many exemptions from it enjoyed by nobles and clerics. The Revolution excited widespread expectations throughout Europe and beyond regarding the prospects of humanity's moral,

36. Louis-Sébastien Mercier, *Fragments on Politics and History*, 2 vols (H. Murray, 1795), 2: 1–2, 4.

37. Mercier, 1: 184; 2: 6. For discussion of the context, see Sonenscher's *Sans-Culottes*, 134–63.

economic, and political improvement. The future was now an exciting proposition, and the past a ladder to be kicked away. The age was pregnant with change: Leibniz's proclamation that "le temps present est gros de l'avenir" was on Mercier's title page.[38]

Yet to some the future bore a more than passing resemblance to the distant past. After 1789 the spirit of Rousseau and Mercier was everywhere evident. As the German philosopher Johann-Gottlieb Fichte enthused in 1794, discussing "the golden age of sensual enjoyment without physical labour which the old poets describe", the general effect of the French Revolution was to bolster the sentiment that "what Rousseau, under the name of the State of Nature, and these poets under the title of the Golden Age, place behind us, lies actually *before us*."[39] An epoch began of what David Landes calls "apocalyptic time" defined by a "messianic hope of regeneration", in which centuries of oppression and inequality seemed to be ending. Justice, the historian Jules Michelet later observed, now appeared "to sit in judgment, on the dogma and on the world. And *that day* of Judgment will be called the Revolution."[40]

The centrality of equality in the unholy trinity of revolutionary ideals meant that the luxury problem haunted its champions from the outset. Efforts like the Law of the General Maximum, passed in September 1793, to restrict commodity prices, and a Law on Hoarding, which in the case of necessities was punishable by death, were widely supported.[41] Various schemes for agrarian laws were also mooted. The Abbé Fauchet's *De la réligion nationale* (1791), for instance, suggested limiting landed estates to 50,000 livres (£3760) annual rent, with a progressive inheritance tax reducing larger estates.[42] Less easily solved was the problem as to whether luxury should be abolished, as in Sparta, or universalised, if possible.[43] The spirit of Rousseau dictated strong measures against it. Amongst the revolutionary leaders, Jean-Paul Marat had declaimed against commerce and luxury in the 1770s, with a view to avoiding poverty.

38. "The present is filled with the future", wrote Leibniz, who added "and laden with the past": G. W. Leibniz, *Discourse on Metaphysics and Other Essays* (Hackett, 1991), 55.

39. Johann Gottlieb Fichte, *The Popular Works of Johann Gottlieb Fichte*, 2 vols (Trübner, 1889), 1: 202–3.

40. Landes, *Heaven on Earth*, 253–54; Jules Michelet, *Historical View of the French Revolution* (G. Bell and Sons, 1896), 26.

41. See Rebecca L. Spang, "What Is Rum? The Politics of Consumption in the French Revolution", in *The Politics of Consumption: Material Culture and Citizenship in Europe and America*, eds Martin Daunton and Matthew Hilton (Berg, 2001), 33–50.

42. A cow cost 100 livres at the time. Thus worth about £564,000 today.

43. On classical references here, see Harold T. Parker, *The Cult of Antiquity and the French Revolution* (University of Chicago Press, 1937), 146–70.

But a prominent disciple of Rousseau, Maximilien Robespierre, told the Convention that the revolution aimed only at "the peaceful enjoyment of liberty and equality", adding, "We do not intend to cast the Republic in a Spartan mold".[44] Other key actors, such as Robespierre's right-hand man, Louis Antoine de Saint-Just (1767–94), were unwilling to offer "any opinion on whether or not luxury is a good thing in itself".[45]

Revolutionary style naturally rejected ostentation. Robespierre's brother claimed he "lived like a Spartan", and in a spirit of personal austerity "spent nothing on himself".[46] Saint-Just, too, disdained aristocratic extravagance and artificiality, and defined the new character-type: "A revolutionary man is uncompromising, but he is judicious, he is frugal; he is simple without displaying luxury or false modesty; he is the irreconcilable enemy of every deceit, every indulgence, every affectation."[47] But the moderate Brissot, while conceding that ancient sumptuary laws might have promoted virtue, doubted whether they suited substantial commercial states like France, and regarded them as too "sublime" for "our little souls".[48] Some Jacobins renounced sugar and coffee to mark their disapproval of colonial commerce. One, Lazare Carnot, told a provincial club that "We shall be the Spartans of the Revolution". But another Jacobin, speaking of Lyons, whose silk weavers had long depended on the luxury trade, insisted that "Sybarites cannot become Spartans."[49] This might well have been, or yet be, a fitting epitaph for the moderns.

Nonetheless, greater simplicity in personal appearance enjoyed what Roche calls a "temporary triumph".[50] For a time, clothing powerfully symbolised the revolution, and luxury in dress beat a hasty retreat.[51] The tricolour

44. Quoted in Patrice Higonnet, *Goodness Beyond Virtue: Jacobins during the French Revolution* (Harvard University Press, 1998), 102.

45. Norman Hampson, *Saint-Just* (Basil Blackwell, 1991), 88.

46. Louis R. Gottschalk, *Jean Paul Marat: A Study in Radicalism* (University of Chicago Press, 1967), 215.

47. Quoted in Whatmore, *Republicanism*, 94.

48. Quoted in H. Parker, *Cult of Antiquity*, 65.

49. Quoted in Higonnet, *Goodness Beyond Virtue*, 141, 138. The Lyonese workers soon stopped supporting the Jacobins.

50. D. Roche, *Culture of Clothing*, 314.

51. Ribeiro, *Fashion*, 60. One British visitor noted "a great affectation of that plainness in dress and simplicity of expression which are supposed to belong to Republicans", having met "a young man of one of the first families in France . . . a violent democrate", at the theatre "in boots, his hair cropt and his whole dress slovenly; on this being taken notice of, he said 'That he was accustoming himself to appear like a Republican'" (70). The *Journal de la mode et du gout* (15 November 1790) noted that the vogue for simplicity in dress led some men to have their hair

cockade became mandatory for men from July 1792, and for all citizens from April 1793. Discussions took place about a national costume, showing, Lynn Hunt suggests, "that some hoped to erase the line between public and private altogether".[52] A minimalist style of dress emerged, defined by the bonnet rouge, the wearing of cockades, sans-culottes (trousers instead of silk breeches) for workers, and even Greek-style togas and sandals, at least for women.[53] But after the reaction of Thermidor, the 1794 coup which overthrew Robespierre, such "maratisme" was condemned as dirty and dishevelled.[54]

The appeal of Sparta nonetheless lived on, and would be transmitted to the next generation by one of the most determined enemies of luxury ever. To the first modern revolutionary communist, François-Noel "Gracchus" Babeuf (1760–97), who read Mercier's *L'an 2440* around 1790, a fair distribution of property was highest priority. Robespierre, he thought, rightly followed Rousseau in aiming to give all enough, but none too much. Nature's bounties were restricted, but dividing the land of France equally among its six million families, based on a natural right to subsistence, would provide each with 14 acres. The context here was thus still overwhelmingly agricultural. Industry and mass production were little developed. Making luxury goods merely distracted from cultivating necessaries, so could be suppressed, along with commerce in "foreign superfluities". Babeuf's egalitarianism was militant and vigorous. Even in clothing, he thought, it was "essential to the happiness of individuals, and to the maintenance of public order, that the citizen should habitually find in all his fellow-countrymen equals, brothers; and that he should nowhere meet with the least sign of even apparent superiority, the precursor of despotism

---

"coupés et frisés comme ceux d'une tête antique" (cut and dressed in the antique style) (146). Helen Maria Williams found that even cleanliness was in decline, and that those who wore a clean shirt were condemned as fops. The uniform of Jacobins was the bonnet rouge, and plain dress for women (83). Discussions were afoot to adopt a specific republican costume, and Jacques-Louis David provided a number of designs, in which classical motifs loom large. A "modest simplicity" (90) in dress peaked as an ideal in 1794—when Robespierre was in power. But by 1795 more-Spartan costumes were giving way to the more luxurious, and "luxury was back in fashion by the late 1790s" (132). The overall effect of the revolution was to banish "the notion that clothing was the symbol of an immutable class system based on unjust privilege" and replace it with the idea that "dress was above all a statement of freedom and an expression of individuality" (141).

52. Lynn Hunt, in Michel Perrot, ed., *A History of Private Life*, 5 vols (Harvard University Press, 1990), 4: 18–19.

53. Restif de la Bretonne, *My Revolution: Promenades in Paris, 1789–1794*, ed. Alex Karmel (McGraw-Hill, 1970), 378–79.

54. Richard Wrigley, *The Politics of Appearances: Representations of Dress in Revolutionary France* (Berg, 2002), 200.

and servile submission." For "The aim of society is the happiness, of all, and happiness consists in equality."[55]

Babeuf's opposition to symbols of inequality lay in the unhappiness they caused the majority, because they could not meet the expectations which resulted:

> The sight of distinctions, of pomp, and of pleasures enjoyed only by a few, was, and ever will be, an inexhaustible source of torments and uneasiness to the mass. It is given to only a few philosophers to preserve themselves from corruption, and moderation is a blessing which the vulgar can no longer appreciate, when once their thoughts have been weaned from it. Do certain citizens create to themselves new factitious wants, and introduce into their enjoyments refinements unknown to the multitude? Simplicity is no longer loved, happiness ceases to consist in an active life and tranquil soul, distinctions and pleasures become the supreme of goods, nobody is content with his station, and all seek in vain for that happiness, the entrance of which into society is debarred by inequality.[56]

Nonetheless, Babeuf reflected,

> Equality and simplicity do not exclude elegance and propriety. There might be, for instance, different colours and forms to distinguish the different ages and occupations; and there is no reason why the citizen should not wear a different costume at the assemblies and festivals, from his ordinary one in the workshop. The girls, too, might be differently attired from the grown women; and it might prove useful, as well as pleasing, that the youth, the adult, the old man, the magistrate, and the warrior, should have each a peculiar and appropriate costume. Indeed, with respect to dress in general, the Insurrectional Committee was of opinion that the main point to consult for was health, and the development of the organs; for fashion and frivolity it had no sympathies.[57]

Hence "all persons who would introduce frivolities and foreign fashions" would be subject to "severe surveillance". To R. B. Rose, Babeuf's scheme was an agrarian, "pseudo-Spartan utopia".[58] His ideal was community of goods, something

55. Philippe Buonarroti, *Buonarroti's History of Babeuf's Conspiracy for Equality*, transl. Bronterre O'Brien (H. Hetherington, 1836), 164; quoted in Hertzler, *History of Utopian Thought*, 189.

56. Manuel and Manuel, *French Utopias*, 251.

57. Buonarroti, *Babeuf's Conspiracy for Equality*, 164.

58. Buonarroti, 191; Lorenz von Stein, *The History of the Social Movement in France, 1789–1850* (Bedminster, 1964), 165; R. B. Rose, *Gracchus Babeuf: The First Revolutionary Communist* (Edward Arnold, 1978), 234.

"more sublime and more just" than a mere agrarian law, without distinctions based on wealth.[59] One of his followers, the "first professional revolutionary" Filippo Buonarroti, subsequently proposed a self-sufficient agrarian republic, where urban development would be discouraged, wealth would be heavily taxed, those guilty of idle or luxurious living would be condemned to hard labour, and gold would be as useless as "sand and stones".[60] We are not far here from Thomas More's ideal. So the Spartan ideal lived on into the next generation, now wedded to what later writers call secular millennialism.[61]

In Britain, the French Revolution drew on long-standing romantic criticisms of urban and commercial society. Neo-Harringtonian and utopian republican sentiments were popularised in the 1790s in the works of, amongst others, Thomas Spence and William Godwin, in the Godwinian *An Essay on Civil Government* (1793), in Thomas Northmore's *Memoirs of Planetes* (1795), and in William Hodgson's *The Commonwealth of Reason* (1795).[62]

The earliest to commence such agitation was the neo-Harringtonian bookseller Thomas Spence (1750–1814), who endeavoured in the 1790s to persuade the public that the limited claims for rights propounded by the much better known Thomas Paine would not stem inequality, and required parish-based public ownership of landed property to succeed. Spence also used the literary form to popularise his ideas, notably in *A Supplement to the History of Robinson Crusoe, Being the History of Crusonia* (1782) and in the *Description of Spensonia* (1795).[63] Unlike some more-primitivist utopians, Spence sought to reconcile traditional republican aims with the attractions of increasing opulence. Despite their agrarian focus, his works still envision town life as central to the ideal society. Spence thought the modern world had done far too much "to support and spread luxury, pride, and all manner of vice". He insisted that "all our boasted civilisation is founded alone on conquest", the savage state having been abandoned solely to benefit plunderers. If all land was owned by parishes,

59. David Thomson, *The Babeuf Plot: The Making of a Republican Legend* (Kegan Paul, Trench, Trubner, 1947), 31.

60. Elizabeth L. Eisenstein, *The First Professional Revolutionist: Filippo Michele Buonarroti (1761–1837)* (Harvard University Press, 1959), 74–75.

61. Frederic J. Baumgartner, *Longing for the End: A History of Millennialism in Western Civilization* (Macmillan, 1999), 139–41.

62. See my *Utopias*, vii–xxviii.

63. A recent account of his works and influence is Alastair Bonnett and Keith Armstrong's *Thomas Spence: The Poor Man's Revolutionary* (Breviary Stuff, 2014).

none would "hunger and thirst after the Riches of the World, to the pernicious degree that is now common. For observe, though they should acquire the Riches of Peru, they could only speculate in fair and honest Trade, and Manufactures." Equality was thus more important than simplicity. So Spence still portrayed his model society in 1782 as "full of superb and well furnished Shops and . . . every Appearance of Grandeur, Opulence, and Convenience, one can conceive to be in a large Place, flourishing with Trade and Manufactures."[64]

Spence's compromises with luxury indicate that Britain's opulence meant that demanding any kind of reversion to primitivism on the grounds of its supposed virtues was a far less appealing strategy than a few decades earlier. There was also the risk that revolution might destroy the economy entirely. This became a central argument in the political debates of the early and mid-1790s, probably tipping the balance amongst the middle classes from the republican Paine to the monarchist Burke. Spence conceded that sliding backwards into barbarism was a possible outcome of radical attacks on wealth. Writing in 1807 to the London physician and social reformer Charles Hall, whose *The Effects of Civilization on the People in European States* (1805) seemingly denounced civilisation as such, he asked whether Hall would "have us all to become again Goths and Vandals and give up every elegant comfort of life?" adding that "You seem to be to be sliding into the System of Sir Thomas More's Utopia wherein he makes every kind of Property the Property of the Nation and the People obliged to work under Gang Masters".[65] In the telling if disingenuously titled *The Restorer of Society to its Natural State* (1801), Spence also distanced himself from Rousseau. In future, he thought, as "all nations, however barbarous or civilized, have naturally a taste for foreign productions and luxuries, and will do anything they can to acquire them, so may we expect this people." They would be well educated, and since "Reading promotes refinement and sensibility, and a taste for elegance in clothes, furniture and every department in life", the gratification of "this multiplication of refined desires" was also only "natural". Yet in *The Constitution of Spensonia* (1803) he still insisted that "Men would learn to moderate their desires, and cease to aspire after boundless wealth, which they could have no means of consolidating."[66] As the eighteenth century drew to a close, commerce was thus ambiguously wedded to agrarian republicanism.

64. Thomas Spence, *The Political Writings of Thomas Spence*, ed. H. T. Dickinson (Avero, 1982), 7.

65. Thomas Spence and Charles Hall, "Four Letters between Thomas Spence and Charles Hall", *Notes and Queries* 28 (1981): 320.

66. Thomas Spence, *Pig's Meat: The Selected Writings of Thomas Spence*, ed. G. I. Gallop (Spokesman, 1982), 63, 75, 145–46.

We encounter similar difficulties in the writings of the leading British utopian republican of the period, and ardent disciple of Rousseau and Fénelon, William Godwin (1756–1836). Godwin is often regarded as the founder of philosophical anarchism, largely because he not only viewed political activity as intrinsically corrupt but also thought many forms of association threatened individual freedom, by bending our opinions towards the outlook of the group and thus thwarting the inner voice of reason. His highly influential *Enquiry concerning Political Justice* (1793) proclaimed the advantages of returning to a more virtuous, simple society based on subsistence agriculture and a morality of "disinterested benevolence" toward all. Here the need for private judgment, forming one's own opinions without group biases, and increasing virtue, would be promoted by minimising dependency on others.[67] Godwin also envisioned self-moving ploughs, and machines generally as the "Helots" of the future; the triumph of reason over sexual desire; the general use of surplus resources for the good of the needy and worthy in particular; and eventually prolonging life.[68] He regarded crime as flowing from inequality and denounced "one man's possessing in abundance that of which another man is destitute."[69] The desire for wealth he thought was driven by "the love of distinction and fear of contempt". The former was the "present ruling passion of the human mind", with "the principal and unintermitting motive to private accumulation" being "the love of distinction and esteem":

> How few would prize the possession of riches, if they were condemned to enjoy their equipage, their palaces and their entertainments in solitude, with no eye to wonder at their magnificence, and no sordid observer ready to convert that wonder into an adulation of the owner? If admiration were not generally deemed the exclusive property of the rich, and contempt the constant lacquey of poverty, the love of gain would cease to be an universal passion.[70]

Yet, thought Godwin, this impulse might well be diverted in a rational society, where the avaricious or luxurious were "contemplated with as much disapprobation as they are now beheld by a mistaken world with deference and respect." As he put it in 1798, he thought

> the allurements that now wait upon costly gratification, would be, for the most part, annihilated. If, through the spurious and incidental

67. The key study here is Mark Philp's *Godwin's Political Justice* (Duckworth, 1986).

68. W. Godwin, *Enquiry Concerning Political Justice*, 2: 846.

69. W. Godwin, 2: 809.

70. W. Godwin, 1: 33; 2: 837.

recommendations it derives from the love of distinction, it is now rendered, to many, a principal source of agreeable sensation, under a different state of opinion, it would not merely be reduced to its intrinsic value in point of sensation, but, in addition to this, would be connected with ideas of injustice, unpopularity and dislike.[71]

So the undue respect accorded to wealth was only a function of public opinion, and if opinion had created it, opinion could also remove it. Luxury was thus a key target here as universally adding further burdens on the poor, and inhibiting the practice of universal benevolence, the maximum sociability of which we are capable. To Godwin, there was "nothing more pernicious to the human mind than the love of opulence". If all were self-subsistent producers, and "If superfluity were banished, the necessity for the greater part of manual industry of mankind would be superseded; and the rest, being amicably shared among the active and vigorous members of the community, would be burthensome to none". Thus in future there would be "no persons employed in the manufacture of trinkets and luxuries".[72]

In 1797 Godwin reiterated that "every man who invents a new luxury adds so much to the quantity of labour entailed on the lower orders of society."[73] Yet he too began to shift his views in the late 1790s, conceding not only that abolishing property might induce barbarism, but that civilisation, rather than simplicity, most conduced to humanity's mental advancement.[74] This retreat clearly indicates just how far radical argument had moved away from Rousseauist primitivism by the turn of the new century. In the mid-1790s, however, Godwin's plea for simplicity won him many acolytes. Samuel Taylor Coleridge, Robert Southey, and others projected a scheme of Pantisocracy in which paradise would bloom again in the North American wilderness, with goods and perhaps wives shared in common.[75] Literary echoes of such sentiments appeared in Godwinian Thomas Northmore's *Memoirs of Planetes; or, A Sketch of the Laws and Manners of Makar* (1795). Makar boasts relative equality of property, estates being divided equally amongst children. Its inhabitants are

71. William Godwin, *Enquiry Concerning Political Justice*, 3rd ed. (1798; Penguin Books, 1976), 706.

72. W. Godwin, *Enquiry Concerning Political Justice*, 2: 806, 821.

73. William Godwin, *The Enquirer* (G. C. and J. Robinson, 1797), 157.

74. See my "From True Virtue to Benevolent Politeness: Godwin and Godwinism Revisited", in *Empire and Revolutions: Papers Presented at the Folger Institute Seminar "Political Thought in the English-Speaking Atlantic, 1760–1800"*, ed. Gordon Schochet (Folger Library, 1993), 187–226.

75. Mrs Henry Sandford, ed., *Thomas Poole and His Friends*, 2 vols (Macmillan, 1888), 1: 98. See J. R. MacGillvray, "The Pantisocracy Scheme and Its Immediate Background", in *Studies in English*, ed. Malcolm W. Williams (Oxford University Press, 1931), 131–69.

"almost total strangers to luxury; they are plain and simple in their diet, living chiefly upon the various fruits of the earth, and eating very little animal food".[76] Utopian republicanism also appeared in the wild communitarian schemes of the eccentric, self-styled "man of nature" "Walking" John Stewart, who envisioned group marriage, community of goods, and abandoning cities, anticipating much that Robert Owen would shortly propose.[77] Other reformers in the period took quite a different tack on the issue of needs, however, even going so far as proposing complete national economic regulation. In William Hodgson's *The Commonwealth of Reason* (1795), a "Committee of Agriculture, Trade and Provisions" ensures that every district has a proper supply of all necessaries at a regulated price, so that each person can "satisfy all the real wants of his nature, and make provision for his old age". Here too people are described as "otherwise naturally friends and brothers", and only "set at enmity with each other, for the enjoyment of paltry titles, that do not *really* distinguish the possessors from the mass of mankind".[78]

Both the Paineite radicals and the Godwinian utopians were satirised in some of the first overtly anti-utopian dystopias to appear in English, where warnings that revolutionary upheaval threatened a reversion to barbarism are common.[79] The success of this loyalist strategy indeed seems to have pushed some other reformers besides Spence and Godwin away from the appeal to primitivism, which was clearly waning by 1800. The years 1795–1805 thus clearly indicate a watershed in British thinking about luxury, inequality, and commercial society. A key figure who marks this shift was the pamphleteer and popular democratic lecturer John Thelwall, the chief orator of the London Corresponding Society, the first mass-scale democratic organisation in Britain. In 1793–94 Thelwall wrote a poem entitled "Simplicity of Manners", and condemned the "noxious weeds" of luxury. In 1795–96 he argued that it was necessary to

> labour to abolish luxury. . . . Let us in our own houses, at our own tables, by our exhortations to our friends, by our admonitions to our enemies, persuade mankind to discard those tinsel ornaments and ridiculous superfluities which enfeeble our minds, and entail voluptuous diseases on the affluent, while diseases of a still more calamitous description overwhelm the oppressed orders of society.[80]

76. Thomas Northmore, in Claeys, *Utopias*, 189–91.

77. See my "'The Only Man of Nature That Ever Appeared in the World': 'Walking' John Stewart and the Trajectories of Social Radicalism, 1790–1822", *Journal of British Studies* 53 (2014): 1–24.

78. William Hodgson, in Claeys, *Utopias*, 231, 225.

79. These are discussed in my *Dystopia*, chap. 4.

80. Gregory Claeys, ed., *The Politics of English Jacobinism: Writings of John Thelwall* (Penn State University Press, 1995), xlii, 67.

But while praising "a love of *virtuous poverty*" as among the indispensable prerequisites of a people "desiring to obtain or to preserve the blessings of liberty", Thelwall like Godwin nonetheless began to shift towards a grudging acceptance of commerce and the utility of buildings, books, paintings, and the like, so long as trade was conducted on "liberal and equal principles."[81]

This was an extremely important concession. Like Godwin, Thelwall in the later 1790s in fact turned sharply away from neo-Spartan republicanism. If commerce and the desire for luxury entailed greater labour for the poor, they were to be condemned. If increasing amenities could be shared equitably, which the newly emerging manufacturing system seemingly implied, this objection to luxury, at least, was neutralised. Charles Hall's *The Effects of Civilization on the People in European States* was amongst the first works to apply these arguments to manufactures. But Hall remained vehemently opposed to commerce, regarding growing wealth as merely a further "increase of poverty in the people, as it subjects them to new and additional demands for the produce of their labour." Terming the rich "capitalists", in one of the first modern usages of the term describing an oppressive relationship, Hall insisted that there was in the employment contract "no voluntary compact equally advantageous on both sides, but an absolute compulsion on the part of masters". Wealth was merely "power over the labour of the poor". He thus offered one of the last great statements of agrarian republicanism.[82] The turning point in this development came with Hall's suggestion that the old republican programme of popular sovereignty and relative social equality, seemingly incarnated in the young United States, did not in fact generate sustained equality, because here commerce concentrated wealth in the hands of the few. By the turn of the nineteenth century, it was becoming evident that economic inequality in the United States was rapidly undermining the egalitarianism of the early republic. Within a generation, radicals would shift from opposing aristocracies and monarchies to seeing capitalism and the factory system as the greatest enemies of the working classes.

In France we can trace a similar pattern of development, as appeals to simplicity paralleled acknowledgements that humanity's progressive improvement rested on a desire for new commodities and experiences. Here the *Esquisse d'un tableau historique des progrès de l'esprit humain* (translated as *Sketch for a Historical Picture of the Progress of the Human Mind*, 1793) by the Marquis de Condorcet (1743–94) became the best-known text to describe the desire for

81. Claeys, *Politics of English Jacobinism*, xlii–xliii.

82. Charles Hall, *The Effects of Civilization on the People in European States* (1805), 94, 268, 73.

novelty as central to humanity's indefinite forward movement. To Condorcet, the "need for new ideas and new feelings" was "the prime mover in the progress of the human mind", the "taste for the superfluities of luxury" being "the spur of industry". In the final stage of progress, inequality between and within nations would disappear, "poverty, humiliation [and] dependence" would end, and the "true perfection of mankind" would occur as all nations reached a uniform, high level of civilisation. Here, however, in a compromise similar to Fénelon's, attitudes towards luxury would alter dramatically. Material and moral progress would advance hand in hand. In "a society enjoying simpler manners and more sensible institutions", wealth would no longer "be a means of satisfying vanity and ambition". All would uphold "principles of strict and unsullied justice, . . . habits of an active and enlightened benevolence, [and] of a fine and generous sensibility".[83] So again, social equality, luxury, emulation, and sociability are understood as inexorably wrapped together, and luxury and vanity are dissevered once and for all. Condorcet thought that "Sumptuary laws are unjust, and hurt industry", and following Turgot proposed instead laws to divide property more equally.[84] Women would also enjoy equal rights with men. But utopia remained an imperial threat, with the warning that savage tribes might be "reduced in number as they are driven back by civilized nations [and] will finally disappear imperceptibly before them or merge into them".[85] So much for the "love of freedom" as the epitome of enlightened progress. Utopia was not yet universal, but would remain subverted by issues of race and empire for more than a century longer.

Condorcet and Godwin were singled out by the Reverend Thomas Malthus as targets in *An Essay on the Principle of Population* (1798), which is generally regarded as the most important anti-utopian tract of the period. Its key argument is that the more opulent a society is, the more rapid would be its growth in population, until a failing food supply reduced numbers again. Condorcet's secretary, Auguste Comte, took a similar line of thinking into the mid-nineteenth century, producing one of the most influential utopian schools of the epoch, Positivism, and invoking Joachim of Fiore as a predecessor in the endeavour.[86]

In the New World utopian republicanism also made its mark in this period. The Scottish emigrant John Lithgow was the anonymous author of the first

83. Condorcet, *Sketch*, 32, 173, 180, and generally 173–202.

84. [Marquis de Condorcet], *The Life of M. Turgot* (J. Johnson, 1787), 319–20.

85. Condorcet, *Sketch*, 177. In the colonies, he thought that Europeans might "either civilize or peacefully remove the savage nations who still inhabit vast tracts of its land" (175).

86. Auguste Comte, *System of Positive Polity*, 4 vols (Longmans, Green, 1875–77), 3: 409. On these developments, see Frank Manuel's *Shapes of Philosophical History* (George Allen and Unwin, 1965, 92–114).

major American literary utopia, the neo-Spartan *Equality: A History of Lithconia* (1802), whose title joins his own name to that of Laconia, or Sparta. This describes a transition from a regime where "nine-tenths of mankind groaned under the most oppressive tyranny, labouring from morning till night for a poor and scanty diet . . . while the other tenth enjoyed every luxury, and rioted in waste and profusion". At present, society was "poisoned by self-interest and worldly pursuits." In Lithconia, "Friendship, love, and a well-regulated industry, are blessings which might amply compensate for all the evils which flesh is heir to". We are asked that where "There may be abundance of wealth, but not much enjoyment . . . Is it living, to exist without friendship? Is life worth the possessing, without love? Does men enjoy themselves [*sic*] either in idleness or in slavery? No."[87] Here community of goods and the regulation of commerce and consumption ensure needs are met. To solve the luxury problem, "Variety of dress and equipage is also unknown. Every citizen having a certain quantity of cloaths distributed to him at stated periods, the whole country appears almost in uniform", though "fanciful variety" exists in self-designed female head coverings. There are no natural inequalities, and "When a person's rank in society is mentioned, it is only by the natural distinctions of infancy, youth, manhood and seniority". Marriage is described as once

> held as sacred here as in other countries; but when property became in common, it fell gradually into disuse—for the children being the property of the state, educated and brought up at the public expence, and as women could live as well single as united young people were seldom at the trouble to make such a contract. Children were born and no one thought it his business or interest to enquire who was the father.[88]

A scheme of universal labour lasts from ages fifteen to fifty, and the nation is directed by its elder members. Lithconia is described as "only one large city upon a grand scale".[89] These proposals, clearly emanating from the revolutionary debate in Britain during the 1790s, indicate yet again that Sparta had not yet lost its appeal, and also that Thomas More was being read consistently as proposing a regime which was practicable. (*Utopia* was reprinted in 1789, in 1795, and twice in 1808.)

87. [John Lithgow], *Equality: A Political Romance; Without Either Kings, Ghosts, or Enchanted Castles* (1802), 33, v–vi.

88. [Lithgow], 16, 18, 6.

89. [Lithgow], 6.

8

Simplicity and Sociability in Nineteenth-Century Utopianism

THE RAPID EXPANSION of industry dominated nineteenth-century Europe. With the end of the Napoleonic Wars came peace, but not stability. The industrial system now attracted hundreds of thousands from the countryside into the new factory towns, first in Britain, then increasingly elsewhere. The new machines were alternately threatening, for those who worked them and became chained to their rhythm, and full of promise, for those who owned them. The factories intensified and speeded up the labour process and brought more onerous discipline, increasingly repetitive and dangerous work, and often wages which barely matched subsistence needs. The new towns and cities were soon overcrowded and by the 1840s had vast slums. The new science of the age, political economy, promised universal affluence. But many saw only relentlessly increasing competition, and a heartless disdain for the poor. Amongst the first great critics of the ethos of commercial society, Thomas Carlyle complained loudly that "We call it a Society; and go about professing openly the totalest separation, isolation. Our life is not a mutual helpfulness; but rather, cloaked under due laws-of-war, named 'fair competition' and so forth, it is a mutual hostility."[1] *Gemeinschaft* seemed the first casualty of the new way of life. Struggle, of race, class, species, nation, and empire, would become the great image of the coming age.

Nostalgia for pre-industrial life and the softer, slower, and mostly less capricious rhythms of the seasons was one response to these developments. Various forms of agrarian radicalism, hostile both to cities and to manufacturing, remained popular until about World War One. In Britain the most influential early nineteenth-century radical, William Cobbett, vehemently opposed the growth of "the great Wen", London. The causes of the nation's degeneration,

1. Thomas Carlyle, *Works*, vol. 10 (Chapman and Hall, 1899), 146.

he thought, included "the drawing of wealth from the many, and giving it to the few" and emulating French habits of luxury.[2] Cobbett's ideal was a prosperous rural peasantry and squirearchy who endorsed unostentatious simplicity. He thought it better if male attire in particular was as cheap "as may be without *shabbiness*". As earlier, this continued to be a gendered debate. Masculinity hinged on what David Kutchka terms "inconspicuous consumption", where the linkage of political corruption, self-indulgence, and the "effeminate luxury" beloved of a French-inspired aristocracy is contrasted to a virile, manly, rural type.[3] What Richard Sennett calls the "corrosion of character" remained a constant concern.[4]

More radical egalitarian principles, now chiefly socialist, were soon revived in this milieu. Resistance to luxury pervaded the socialist and utopian traditions for most of the coming century. Early socialism, we have seen, was at least initially mostly hostile to luxury as tending to promote inequality. Soon, however, socialists would begin to explore the prospect that machinery could provide a much higher standard of living for all, and thus decouple luxury from exploitation. This transition would eventually indicate a marked shift away from the ideal of a simpler society as we reach the twentieth century.

Early Nineteenth-Century Literary Utopianism

More than ever before, nineteenth-century authors of literary utopias were aware that attaining simplicity was easier in projected new colonies well removed from the temptations of civilised urban life, where luxuries were scarce, or in rediscovering imaginary communities which history had passed by, compared to describing how corrupt societies could restore virtue. The remote island colony or lost mariner tropes, and the hopes they inspired, began, however, to weaken in appeal throughout the nineteenth century, as the world shrank through conquest and exploration. The subgenres remained, but more as entertainment than social theory. Increasingly they represented not the

2. William Cobbett, *Advice to Young Men* (1829; Henry Froude, 1906), 208.

3. Cobbett, *Advice to Young Men*, 21. See David Kuchtam, "The Making of the Self-Made Man: Class, Clothing, and English Masculinity, 1688–1732", in *The Sex of Things: Gender and Consumption in Historical Perspective*, ed. Victoria de Grazia, with Ellen Furlough (University of California Press, 1996), 54–78.

4. Richard Sennett, *The Corrosion of Character: The Personal Consequences of Work in the New Capitalism* (W. W. Norton, 1998).

possibility of an alternative reality awaiting discovery, but either a psychological alter ego or a variation on imperial occupation, particularly, as Duncan Bell shows, in an Anglo-American context.[5] The tradition of locating ideal societies in unknown parts of the earth did not die out until the planet was almost fully explored, around 1900. Then this narrative largely shifted to space, and the genre to science fiction. So imaginative projections in the Americas, Africa, and Australia continued throughout this period where ideals of an alternative simplicity appear. A few brief examples illustrate this theme.

Archbishop Richard Whately's *An Account of an Expedition to the Interior of New Holland* (1837) describes a European colony called Southland, in what is now Australia, after three hundred years of settlement. Here, excessive bodily ornamentation is condemned where merely demonstrating superfluity is intended. So too:

> The cookery among the higher classes is for the most part plain and simple, and the few who have refined much upon the luxuries of the table are exposed to something of the same sort of contemptuous ridicule that the being called a dandy incurs among us.

Luxuries are still permitted where their functionality is evident, such as

> handsome and costly gold brooches and buckles, buttons made of jewels, embroidered garments, inlaid tables, and other such ornamental articles; but you will see no article that is merely an ornament. A gold brooch or button served as a fastening, not better indeed, but as well, as an iron or brass one. Its beauty is superfluous, but it is not itself superfluous, and destitute of all ostensible use. So, also, a silver goblet serves to drink out of, and an embroidered gown to cover one, no less than plain ones.[6]

More was known about the United States than Australia in this period, and perhaps for this reason utopian literary projections were less fanciful here. The United States also attracted large number of radical emigrants in the 1790s and subsequently. In *New Britain* (1820), an imaginary republican settlement in the American West, maintaining small-scale communities, valuing individual and martial virtue, preserving social equality, and inhibiting the desire for power are all intimately intertwined. Here, we are told, "We do not consider it just for a man to acquire all he can: such conduct would tend to deprive his neighbours of their rights, and to enslave his equals. Prevailing luxury and general

5. See Duncan Bell, *Dreamworlds of Race: Empire and the Utopian Destiny of Anglo-America* (Princeton University Press, 2020).

6. Richard Whately, in Claeys, *Modern British Utopias*, 7: 269, 290.

happiness are the opposing scales of human life." New Britons "consider true riches to consist in exemption from want: they think those who have the fewest wants, not those who have the largest possessions, to be the richest." They believe commerce "creates more wants than it supplies", "seldom brings a return adequate, in usefulness, to the goods and the labour", and undermines "that equality of condition which is requisite to real liberty." Similarly they are hostile to large cities:

> We consider it wise to live upon the spot where our necessaries of life are produced; and not to congregate in great cities, and thus have to fetch them from a distance. We by this, prevent a few from rising into improper consequence, and the many from sinking into servility; the inevitable result of their different employments in such large collections of men; which appear to us to be large masses of human corruption. We also thus prevent that degradation of moral character, which would be alike despicable in those who would be so unjust as to grasp at more than what is needful of property or power, and in those who would be mean enough to submit to servility, or to the privation of any part of their just and equal rights. We read attentively the histories of other nations; and we consider that we have discovered the true source of all their misfortunes, collectively and individually, to be, the love of unreasonable wealth and power. Be not then surprised that we most determinedly avoid in ourselves, and prevent in others, these, as the greatest evils, which the people of other countries eagerly pursue as their greatest good.[7]

A similar plea for simplicity defines a domestic American utopia, *Symzonia* (1820). Here "by dressing alike, they maintain a perfect equality in their wants in that respect", for "They wear garments because they defend the body, and are necessary to decency; but it never occurred to their simple minds, that the fairest work of an Infinite Being could be improved by trinkets and fripperies of man's device." Trade "is practised only for the common convenience of society. The accumulation of wealth, and indulgence in luxury, being disreputable, and a bar to admission to the distinguished orders, an overreaching and avaricious spirit is not generated by traffic". When anyone amasses "beyond his wants, the surplus is in general voluntarily devoted by him to the use and benefit of his fellow-beings". We find also the observation that "men feeding upon animal food and costly drinks, and given to the indulgence of inordinate passions, must of necessity become very unequal in their condition, depraved

7. Claeys, *Modern British Utopias*, 6: 174, 177, 298, 301.

in their appetites, and miserable in proportion to their aberrations from the strictest temperance, virtue, and piety."[8]

Utopian Social Theory

Karl Marx

Turning from literature to the practical utopian reformers of the early and mid-nineteenth century, we recall that by the 1830s socialist hostility towards luxury and the multiplication of needs began to give way to visions of more egalitarian opulence, often described as public luxury. For the later moderns, Karl Marx is to utopia what George Orwell is to dystopia: the necessary starting point, in terms of sheer influence.[9] But positioning Marx within this tradition is highly controversial, since he is rarely classed as a utopian.[10] His credentials as a proto-environmentalist thinker can also be disputed.[11] Marx's chief statement of his utopian aspirations, the programmatic *Manifesto of the Communist Party* (1848), offered a detailed account of future organisation and clearly extended Thomas More's original ideal, to abolishing exploitation as such. But he was wary of, and indeed hostile towards, the "utopian" label.

One reason for this was a desire to break explicitly from the tradition. Both Marx and Engels were engaged in the early 1840s with the ideas of Owen, Fourier, Cabet, Proudhon, Hess, and Weitling. By 1844–45 they began to refer to these derogatorily as "utopian socialists" who ignored the need for class struggle and a violent proletarian revolution to usher in the new society, and presumed instead that the bourgeoisie could be persuaded rationally to reform society on the basis of "duodecimo editions of the New Jerusalem", fanciful models of the ideal community.[12] Engels flirted briefly with Owenism in the mid-1840s, but around late 1847, after beginning his partnership with Marx, largely abandoned proposals to house unemployed workers in "palaces" in the

8. [John Cleve Symmes ?], *Symzonia: A Voyage of Discovery* (1820; Scholars' Facsimiles and Reprints, 1965), 159, 162–63, 151.

9. It might be claimed that the Bible is the single and still more influential source of both concepts.

10. The case for seeing him as one is presented in my *Marx and Marxism* (229–40). See also Axel van den Berg, *The Immanent Utopia: From Marxism on the State to the State of Marxism* (Transaction, 2003), 43–77.

11. The case is restated in Kohei Saito's *Karl Marx's Ecosocialism: Capital, Nature, and the Unfinished Critique of Political Economy* (Monthly Review, 2017).

12. Marx and Engels, *Collected Works*, 6: 516. Here, see generally David Leopold's "The Structure of Marx and Engels' Considered Account of Utopian Socialism", *History of Political Thought* 26 (2005): 443–66.

countryside. Thereafter he rarely referred to such ideals.[13] Small-scale group life never appealed to the temperamental Marx, whose one brief attempt at communal living in Paris with the poet Georg Herwegh and another Young Hegelian, Arnold Ruge, ended badly. Marx also clearly disliked "barracks-communism" (*Kasernenkommunismus*) in principle, describing it, vis-à-vis the Russian nihilist Sergei Nechayev, as "an absurd plan of practical organisation" of "communal eating, communal sleeping, assessors and offices regulating education, production, consumption, in a word, all social activity".[14]

Nor was Marx disposed to asceticism. When funds permitted, which was rarely the case for many years, his own lifestyle was certainly bourgeois, even sybaritic, and Engels was the paradigmatic champagne socialist, this being perhaps his favourite tipple. Marx's early writings acknowledge the profoundly corrosive effects of commercial society upon character. The "Paris Manuscripts" of 1844 indicate a utopia of psychological wholeness as well as economic justice, where all forms of alienation would be superseded under communism, and humanity reunified with nature, which implies a non-exploitative relationship. Marx described commerce as a system which bred alienation, where money led each to regard "his will, his activity and his relation to other men as a power independent of him and them". In commerce, Marx suggested, people did not see each other as human, so "things lose the significance of human, personal property". Exchange, however, might be made "equivalent to *species-activity* and species-spirit, the real, conscious and true mode of existence which is *social* activity and social enjoyment". Money need not corrupt relationships: "Assume *man* to be *man* and his relationship to the world to be a human one: then you can exchange love only for love, trust for trust, etc."[15] A famous, if possibly satirical, passage in the *German Ideology* also alluded to a largely Fourierist solution to the division of labour which underpinned private property:

> in communist society, where nobody has one exclusive sphere of activity but each can become accomplished in any branch he wishes, society regulates the general production and thus makes it possible for me to do one thing today and another tomorrow, to hunt in the morning, fish in the afternoon, rear cattle in the evening, criticise after dinner, just as I have a mind, without ever becoming hunter, fisherman, shepherd or critic.[16]

13. See "Principles of Communism", in Marx and Engels, *Collected Works*, 6: 341–57, here 351.

14. "The Alliance in Russia", in Marx and Engels, 23: 543. He associated such ideas with Auguste Willich, a revolutionary rival amongst the German exiles, as well as Bakunin.

15. Marx and Engels, 3: 212–13, 216–17 ("Comments on James Mill, *Elemens d'economie politique*"), 326 ("Economic and Philosophic Manuscripts of 1844").

16. "The German Ideology", in Marx and Engels, 5: 47.

Like More's Utopia, Marx and Engels' scheme in the *Communist Manifesto* was national in scope, and was as such vastly more ambitious than the small-scale communitarian plans of the supposed "utopians". Again we return to the question of scale. By definition, nearly all the problems utopians aim to eliminate, like crime, violence, and war, are more difficult to resolve the larger the number of people involved. Utopia might be attainable in small groups, but impossible for larger numbers. Marx and Engels, however, united the same set of assumptions which wedded most of the early socialists and the anarchists vis-à-vis the future state. Most suffering and misery is traced to private property, which once abolished would induce a profound amelioration of behaviour generally. The scale now seems unimportant.

A more justly distributed plenitude was central to this vision. Many more would have their needs satisfied in the future. From 1848 onwards, Marx insisted that a high level of industrialisation could alone provide this opulence, and ensure that "all-round development" and universal education resulted. Just how the produce of industry would be consumed he did not consider. He obviously assumed that the excessive egoism and materialism associated with capitalist consumption would disappear. The working classes mainly lacked decent food and housing. Marx evidently thought that their needs would expand as their standard of living rose. The *Manifesto of the Communist Party* (1848) observed that "In place of the old wants, satisfied by the productions of the country, we find new wants, requiring for their satisfaction the products of distant lands and climes."[17] The *Grundrisse* (1857–58) noted in passing that "what earlier appeared as luxury is now necessary, and . . . so-called luxury needs appear e.g. as a necessity for the most natural industry rooted in the most basic natural need."[18] There was no reason to presume this pattern would alter under socialism. Marx was, moreover, a stalwart anti-Malthusian, and never conceded that the world could even in principle be overpopulated vis-à-vis its natural resources. So communism implied a growth in demand, production, and population.

The presumption of expanding needs, however, meant that the eventual goal of rewarding people on the principle drawn from Morelly and Babeuf via the Saint-Simonians and Cabet, "from each according to his abilities, to each according to his needs", is only superficially simple.[19] Marx acknowledged that individual needs varied. For "one worker is married, another not; one has

17. Marx and Engels, 6: 488.

18. "Outlines of the Critique of Political Economy", in Marx and Engels, 28: 451.

19. Marx first explored this in "The German Ideology" (Marx and Engels, 5: 537), which states that a different form of labour does not "justify *inequality*, confers no *privileges* in respect of possession and enjoyment".

more children than another, etc., etc. Thus, given an equal amount of work done, and hence an equal share in the social consumption fund, one will in fact receive more than another, one will be richer than another, etc."[20] An expansion of needs would also be driven by other factors. The "all-round development" of human character implied cultivating higher and more cultural sides of the self—indeed, for the majority, well beyond the norms of capitalism. This was also an essentially urban ideal, as Marx's caustic comments on the bourgeoisie rescuing "a considerable part of the population from the idiocy of rural life [*Idiotismus des Landlebens*]" indicate.[21] But in cities needs multiplied much more rapidly than on the land.

Marx's followers thought his system heralded great improvements in both morality and intellectual capacity. His aspirations for humanity's advancement were captured in Leon Trotsky's famous boast that in future, at least eventually, "The average human being will rise to the heights of an Aristotle, a Goethe, or a Marx".[22] But this ideal was rarely linked to the expectation that fewer material goods would be desired. The assumption that behaviour would be more peaceable and orderly and that people would be less disposed to violence and cruelty where relative social equality existed does underscore the idea that a "higher morality" was a key aim of the future society. Yet there is little consensus with respect to how needs would fit into this ideal. Some think Marx and Engels viewed the ideal proletarian as a "paragon of virtue . . . a hero of virtue, who did not indulge in the imminent gratification of his needs."[23] This implies an ascetic, even Spartan, component in his inheritance. To Hans Jonas, however, material prosperity has been "a causal condition for Marxist utopia."[24] Then there is the question of public luxury. Like the early socialists, many later Marxists thought that economy of effort would bring about large-scale communal kitchens, where, the German Social Democrat August Bebel, for example, surmised, "co-operative cooking, with a large central kitchen and machinery" would liberate millions of women from private drudgery.[25]

20. "Critique of the Gotha Programme", in Marx and Engels, 24: 87.

21. "Manifesto of the Communist Party", in Marx and Engels, 6: 488; Karl Marx and Friedrich Engels, *Werke*, 43 vols (Dietz Verlag, 1969), 4: 466.

22. Leon Trotsky, *Literature and Revolution* (1925; University of Michigan Press, 1975), 256.

23. Timo Vihavainen, "Consumerism and the Soviet Project", in *Communism and Consumerism: The Soviet Alternative to the Affluent Society*, eds Timo Vihavainen and Elena Bogdanova (Brill, 2016), 29.

24. Jonas, *Imperative of Responsibility*, 160.

25. August Bebel, *Women under Socialism* (1883; Schocken Books, 1971), 185. The antisocialist Eugene Richter, however, parodied such proposals in deriding the "caprice of fashion", and floated the idea that "in place of having any variety in goods which are destined to fulfil the

Marx was often seen, as a San Francisco branch of the International put it, as opposing "luxurious idleness", not luxury as such.[26] Luxury could be collective and extended to all. As Kristin Ross shows, some of these ideas were tentatively explored during the brief experiment of the Paris Commune of 1871, defined not just by social equality, but a "communal luxury" (the phrase itself was used), which entailed "transforming the aesthetic coordinates of the entire community."[27] This complex legacy, we will see below, was played out during the history of the USSR.

John Stuart Mill and the Stationary State

The leading mid-nineteenth-century liberal John Stuart Mill (1806–73) is not included in histories of utopianism on the mistaken presumption that liberalism as such is never utopian, but focuses on only the virtues of the market, economically, and on democracy, politically. Mill was, however, a utopian writer in several key senses, and for our purposes one of the most important.[28] The Saint-Simonians taught him that society progressed through distinctive "organic" and "critical" stages, implying that the present, too, might be merely a "crisis of transition" to a new order.[29] From his own observation, and from Tocqueville, he learned that what the latter called the "hypocrisy of luxury", "the maintaining an appearance beyond one's real expenditure", was when taken to excess a peculiarly English and American vice.[30] Crucially for us, in chapter 6, book 4, of his *Principles of Political Economy* (1848), acknowledging that

same purpose, all such articles were limited to a few patterns, or, better still, if they were all made on one single pattern." He also suggested that:

We do not contemplate carrying equality in dress to such a length that all diversities will be entirely abolished. On the contrary, we suggest the wearing of various badges as marks whereby the ladies and gentlemen of the different provinces, towns, and trades, may readily be distinguished from each other at a glance [and to] materially facilitate the surveillance of individual persons by "checkers" (*Pictures of the Socialistic Future* (1893; George Allen, 1912), 105–6).

26. Robert V. Hine, *California's Utopian Colonies* (Yale University Press, 1969), 79.

27. Kristin Ross, *Communal Luxury: The Political Imaginary of the Paris Commune* (Verso, 2015), 39.

28. See my *Mill and Paternalism* (Cambridge University Press, 2013), 137–72, and *John Stuart Mill: A Very Short Introduction* (Oxford University Press, 2022). An argument along similar lines is F. L. van Holtoon's *The Road to Utopia: A Study in John Stuart Mill's Social Thought* (Van Gorcum, 1971).

29. John Stuart Mill, *The Spirit of the Age* (University of Chicago Press, 1942), 93.

30. John Stuart Mill, *Dissertations and Discussions*, 2nd ed., 2 vols (1862), 2: 66.

"the increase of wealth is not boundless", Mill proposed the first positive vision of a "stationary state". Here the qualitative improvement of life would take priority over economic growth.[31] Hitherto the stationary state had been a kind of bogey for economists, implying stagnant wages, demand, and population growth. Mill projected instead a condition where wealth, production, and population no longer increased, terming this

> a very considerable improvement on our present condition. I confess I am not charmed with the ideal of life held out by those who think that the normal state of human beings is that of struggling to get on; that the trampling, crushing, elbowing, and treading on each other's heels, which form the existing type of social life, are the most desirable lot of human kind, or anything but the disagreeable symptoms of one of the phases of industrial progress.[32]

Continuing, Mill wrote, "the best state for human nature is that in which, while no one is poor, no one desires to be richer, nor has any reason to fear being thrust back, by the efforts of others to push themselves forward." Why, he asked, was it a

> matter of congratulation that persons who are already richer than any one needs to be, should have doubled their means of consuming things which give little or no pleasure except as representative of wealth; or that numbers of individuals should pass over, every year, from the middle classes into a richer class, or from the class of the occupied rich to that of the unoccupied[?][33]

Instead, under a better distribution of property, where inheritance would be sufficiently limited to require all to work, a liberal utopia was possible.[34] Here,

> society would exhibit these leading features: a well paid and affluent body of labourers; no enormous fortunes, except what were earned and accumulated during a single lifetime; but a much larger body of persons than at present, not only exempt from the coarser toils, but with sufficient leisure,

31. J. S. Mill, *Collected Works*, 3: 752. David Ricardo had used the phrase "stationary state" to describe what might well be a condition of "universal poverty", without further "progressive prosperity", and where wages in particular were contingent solely on population (*The Works of David Ricardo* (John Murray, 1846), 59, 379).

32. John Stuart Mill, *Principles of Political Economy*, 2 vols (John Parker, 1848), 2: 308.

33. J. S. Mill, *Principles of Political Economy*, 2: 309–10.

34. On Mill's assessment of whether even a "moderate independence" was too much, and what this might consist of, see my *Mill and Paternalism* (72–76, 159–60).

> both physical and mental, from mechanical details, to cultivate freely the graces of life, and afford examples of them to the classes less favourably circumstanced for their growth. This state of things, which seems, economically considered, to be the most desirable condition of society, is not only perfectly compatible with the stationary state, but, it would seem, more naturally allied with that state than with any other.[35]

This is, thus, a kind of middle-class utopia—one amongst many—where classes and private property still exist but all are relatively well off. Reflecting that "it is questionable if all the mechanical inventions yet made have lightened the day's toil of any human being", Mill added that here "industrial improvements would produce their legitimate effect, that of abridging labour." He then tackled the issue of voluntary population restriction. Besides tending to force wages down, an overly dense population removed much of what was valuable in human life:

> A population may be too crowded, though all be amply supplied with food and raiment. It is not good for man to be kept perforce at all times in the presence of his species. A world from which solitude is extirpated, is a very poor ideal. Solitude, in the sense of being often alone, is essential to any depth of meditation or of character; and solitude in the presence of natural beauty and grandeur, is the cradle of thoughts and aspirations which are not only good for the individual, but which society could ill do without. Nor is there much satisfaction in contemplating the world with nothing left to the spontaneous activity of nature; with every rood of land brought into cultivation which is capable of growing food for human beings; every flowery waste or natural pasture ploughed up, all quadrupeds or birds which are not domesticated for man's use exterminated as his rivals for food, every hedgerow or superfluous tree rooted out, and scarcely a place left where a wild shrub or flower could grow without being eradicated as a weed in the name of improved agriculture. If the earth must lose that great portion of its pleasantness which it owes to things that the unlimited increase of wealth and population would extirpate from it, for the mere purpose of enabling it to support a larger, but not a better or a happier population, I sincerely hope, for the sake of posterity, that they will be content to be stationary, long before necessity compels them to it.[36]

To achieve these ends, Mill promoted workers' co-ownership in the means of production, along with land nationalisation, preferring this to the more

35. J. S. Mill, *Principles of Political Economy*, 2: 310.

36. J. S. Mill, 2: 311–12.

statist schemes associated with Saint-Simon and the communitarianism of Owen and Fourier. Competition would still exist, for Mill feared the consequences of monopoly arising in its stead. He was in fact the most important nineteenth-century theorist to write sympathetically on the need to blend both considerable liberty and increasing equality in the ideal society—Marx was much weaker on the institutional and constitutional mechanisms required to secure and maintain liberty. This happy meeting ground between the extremes of libertarian capitalism and statist communism might thereafter be considered as one of the more realistic utopian alternatives which the twentieth century inherited from the nineteenth. And Mill was also considerably ahead of Marx on the question of feminism, championing with his wife Harriet Taylor Mill the cause of female enfranchisement, most notably in *The Subjection of Women* (1869), and indeed making social progress effectively contingent on full social and political equality for women.

Anarchism and Luxury

Anarchist writers have often focused more upon the evils of state power than on the corrosive effects of luxury and the virtues of a simpler life. Most blame the luxury of the idle few for the destitution of the poor. Anarchists too have usually been divided into those who accept the prospect of expanding needs, and the legitimate desire of the masses for luxuries, and those who prefer greater simplicity.

The two main nineteenth-century anarchists were Michael Bakunin (1814–76) and Peter Kropotkin (1842–1921). Both addressed the problem of luxury at various points. Bakunin condemned "a life of luxurious idleness" as leading to "moral and intellectual degeneration". In future, he thought, communal banks might distribute goods "in accordance with the needs of the consumers". After necessities had been met, "Other commodities, formerly scarce and today considered luxuries, will in a reasonable length of time be produced in great quantity and will no longer be rationed." But "rare and useless baubles, such as pearls, diamonds, certain precious metals, etc., will cease to have the value attributed to them by public opinion", and would be used only for research purposes. This could take some time, however.[37] Following the revolution, "luxury and all the refinements of life will have to disappear from society for a long time, and they will not be able to reappear until society finds the necessities of life for everyone, when luxuries no longer will be exclusive

37. Michael Bakunin, *Bakunin on Anarchy*, ed. Sam Dolgoff (George Allen and Unwin, 1973), 91, 368.

delights but will ennoble everyone's life."[38] So, again, luxury might reappear once the problem of raising the general standard of living had been solved.

To Kropotkin, once the people had "made a clean sweep of the Government, they will seek first of all to ensure to themselves decent dwellings and sufficient food and clothes". Thereafter most trades and crafts could focus on necessities, and "abstain from producing mere luxuries". Yet in the future "Steam, electricity, the heat of the sun, and the breath of the wind, will ere long be pressed into service."[39] Fewer workers would cultivate food. Necessities might be furnished from public stocks, while

> Not being able to offer to each man a sable-lined coat, and to every woman a velvet gown, society would probably distinguish between the superfluous and the necessary, and, provisionally, at least, class sable and velvet among the superfluities of life, ready to let time prove whether what is a luxury to-day may not become common to all to-morrow.

Kropotkin admitted that "the conception of the necessaries of life can be so extended as to include luxuries". A chapter in *The Conquest of Bread* tellingly entitled "The Need for Luxury" detailed how "As soon as his material wants are satisfied, other needs, of an artistic character, will thrust themselves forward the more ardently. Aims of life vary with each and every individual; and the more society is civilized, the more will individuality be developed, and the more will desires be varied."[40] When

> all can eat sufficiently, the needs which we consider luxuries to-day will be the more keenly felt. And as all men do not and cannot resemble one another (the variety of tastes and needs is the chief guarantee of human progress) there will always be, and it is desirable that there should always be, men and women whose desire will go beyond those of ordinary individuals in some particular direction.

The anarchist commune would thus "strive to satisfy all manifestations of the human mind." Many more would write, invent, paint, or sculpt:

> In short, the five or seven hours a day which each will have at his disposal, after having consecrated several hours to the production of necessities, will amply suffice to satisfy all longings for luxury, however varied. Thousands of associations would undertake to supply them. What is now the privilege of an insignificant minority would be accessible to all. Luxury, ceasing to

38. Bakunin, *Out of the Dustbin*, 116.

39. Peter Kropotkin, *The Conquest of Bread* (Chapman and Hall, 1906), 65, 96, 101.

40. Kropotkin, 119, 123, 134.

> be a foolish and ostentatious display of the bourgeois class, would become an artistic pleasure.[41]

Once again, thus, we see luxury universalised rather than condemned as such.

But there were notable exceptions to this trend. Amongst the later anarchists, Count Leo Tolstoy (1828–1910) moved markedly towards simplicity in the 1870s, after finding that, in comparing his own considerable wealth to the poverty of the many, "all the indulgencies of a luxurious existence, in which I had formerly delighted, began to torment me." He now embraced an ideal of Christian self-renunciation, and condemned idleness and luxury as entailing slavery and misery on others. He discovered that "my most expensive demands on life, the demands of vanity and for distraction from ennui, were directly due to an idle life."[42] The "conditions of luxury in which we live", he reflected, "deprive us of the possibility of understanding life", especially that of the "simple working people." "Luxury cannot be obtained other than by enslaving other people", he concluded.[43] Tolstoy did not reject modern technology as such, but came to believe that it would be better to relinquish "book-printing, railroads, ploughs and scythes . . . [until] we can learn to get them without destroying the happiness and life of men." No true utopia, he reflected, could depend on exploiting others. So he derided the "augmentation of wants", and warned of educating children in "habits of effeminacy, physical idleness, and luxury".[44] The trend of modern life, he thought, was for increasing numbers to work excessively to support the "luxurious lives of the ever increasing number of idle people". Since there was "no end to the caprices of men when they are met not by their own labour but by that of others, industry is more and more diverted to the production of the most unnecessary, stupid, depraving products, and draws people more and more from reasonable work." Even "the people of distant nations are themselves becoming depraved" by "the sale of unnecessary and injurious articles." At least for Russia, this malaise could be avoided only by remaining a chiefly agricultural nation, while renouncing private property in land, military service, and taxation, and serving God rather than an immoral state.[45]

Tolstoy inspired Mohandas Gandhi (1869–1948), who was also indebted to the British social critic John Ruskin, who taught him that "a life of labour,

41. Kropotkin, 135–37, 154.

42. Leo Tolstoy, *What Shall We Do?* (Free Age, 1910), 14–15, 211.

43. Leo Tolstoy, *A Confession and Other Religious Writings* (Penguin Books, 1987), 67, 103.

44. Quoted in Rob Knowles, *Political Economy from Below: Economic Thought in Communitarian Anarchism, 1840–1914* (Routledge, 2004), 282.

45. Leo Tolstoy, *The Russian Revolution* (Free Age, 1907), 13, 15–16, 20, 22.

i.e., the life of the tiller of the soil and the handicraftsman, is the life worth living."[46] Describing himself in 1928 as "a passionate devotee of simplicity in life", Gandhi came to believe that multiplying bodily desires beyond a minimum was a spiritually degrading sensual indulgence. The wealthier people were, he observed, "the greater was their moral turpitude." And so too with nations: "Western nations are today groaning under the hell of the monster-god of materialism. Their moral growth has become stunted." Every attempt "to gain American wealth but avoid its methods" was "foredoomed to failure." Rejecting much modern machinery, including transport and medicine, he adopted the self-sufficient, self-governing village as his basic social model. Gaoled for resisting South Africa's race laws in 1908, Gandhi took up fasting, and became a vegetarian in the belief that "the carnal mind always lusts for delicacies and luxuries." Slowly, simplicity became his central precept, the ideal being "not to possess anything which the poorest on earth do not". He now rejected jewellery and household ornaments and began baking his own bread. Self-reliance, or *Swadeshi*, became a "rule of life", which Gandhi described as "that spirit within us which restricts us to the use and service of our immediate surroundings to the exclusion of the more remote."[47] Civilisation, he came to believe, consisted not in machines and luxury but in "that mode of conduct which points out to man the path of duty." Here, he thought, India had "nothing to learn from anybody else." Indians had discovered that "the mind is a restless bird; the more it gets the more it wants, and it still remains unsatisfied. The more we indulge our passions, the more unbridled they become. Our ancestors, therefore, set a limit to our indulgences." Indians had not invented machines because "our forefathers knew that, if we set our hearts after such things, we would become slaves and lose our moral fibre." Thus they "decided that we should only do what we could with our hands and feet. They saw that our real happiness consisted in a proper use of our hands and feet."[48] For similar reasons they preferred small villages to large cities, and traditional to modern medicine. India, thought Gandhi, could "realise truth and non-violence only in the simplicity of village life."[49] So he gave away everything superfluous. At his death he owned ten things.

46. M. K. Gandhi, *An Autobiography* (Phoenix, 1949), 250.

47. M. K. Gandhi, *The Moral and Political Writings of Mahatma Gandhi*, ed. Raghavan Iyer, 3 vols (Clarendon, 1986), 1: 354, 357, 359; 2: 616; 3: 326; Gandhi, *Autobiography*, 275, 184.

48. M. K. Gandhi, *Hind Swaraj and Other Writings* (Cambridge University Press, 1997), 67–69.

49. Gandhi, *Hind Swaraj*, 150.

Luxury and Simplicity in Later Nineteenth-Century Literary Utopianism, 1880–1917

Late nineteenth-century literary utopianism was torn between traditional appeals to simplicity and the "natural life" and the increasingly dazzling vision of a new world of scientific and technological innovation. The first great age of modern utopian writing, in terms of the number of texts produced and their popularity, stretched from the 1880s to World War One. Often addressing new socialist alternatives to capitalism, utopian literature now addressed the apparent collision of this promise of scientific and technological advance with the threats which the new techniques of production and some of their results, especially weapons, posed to humanity and the prospect of sharply increasing social inequality, which provoked the rise of the modern labour movement and again raised the prospect of revolution. The contrast of great wealth and deep poverty was regarded as the definitive paradox of the age. As the influential land reformer Henry George (1839–97) put it, in new countries, "though you will find an absence of wealth and all its concomitants, you will find no beggars. There is no luxury, but there is no destitution." In older countries, however, "that is to say, the countries where material progress has reached later stages . . . widespread destitution is found in the midst of the greatest abundance."[50]

In assessing the problem of needs, later nineteenth-century literary texts continued the three main trends noted earlier: asceticism, public luxury, and the acceptance of more or less unlimited expansion of needs. Here we will introduce a few texts representing each trend.

Opposition to luxury was common amongst those who saw it as "robbing the labourer of his hardly earned gains", as Henry Crocker Marriott Watson's *The Decline and Fall of the British Empire* (1890) put it.[51] Fashion was thus condemned by various well-known writers. In Frederick Hayes's *The Great Revolution of 1905* (1893), we are told that in the future society

> The impersonal interests of the home circle became associated with more artistic dress, furniture, and domestic appointments, in place of the vulgarity of mere costliness. Fashion became daily less and less of a tyrant among women, and with men ceased to have any influence. No one having anything to gain by imitating or consulting other people in attire, style, habits, or fads, or to lose by indulging their own preferences therein, a diversity

50. Henry George, *Progress and Poverty* (Wm M. Hinton, 1879), 4.

51. Henry Crocker Marriott Watson, in Gregory Claeys, ed., *Late Victorian Utopias*, 6 vols (Pickering and Chatto, 2009), 3: 171.

> and individuality hitherto unknown and undreamed-of began to characterise the mass of the community.[52]

Hayes envisioned detaching the growth of wealth from its deleterious psychological effects. In the old society "very many men, and almost all women, sought further riches, not so much to add to their own personal comfort or luxury as to exercise tyranny, and by parade and ostentation to excite a satisfactory degree of envy and jealousy amongst their acquaintances." After the revolution occurred, however,

> While there would be an increase of the consumption of the necessaries and comforts of life to an entirely unprecedented extent, it was easy to foresee that the whims and fads of reckless extravagance, the costly novelties of ephemeral fashion, the lavish ostentation of a fabulously wealthy handful of the community, would undergo gradual extinction.[53]

Another advocate of frugality was Frank Perry Coste, whose *Towards Utopia* (1894) advocated an "*earnest and beautiful simplicity of life*". But Coste insisted that

> *our* simplicity is very different from that of Sparta or of the Cynics—as different as is day from night. For us, the Spartan life, with its incessant hardship, rigid discipline, ceaseless governmental interference, and cruel inhumanity, its narrowness, bareness, and unamiability, has none of the attractions which it has offered to some foolish sentimentalists who profess to long for such "wholesome" living: to us Lycurgean Sparta rather suggests devildom let loose on a saturnalia of repression; whilst the coming life, that we picture, is warm, glowing, bright, *human*, cultured, refined, beautiful, and unrestrained, beyond anything that most of us to-day can realise . . . If champagne and madeira, winter-pine-apples and peaches, were cheap as daisies, mankind would be very absurd not to enjoy such delicacies; but we do contend that since these luxuries can be had only at a great expense of toil and treasure, since the pleasure afforded by them is essentially fleeting and of the moment—the pleasures of taste having least of all our pleasures any ideal persistence . . . the consumption of such luxuries means the mere emptying down our gullets of so much wealth and labor that would otherwise—Proteus-like—have appeared in the form of esthetic intellectual or material gains, or prolonged leisure.[54]

52. Frederick Hayes, in Claeys, 4: 413.

53. Frederick Hayes, in Claeys, 4: 217, 208.

54. [Frank Perry Coste], *Towards Utopia (Being Speculations in Social Evolution)* (Swan Sonnenschein, 1894), 146–47.

Some texts link simplicity positively to a new ethos of barbarism and Spartanism. Charles Wicksteed Armstrong's *The Yorl of the Northmen* (1892) tells us that "we count it effeminate to love luxury and idleness and manly to be ready to face hardships" (and they are called "this new race of Spartans, or rather, I should say, these revivers of the customs of old English chivalry").[55] Most celebrate moderation and restraint. In *Etymonia* (1875) labour is universal, idleness has disappeared, and no one consumes more than the value of what he or she produces. Adults each have a two-room apartment and draw their sustenance from communal stores. Here "the charge of luxury was a serious one". Every Etymonian satisfies "his bodily wants without stint, but at the same time without imagination; if he does not deny his appetites, at least he does not excite them." Here "All the men dress, alike, without ostentation, there being but one fixed and invariable fashion, as becoming men of one form and of one need." As for the women, "Coquetry is with them a thing quite unknown. In fine—they talk, walk, and act like reasonable creatures. The only ornament they wear, which, however, it must be said, is sufficient, is their own beauty." Overpopulation is treated as a problem, and the question is asked, "Shall we improve our system of cultivation, and sacrifice our follies and our luxuries to the wants and necessities of our fellow-creatures?" But this is not seen as a "permanent solution" to the issue.[56] Amongst the frugal inhabitants of *Pyrna* (1875), a commune discovered beneath a Swiss glacier, too, women are equal to men. Here "This subject of dress, so important to our women, was to them a matter of no account at all", though "they never used rouge or cosmetics, and only the most delicate perfumes were tolerated, and those used alike by men and women". One result was that

> there was affection and devotion among this people of the highest order. If love, such as it is here, was banished from amongst them, with its good as well as its evil, a pure and holy sentiment of devotion and attachment transcending anything we know of was substituted. All were one family; each one was prepared to make any sacrifice for his or her neighbour's benefit.[57]

One of the best accounts of this type was Henry Wright's *Mental Travel in Imagined Lands* (1878). In its ideal society, Nommuniburgh, all "by their unselfishness, their thorough appreciation of the maxim that 'it is more blessed to give than receive,' are enabled amicably to settle the ownership, and to distribute, without competing for, 'luxuries.'" All "eat to live" rather than "live to

55. Charles Wicksteed Armstrong, in Claeys, *Late Victorian Utopias*, 3: 347, 366.

56. Claeys, 1: 240.

57. Claeys, 1: 30.

eat", and though they are satisfied, "creature comforts do not enter to any great extent into our scheme of happiness." The belief that constant improvement provides "the true solace and greatest happiness mankind is capable of experiencing" "saves us from giving way to luxury and easy living, to which our mode of life makes us so especially liable." Thus "as it is idleness and self-satisfaction, and not prosperity, which lead to the degeneracy of the individual man, so is it the effeminacy and luxury which are the outcome of these which bring the downfall of a nation—and not peace." Here are "none of the drawbacks which tend to deteriorate the higher classes in other countries, such as the temptations to idleness and luxury and the overindulgence of the senses; for while the law prevents the first evil, it is the care of education—the vital education—to prevent the latter."[58] Here too "Such a thing as loneliness, or being without friends, is impossible, while the chances of being bored by acquaintances is equally so; for as it takes two to make a quarrel, so it takes two to make a friendship; and other relationship there is none recognised or required". This is possible for two reasons—because

> all interests should be identical,—thereby taking away any cause for what are called "interested friendships;"—and, secondly, that the moral tone should be sufficiently high,—or, in other words, that the people should be all true ladies and gentlemen,—so that no base advantage should be taken of this freedom of intercourse, especially where the difference of sex might tempt it.

Friendships here are thus based on "affinity", rather than being "formal and conventional". In "Fortuneland" (modern Britain), "King Fashion" is the "semi-autocratic" ruler who dictates, for example, that all wear shoes which are too small. By contrast, "It is said at Nomunniburgh that *overfeeding the sensual organs comes of starving the spiritual.* Thus, for instance, if a man drinks too much, it is because his mental faculties have been neglected".[59] A similar sentiment is expressed in J. H. Levy's little-known *An Individualist's Utopia* (1912), where

> Life was full of an undreamt-of happiness; but it was also in many respects much simpler. People had less need of sensual luxury. When poverty and bad social arrangements had passed away, the *raison d'etre* of drinking alcoholics and smoking tobacco was gone. The pleasures of ostentation came gradually into public contempt; and, as their only object was to draw admiration, they ceased when the reverse of admiration was awarded to them.[60]

58. Claeys, 2: 65.

59. Claeys, 2: 27.

60. J. H. Levy, *An Individualist's Utopia* (Lawrence Nelson, 1912), 11.

Social harmony was of course a key theme in most such texts, and inequality of property was usually deemed its greatest enemy. Such harmony was sometimes portrayed specifically in terms of an ideal of friendship. A good example is the Danish American lawyer Laurence Gronlund's *Our Destiny* (1890), where friendship is described as a key reward of the new social system:

> when, under Socialism, the chasm now dividing the classes is filled, then we shall have everywhere the sympathy, greater than that with pain, greater than that with pleasure: sympathy with the thoughts and purposes of others—*friendship*. Friendship is the *bouquet* of morality—the distilled flavour of morality; the important link that will make sexual love evolve love for fellow-men. The future moralised society will be constituted of groups of friends, each group formed out of men and women from various callings and departments. We know very little of true friendship now, and it is again the Established Order that is at fault, which is the cause of the fact that we have only cliques. Friendship, being sympathy with thoughts and purposes, of course, demands community of sentiments, but it is equally a law that true friendships are formed out of diversities of character—such diversities as are found in people of various callings. It is this diversity that creates admiration. We should thus expect model friendships between literary men and workingmen; but they are at present separated by a yawning chasm. We have therefore now comradeships where the mutual influence is by no means always salutary, while true friendships, with their mutual confidences, always have the moral advantage of conferring personal dignity on the parties.[61]

Gronlund even imagined that servants would continue under socialism, except that:

> Under Socialism we undoubtedly shall not be without "helpers" in our private houses—those in the public establishments will, of course, be just as much public functionaries as the guests who have their wants attended to. But the relations of these private attendants to their principals will be very different from what it is now—it will be a sympathetic, not a pecuniary one. These attendants will attach themselves to our persons because attracted by our personal qualities, and on the condition of being incorporated into our families as members thereof—something like the pages of mediaeval households; they will hardly accept such positions on other terms.[62]

Utopias of this period often assume that reducing inequality would virtually eliminate crime. Chauncey Thomas's *The Crystal Button* (1891), for

61. Laurence Gronlund, *Our Destiny* (Swan Sonnenschein, 1890), 116.

62. Gronlund, 110–11.

instance, tells us "we have so reduced, where we have not entirely removed, the chief inducements to crime, including poverty, excess of wealth, injustice, and ambition for undeserved power, inevitably leading to tyranny, that it is now infrequent."[63] George Griffith's *The Angel of the Revolution* (1895) similarly relates that

> One usually conspicuous feature in similar ceremonies was entirely wanting. There were no wedding presents. For this there was a very sufficient reason. All the property of the members of the Inner Circle, saving only articles of personal necessity, were held in common. Articles of mere convenience or luxury were looked upon with indifference, if not with absolute contempt, and so no one had anything to give.[64]

In David Hilton Wheeler's *Our Industrial Utopia and Its Unhappy Citizens* (1895), we learn that

> Display for the sake of display, expense for the sake of expense, a large staff of servants as a badge of wealth; these things are out to stay out. A few light-headed people will try to perpetuate them; but they are not socially required nor are they more than tolerated. Gaping admiration of them is dead in Utopia.[65]

Fashion was, of course, commonly targeted in these texts. In Joseph Carne-Ross's *Quintura* (1886), the numbers of the idle had grown greatly in the old society, such that "there could not be found in the whole empire a creature more shameless, sensual, and selfish than a Quinturian lady of fashion". Accordingly it was decided to introduce a law that within two years

> every woman of twenty years and upwards who had no ostensible means of subsistence beyond the charity of her friends, should be ranked in that class that was under especial police supervision. And that similarly every man of twenty years and upwards who had no occupation, or ostensible means of gaining a livelihood, should be at once proceeded against as a rogue and a vagabond.

Then an agrarian law was introduced limiting landholdings to 1000 acres, and inheritance was restricted to £5000. And for the last 400 years there has been

63. Chauncey Thomas, *The Crystal Button; or, Adventures of Paul Prognosis in the Forty-Ninth Century* (Houghton, Mifflin, 1891), 58.

64. George Griffith, *The Angel of the Revolution* (Tower, 1895), 199.

65. David Hilton Wheeler, *Our Industrial Utopia and Its Unhappy Citizens* (A. C. McClurg, 1895), 42–43.

neither millionaire nor pauper in Quintura.[66] In Henry Olerich's *A Countryless and Cityless World* (1893), which is set on Mars, we are told: "Our clothing is not made and worn so wastefully, and our fashions are not so changeable as yours". "All this display of unnatural costumes and continuous change of fashion requires an immense amount of labor, which has to be performed by the man as well as by the woman."[67] William Wallace Cook's *A Round Trip to the Year 2000* (1903) states simply that "Fashions in clothes have gone out of fashion. There's one style for everybody."[68] King Gillette's *The Human Drift* (1894), too, relates that

> Our whole system of exchange with foreign countries will undergo a change. We will only produce an excess of goods beyond our needs to an amount sufficient to answer purpose of exchange for such foreign products as we do not produce. Thus we can exchange wheat, corn, cotton, etc., for teas, coffees, spices, etc. We shall not import any foods or clothing which it is possible for us to produce ourselves.[69]

It was not only clothing that was at issue where simplicity was demanded. J. W. Roberts's *Looking Within* (1893) describes the rooms of the future as simply furnished: "There was nothing attractive about them; no display; no evidence of an effort to outrival some other house. The furniture was plain and substantial; good enough as far as it went, but neither profuse nor ornamental. It was quite unlike the fashionable hotels of a hundred years ago."[70] In the last chapter of his very popular *Caesar's Column* (1890), Ignatius Donnelly portrayed an agrarian republican utopia, where a colony of five thousand, armed to the teeth, is surrounded by a 30-foot wall, disdaining all trade with any poorer nations in order to keep wages up, and free, like the Swiss, of "the luxuries and excesses that had wrecked the world".[71] Donnelly's description of Plato's Atlantis betrays a similar sentiment:

> They despised everything but virtue, not caring for their present state of life, and thinking lightly on the possession of gold and other property, which seemed only a burden to them; neither were they intoxicated by luxury; nor did wealth deprive them of their self-control; but they were sober, and saw clearly that all these goods are increased by virtuous

66. Joseph Carne-Ross, in Claeys, *Late Victorian Utopias*, 3: 18–19.
67. Henry Olerich, *A Countryless and Cityless World* (Gilmore and Olerich, 1893), 146, 308.
68. William Wallace Cook, *A Round Trip to the Year 2000* (Street and Smith, 1903), 74.
69. King Gillette, *The Human Drift* (New Era, 1894), 70–71.
70. J. W. Roberts, *Looking Within* (A. S. Barnes, 1893), 149.
71. Ignatius Donnelly, *Caesar's Column* (Sampson, Low, Marston, 1891), 351.

> friendship with one another, and that by excessive zeal for them, and honor of them, the good of them is lost, and friendship perishes with them.[72]

Even hobbies and pastimes would be affected. In his *New Utopia; or, England in 1985* (1885) the Reverend W. Tuckwell speaks of

> the simpler habits of our day, not only in the cases which I just now mentioned, but by the disappearance of the idle classes, the extinction of sport and racing—both relics surely of primaeval savagery—by the great limitation of domestic servants, the abandonment of all trades ministering to luxury and ostentation.

Here the towns are now all suburbs: "an English town is now a ring of villages enclosing a nucleus of factories, penetrated, entwined, and refreshed by green spaces of public recreation or private garden ground."[73]

A second group of texts developed the theme of public luxury which we saw slowly emerging in both literary utopias and socialist texts earlier in the century. George Read Murphy's *Beyond the Ice* (1894), for example, relates that "extra luxuries here would encourage people to be ostentatious, and tend to create envy, or discontent". So,

> In place of domiciles, replete with selfish, wasteful splendour, or filthy crowded dens of misery, with every grade between, had risen tall, handsome, regular buildings, giving reasonable luxury and comfort to all. In place of the idle plethoric wealth and the pinched miserable existences of perpetual toil had come wealth, with duties that could not be shirked, and a comfortable competence for all. In place of carriages with two horses, and two or more men, to carry two or three healthy people, as if they were weak-spined cripples, huge, slow, overcrowded vehicles, and slow traffic with occasional blocks, were splendid trams which ran down every street, and provided rapid, comfortable, and economical transit for all; while the dirty, ineffectual horse traffic was replaced by infinitely superior vehicles, using electricity or compressed air.[74]

The theme of public luxury overlaps with a third group of works which described the ideal of the future in terms of expanding needs. In W. D. Howells's vision of Altruria, which has "realized the Utopian dream of brotherly equality", apartments were not in decorated in "the spirit of profuse and

72. Ignatius Donnelly, *Atlantis* (Harper and Bros, 1885), 20.

73. Reverend W. Tuckwell, *The New Utopia; or, England in 1985* (Hudson and Son, 1885), 13.

74. G. Read Murphy, *Beyond the Ice* (Sampson Low, Marston, 1894), 35, 275.

vulgar luxury which it must be allowed once characterized them". Nonetheless, as *Through the Eye of the Needle* put it,

> there are many that are fine in a good taste, in the things that are common to the inmates. Their fittings for housekeeping are of all degrees of perfection, and, except for the want of light and air, life in them has a high degree of gross luxury. They are heated throughout with pipes of steam or hot water, and they are sometimes lighted with both gas and electricity, which the inmate uses at will, though of course at his own cost.[75]

In C. A. Steere's *When Things Were Doing* (1907), "everybody was entitled to make requisition upon the general stores for anything in reason he or she could use", such "that the word luxury was stricken from the lexicon, whatever was desired being regarded as a necessity, up to the capacity of the commonwealth for turning it out."[76] Some went even further down the road to the bourgeois paradise. In the sumptuous future described in Paul Devinne's *The Day of Prosperity* (1902), everyone lives in ten-storey hotels, which include "Marble steps and columns, statues, luxurious carpets, rich wall decorations, gold-framed mirrors, rare pictures, wood and ivory carvings, proud palms, and fragrant flowers".[77] In William Delisle Hay's *Three Hundred Years Hence* (1881), even provincial houses possess "every comfort and luxury . . . that would be deemed requisite in, say, Londinova itself." Here, the house of a wealthy citizen "was magnificently furnished; . . . gorgeous silken draperies of every hue . . . adorned the ladies' rooms":

> You are bewildered now by the splendid luxury that surrounds you, as well you may be, for there is nothing your time could produce in the way of decoration or of furniture, that may not be seen here in for more beauty and perfection of material and construction. If it be night you do not know it, for light springs round you, not from lamps, but from a golden network on the roof or walls, from which it is diffused from numberless glittering points that cannot individually distress the eye. There is warmth, there is changing perfume, there is soft music to enchant the senses. There are paintings, and carvings, and decorations, all the glories of art spread round you, like a book whose many pages would take long time to read.[78]

75. William Dean Howells, *A Traveller from Altruria* (David Douglas, 1894), 225; Howells, *Through the Eye of the Needle* (Harper and Bros, 1907), viii, 10.

76. C. A. Steere, *When Things Were Doing* (Charles H. Kerr, 1907), 182.

77. Paul Devinne, *The Day of Prosperity* (G. W. Dillingham, 1902), 80.

78. W. D. Hay, *Three Hundred Years Hence; or, A Voice from Posterity* (Newman, 1881), 307.

Yet the evils of folly and fashion are gone. No servants are seen, and machinery does much of their former work, so that "You seem merely to wish and a thing is done; a look or a word compels the obedience of inanimate things, or seems to you to do so." But then we see what price must be paid for this. To accommodate the needs of a vastly expanded population, large parts of other continents have been conquered. Utopian imperialism was once again on the march. "Extermination" frees Africa and China of "the inferior races" for white settlement, and in Africa "extinction" follows indigenous rebellions.[79] These themes were fairly common in this period: many such utopias were imperial fantasies which were inevitably other people's dystopias, because they depended on domination and exploitation. A refusal to accept this ideal of utopia as imposing drudgery or worse on others, however, we will also see became a key element in late Victorian visions of the simple life.

A continuation of modern luxury, but without moral contamination, is also suggested in one of the better-known texts of this period, Edward Bulwer-Lytton's *The Coming Race* (1871), though it is strictly speaking closer to science fiction than the utopian genre. Here, residences in the future conform "to modern notions of luxury, and would have excited admiration if found attached to the apartments of an English duchess or a fashionable French author." Poverty is unknown, but so too "a fortune much above the average is a heavy burden." While there is "no difference of rank or position between the grades of wealth or the choice of occupations, each pursues his own inclinations without creating envy or vying; some like a modest, some a more splendid kind of life; each makes himself happy in his own way." The new race, the Vril-ya, still have "a considerable commercial traffic with other states, both near and distant", which includes "articles more of luxury than necessity", such as birds which sing delightfully.[80] "In their own way", we are told, "they are the most luxurious of people, but all their luxuries are innocent." "Every room has its mechanical contrivances for melodious sounds", and they have a taste for saunas and perfumes. They eat no animal food but drink milk. They abstain from alcohol but still have great banquets, which are "not without a certain resemblance to those we read of in the luxurious age of the Roman empire." Here, then, was a model to be emulated, though Lytton observed that "it would be utterly impossible for us to adopt the modes of life, or to reconcile our passions to the modes of thought, among the Vril-ya".[81] Once again, thus,

79. Hay, 257.

80. Edward Bulwer-Lytton, *The Coming Race* (George Routledge and Sons, 1874), 39, 190, 59–60, 98.

81. Bulwer-Lytton, 102, 201, 229.

the impossibility of creating utopia in a corrupt society is suggested. But the suggestion is a dare, too.

Many other utopias of this period present a future defined by scientific and technological innovation. Alvarado M. Fuller's *A.D. 2000* (1890) typifies these urban and techno-centric visions. Land ownership is restricted to one square mile. Fast electric trains connect the great cities, and submarines and ships powered without coal or wood silently ply the oceans. The new cities are designed on a grid system, with each square having its own shops and stores, and only one of each particular trade allowed per neighbourhood, with accompanying restaurants, club rooms, libraries, and churches. So well organised is this system, we are told, that "the residents have little need of ever going outside their square to have their wants properly attended to."[82] In the shops the highest quality goods are available, with prices controlled by the authorities. Universal plenitude was often envisioned as not entailing more labour. William Stanley Child's *The Legal Revolution of 1902* (1898) speculated that "When all are engaged in productive enterprises the output will be simply fabulous. With but a few hours' work a day everything the mind can crave will be satisfied". Thus it came to pass that "The people have long since earned sufficient so they can have all the luxuries that beautify the home and make life cheerful, aside from all the substantial necessities. This at once wonderfully increased the consumption of goods, and enormously multiplied the demand on the factories."[83] Often government provides these bounties. In Thomas McGrady's *Beyond the Black Ocean* (1901), "The government produces quantities sufficient to keep the people in luxuries".[84] In Herman Hine Brinsmade's *Utopia Achieved* (1912), they make it possible for people to "live to get the maximum of efficiency and happiness out of existence. They have, thanks to the federal bureau of health, learned to eat the proper things—far less expensive and far more wholesome than those things which years ago they ate at a far greater expense."[85] Governments thus often shape desire. In Cosimo Noto's *The Ideal City* (1903), commodities which previously "were not only luxuries, but were dangerous to health, cigars and strong liquors, for instance . . . are much more expensive than formerly."[86]

Finally, we should consider how these authors treated sociability in relation to luxury. Inequality continued to be recognised as a threat to sociability as

82. Alvarado M. Fuller, *A.D. 2000* (Laird and Lee, 1890), 287.

83. William Stanley Child, *The Legal Revolution of 1902* (Charles H. Kerr, 1898), 257, 316.

84. Thomas McGrady, *Beyond the Black Ocean* (C. H. Kerr, 1901), 301.

85. Herman Hine Brinsmade, *Utopia Achieved* (Broadway, 1912), 44.

86. Cosimo Noto, *The Ideal City* (1903), 187.

such. The latter propensity, however, most regarded as a central feature of utopian life. In Theodor Hertzka's *Freeland* (1891), we read:

> Social life here is very bright and animated. Families that are intimate with each other meet together without ceremony almost every evening; and there is conversation, music, and, among the young people, not a little dancing. There is nothing particular in all this; but the very peculiar, and to the stranger at first altogether inexplicable, attraction of Freeland society is due to the prevailing tone of the most perfect freedom in combination with the loftiest nobility and the most exquisite delicacy.[87]

In William Dean Howells's *A Traveller from Altruria* (1894), too, we are told that:

> The privacy of the family is sacredly guarded in essentials, but the social instinct is so highly developed with us that we like to eat together in large refectories, and we meet constantly to argue and dispute on questions of aesthetics and metaphysics. We do not, perhaps, read so many books as you do, for most of our reading, when not for special research, but for culture and entertainment, is done by public readers, to large groups of listeners. We have no social meetings which are not free to all; and we encourage joking and the friendly give and take of witty encounters. . . .
>
> . . . We make our pleasures civic and public as far as possible, and the ideal is inclusive, and not exclusive. There are, of course, festivities which all cannot share, but our distribution into small communities favors the possibility of all doing so. Our daily life, however, is so largely social that we seldom meet by special invitation or engagement.[88]

And Arthur Bird's *Looking Forward* (1899) relates that:

> About the period of 1925 a radical change was effected. Upon meeting in public places, it was no longer customary for the gentleman to uncover, or for the lady to cast a glance in acknowledgement of his salutation. The mode was simplified. Ladies and gentlemen saluted one another in precisely the same manner. Each one, upon approach, raised their right hand in military salute, touching the hat, and by a quick movement, letting the hand drop to the side. This new custom placed both sexes upon equal and exact terms.[89]

A connection between restraining personal luxury and enjoying greater sociability was thus clearly made by some authors. One work more than any other links the new sociability of the future to relinquishing consumerism,

87. Theodor Hertzka, *Freeland: A Social Anticipation* (Chatto and Windus, 1891), 285.

88. Howells, *Traveller from Altruria*, 227, 238.

89. Arthur Bird, *Looking Forward* (L. C. Childs and Son, 1899), 216.

Gabriel Tarde's *Underground Man* (1905). This is unsurprising: as we have seen, Tarde was an accomplished sociologist who particularly stressed the importance of social relations to well-being. Emulation was central to his social theory, rooted in the belief that "Mutually to ape one another, and by dint of accumulated apings diversely combined to create an originality is the important thing." His utopia asserted that "If it has been possible for us to realise the most perfect and the most intense social life that has ever been seen, it is thanks to the extreme simplicity of our strictly so-called wants." Food was now supplied freely by machinery, while "The need of clothing has been pretty nearly abolished by the softness of an ever constant climate, and, we must also admit it, by the absence of silkworms and of textile plants." The old division of labour in manufactures had created "a non-social, an almost anti-social relationship" amongst classes, which has been superseded. Tarde criticised the "ancient visionaries called socialists" because of "their failure to see that this life in common, this intense social life, they dreamt of so ardently, had for its indispensable condition the aesthetic life and the universal propagation of the religion of truth and beauty." The latter principle, he thought, assumed "the drastic lopping off of numerous personal wants", but the socialists, "in rushing, as they did, into an exaggerated development of commercial life", had been "marching in the opposite direction to their own goal." The new sociability would also be attained by improving urban life:

> Man in becoming a town dweller has become really human. From the time that all sorts of trees and beasts, of flowers and insects no longer interpose between men, and all sorts of vulgar wants no longer hinder the progress of the truly human faculties, every one seems to be born well-bred, just as every one is born a sculptor or musician, philosopher or poet, and speaks the most correct language with the purest accent. An indescribable courtesy, skilled to charm without falsehood, to please without obsequiousness, the most free from fawning one has ever seen, is united to a politeness which has at heart the feeling, not of a social hierarchy to be respected, but of a social harmony to be maintained. It is composed not of more or less degenerate airs of the court, but of more or less faithful reflections of the heart . . . No unsociableness, no misanthropy can resist it. The charm is too profound.[90]

In the late nineteenth and early twentieth-century historic utopian pleas for simplicity were rendered considerably more complex by a growing infatuation

90. Gabriel Tarde, *Underground Man* (Duckworth, 1905), 117–18, 120–21, 124, 141–42.

with science and technology.[91] Within a few decades, gas lighting and heating, electricity, the telegraph, the phonograph, refrigeration, radiation, the internal combustion engine, human flight, the radio, and much else were discovered, invented, or greatly improved and more widely disseminated. The new technologies promised to solve the age-old problem of requiring extensive manual labour to supply needs. Twentieth-century intoxication with progress dates from this period, as technology became central to the assumption of plenty which some utopian authors entertained. Now, in principle, no future utopia need rely on slavery or coerced labour, and the requirement of extreme frugality to avoid this was virtually removed. But to compensate, in the real world, the new machines threatened to assist in forming an even more powerful form of tyranny over the majority of workers, and thus to oppress them still further. One of the most prescient warnings of this type was *Erewhon; or, Over the Range* (1872), by Samuel Butler (1835–1902).[92] This speculated, very dramatically for its own times and no less so for ours, that if mankind had, according to Darwin, evolved from the apes, so might an advanced artificial intelligence proceed from modern technology. No Luddite, Butler did not recommend abolishing machines, only reducing them to a minimum.

Edward Bellamy and the Shift to Public Luxury

As we now usually view it, perhaps owing to the way the subject is often taught, later Anglo-American nineteenth-century utopian literature is dominated by two key texts, Edward Bellamy's *Looking Backward 2000–1887* (1888), and William Morris's critique of it, *News from Nowhere* (1890). At the time Bellamy was far more influential, selling perhaps 800,000 copies over half a century. William Dean Howells, indeed, thought that he had "revived throughout Christendom the faith in a millennium".[93] Today, however, few read the somewhat stilted prose of Bellamy, and our tastes run mostly to the more romantic, as well as more Marxist, Morris. But time may again reverse this trend.

The issue of luxury, how to generate and share it, even how to define it, was central to the differences between the two writers. Bellamy had a powerful sense of the liberating potential of the new technologies, of the burgeoning demand for affluence by the majority, and of the appeal of life in great cities.

91. See Nell Eurich, *Science in Utopia: A Mighty Design* (Harvard University Press, 1967).

92. A recent study is Peter Raby's *Samuel Butler: A Biography* (Hogarth, 1991).

93. Quoted in W.H.G. Armytage, *Yesterday's Tomorrows: A Historical Survey of Future Societies* (Routledge and Kegan Paul, 1968), 83. On British utopias in this period, see Matthew Beaumont's *Utopia Ltd: Ideologies of Social Dreaming in England, 1870–1900* (Brill, 2005).

His vision is both utopian and millenarian. The system he proposes is not psychologically demanding; there are no parties or politicians, no demagoguery or corruption. This prompts the narrator to comment, "Human nature itself must have changed very much," which receives the reply, "Not at all . . . but the conditions of human life have changed, and with them the motives of human action."[94] With no private property to speak of there is little crime. The fact that people now "live together like brethren dwelling in unity, without strifes or envying, violence or overreaching" is likened to earlier ideas of heaven and the millennium, "and the theory from their point of view does not lack plausibility."[95] Bellamy called his system—"Nationalism"—"a religion," "a Judgment Day," or "God's kingdom of fraternal equality." Publicly he termed its underlying principle "human brotherhood", and implied the Second Coming of Christ would now occur in the hearts of men.[96] In a dialogue he proclaimed "that the world is upon the verge of the realization of the visions of universal peace, love and justice, which the seers and poets of all ages have more or less dimly foreseen and testified of."[97]

So at one level, Nationalism was clearly a secular millenarian project. At another, however, it was simply a new theory of sociability. "Individualism, which in your day was the animating idea of society", he insisted, "not only was fatal to any vital sentiment of brotherhood and common interest among living men, but equally to any realization of the responsibility of the living for the generation to follow."[98] What Bellamy termed the "passion for losing ourselves in others or for absorbing them into ourselves, which rebels against individuality as an impediment" was "the expression of the greatest law of solidarity". His essay "The Religion of Solidarity" concluded it was "the only rational philosophy of the moral instincts. Unselfishness, self-sacrifice, are the essence of morality".[99]

Looking Backward rejected the "colony" approach to the ideal society, and proposed instead a highly centralised, urban, statist organisation of labour, with mandatory industrial armies for most, and equal wages. A debt to Carlyle is likely here, and, in the Saint-Simonian spirit, industry is at the centre of human activity.[100] Many inventions, like music piped into houses—he had

94. Bellamy, *Looking Backward*, 36.

95. Bellamy, 120, 169–70.

96. Bellamy, *Selected Writings*, xxvi; Sylvia E. Bowman, *Edward Bellamy* (Twayne, 1986), 88.

97. Edward Bellamy, *Talks on Nationalism* (Books for Libraries, 1938), 67.

98. Bellamy, *Looking Backward*, 157.

99. Bellamy, *Selected Writings*, 18, 22.

100. Edward Bellamy, *Edward Bellamy Speaks Again* (Peerage, 1937), 77. Bellamy mentions Saint-Simon at various points: see Sylvia E. Bowman et al., *Edward Bellamy Abroad: An American Prophet's Influence* (Twayne, 1962), 55.

read Cabet—ameliorate urban life. But private luxury has given way to that of the public, and here we encounter a powerful exposition of how this might work. Now "the splendor of our public and common life as compared with the simplicity of our private and home life" is presented as "the contrast which, in this respect, the twentieth bears to the nineteenth century. To save ourselves useless burdens, we have as little gear about us at home as is consistent with comfort, but the social side of our life is ornate and luxurious beyond anything the world ever knew before." The narrator is told that the population

> are pretty well off. The rate at which we live is as luxurious as we could wish. The rivalry of ostentation, which in your day led to extravagance in no way conducive to comfort, finds no place, of course, in a society of people absolutely equal in resources, and our ambition stops at the surroundings which minister to the enjoyment of life . . . We might, indeed, have much larger incomes, individually, if we chose so to use the surplus of our product, but we prefer to expend it upon public works and pleasures in which all share, upon public halls and buildings, art galleries, bridges, statuary, means of transit, and the conveniences of our cities, great musical and theatrical exhibitions, and in providing, on a vast scale, for the recreations of the people . . . At home we have comfort, but the splendor of our life is, on its social side, that which we share with our fellows.

While, in the past, "What little wealth you had seems almost wholly to have been lavished in private luxury. Nowadays, on the contrary, there is no destination of the surplus wealth so popular as the adornment of the city, which all enjoy in equal degree." Private houses are much less sumptuous than public buildings. Cooking is done in public kitchens, while guilds and societies have "club houses" for their members. The word "menial" is obsolete, as is contempt for service roles. For "nowadays it is an axiom of ethics that to accept a service from another which we would be unwilling to return in kind" is impermissible. Now

> there is recognized no sort of difference between the dignity of the different sorts of work required by the nation. The individual is never regarded, nor regards himself, as the servant of those he serves, nor is he in any way dependent upon them. It is always the nation which he is serving. No difference is recognized between a waiter's functions and those of any other worker.

Fear of poverty and love of luxury have given way to the higher motives of "service of the nation, patriotism, passion for humanity". Thus, "The equal wealth and equal opportunities of culture which all persons now enjoy have simply made us all members of one class, which corresponds to the most fortunate class with you. Until this equality of condition had come to pass, the idea of the solidarity of humanity, the brotherhood of all men, could never have become the real conviction and practical principle of action it is nowadays."[101]

This process is aided by the fact that property cannot be converted into money:

> Under the present organization of society, accumulations of personal property are merely burdensome the moment they exceed what adds to the real comfort. In your day, if a man had a house crammed full with gold and silver plate, rare china, expensive furniture, and such things, he was considered rich, for these things represented money, and could at any time be turned into it. Nowadays a man whom the legacies of a hundred relatives, simultaneously dying, should place in a similar position, would be considered very unlucky. The articles, not being saleable, would be of no value to him except for their actual use or the enjoyment of their beauty.[102]

So, most such accumulations are given to the nation as "common stock". Consumption is rationalised, with each ward of the city having its own shop. Advertising is banned, evidently to suppress demand:

> There was no display of goods in the great windows, or any device to advertise wares, or attract custom. Nor was there any sort of sign or legend on the front of the building to indicate the character of the business carried on there; but instead, above the portal, standing out from the front of the building, a majestic life-size group of statuary, the central figure of which was a female ideal of Plenty, with her cornucopia.

We also learn that minority tastes are catered for, and the "administration has no power to stop the production of any commodity for which there continues to be a demand", though items are priced according to cost.[103] A system of setting work credits against demand for goods would balance production against consumption. Some of Bellamy's imitators, like Bradford Peck's *The World a Department Store: A Story of Life under a Cooperative System* (1900), insisted that providing a universal income would give all access to luxuries.[104] Some critics warned, however, that the luxury Bellamy's system promised might well prove its undoing.[105]

101. Bellamy, *Looking Backward*, 24–25, 93, 57, 91–92, 143.

102. Bellamy, 69.

103. Bellamy, 60, 108.

104. Bradford Peck, *The World A Department Store: A Story of Life under a Cooperative System* (Bradford Peck, 1900), 81–82.

105. E.g., W. W. Satterlee, *Looking Backward, and What I Saw* (Harrison and Smith, 1890), 135.

Bellamy's sequel to *Looking Backward, Equality* (1897), developed his account of consumer behaviour. He described the nineteenth century's "fashion mania" as

> the natural result of a disparity of economic conditions prevailing in a community in which rigid distinctions of caste had ceased to exist. It resulted from two factors: the desire of the common herd to imitate the superior class, and the desire of the superior class to protect themselves from that imitation and preserve distinction of appearance.

The gendered dimensions of this debate are revealed more clearly here. Gems and precious metals have lost their value as symbols of social inequality. So "the wearing of jewelry had been virtually an obsolete custom for a couple of generations if not more". The "ancient practice of using paints on our faces and bodies" has ceased. Now "women give much less thought to dress than in your day and men considerably more", for neither sex seeks attractiveness more than the other. Again, equality is the cause: "from the moment that equality became established between them it ceased to be a whit more the interest of women to make themselves attractive and desirable to men than for men to produce the same impression upon women."[106] So the myth of greater female vanity is disposed of: an economic cause is responsible.

William Morris's News from Nowhere *(1890): Beauty and Creativity*

In Europe, Bellamy's main critic was a leading late Victorian socialist, William Morris (1834–96), whose philo-medievalist utopia *News from Nowhere* makes beauty and nature integral to the ideal life.[107] As for Rousseau, "modern" meant "inauthentic" to Morris, who called "hatred of modern civilisation" the "leading passion of my life". Inspired by Richard Jefferies's post-apocalyptic *After London* (1885), which celebrated a reversion to "barbarism", he wrote in 1885 "how often it consoles me to think of barbarism once more flooding the world, and real feelings and passions, however rudimentary, taking the place of our wretched hypocrisies."[108] Morris thought "the elaboration of machinery . . .

106. Edward Bellamy, *Equality* (1897; D. Appleton Century, 1937), 60, 125–28.

107. The classic study remains E. P. Thompson's *William Morris*. Morris's review of Bellamy is reprinted in Claeys and Sargent's *Utopia Reader* (315–20).

108. *How I Became a Socialist: A Series of Biographical Sketches* (Twentieth Century, 1896), 20; William Morris, *The Collected Letters of William Morris* (Princeton University Press, 1987), vol. 2, pt 2: 436. He did not of course mean by this a hatred of machinery as such, a view

will lead to the simplification of life, and so once more to the limitation of machinery". He considered fashion "a strange monster born out of the vacancy of the lives of rich people", and demanded "the extinction of luxury" in the sense of the grand clubs of bourgeois London.[109] Morris's view of luxury and modernity generally was much indebted to John Ruskin, who wanted most of the railways destroyed and a mass return of workers to the land to relearn traditional ways of life.[110] To Ruskin, "useless luxury" was anything men were "degraded in providing", again implying that "luxury" did not mean things themselves, but the social consequences of supplying them. "Luxury is indeed possible in the future", he wrote, "innocent and exquisite; luxury for all, and by the help of all; but luxury at present can only be enjoyed by the ignorant; the cruelest man living could not sit at his feast, unless he sat blindfold." So "severe limitations" should be placed on this process.[111] In his experimental land settlement, the "Guild of St. George", female members were encouraged to "set an example in frugal living, avoiding luxuries and wearing plain clothes."[112]

Unsurprisingly, Morris found Bellamy's approach to labour pedestrian and his infatuation with technology reprehensible. He objected to *Looking Backward*'s over-centralisation, its neglect of the role of beauty in human life, the implication that enormous cities were desirable, and the immense concentration of power in the government, which other detractors also thought had authoritarian overtones. With a nod towards J. S. Mill's idea of the stationary state, Morris imagined a future Britain where heavy industry was minimised, large cities shrunken, and a balance of creative and necessary labour achieved. A love of craftsmanship lay at the centre of his ideal. Much of his own vision of future luxury is, however, curiously private and individual: we make beautiful things for ourselves in metal, leather, and other materials, and decorate

---

sometimes still imputed to him. For while *News from Nowhere* does describe machine printing as "beginning to die out" (20), we are told unequivocally that "All work which, would be irksome to do by hand is done by immensely improved machinery; and in all work which it is a pleasure to do by hand machinery is done without" (108).

109. William Morris, *Signs of Change* (Reeves and Turner, 1888), 33; Morris, *The Political Writings of William Morris*, ed. A. L. Morton (Lawrence and Wishart, 1984), 193.

110. On some results thereof, see Vicky Albritton and Fredrik Albritton Jonsson's *Green Victorians: The Simple Life in John Ruskin's Lake District* (University of Chicago Press, 2016). On Ruskin's version of this, see Jeffrey R. Spear's *Dreams of an English Eden: Ruskin and His Tradition in Social Criticism* (Columbia University Press, 1984).

111. John Ruskin, *The Works of John Ruskin*, 11 vols (George Allen, 1882), 5: 144–45; Ruskin, *Unto This Last* (Smith, Elder, 1862), 173; Ruskin, *Works*, 11: 161.

112. Quoted in Albritton and Jonsson, *Green Victorians*, 79.

ourselves with them. Morris spent much of his working life designing expensive furnishings and interior decorations for the well-to-do, but told his friend Edward Carpenter, whom we will shortly meet, that "I am inclined to think that sort of thing is mostly rubbish, and I would prefer for my part to live with the plainest whitewashed walls and wooden chairs and tables."[113]

News from Nowhere projects an ideal society which is, like that of many early socialists, a kind of halfway house between city and countryside. As with Mill, a passionate love of nature from childhood onwards helped drive Morris in this direction. Environmentalists today regard him amongst the first to portray reduced air and water pollution, and the cultivation of green spaces within cities. Like Marx and Engels, at least at points, Morris conceived that the model human character consisted in a merging of rural and urban occupations and outlooks. So the people "flocked into the country villages" at the outset of the revolutionary period of transition, "but the invaders, like the warlike invaders of early days, yielded to the influence of their surroundings, and became country people." Thus the "difference between town and country grew less and less", and the "world of the country vivified by the thought and briskness of town-bred folk" now produced a "happy and leisurely but eager life".[114] England, one of the most famous passages in the book tells us, "is now a garden, where nothing is wasted and nothing is spoilt, with the necessary dwellings, sheds, and workshops scattered up and down the country, all trim and neat and pretty."[115] In its economics, the future society discourages "centralisation all we can, and we have long ago dropped the pretension to be the market of the world."[116]

This pretty portrait is, however, lacking in some essential details. Though Morris hints that rare materials like gold are readily available, it is unclear how supply and demand function in the future. In one key example he gives us, the narrator goes shopping for a pipe. The article provided is indeed luxurious, being "carved out of some hard wood very elaborately, and mounted in gold sprinkled with little gems".[117] But it costs nothing. Then there is the question of drink. A "rougher kind of wine" is sold alongside fine Rhine wines, and the narrator is indeed puzzled as to "how they managed to make fine wine when there were no longer labourers compelled to drink rot-gut instead of the fine wine which they themselves made." With respect to fashion, the women are described as "decently veiled with drapery, and not bundled up with millinery; . . .

113. Edward Carpenter, *My Days and Dreams: Being Autobiographical Notes* (George Allen and Unwin, 1916), 217.

114. William Morris, *News from Nowhere* (1890; Longmans, Green, 1899), 79.

115. Morris, 80.

116. Morris, 74.

117. Morris, 40.

they were clothed like women, not upholstered like armchairs, as most women of our time are. In short, their dress was somewhat between that of the ancient classical costume and the simpler forms of the fourteenth century garments."[118]

To late nineteenth-century Britons, at least, *News from Nowhere* established the great utopian divide as being between Bellamy's gleaming, machine-driven modernity and Morris's simpler, craft-centred ideal. With respect to "the Great Change" which would inaugurate the new society, Morris was also a revolutionary where Bellamy was not. Already in 1888 Morris argued that "it is utopian to put forward a scheme of gradual logical reconstruction of society which is liable to be overturned at the first historical hitch it comes to."[119] By contrast, Bellamy, he thought, conceived "of the change to Socialism as taking place without any breakdown of that life, or indeed disturbance of it, by means of the final development of the great private monopolies which are such a noteworthy feature of the present day. He supposes that these must necessarily be absorbed into one great monopoly which will include the whole people and be worked for its benefit by the whole people."[120]

Amongst the late Victorian socialists, William Morris was not alone in his antipathy to luxury. The most influential writer to address the theme was the leader of the Clarion movement, Robert Blatchford (1851–1943), whose *Merrie England* (1893) sold over two million copies. Blatchford was a pro-imperialist Little Englander whose support for the Boer War and opposition to female suffrage have left him marginalised in histories of socialism. On our central theme here, he is in many respects the heir to William Cobbett. Rejecting the view that the "consumption of luxuries by the rich finds useful employment for the poor", Blatchford asserted that "the luxury of the rich is a direct cause of the misery of the poor". He described his ideal of life as "*frugality of body and opulence of mind*. I suggest that we should be as temperate and as simple as possible in our use of mere bodily necessaries, so that we may have as much time as possible to enjoy pleasures of a higher, purer, and more delightful kind". "I propose", he argued, "to make our material lives simple; to spend as little time and labour as possible upon the production of food, clothing, houses, and fuel, in order that we may have more leisure". This would involve growing and producing more at home, on the principle "that the people should make

118. Morris, 36, 41, 14.
119. Morris, *Political Writings*, 341.
120. Morris, 421.

the best of their own country before attempting to trade with other countries", thus restraining the factory system, while stopping "the smoke nuisance by developing water power and electricity".[121] Britain should thus no longer be the "Workshop of the World". In *Britain for the British* (1902), meaning the working population rather than the wealthy few, he argued that "it is very clear that the more luxuries a people produce, the fewer necessaries they will produce". Here he concluded that "In a well-ordered State no luxuries would be produced until there were enough necessaries for all", which does not, however, imply antipathy to luxury as such.[122] His utopia, *The Sorcery Shop* (1907), condemned "wasteful extravagance and luxury, which bring but little pleasure to those who indulge in them", and he suggested that if it were possible to convince people "that there are better things than luxury, and power, and applause, and ease, and that these more desirable ends could be attained without money, they would cease to care for money." He added:

> when one has physical health and comfort one can be happy without luxury and waste. And these people recognise that fact. Real happiness comes by higher paths. Real happiness comes by service, by achievement, by purity of morals, and loftiness of mind. Give a man health, respect, food, a home, a wife and children, congenial recreation and congenial work, and he is happier than a millionaire or an emperor.[123]

This autarkic, nationalist, humanitarian ideal—effectively socialism in one country—was immensely influential, as Blatchford's Clarion movement, which included cycling clubs, theatrical groups, and choral societies, attracted tens of thousands of adherents.

A trend more often associated with Morris was the Simple Life movement, as it came to be known, which expanded in the 1890s to comprise ideals of improved furniture, dress, art, craftsmanship, manners, and morals.[124] It wedded no small element of pastoral nostalgia to social and political radicalism and a passionate dedication to the aesthetic. Linked to the Arts and Crafts movement, and with initiatives to preserve older buildings and monuments, amongst other things, its ideals were generally regarded as leading progressive thinking at least through the 1930s. Its more extreme exponents were

121. Robert Blatchford, *Britain for the British* (Clarion, 1902), 63; Blatchford, *What's All This?* (Labour Book Service, 1940), 57; Blatchford, *Merrie England* (Clarion, 1908), 237, 16, 47. Here he noted, "read 'Walden'" (45).

122. Blatchford, *Britain for the British*, 66.

123. Robert Blatchford, *The Sorcery Shop: An Impossible Romance* (Clarion, 1907), 184, 112–13, 194.

124. For a literary portrayal of the movement, see Daniel Chaucer's *The Simple Life Limited* (Bodley Head, 1911).

very hostile to urban life. Some aimed to escape the city entirely. Others, like the evangelical clergyman W. J. Dawson (1854–1928), who wrote a play about Savonarola, longed at least for "the life of pleasant alternation between town and country", if only because there was no work in rural areas. This is similar to what Morris hoped to institutionalise. Cities were overrated, and to Dawson their inhabitants, far from being more gregarious, were actually largely "indifferent to social intercourse". Friendships were strained by the distance required to meet and the complexity of arranging complex timetables in a frenetic schedule. Dawson thought that even the wealthy usually failed to enjoy their riches, and hoped an ideal state would place a ceiling of £10,000 per annum (*c.* £200,000 today) on income, to reduce avarice and "vulgar ostentation". But when he finally fled to a cottage in the countryside, Dawson found that "there was universal restlessness among the people. The common ambition of all the younger generation was to get to London by almost any means, and in almost any capacity . . . The universal complaint was that life was dull". So the answer to urban malaise seemed rather to lie in redesigning the city to make it more like the country. Let us suppose, Dawson wrote in 1903,

> a reconstructed London, devised upon the broad principle of ample space and air according to population; of congregated and contiguous cities under a common government; of public buildings of utility and beauty equally distributed; and it is easy to imagine a London that should combine all the charm of the country with the advantages of the metropolis. The splendid streets, which are the main arteries of traffic, would remain, but the squalid tenements and alleys which are packed away behind them would disappear. A long chain of parks and gardens would unite the West and East, taking the place of a host of rotten rabbit-warrens, which are a disgrace to any civilised community. There would be no quarter of the town relinquished to the absolutely poor . . . Manufactories would be distributed as well as mansions. The various trades would not be huddled together in narrow inconvenient corners of the metropolis . . . The artisan would thus work within sight of his house, and that entire dislocation of home-life, involved by present conditions of labour, would disappear. And each of these townships would have its baths, libraries, and technical schools, not dependent on local enterprise or generosity, but administered by a central body . . . and each would be saved from the narrow spirit of suburbanism by the proud sense of its corporate unity with London.[125]

125. W. Dawson, *Quest*, 266–68. Dawson's short play is *Savonarola: A Drama* (Grant Richards, 1900).

This vision of the "Regenerated City", which overlaps with Morris's proposals, we must return to later.

Others, however, made the transition to rural simplicity more successfully. Chief amongst those craving a clean break from city life was Edward Carpenter (1844–1929), a pioneering gay theorist of sexual liberation, nudism, and vegetarianism who condemned the "civilised" view of sex as "thoroughly unclean", and thought that to "live in opulent and luxurious surroundings is to erect a fence between yourself and the mass-world which no self-respecting manual worker will pass. It is consequently to stultify yourself and to lose some of the best that the world can give." In middle age, Carpenter plunged himself into the "Thoreau ideal" of rural isolation, living in a hut on a seven-acre plot at Millthorpe near Sheffield with his male companion, growing or raising most of his own produce and giving his dress clothes away in favour of plainer garb. He even had a "Saxon" tunic made for him, though apparently never wore it. In *England's Ideal* (1887), taking a leaf from Thoreau, Carpenter described his goal as "the simplification of life", defined by moderation in eating, furnishing, clothing, and adornment, and commending all this as the socialistic norm towards which society was tending. This was the only means, he thought, to live, "if you do not want to be a vampire and a parasite upon others"—the utopia/dystopia problem again. It also involved reducing foreign imports, especially of food, and avoiding the "*feverish* work" of modern life.[126] In *Civilization: Its Cause and Cure* (1889) he contended for the "life of the open air, familiarity with the winds and waves, clean and pure food, the companionship of the animals—the very wrestling with the great Mother for his food", and claimed that "all these things will tend to restore that relationship which man has so long disowned".[127]

Carpenter was involved in a forerunner of the Fabian Society, the Fellowship of the New Life, formed in 1883, which held "rustic gatherings" in the summer, one of which addressed "The Return to Nature" and "truth, beauty and simplicity of life." Though self-proclaimed "scientific socialists" like H. M. Hyndman derided them as "sentimentalists", many aimed to balance urban and rural life.[128] The best-known result was the Garden City movement, whose leader, Ebenezer Howard, deeply inspired by Bellamy, recognised that

126. Edward Carpenter, *England's Ideal and Other Papers on Social Subjects* (Swan Sonnenschein, 1902), 79, 84; Fiona McCarthy, *The Simple Life: C. R. Ashbee in the Cotswolds* (Lund Humphries, 1981), 12.

127. Edward Carpenter, *Civilization: Its Cause and Cure* (George Allen and Unwin, 1921), 60; Edward Carpenter, *Sex-Love, and Its Place in a Free Society* (Labour Press Society, 1894), 16; Carpenter, *My Days and Dreams*, 166, 147–49; Carpenter, *England's Ideal*, 79–105.

128. Quoted in Sheila Rowbotham and Jeffrey Weeks, *Socialism and the New Life: The Personal and Sexual Politics of Edward Carpenter and Havelock Ellis* (Pluto, 1977), 45.

"lack of amusement"—today called "no broadband"—was draining the countryside. He proposed an alternative scheme for a circular "social city" of thirty-two thousand.[129] One of the model towns constructed, Letchworth, begun in 1903, had a Simple Life Hotel, and drew reformers of all stripes like a magnet. Its male inhabitants were mocked as bearded, long-haired, sandal-wearing, fruit-and-nut Healthy Life cranks, forerunners in other words of the Beatniks and Hippies we consider below.[130] The Garden City movement, indeed, represented a major episode in the Simple Life epoch, and attracted large numbers of Quakers and similarly minded puritans seeking "elegance rather than luxury: and refinement rather than fashion".[131]

Another of Carpenter's friends was Henry S. Salt (1851–1939), who first interested Gandhi in Thoreau and became a leading advocate of animal rights and vegetarianism. (The Vegetarian Society, which still exists, had been founded in 1847, and interest in the subject expanded greatly in the 1880s and 1890s.)[132] With a nod to Rousseau, Godwin, and Thoreau, Salt insisted that since "luxury on the part of one man must involve drudgery on the part of another", "simplification of life" was the necessary corollary.[133] Parallel movements at the time included the Rational Dress Society, founded in 1881, which protested against "constant changes in fashion" and constricting dress for women in particular.[134] At the turn of the century the Simple Life Press was founded by the Ruskinian Godfrey Blount. Proclaiming a "Utopia the world must have", Blount called for a "New Crusade" in the form of a "plea for The Country Life in the ideal purity of its simplicity".[135] This implied a "return to simpler methods of manufacture, and the simpler methods of living which such return would involve", and renouncing "the mere accumulation of luxuries and the opportunity of indulging in them".[136] Blount understood this reaction in terms of antipathy to machinery as such as setting a standard of work which human beings felt compelled to follow:

129. Ebenezer Howard, *Garden Cities of Tomorrow* (1898; Faber and Faber, 1945), 142.

130. Marsh, *Back to the Land*, 103, 233.

131. Dugald Macfadyen, *Sir Ebenezer Howard and the Town Planning Movement* (Manchester University Press, 1933), 35.

132. Ideas of a vegetarian utopia have existed since the classical world: Colin Spencer, *Vegetarianism: A History* (Grub Street, 2000), 93. See also Rod Preece, *Sins of the Flesh: A History of Ethical Vegetarian Thought* (UBC, 2008).

133. Henry S. Salt, *Seventy Years among Savages* (George Allen and Unwin, 1921), 73.

134. Marsh, *Back to the Land*, 189.

135. Godfrey Blount [pseud. Hans Breitman], *For Our Country's Sake* (A. C. Fifield, 1905), 6, 30.

136. Godfrey Blount, *The Gospel of Simplicity: A Plea for Country Life & Handicrafts* (Simple Life, 1903), 7.

> The deadly perfection of what the machine does, besides setting a false standard of finish before the workman and destroying his imagination and pride, exerts a peculiar fascination over the mind of the consumer who has had no share in the production, and tempts him to believe that the triumphant tendency of our civilization in the greater and greater perfection of the machines, whereby with the nationalization of the means of production and the reduction of human labour to a minimum, he is confident that we should all be free to educate and amuse ourselves to our hearts' content.

This plea for "leisure", however, he felt was "divorced from our feeling of pleasure from work".[137] Here, thus, we see the chasm separating Ruskin and Morris from many other socialists of the period.

Outside Britain, as society became more complex and frenetic in the 1880s and 1890s, the appeal to simplicity assumed various forms. In France, the great post-impressionist painter Paul Gauguin (1848–1903) once again popularised a Tahitian ideal of languid simplicity, as Henry Melville had done a generation earlier in *Typee: A Peep at Polynesian Life* (1846), Bougainville and Diderot seventy years before that, and James Michener did seventy years later, describing it as "the most lasting vision of the earthly paradise . . . the symbol of hedonism".[138] Gauguin first reached Tahiti in 1891. Seeing it as the direct opposite of a money-obsessed bourgeois French culture, he relished the prospect of "turning native" there.[139] "Bit by bit civilisation is leaving me", he exulted, as he settled into his hut:

> I'm beginning to think simply and to lose hatred for my neighbour—or even better, to love him—I live a free life, enjoying animal and human pleasures. I escape from the factitious, I identify myself with nature, knowing that the next day will be like the present one, just as free, just as beautiful. Peace is taking possession of me.[140]

He revelled, he told the writer August Strindberg, in "a barbarism which spells rejuvenation for me", and hoped to display in his paintings of native women a new "Eve of primitive times" foreign to European misogyny.[141] Today we perceive much ambiguity and self-delusion in these reflections. By this time, nearly 90 per cent of Polynesians had died of disease and the western way of life.

137. Godfrey Blount, *A New Crusade: An Appeal* (Simple Life, 1903), viii.

138. James Michener, *Return to Paradise* (Secker and Warburg, 1951), 45.

139. Paul Gauguin, *The Letters of Paul Gauguin to Georges Daniel de Monfreid* (William Heinemann, 1923), 25. See Gauguin, *Noa Noa* (Nicholas L. Brown, 1919), 40.

140. Quoted in Lawrence Hanson and Elisabeth Hanson, *The Noble Savage: A Life of Paul Gauguin* (Chatto and Windus, 1954), 214.

141. Paul Gauguin, *Letters to His Wife and Friends* (World Publishing, 1949), 197.

Gauguin's trysts with young Tahitian maidens parody paradise, and smack of the seduction of Eve by Satan, its women indeed inviting what Gauguin himself called "a longing to rape", and a dystopian fantasy of conquest which produced a miniature empire of at least ten of his own children.[142] And there is doubt that he ever really sought to renounce bourgeois society, as opposed to merely coquetting with what he took to be its opposite, but which might, for a habitué of brothels, also appear to be an extension of their libertine fantasies.

Nonetheless, such avowals of primitivism struck a chord at many levels. Many, indeed, were disenchanted with so-called "civilisation". In France, the Lutheran writer Charles Wagner argued that the unlimited multiplication of material needs had not brought happiness, social peace, an increase of independence, or any higher standard of morality. Instead, and especially among the rich, "the more goods a man has, the more he wants", and the "more desires and needs a man has, the more occasion he finds for conflict with his fellowmen".[143] In central Europe, Carl Jung, Isadora Duncan, and other "spiritual rebels" took refuge during 1900–1920 in Ascona, Switzerland, where vegetarianism, naturism, feminism, and paganism were amongst the trends explored.[144] In the United States, in the same period, referring to Whitman, Thoreau, Carpenter, and Morris, Leonard Abbott predicted that in the future "All kinds of luxury will surely disappear . . . This tendency towards simplicity will probably make itself felt in every department of our life."[145] A few decades later, the Tarzan films exemplified a similar juxtaposition between the civilised and primitive, the (literally) noble savage, Lord Greystoke, rejecting civilisation for the life of the forest. Anthropologists like Margaret Mead stimulated further interest in the attractions of more-primitive societies. Marshall Sahlins famously described a condition of "original affluence" where "all the people's material wants are easily satisfied" and none desire more.[146] And the respectfully protective attitude of many indigenous peoples towards nature rightly continues to inspire environmentalists today.

H. G. Wells

One later writer, who read Bellamy carefully, markedly helped to tilt utopianism away from such appeals to the primitive.[147] This was H. G. Wells (1866–1946), who more than any other writer defined the future-oriented zeitgeist

142. Nancy Mowll Mathews, *Paul Gauguin: An Erotic Life* (Yale University Press, 2001), 181.

143. Charles Wagner, *The Simple Life* (1897; Ibister, 1903), 49, 52.

144. Ian Marchant, *A Hero for High Times* (Jonathan Cape, 2018), 22.

145. Leonard Abbott, *The Society of the Future* (J. A. Wayland, 1898), 15.

146. Sahlins, *Stone-Age Economics*, 1.

147. Wells parodies Bellamy in *When the Sleeper Wakes* (1899).

of the early twentieth century in terms of the speed and promise of technological and scientific innovation.[148] Wells had a background in biology and, like the Marxists, imagined science to be firmly on his side. He railed at those for whom "Socialism has got mixed up with Return-to-Nature ideas, with proposals for living in a state of unregulated primitive virtue in purely hand-made houses, upon rain water and uncooked fruit". Instead he insisted on the need "to disavow, with all necessary emphasis, that gibing at science and the medical profession, at schools and books and the necessary apparatus for collective thinking, which has been one of our little ornamental weaknesses in the past."[149] But he also rejected the idea that revolutionaries ought to be given a "blank cheque" to "put everything on a scientific basis".[150]

Wells was adamant that any new utopia could not be fixed in time and space. In his view, the only future possible after Darwin had to be "kinetic", aiming not at a "permanent state but at a hopeful stage, leading to a long ascent of stages". The problem of excessive regimentation, which made so many earlier utopias unappealing to the later moderns, also needed to be confronted. "Compared with the older writers", Wells thought, "Bellamy and Morris have a vivid sense of individual separation, and their departure from the old homogeneity is sufficiently marked to justify a doubt whether there will be any more thoroughly communistic Utopias for ever."[151] So utopia had to cultivate "individualities". Wells doubted whether "anyone could stand a month of the relentless publicity of virtue planned by More". He insisted against an "intolerable continuity of contact" that "No one wants to live in any community of intercourse really, save for the sake of the individualities he would meet there."[152] And utopia could no longer be confined within national boundaries: its aims required not only powerful centralised governments but a world-state. An enlarged patriotism could wed ever greater numbers of people together in a sense of communality, for Wells still thought it "necessary to his moral life, that a man should feel himself part of a community, belonging to it, and it belonging to him. And that this community should be a single and lovable reality, inspired by a common idea, with a common fashion and aim."[153]

148. On Wells's intellectual development, see John S. Partington's *Building Cosmopolis: The Political Thought of H. G. Wells* (Ashgate, 2003).

149. H. G. Wells, *Socialism and the Family* (A. C. Fifield, 1906), 26–27.

150. H. G. Wells, *The Bulpington of Blup* (Hutchinson, 1933), 104.

151. Wells, *Modern Utopia*, 87.

152. Wells, 5, 37, 10.

153. H. G. Wells, *The Salvaging of Civilisation* (Cassell, 1921), 69. See further Maxim Shadurski, *The Nationality of Utopia: H. G. Wells, England, and the World-State* (Routledge, 2020).

Critics like the Catholic philosopher H. G. Chesterton of course denied that this was psychologically possible.[154]

Like Bellamy's, Wells's utopian vision involved a future often defined by America, of "growth invincible" and "inexhaustible supply" of concrete, glass, Formica, neon, and steel, and plenty for all.[155] High-rise apartments, aeroplanes, automobiles, and space travel would come to epitomise the new ideal of utopian modernity far more than the whimsical musings of nostalgic dissidents like Morris or Carpenter. Manhattan or, today, the gleaming towers of Dubai, not garden cities like Letchworth, would come to embody most people's idea of the future, even if in reality they represent "evil paradises" which mask large-scale criminality and exploitation.[156] Wells was acutely aware of the shortcomings of the American Dream, and the growing chasm between rich and poor; he was a socialist, after all, who believed in "the establishment of a new and better order of society by the abolition of private property in land, in natural productions, and in their exploitation."[157] A key problem, he suggested, was achieving "the use and consumption of material goods without the burthen of ownership", which would then permit unlimited luxury.[158] This Wells thought was an important step beyond Bellamy. In *When the Sleeper Wakes* (1899) he confronted Bellamy directly, imagining his world view realised, and proposing that the result would be "no Utopia, no Socialistic state", for "the ancient antithesis of luxury, waste and sensuality on the one hand and abject poverty on the other, still prevailed." For here, capitalism improved turns out to be only a new and more deadly authoritarianism, as indeed some of Bellamy's critics had warned.[159]

Wells's *A Modern Utopia* (1905) moved beyond the static and collectivist end points of much of the earlier tradition, but did not presume to "make the whole race wise, tolerant, noble, perfect".[160] It is urban centred and replete with the amenities beloved of the Edwardian middle classes, if considerably

154. Cf. G. K. Chesterton, *Twelve Types* (Arthur L. Humphreys, 1910), 128–29: "the cosmopolitan is basing his whole case upon the idea that man should, if he can, become as God, with equal sympathies and no prejudices, while the nationalist denies any such duty at the very start, and regards man as an animal who has preferences, as a bird has feathers."

155. H. G. Wells, *The Future in America* (Chapman and Hall, 1906), 60.

156. Mike Davis and Daniel Bertrand Hunt, eds, *Evil Paradises: Dreamworlds of Neoliberalism* (New Press, 2007), xiv. This volume offers many other examples of what might be termed hetero- or micro-dystopias.

157. H. G. Wells, *This Misery of Boots* (Fabian Society, 1907), 37.

158. H. G. Wells, *The Shape of Things to Come* (Hutchinson, 1933), 413.

159. H. G. Wells, *When the Sleeper Wakes* (Harper and Bros, 1899), 69.

160. Wells, *Modern Utopia*, 7.

less individualistic. Wells thought most people would like living in hotel-like apartments, which seemed to epitomise modern sociability, efficiency, and convenience, somewhat on the utopia-as-holiday model, with common dining the norm. Ebenezer Howard's schemes he thought too confined, with work and living in overly close proximity. The more affluent would live in "clubs", with private libraries, studies, and gardens. We encounter few solitary houses typical of the rapidly expanding Edwardian suburbs. In the new society, inequality of property would be reduced by taxation on inheritance. Machinery would abolish unnecessary and burdensome labour. There would be "faultless roads and beautifully arranged inter-urban communications, swift trains or motor services or what not, to diffuse its population", though he derided "smoke-disgorging steam-railway trains" as "already doomed on earth".[161] The exploitation of nature would necessarily continue. Wells thought that "in Utopia there will be wide stretches of cheerless or unhealthy or toilsome or dangerous land with never a household; there will be regions of mining and smelting, black with the smoke of furnaces and gashed and desolated by mines, with a sort of weird inhospitable grandeur of industrial desolation, and the men will come thither and work for a spell and return to civilisation again". A "great multitude of selected men" would struggle with nature. They would help ensure that electricity was generated "by water power, by combustion, by wind or tide or whatever other natural force is available", though coal is still used. Waste would be vastly reduced. Every possible advance would be contemplated and studied, and thus "Bacon's visionary House of Salomon will be a thing realised".[162]

Wells saw "no limit to the invasion of life by the machine" and thought indeed that the great achievement of nineteenth-century utopias lay in recognising that "the social fabric rests no longer upon human labour", which removed "the last base reason for anyone's servitude or inferiority". With a nod to the Simple Lifers, however, there are no "changing fashions" in clothes, no "disorderly conflict, of self-assertion qualified by the fear of ridicule."[163] In Wells's later works, the world-state is routinely associated with public luxury. In 1921 he assured readers that even a village school there would be "in a beautiful little building costing as much perhaps as a big naval gun or a bombing-aeroplane costs to-day", adding that "I know this will sound like shocking extravagance to many contemporary hearers, but in the World State the standards will be

161. Wells, 216, 42, 45.

162. Wells, 49, 60. On Wells's conception of the individual, see R. D. Haynes's *H. G. Wells: Discoverer of the Future* (Macmillan, 1980), 163–96.

163. Wells, *Modern Utopia*, 98, 100, 102, 296, 228.

different."[164] And all would be made as beautiful as possible by artist-craftsmen who were also builders and designers. Travel, mainly by train, would be cheap, and the population would move frequently. In the Depression, when under-consumption was regarded as a key problem, Wells subscribed to "the ideal of abundant production not only for immediate consumption but for a continual extension of human activities and a continual raising of the individual's standard of life.[165]

The control of population through regulating marriage was another part of Wells's image of the future, and here he largely followed John Stuart Mill. Regarding population, there is a hint, at least for a time, of eugenic regulation: "No longer will it be that failures must suffer and perish lest their breed increase, but the breed of failure must not increase, lest they suffer and perish, and the race with them."[166] The right of marriage was to be regulated, and, like Mill, whom he quoted on marriage restrictions,[167] Wells thought the state could legitimately claim that:

> before you may add children to the community for the community to educate and in part to support, you must be above a certain minimum of personal efficiency, and this you must show by holding a position of solvency and independence in the world; you must be above a certain age, and a certain minimum of physical development, and free of any transmissible disease. You must not be a criminal unless you have expiated your offence.[168]

In addition, "prolific marriage" would be confined to those "sufficiently intelligent and energetic to have acquired a minimum education. The man at least must be in receipt of a net income above the minimum wage, after any outstanding charges against him have been paid." The future society would thus "control the increase of its population. Without the determination and ability to limit that increase as well as to stimulate it whenever it is necessary, no Utopia is possible. That was clearly demonstrated by Malthus for all time."[169] Wells also questioned whether, when "all the people are Samurai", as in *Men Like Gods*, it would be "necessary to admit the necessary survival of inferior types."[170] None of this was new, but none of it was popular either.

164. Wells, *Salvaging of Civilisation*, 90.

165. H. G. Wells, *The New America* (Cresset, 1935), 67.

166. Wells, *Modern Utopia*, 137.

167. H. G. Wells. *Mankind in the Making* (Chapman and Hall, 1911), 101.

168. Wells, *Modern Utopia*, 184.

169. Wells, 191, 152.

170. Quoted in W. Warren Wagar, *H. G. Wells and the World State* (Yale University Press, 1961), 215.

Nor at points was Wells particularly sympathetic to racial integration.[171] After 1903, however, he backed away from eugenics and ideas of racial purity, and declared himself "inclined to discount all adverse judgments and all statements of insurmountable differences between race and race", and questioned the existence of any "all-round inferior race".[172]

Like Plato, Wells proposed creating an elite group to offer moral governance for mankind, the quasi-Platonic voluntary nobility called the Samurai, and indeed *A Modern Utopia* was often referred to as "the Samurai book". The Samurai are mostly administrators who occupy key positions of authority. They follow "the Rule", which includes a meatless diet, and no tobacco, alcoholic drinks, or narcotics. They epitomise an ideal of self-sacrifice. Though some are wealthy, they cannot lend money at interest, or run hotels, act or sing professionally, gamble, or play or watch competitive games. They are chaste, though not celibate. All, men and women, have a peculiar dress.[173] Here we see Wells's attempt to balance the demands of public luxury with those of the need for austerity and simplicity amongst the governing group. As late as 1932 Wells was still insisting that "I am asking for a Liberal Fascisti, for enlightened Nazis; I am proposing that you consider the formation of a greater Communist Party, a Western response to Russia."[174] But by 1939 he admitted to being "taken in by" the Samurai idea. He worried that in his portrayal "they anticipated the Communist Party commissars very strikingly", and apologised for "thinking with my generation".[175] Wells also came eventually to query the results of mass production, writing in 1941 that "Chain-shops, controlled stores and standardised production" had reduced "mankind to the same dead level of everyday living. They live in the same sort of houses, wear the same sort of clothes, eat the same flabby foods and upset themselves with the same advertised medicines."[176] But though he knew Morris's works well enough, he regarded their author as "profoundly reactionary", and there is no turn towards simplicity in his own writings.[177]

Wells's modernist portrait of utopia did not overlook sociability. In various works, notably *Men Like Gods* (1923), he portrayed a future in which "there was much love and laughter and friendship in Utopia and an abundant easy

171. He termed the "coloured" population possibly "unassimilable labour immigrants" (*Future in America*, 200).

172. Wells, *Modern Utopia*, 334, 338–39.

173. Wells, 49, 296, 279, 286, 289, 293.

174. H. G. Wells, *After Democracy* (Watts, 1932), 24.

175. H. G. Wells, *The Fate of Homo Sapiens* (Secker and Warburg, 1939), 217.

176. H. G. Wells, *Science and the World Mind* (New Europe, 1942), 19.

177. H. G. Wells, *An Englishman Looks at the World* (Cassell, 1914), 107.

informal social life." "The friendship was all the franker and closer", we are told, "because of that lack of power, and all the easier because age for age the Utopians were so much younger and fresher-minded than Earthlings."[178] For a time Wells thought he had discovered moral order in the universe, and called it "God", writing of "belief in God as the Invisible King", and that "Each believer, as he grasps this natural and immediate consequence of the faith that has come into his life will form at the same time a Utopian conception of this world changed in the direction of God's purpose".[179] But this phase too passed. The project and assumptions of biological progress to which Wells, like so many others, was wedded in the early twentieth century mostly leave us cold today, though we still pay lip-service to their objectives. "Wellsian" and "welfare state" were long synonymous. At Wells's death in 1946 it was harder to project a rationalist vision of humanity's reorganisation and moral reawakening than it had been in 1900. His Marxist critics insisted that this was inevitable: he had misunderstood the agency of revolution—who would turn the bourgeois intelligentsia into the misnamed "Samurai"?[180] Indeed, Wells had reached a similar conclusion. But he had made an indelible mark on a generation and more.

Summary of the Historical Argument concerning Utopia and Luxury to the 1930s

Literary and practical utopians alike treated the themes of luxury and simplicity in three ways from the eighteenth until the early twentieth century. Early on, many applauded a Spartan simplicity as the only condition conducive to virtue and equality. As the Manuels summarise it, "In virtually all communist utopias from More to the nineteenth century, repression of desires makes egalitarian commonality possible."[181] This option became much less common by the late nineteenth century, however, and in any case few offered any firm plan for returning corrupted Europeans to a simpler and more virtuous life. Neither of the chief options, returning to barbarism following civilisation's collapse, or a despotic intervention to end corrosive luxury, had much appeal. Others, secondly, plumped for public over private luxury, contending that this

178. H. G. Wells, *Men like Gods* (Cassell, 1923), 287, 344.
179. H. G. Wells, *God the Invisible King* (Cassell, 1917), 129.
180. E.g., Christopher Caudwell, *Studies in a Dying Culture* (Bodley Head, 1949), 73–95.
181. Manuel and Manuel, *Utopian Thought*, 176.

both facilitated sociability and permitted a reasonable amount of indulgence without entailing additional burdens on the poor. From the early nineteenth century, a third group contended that machinery made virtually unlimited luxury for all possible, while still generally arguing that necessities had to be prioritised over other desires. Through Bellamy and Wells, in particular, this view came to dominate much utopian thought in the following century.

There were variants within these three main positions, which began to confront one another during the second half of the eighteenth century, when a marked shift in utopian projections from the past and primitive to the future and complex occurred. The advent of commercial and consumer society now revolutionised ideas of need and want. Nonetheless, the Spartan tradition and the warning that luxury generated great social inequality remained prominent in social and political thinking in this period, notably through Jean-Jacques Rousseau. After 1800, the image of a simpler society was central to some socialists, like Robert Owen and Charles Fourier, then William Morris; and to the anarchism of William Godwin, Peter Kropotkin, and later Mohandas Gandhi. Few liberals took up these themes, though John Stuart Mill hinted at the advantages of a stationary state, and indeed insisted that the "superior worth of simplicity of life" as described by Rousseau needed at present "to be asserted as much as ever".[182]

Luxury was condemned by utopian writers because it entailed greater labour on the poor, because it reduced the number of cultivators, because it involved costly foreign imports, and because it was personally corrupting and undermined civic virtue. From the ancients until the nineteenth century, the idea of an egalitarian society in which luxury proliferated was unimaginable. By the 1880s, however, modern industry made it possible to envision universal plenty because extra labour could be done by machines. Carpenter and Thoreau notwithstanding, frugal or neo-Spartan utopianism now largely fell out of fashion. Appeals to the virtues of simplicity and the primitive occurred sporadically every twenty years or so from the 1750s onwards. But by 1900 these were largely displaced by a greater focus on public luxury, and on a more equal sharing of pleasures and leisure pursuits. An immensely optimistic technocratic vision of the future prevailed until 1914, as utopianism edged closer than ever to science fiction in its projections of colonising the universe.[183] Then came the colossal crash of World War One, and the realisation that humanity's destructive capabilities had surpassed its

182. John Stuart Mill, *On Liberty* (John W. Parker and Son, 1859), 85.

183. For a recent account of these trends, see Peter J. Bowler's *A History of the Future: Prophets of Progress from H. G. Wells to Isaac Asimov* (Cambridge University Press, 2017).

ability to maintain peace. Until 1945, much of the promise of universal luxury was put on hold as industry manufactured armaments instead of domestic consumer goods, or nothing at all, when the Great Depression closed factory gates.

In an age when strict canons of interpersonal propriety, stiff formality, and conventional manners and dress demarcated by class were common, utopianism also represented a romantic reaction to artificiality in favour of "authentic" modes of interacting. The themes of luxury versus simplicity were intermixed with discussions of sociability from the outset, since it was widely recognised that social inequality inhibited the open, honest, frank interactions usually associated with friendship. Utopian writers first began to imagine what a post-commercial sociability might look like both because they nostalgically recalled or invented pre-commercial mentalities and because they realised the deficiencies of commercial, unsocial sociability, especially in placing selfishness, or at least commercial self-interest, ahead of either pity or the duties of one's station. By the late nineteenth century, sociologists in particular came to see *Gemeinschaft* as invoking an image of pre-commercial community, compared with the unsocial sociability of *Gesellschaft*.[184] In the latter, needs were unrestrained and individuals were predominantly seen as defined by their satisfaction. In the former, needs were limited, and priority was given to maintaining equality and group identity, suppressing competition based on greed and vanity, and supplying whatever deficiencies humanity's lack of natural sociability evidenced by artificial restraints and incentives. Those societies were happiest which exalted love, gratitude, friendship, and esteem, as even Adam Smith, who advocated state support for public festivals, had acknowledged.[185] Wherever injustice, usually generated by inequality, was prevalent, commercial society eroded these qualities, eventually destroying them.[186] Its defenders usually claim that unsocial sociability is the best that humanity is capable of. This is the chief moral argument we will need to confront in part IV of this book.

By 1900, then, three major arguments against luxury had largely gone into remission: that it was intrinsically harmful to personal moral as well as civic public virtue; that it reduced the amount of agricultural labour unduly, thus

184. On the origins of the distinction, see Istvan Hont's *Politics in Commercial Society: Jean-Jacques Rousseau and Adam Smith* (Harvard University Press, 2015).

185. He thought the state might help "amuse and divert the people by painting, poetry, music, dancing, by all sorts of dramatic representations and exhibitions, [which] would easily dissipate, in the greater part of them, that melancholy and gloomy humour which is almost always the nurse of popular superstition and enthusiasm" (*Wealth of Nations*, 2: 381).

186. Hont, *Politics in Commercial Society*, 9–13.

producing want and famine; and that it entailed too much additional labour on the poor. Machines, Marx, Wells, and many others surmised, could now take up the most onerous tasks. Consuming luxuries was not dangerous as such, provided none were rendered poorer or had to work harder as a result. In addition, utopia was now firmly projected into the future and conceived as realisable, even inevitable. Mill, Morris, and a few others had begun to posit a "limits to growth" hypothesis. But this idea, we will now see, did not begin to bear fruit for a century.

PART IV

Modern Consumerism and Its Opponents

9

Twentieth-Century Consumerism and the Utopian Response

IF THE FIRST stage of modern consumerism emerged in northern Europe, and especially France and Britain, from the late seventeenth century onwards, a second stage began on a larger scale in the mid-nineteenth century. The third and most familiar commenced after World War Two and was dominated by the United States, our focus in this section.[1] By the late twentieth century, it became evident that aspiring to a universal high standard of living threatened to deplete resources and degrade the environment. The seventy years from 1945 to 2015 thus represent the apogee of modern materialism and consumerism, and of the now unsustainable ideal of the indefinite progress of population, production, and consumption. This was, and continues to be, the age of the greatest excess and waste of any in history. Only by understanding how status and consumerism came to be fused will we be able to disconnect them.

The second phase of modern consumerism was defined by huge new department stores, which soon became the hearts of cities. Entering them was one of the great novelties and delights of the age. A policy of free entry permitted browsing without having to buy, so shopping now became a leisure activity valued for the experience itself, with acquisition often secondary to titillating desire. These temples of opulence catered to every imaginable desire. In London, the five floors of Maple's furniture store on Tottenham Court Road

1. For an international overview of this process, see Hartmut Berghoff and Uwe Spiekermann's. *Decoding Modern Consumer Societies* (Palgrave Macmillan, 2012) and Roberta Sassatelli's *Consumer Culture: History, Theory and Politics* (Sage, 2007). On the United States, see Richard L. Bushman's *The Refinement of America: Persons, Houses, Cities* (Alfred A. Knopf, 1992).

"stretched the length of twenty-five houses" by 1890, and employed some two thousand people.[2] Here too in 1898, Harrods, now the largest department store in Europe, introduced the first escalator; customers were offered brandy when they reached the top to compensate for the thrill, as if they had conquered some fabled peak. In Brussels, one of the first shopping arcades, the sumptuous Galeries St Hubert, opened in 1847. In France, where the concept was first floated in the 1850s in the Bon Marché of Aristide Boucicaut, occupying a city block, Emile Zola portrayed a *grand magasin* as *The Ladies' Paradise* (1883). The Galeries Lafayette followed in 1912. In New York, Macy's opened to great fanfare in 1878. Saks Fifth Avenue would appear in 1924.

The new stores represented the new prosperity like nothing else, and were a world unto themselves. They defined the mode, announcing the new fashions, alluring and overwhelming the senses with a thousand different stimuli, and encouraging the imagination to run riot with the multitudinous possibilities of being and possessing. Shopping now became an exhilarating, tantalising process of immersion in a sea of things. Inflamed by advertising—the first ad agency was founded in 1864—desires multiplied constantly, especially in towns and cities. By 1940, in the United States, a dollar was spent on advertising for every 70 cents invested in all forms of education from primary school to university.[3] In 2020 about $700 billion was spent on public education, and $239 billion on advertising. And so the modern consumer was born, defined by the constancy of relentless desires, with novelty-as-renewal representing life itself. Increasingly this became the dominant type of modern identity.

The most accelerated form of the quest for incessant novelty is an intensively American phenomenon.[4] By 1900, the United States was defined by the desire for riches and the ideal of unlimited abundance like no nation previously.[5] The new consumerism rapidly wormed its way into every nook and cranny of private life. Advertisers commercialised many traditional events, insisting that there was no experience worth having, and some which might be well avoided, that wouldn't be made better by acquiring something symbolic to commemorate it. Memorable occasions were redefined by buying things, or by cards: Christmas, Mother's Day (1908), Valentine's Day, Halloween. Events which notionally marked our love and loyalty to others were made ostensibly more meaningful by multiplying symbolic gifts. To consume became an emotional

2. Albritton and Jonsson, *Green Victorians*, 23.

3. Stuart Chase, *The Tragedy of Waste* (Macmillan, 1942), 124.

4. See Michael North, *Novelty: A History of the New* (University of Chicago Press, 2013); Sanford, *Quest for Paradise*, 94–113.

5. See Stanley Lebergott, *Pursuing Happiness: American Consumers in the Twentieth Century* (Princeton University Press, 1993).

obligation. Voraciousness came to match plenitude. Soon the movies, and then television, became the greatest advertisers of all, as all the world viewed on screens large and then small an aspirational American lifestyle, and sought to emulate it in local variants of the American Dream. Middle-class women in particular came to play a central role in the rituals of consumption and were soon portrayed as the consummate shoppers, "the unresisting victims of manipulative advertising and vulgar, alluring displays". For them, it was assumed, shopping was a social as well as a self-validating and self-empowering experience.[6]

The third great shift in shopping patterns was the transition from department store to warehouse-style supermarket, and from a focus on middle-class luxury to mass-market consumption of a vast array of goods, piled high and sold cheap. Novelty and status were now interwoven as never before. Bustle and haste had long defined America. Now, as the pace of life intensified, so, already by the 1920s, did the emphasis on "fast" and "instant", time- and labour-saving products. When the frontier ended in the 1890s, mobility both social and spatial was redefined. The abundance of free or cheap land ended. It became harder to move on, and more necessary to move upwards. Status consciousness came increasingly to define personal aspiration. And status meant things. By the 1930s, note Robert and Helen Lynd in their classic study of American life in "Middletown" (Muncie, Indiana), the pressure to prove one's status was particularly evident in two areas: women's clothing and "the deliberately instigated vogue of the annual new model of each make of automobile".[7]

In the land of constant self-renewal the mentality of novelty was all-pervasive. Replacing things rapidly became a sign of status. The pioneer virtues of thrift and conservation swiftly melted away. Modern life left no time to save and mend. Doing so was seen as fussy, old-fashioned, clumsy, old-worldly, *déclassé*—in a word, un-American. Mending things was what elderly poor people and immigrants from the Old Country did, a necessity, not a virtue. This was the mentality people had fled from. Waste became obligatory: real Americans threw things away to prove how far they had risen. "Throw-away living", as *Life* magazine termed it in 1955, was born as the incentive to buy the new and discard the old became both socially and economically compelling.[8] Every event had to be performed quickly to prove one's devotion

6. Rachel Bowlby, *Carried Away: The Invention of Modern Shopping* (Faber and Faber, 2000), 6, 225.

7. Robert S. Lynd and Helen Merrell Lynd, *Middletown in Transition: A Study in Cultural Conflicts* (Constable, 1937), 268.

8. Alan Weisman, *The World without Us* (Thomas Dunne, 2007), 119.

to the work ethic. Today, millions parade disposable plastic coffee cups through the streets held aloft like torches of liberty to emphasise their busyness and lack of leisure.

In the fourth stage of this process, in the late twentieth century, the most visible symbol of advancement was the mall, the new temple of the shopping religion, which soon spawned the mega-mall, most of which are now in Asia. The largest of these, in Dongguan, China, is some 900,000 square metres. In the United States, their growth followed white flight to the suburbs from the 1950s onwards. Mostly located outside cities, and readily accessible only by car, malls swiftly became the focus of communal life in many areas, the place to "hang out", to see and be seen. They soon sucked the life out of town and city centres, as great chains overwhelmed family businesses, turned millions into wage-labourers, and transformed downtowns into wastelands.[9] By 1985 there were more malls than cities, four-year colleges, or television stations, and Americans spent more time in them than anywhere else except at work and at home.[10] Shopping had become a way of life, and a definitive statement of existence.

Malls are commercial heterotopias which aim to project a purified, bourgeois way of life, a "safe space" defined by opulence, consumption, and a reassuring sense of certainty. Miniature idealised cities, they are controlled environments: weatherproof, protected, policed, mostly white or partially ethnically and class-cleansed spaces, a confined and ordered mixture of sociability and shopping experience which extends the fantasy of the suburban ideal, a new "Paradise Enclosed".[11] Within, every device in the psychologist's armoury is deployed to weaken resistance to the urge to consume. Overwhelming our senses is essential to the task. A constant barrage of sensations includes all-pervasive inoffensive muzak, the bright lights and glitter of glowing screens full of action and movement, the alluring smells of air freshener and sugar and fat junk-food products, all designed for the "feel-good" impression, and to whet the shopping appetite.

These stimuli also aim to fill the void that results from our supposed fear of being alone, plagued by the self-doubt and anxiety of being forced to confront our own thoughts which silence reveals, and the sense of inner emptiness which results. Studies show, Christopher Lasch relates, that "shopping serves as a means of 'alleviating loneliness,' 'dispelling boredom,' and 'relieving depression.'" Many "don't really need what they are shopping for. Often they

9. See Jane Jacobs, *The Life and Death of Great American Cities* (Penguin Books, 1974).

10. William Severini Kowinski, *The Malling of America: Travels in the United States of Shopping* (Xlibris, 2002), 43, 45.

11. Kowinski, *Malling of America*, 523.

don't even know what they're after." Only 25 per cent of shoppers in malls come to buy a particular item. The rest are there for compensatory psychological relief, which they may not find, or at least not for long enough.[12] But people who sell us things are paid to pretend to be nice to us, and we can pretend they mean it. They are subservient, which makes us feel superior. We are made to feel welcome, so long as we spend. Stores (or "retail outlets") always have their doors open, and approximate a Disneyfied, television version of the American Main Street in the 1950s, a fantasy within a fantasy.[13] "Disneyfication" implies extending a quasi-magical concept of "re-enchanting" everyday space, a kind of consumer imperialism which conquers everyday space for the middle class, for this is the epitome of one variant on the bourgeois utopia, while ensuring the low-wage workers skulk past invisibly or wait in uniformed politeness behind the counter, and that the vast waste this cornucopia creates is hidden.[14] "McDonaldisation", the mass branding experience, ensures that chain after chain brings reassuring similarity and reliability. Malls became the great symbol of American life in the late twentieth century, and then of American decline. At the turn of the twenty-first century, mass consumption shifted once again, to the internet, threatening the final obliteration of town centres but now also malls everywhere, and production shifted to cheap labour economies, spelling the doom of American manufacturing.

During the post-war prosperity of the 1950s, the new consumerism rapidly became an addictive mindset. A shift from valuing relationships to prizing things in assigning meaning is readily discernible. The high-consumption ideal, warned William Leiss some forty years ago, "tends to orient all aspects of an individual's striving for personal satisfaction towards the realm of commodities."[15] The poor too were easily drawn into the mentality. Immigrants who after arriving lived in dense, low-income, often ethnically defined communities, later remembered that "In the old days, there were no cars or luxuries, no one had anything . . . people were communal".[16] Children played

12. A starting point here is Christopher Lasch's *True and Only Heaven* (quotes at 522).

13. Kowinski, *Malling of America*, 101.

14. George Ritzer, *Enchanting a Disenchanted World: Continuity and Change in the Cathedrals of Consumption* (Sage, 2010), 9.

15. William Leiss, *The Limits to Satisfaction: On Needs and Commodities* (Marion Boyers, 1978), 60.

16. Said of an Italian-American neighbourhood in Boston: Herbert J. Gans, *The Urban Villagers: Group and Class in the Life of Italian-Americans* (Free Press, 1962), 223.

in the streets, on doorsteps and front porches, and in public parks. This made for "community". Later generations assimilated, Americanised, lost the language of their ancestors, and showed they had "made it" by moving to individual homes in the suburbs. Here class was more important, and status competition intensified, as a range of new goods became available: cars, washers, home freezers, lawnmowers, air conditioners. Wanting and then flaunting it became the norm.

As key symbols of freedom and mobility as well as status, cars were vital to this process. Planned obsolescence is often associated with Alfred P. Sloan's introduction of the "annual model" change at General Motors in the mid-1950s, which he touted as "the spur to which the organization must respond or die."[17] In a 1954 speech the industrial designer Brooks Stevens described this ritual as aimed at "Instilling in the buyer the desire to own something a little newer, a little better, a little sooner than is necessary".[18] Here, one group set the standard: corporate executives. As William Whyte stresses in his classic account of its emergence, "*It is the group that determines when a luxury becomes a necessity*." This group set the norms for consumption and defined the gradual escalation of acquisition. It established when "the nonpossession of the item becomes an almost unsocial act", and thus when it becomes impossible to resist the expenditure. Then, "Item by item, the process is constantly repeated, and the norm never stays still."[19] Here, the social pressure is great but not immense, and the economic damage is sustainable. But when these expectations leapt outside of the relatively equal group they became much more threatening. From the outset of consumer society, as we have seen, people commonly aspired to possess things they could not afford, but without which they felt inferior. This process now accelerated even further by being extended to a wider range of goods. With the "new consumerism" of the 1980s onwards in the United States, as Juliet B. Schor terms it, even poorer Americans came to view others with much higher incomes—the upper middle class—as a model, and began to desire, acquire, and consume accordingly.[20]

For millions this was a fatal leap in aspiration. By the 1970s real incomes were declining, and many jobs disappeared through globalisation. The

17. Alfred P. Sloan, *My Years with General Motors* (Sidgwick and Jackson, 1966), xxxv ("respond or die"), 238–47 (annual model change). Ford himself had begun with a "static model" concept (152). The process of transformation began around 1925, and became "regularized" in the 1930s (167). There was even a General Motors "autopia" exhibit at the 1939 World's Fair.

18. Michael Schudson, "Delectable Materialism", in Crocker and Linden, *Ethics of Consumption*, 255. See Sloan, *My Years*, 238–47: "The Annual Model Change".

19. Whyte, *Organization Man*, 314.

20. Schor, *Overspent American*, 4–5.

post-war ideal of one male income sufficing to raise a middle-class family no longer held. Keeping up became increasingly stressful, and at the same time status competition if anything intensified as the proliferation of styles reached ever-greater numbers. Standards and expectations rose relentlessly, especially for the young, for whom the right clothes and shoes were vital to social recognition. From the 1950s to 2006 consumption increased about threefold.[21] From the 1980s onwards, in particular, consumer spending rocketed, aided in part by conservative-led tax breaks. In only six years, 1990–96, credit card debt in the United States doubled. Household debt reached an all-time high in 2019. The result was growing frustration, stress, and overwork for many. As incomes declined, the "lower-middle and working classes fell even farther behind", with the least well-off the chief victims of the system.[22] The upper middle classes, however, continued to prosper. In the late twentieth century, their extravagance reached its apogee, and an entire new group, then called yuppies (young upwardly mobile professionals), epitomised new styles of consumption and became "known almost entirely through their display of goods", expensive watches, cars, clothes, computers, apartments, houses, holidays, and taste generally.[23] This trend is also associated with the gentrification of decaying urban areas, and the advantages and problems this brings.[24]

By the late twentieth century, keeping up with the Joneses by demonstrating our superiority became a veritable international obsession, except that keeping up with "celebrities" was now equally important. The middle classes now spent to impress the Joneses, or to ape the Kardashians, to impress themselves, to impress their children, and so that their children could impress others. Driven by advertising and by brand identification, happiness became identified with consumption, especially of luxury goods, as never before. Increasingly, too, this was now an international phenomenon. In the four main areas of luxury—sustenance (food and drink), shelter, clothing, and sufficient leisure—the race was on.[25] Gucci, Prada, Dior, Chanel, Louis Vuitton, Ralph Lauren, Tommy Hilfiger, Calvin Klein, Giorgio Armani, Rolex, BMW, Jaguar,

21. Avner Offer, *The Challenge of Affluence: Self-Control and Well-Being in the United States and Britain since 1950* (Oxford University Press, 2006), 278.

22. Schor, *Overspent American*, 14.

23. James B. Twitchell, *Living It Up: Our Love Affair with Luxury* (Columbia University Press, 2002), 76.

24. A good start with this theme is David Harvey's *Spaces of Hope* (Edinburgh University Press, 2000), esp. 133–81, for the example of Baltimore.

25. Twitchell, *Living It Up*, 59. The first three took up about 80 per cent of US disposable income in 1918, compared with 50 per cent in 2000 (p. 61).

Rolls-Royce, Apple: the list of luxury brands is lengthy. (About half of the luxury goods market is French.) Products with a luxury label are valued more highly than identical products which lack one.[26] Their symbols are instantly recognised by millions. The brands are often worth far more than the economic cost of the materials involved. We even name our children after them.[27]

Obsession with such goods is now nearly universal. By 2005 more than 90 per cent of women in Tokyo owned a Vuitton or Gucci item, and more than half something from Prada or Chanel.[28] In 2007 the luxury industry was worth $157 billion, with thirty-five major brands controlling 60 per cent of the business; by 2019 it amounted to $306 billion. Upper-income women aged thirty to fifty are the dominant consumers, with Japan accounting for 41 per cent of sales in 2004. China was at 11 per cent in 2005, and 20 per cent by 2015. The Russian and Indian markets are growing fast. Luxury goods are more accessible than ever, being mass produced, sometimes as fakes, with inferior materials. To their defenders, such as Christian Dior, sometimes regarded as the father of modern fashion, such goods "shield us against the shabby and humdrum", and reflect an inner yearning to express ourselves.[29] To their possessors they may represent the "foreign", the "other", the "different", the superior. To their critics, they epitomise falsehood, pretentiousness, superficiality, vanity, and class arrogance. They represent the mentality that we don't want others to have things we have, because this detracts from the value of our possessions. And worse, modernity's own monster, the all-desiring consumer, crafted from the detritus of airport duty-free zones and malls, had become a Shiva-like destroyer of the planet.

These trends, exacerbated by economic and pandemic crises, continue inexorably in the present. Inequality is more extreme than ever, and the price paid for being poor and/or marginally employed has become much greater. With growing poverty, debt, rates of crime, incarceration, drug use, obesity, poor health, and infection from contagious diseases have risen steadily. Well-paid working class jobs have become scarce. Now middle class jobs are also rapidly disappearing. We are all clearly markedly less happy, and more chained to the treadmill than ever. The consumerist paradigm has reached its extinction point.

26. Schor, *Overspent American*, 95.

27. See David Boyle, *Authenticity: Brands, Fakes, Spin and the Lust for Real Life* (Flamingo, 2003).

28. McNeil and Riello, *Luxury*, 245.

29. Dana Thomas, *Deluxe: How Luxury Lost Its Lustre* (Allen Lane, 2007), 8.

Explaining Waste: Veblen and Conspicuous Consumption

Warnings of such consequences emerged early on in the process. Though consumerism reached its apogee in the past few decades, it was a definitive late nineteenth- and twentieth-century phenomenon. Its greatest early chronicler was the sociologist Thorstein Veblen (1857–1929), whose *Theory of the Leisure Class* (1899) became an instant classic and guided several generations of American critics, including David Riesman, Vance Packard, and Richard Sennett.[30] Veblen's catch phrase, "conspicuous consumption", was coined at the height of America's Gilded Age (1870s–1900), when the vast fortunes of the Rockefellers, Vanderbilts, Goulds, and others came to define the profligacy of the new upper class, and agrarian democracy was rapidly giving way to a corrupt plutocracy and the economic predominance of "trusts", or great monopolistic corporations.[31]

Veblen's analysis of this age of excess owed much to the anthropologist Franz Boas's account of the Kwakiutl, a north-western Canadian indigenous, or First Nations, tribe. Their practice of potlatch, or the public waste of wealth at feasts, such as burning blankets, to the point of bankrupting those who were forced by the rules of hospitality to imitate their hosts, presented a clear prototype of modern consumption.[32] Like Mandeville, Hume, and Smith, Veblen thought a desire for "the esteem and envy of one's fellow men" crucially motivated such behaviour. Emulation, "the stimulus of an invidious comparison which prompts us to outdo those with whom we are in the habit of classing ourselves", was a "pervading trait of human nature", and "probably the strongest and most alert and persistent of the economic motives proper."[33] In an "industrial

30. On Veblen, see David Riesman's *Thorstein Veblen* (Charles Scribner's Sons, 1953), John P. Diggins's *The Bard of Savagery: Thorstein Veblen and Modern Social Theory* (Harvester, 1978), Rick Tilman's *Thorstein Veblen and His Critics, 1891–1963* (Princeton University Press, 1992), and Charles Camic's *Veblen: The Making of an Economist Who Unmade Economics* (Harvard University Press, 2020).

31. The classic critique is Henry Demerest Lloyd's *Wealth against Commonwealth* (Harper and Bros, 1894).

32. John Brooks, *Showing Off in America: From Conspicuous Consumption to Parody Display* (Little, Brown, 1979), 8–11. An 1884 amendment to the Indian Act of 1876 made potlatch illegal, and was only repealed in 1951. The practice originated in a period of slavery, when the results of a slave's labours were distributed amongst the community (W. M. Halliday, *Potlatch and Totem* (J. M. Dent and Sons, 1935), 3–4). For an updating of the argument in a comparative context, see Brian Hayden's *The Power of Feasts: From Prehistory to the Present* (Cambridge University Press, 2014).

33. Thorstein Veblen, *The Theory of the Leisure Class* (1899; Allen and Unwin, 1924), 32, 110. We would today give greater stress to their role in group self-definition and empowerment.

community", status and reputation were defined by conspicuous leisure and conspicuous consumption.[34] The point of wealth was to impress others through expressing one's social power in display and ritual consumption. So, as another anthropologist, Mary Douglas, put it, we desire goods primarily not for their intrinsic worth or pleasure-inducing capacity, but "for mobilizing other people".[35] Prestige trumped both utility and aesthetics at every step. Cost alone was important in consumption, so "a beautiful article which is not expensive is accounted not beautiful."[36] This accorded neatly with the demands of mass production, where overproduction was always a risk, and imitation had to be stimulated by advertising. Here, as an early critic put it, "Artificial obsolescence was attempted by appeals to the invidious sentiment, the snobist and emulative passions". "Salvation by advertising" was thus the response to "plenty as disaster".[37] To save capitalism it was necessary to destroy as many commodities as possible.

The central rationale for capitalism—that entrepreneurial talent, industry, and drive produced the accumulation which employed others and distributed wealth throughout the society—seemed greatly undermined by this process. Now the want and privation of the many was juxtaposed not to hard work but to wantonly destroying wealth through waste. Emulation focused increasingly on acquiring and consuming material goods to bolster social standing and "command the deference of the community". Failure to attain these expectations, which were heightened in cities, invited the "pain of losing caste": people looked down on you.[38] This was in some respects a peculiarly American phenomenon. In an era of "new money" and a milieu which was superficially vastly more egalitarian than the old world, it was especially important to display one's possessions quickly and to gain the widest possible recognition and impact, to show that one had "arrived" and achieved the "American Dream". The more lavish and splendid the show, the better: "In order to be reputable it must be wasteful", insisted Veblen.[39]

The Gilded Age epitomised such extravagance. At a party in February 1897, the Bradley Martins of New York City transformed the Waldorf Astoria into a replica of Versailles at a cost of $369,000, or perhaps $20 million today. (A carpenter made around $1000 annually at this time.) One guest came in a

34. Veblen, *Leisure Class*, 32.

35. Mary Douglas, *In the Active Voice* (Routledge and Kegan Paul, 1984), 24.

36. Veblen, *Leisure Class*, 132.

37. Horace M. Kallen, *The Decline and Rise of the Consumer* (Packard, 1936), 80–81.

38. Veblen, *Leisure Class*, 71, 88.

39. Veblen, 96.

gold-inlaid suit of armour costing $10,000.[40] Public outcry at this ostentation forced the Martins to flee to Europe. But the mentality displayed at such events soon extended itself rapidly and widely. Disciples of Veblen pointed to the cost of fine ladies changing their clothes five times daily. By 1919 "illth", or conspicuous waste, in Stuart Chase's term, borrowed from Ruskin, comprised almost a third of US consumption.[41] If we include food, and avoidable waste generally, the figure is probably nearer 40 per cent today. Not all consumption, however, is "conspicuous". Besides showing off, or asserting our superiority, we consume more, writes Judith Lichtenberg, to "achieve a certain kind of equality that is essential to self-respect".[42] But our sense of our needs is also constantly expanding. The net effect on resources remains the same regardless of our motive.

Modern Consumerism Defined

The linkage between mass consumption and personal identity has been called "the most important subject for our times"; in the context of environmental degradation this is readily conceded.[43] The progress of modern consumerism has involved a relentless transformation not only in what we consume but in how we feel as consumers, and how consumerism has come to define our very selves. Amongst the three main elements in the psychology of consumption defined by Maxine Berg—the roles played by the senses, by novelty, and by imitation—the position of novelty comes to the fore here.[44] As we have seen, our desire for the new most clearly differentiates us from earlier societies and shows how exceptional the lives of the most affluent later moderns are. In traditional societies, needs are simple and fixed for the vast majority. Merely gaining a subsistence occupies most of people's time. They may move to follow game, or their animals, or to cultivate new fields, but their assumptions about their rank and standard of living rarely alter substantially during their lives. Guided by custom, they usually aspire to keep things exactly as they have always been.

But the inflammation of desire is like a spark set to dry grass. Wants once awakened can scarcely ever be dampened. To set consumerism in motion requires only introducing one new and unusual commodity—say, a brass bed

40. J. Brooks, *Showing Off in America*, 13.

41. Chase, *Tragedy of Waste*, 87–88. See Ruskin, *Unto This Last*, 126.

42. Judith Lichtenberg, "Consuming because Others Consume", in *Ethics of Consumption: The Good Life, Justice, and Global Stewardship*, eds David A. Crocker and Toby Linden (Rowman and Littlefield, 1998), 171.

43. Peter K. Lunt and Sonia M. Livingstone, *Mass Consumption and Personal Identity* (Open University Press, 1992), 2.

44. Berg, *Luxury & Pleasure*, 249.

brought to a small tropical island, which, in 1942, upset the social equilibrium when it "set the other women to longing for one like it—for show."[45] Now we want things simply because others have them, rooted, mostly likely, in a profound desire for equality. The process of constructing a "commercialized self", and defining ourselves by novel things, even has a name, "the Diderot effect", after an essay by the French Enlightenment philosopher we have already encountered. He reflected how acquiring a new dressing gown led to want to improve everything else around him, to mirror its superiority, an interesting obsession in someone supposedly enamoured of primitivism.[46] So novelty is imperious: it demands the remaking of the world according to its own standard.

This gives us an important clue with respect to the intrinsically pathological nature of the consumer personality type. Desiring more, better, faster, and more stimulating things becomes an addiction which eventually hinders our sane self-development and makes us gravely imbalanced. Our adrenaline pumps ever more rapidly, and our bodies demand still more arousal. Anxiety results: we live constantly in anticipation, especially of the desired but not-yet-acquired, which forever beckons us to hasten. People often relax in traditional societies. The moderns demand incessant stimulation. They fidget, are never satisfied, get bored quickly, have short attention spans, are endlessly restless and uneasy, and are constantly driven by vanity, greed, ambition, and the need to appear to be in a hurry, and in demand. They rush so fast to see and do that they forget the value of the moment, and of merely being. Aldous Huxley noted that "Savages can spend a great deal of time simply sitting still; and it is possible that the white races of the West may in time be educated into being less restless than they are at present. This restlessness is encouraged by manufacturers for their own profit."[47] Avner Offer writes that "affluence is driven by novelty and novelty unsettles . . . Affluence breeds impatience, and impatience undermines well-being."[48] All pleasure is seductive, and the stimulation involved in every aspect of consuming is highly addictive. Eating makes us want to eat more, drinking to drink more, gambling to gamble more, buying to buy more. The "salted-nut" syndrome described by Tibor Scitovsky thus produces a "mini-addiction" of many types, when we continue consuming

45. James Norman Hall, *Lost Island* (Little, Brown, 1944), 180.

46. Twitchell, *Living It Up*, 84–85. Diderot's essay, "Regrets on Parting with My Old Dressing Gown" (1772), is reprinted in *Rameau's Nephew*, 325–33. The paradox that doing so degrades the status of the first-acquired object alone makes this case worth studying. Try painting one wall of a room to achieve this effect.

47. A. Huxley, *Hearst Essays*, 102.

48. Offer, *Challenge of Affluence*, vii, 1. This is the best summary of trends in modern consumption and their implications for both countries.

beyond any reasonable need, simply to sustain the experience.[49] Physical obesity becomes common, but so does mental obesity, as we cram our minds to bursting with images of unsatisfied and exaggerated desires and bloat our egos with the sense of domain over our possessions. The process by which novelty becomes addicting was described in 1930 by Waldo Frank vis-à-vis children. Initially, "A fresh plaything renews the child's opportunity to say: this is mine", and "prizes it as part of himself". But then as "toys become more frequent, value is gradually transferred from the toy to the toy's novelty . . . The arrival of the toy, not the toy itself, becomes the event."[50] Desire surpasses its objects and becomes itself the key source of exhilaration. And so it masters us, and we cower before its demands.

By such means the goalposts of self-satisfaction are constantly being moved, and our consequent unhappiness is virtually guaranteed. The tremendous impetus given to the natural desire to imitate by visual images in film, television, or social media intensifies this want. Now the world of appearance threatens to overtake any other version of "reality". "Real" is real *because* it appears on a screen, which validates others' opinions of what "reality" consists of, since we are so doubtful ourselves of our grasp of it without others' agreement. Unattainable ideals of the good life are dangled tantalisingly before us by internet "influencers" and media or sports "celebs". In the case of physical appearance and the body, the more beautiful, the younger, the sexier, the fitter the ideal becomes, to take some later images, the greater the dismay of all who fail to meet it—necessarily the great majority. But even the successful have to struggle constantly to stay in place. The pace of frenzied emulation is relentless, and the price of failure high.

The experience of consumption, then, involves an expectation of pleasure, usually initiated by advertising, or by seeing others consuming things we lack, whom we thus envy and desire to imitate in order to feel at least their equals. These pleasures extend from the moment a desire is conceived. We are aroused by the thought of the product, and in anticipating its consumption. Then we glimpse, smell, unpack, touch, devour, turn on, the object of our desire. Many tactile pleasures now come into play. Our experience is multisensory and multidimensional, at once, in Rachel Bowlby's phrase, "ecstasy and waste: not just the packages but also the things they contain are for throwing out or passing on, but meanwhile there are exorbitant private pleasures to be got from sniffing and seeing and touching."[51]

49. Tibor Scitovsky, *The Joyless Economy: An Inquiry into Human Satisfaction and Consumer Dissatisfaction* (Oxford University Press, 1976), 62–63.

50. Waldo Frank, *The Rediscovery of America* (Charles Scribner's Sons, 1929), 117.

51. Bowlby, *Carried Away*, 19.

Branding

Symbols play a profoundly important role in the process by which we become prisoners of our desires. Much of our desire to consume comes less from a lust for objects than a wish to identify ourselves with the symbols behind them. This is revealed most clearly in branding. Brands represent symbolic groups. Through the power of suggestion, our extreme gullibility and susceptibility to flattery, and our perennial weakness for strong symbols, brands succeed when we come to identify with the ideas we regard as embodied in things as much as the things themselves, and to have them stamped or branded on us as group markers for all the world to see. When the symbolic value of objects surpasses their capacity to gratify in the act of consumption, and even our desire to have them do so, the idea becomes more important than the thing. In Jean Baudrillard's terms, signifiers win out over things, and affluence now means accumulating the signs of happiness.[52] As opposed to word magic, we might call this sign magic.

To its critics this process is insidious. Our capacity for individuality is hampered by being moulded by a limited number of types. Our minds are incessantly crowded with symbols which refer only to consumption, shutting out the non-acquisitive parts of our personalities. Their prevalence also constantly threatens whatever free will or sense of self-control we may imagine we possess. We become puppets who dance to the advertiser's tune. We fall easily into a state which Rachel Bowlby calls "unconscious imprisonment", at best vaguely aware that our "desires are for worthless or superfluous things, or that they are shaped, if not entirely created, by the skills and tricks of advertising and other forms of presentation."[53] But the process is only partly un- or semiconscious. For we commonly collude in this seduction and play along with the game, knowing it to be such. The symbols give us strength and sustenance. If we cannot have the real thing, give us a counterfeit: it does not really matter, and most can't tell the difference. If this is a Potemkin or papier-mâché world, who cares what lies behind the false fronts? Appearance is what counts; the point is that we bond with the symbol, so the logo and brand define our peer group, our status, our ideal self. For the brand, too, constitutes a group. Belonging to it lifts us out of ourselves and makes us great. As Andrew Wernick writes, the object you desire to possess "is your own essence, and by consuming it you would be indicating your membership in the communion of all those who share it with you. The narcissistic binding of ego to product, as an act of individual and collective self-celebration, converts

52. Jean Baudrillard, *The Consumer Society: Myths and Structures* (Sage, 1998), 88, 31.

53. Bowlby, *Carried Away*, 3.

consumption into a sacrificial rite."[54] Now what Juliet B. Schor calls the "identity-consumption" relationship, the commodification of the self, is complete.[55] We are the thing, and the thing is us.

This process, which is doubtless linked to the decline of more traditional forms of identity based on class, location, gender, and religion, began relatively recently. Corporate logos appeared in the late nineteenth century, with mass-produced foods like Campbell's Soup, Quaker Oats cereal, and Coke, the most successful of them all. In the twentieth century, branding became central to the strategy of standardising and marketing commodities, the result, Naomi Klein explains, largely of a single management decision that "successful corporations must primarily produce brands, as opposed to products."[56] Entire epochs and ways of life came to be reflected in the image of the brand, famously epitomised in Andy Warhol's picture of a Campbell's Soup can, symbolising the ubiquitous normality of mass production. Blue jeans or Levis played the same role in the 1960s. It was soon realised that lifestyles and idealised selves were being sold with logos. The brand became, as John Seabrook put it, "your price of admission" to a particular subculture.[57] Gender roles, and definitions of manliness and femininity, were key to manipulating identity. Image and symbol became more important than content and quality. A century later, as the power of corporations grew ever stronger, and increasingly global, they came to define the entirety of commercial culture and its inhabitants. In the internet age, tailored or targeted marketing provides much greater opportunities to profile consumers through algorithms based on shopping patterns. Studies show that men who display large logos are less interested in long-term relationships, identification with the brand evidently satisfying some desire for belonging.[58] Personal branding has also been added to the list of varieties of commodification.

The interaction of brand identities with our sense of self is complex. When we link moral qualities and personal attainments to the image of the object, we believe we are augmenting both our virtue and our status by acquiring it. Here, luxury goods in particular elevate us, William Whyte observes, by dangling before us a self-definition above our normal expectations.[59] They can

54. Andrew Wernick, *Promotional Culture: Advertising, Ideology and Symbolic Expression* (Sage, 1991), 34.

55. Juliet B. Schor, *Overspent American*, 57.

56. Naomi Klein, *No Logo* (Picador, 2000), 3.

57. John Seabrook, *Nobrow: The Culture of Marketing, the Marketing of Culture* (Methuen, 2000), 163.

58. *Guardian*, 1 May 2021.

59. Whyte, *Organization Man*, 316.

provide what James B. Twitchell calls "an almost transcendental experience", a kind of substitute religion, a feeling close to "an epiphany": "The appeal of luxury is that nothing is higher".[60] (A parallel with a medieval proximity to saints' bones seems evident.) At an everyday level, by associating things with virtues and ideals, say soap with psychological as well as physical cleansing, sex appeal, and social acceptance, many desires are fulfilled by consuming the product.[61] Packaging adds "shelf appeal", with both the pun on and the overlap with "sex appeal" being pretty clear.[62] To penetrate still more deeply, add an asinine slogan and a catchy jingle which gets stuck in your head. ("Created just for you"—*tra la la, tra la la*: hum it, I dare you.) The associated impressions become indelibly etched into your subconscious. When the symbol appears, you feel the product to be part of what you are, because it is.

Advertising is thus key to the success of branding. It aims to get us to want to identify with the self-image associated with the product, and to aspire to be the people in the advertisements. This works for anything: the jaunty cigarette smoker, the motorist exuding freedom and independence, the happy housewife oozing with delight at her laundry whiter than white, the cool jeans wearer, the mouthwash-using sex-appeal-incarnate model, the IKEA-bookshelf-constructing couple, the healthier orange-juice drinker, and so on. Every product we consume, we are assured, makes us better and stronger, longer-lived and more attractive, more like these happy figures whose lives are evidently made so much richer, deeper, and more meaningful by their act of consumption, whose highest form of self-expression exists in the moment they submit to and merge with the commodity. A higher aim still is to get us to identify the experience with the product by name. So we buy not a drink but a Coke, smoke not a cigarette but a Marlboro, eat not a meal but a Big Mac, are not teenagers but belong to the Pepsi Generation. Verbs are even better than nouns: we don't make photocopies, we Xerox; we do not search the internet, but Google things. The same holds for some places. We enter Hard Rock cafés not to eat or listen, but to buy the T-shirt with the logo.[63] This shows that we are "Hard Rock" people, and to be reckoned with in the great contest of life.

In the "commercial utopia", in Naomi Klein's term, we live, think, breathe, and dream in and through logos.[64] Every brand provides a multiplier effect

60. Twitchell, *Living It Up*, xv, 38, 57.

61. The example is famous: see Eric Clark, *The Want Makers: Lifting the Lid off the World Advertising Industry* (Guild, 1988), 73.

62. Bowlby, *Carried Away*, 84–91.

63. So claims George Ritzer in *Enchanting a Disenchanted World* (19).

64. Klein, *No Logo*, 143.

for even our most basic pleasures and invoking it enhances the consumption greatly, indeed transforms it into a more meaningful experience. Sometimes our bond with a single brand becomes all-consuming and all-defining, swallowing us up in the obsession to belong or to be cool. If I belong to the wealthy minority, for example, I may see myself as a "Porsche" person: affluent, cool, sexy, tough, part of an elite, in the fast lane in life: *vroom, vroom*! Look at me![65] If I can't afford the car I buy the hat with the logo, or the T-shirt, to show where my heart is and experience the car pleasure vicariously—and yes, there is an online shop which sells all these, including bags, golf balls, chairs, and watches—and now there are even face masks. Or take the example of tobacco. To smoke Marlboro cigarettes, in one of the most famous ads of the 1950s and 1960s, was to participate in the Marlboro cowboy's daring machismo, to ride through beautiful Western scenery: freedom, the most saleable modern idea, in a word.[66] (How many millions died of lung cancer for this freedom?) Normally we float in a sea of logo images, grasping at this or that straw to bolster our egos. Where the reinforcement is sufficiently strong and repetitive, they worm their way into our unconscious, subliminal thought patterns, lurking in these shadows with constant whispering reminders of our desires, the jingle no doubt still resonating even as we sleep, a little monster buried within us. When you wake up with one of these jingles reverberating around your brain (and who has not?), you know you have been captured. *Tra la la*, hum it again.

This system of understanding and structuring desire, and of self-definition by commodities, became ubiquitous in the television age, the late twentieth century. This new medium came to define reality as none before had ever done. The sense arose, indeed, that nothing was real unless it was on TV, as if the imprimatur of the small screen validated our lack of confidence in our native grasp of reality. (YouTube videos play the same role now.) In the United States, particularly, advertising became so pervasive and its definition of meaning so powerful that, as James B. Twitchell puts it, it is "not just the way the characters know other humans, it is how they know reality."[67] The line between the reality of the commercialised image and any other now became increasingly blurred. Information and entertainment became indistinguishably "infotainment", often interwoven with propaganda promoting the broadcasters'

65. Viannis Gabriel and Tim Lang, *The Unmanageable Consumer: Contemporary Consumption and Its Fragmentation* (Sage, 1995), 86–88.

66. Vance Packard, *The Hidden Persuaders* (Longmans, Green, 1957), 51, 96. Camels and Lucky Strikes were "masculine" and Marlboro "virile" in the 1950s, Packard observed.

67. James B. Twitchell, *Lead Us into Temptation: The Triumph of American Materialism* (Columbia University Press, 1999), 87.

corporate backers. Product placement and commercial sponsorship became ubiquitous, and the assumption that commercial advertising is "normal" in schools, hospitals, and other "public" spaces, and that all activities should involve profit to someone, and be branded, became more widely accepted. So we move constantly from one logo or product image to another, all markers on our map of life. This process works as effectively in politics, where branding is incessant, image as a rule trumps content, and the appeal is routinely to the self-interest of the political "consumer" concerned about low taxes and cheap commodities rather than the public good, public services (if primarily for the poor), or decent wages for workers. The desire to consume now flourishes as never before, and consumer indebtedness is correlated with excessive television viewing and exposure to advertising.[68] Post-war America thus created what Lizabeth Cohen calls a "consumers' republic", where satisfying personal desires became identified with the national interest.[69]

Accepting this system occurred so easily because it suits us. To an impressive degree we connive as co-conspirators in this grand deception of selling illusions. We know that when the ad says "just for you" it cannot be true, for millions of others will see it and buy the product. But we *want* to be deceived, so long as lies make us feel better than the truth, which is why propaganda is so successful. Much more than, say, religion, where we may have deeply felt beliefs in what we think is true, we know or at least suspect that much of brand advertising is not. But we would like it to be. As external reality constantly shakes our self-esteem, we desperately want to be flattered and told how great we are (or, in politics, our country or dominant group is, or was). We want the products we acquire to have the qualities advertisers associate with them. So we are happy to be lied to so long as the liars make us feel good. We want to be told that our aspirations to happiness will succeed. Yet we pay a heavy price for this diabolical pact. Every product sold (including political systems) to the general population is retailed at the lowest common intellectual level, to permit the greatest sales, dragging everyone down to the level of crude slogans. As literacy standards fall and the ability to distinguish among advertisement, propaganda, and reality declines, our capacity to resist this process diminishes. Complexity is perplexing and involves effort. So paragraphs shrink, and long words are banned. Those with more than six syllables get chucked into the waste bin, to be retrieved by homeless philosophers who quibble over their meaning. Then we find that images are easier to comprehend than words. They

68. Schor, *Overspent American*, 81.

69. Lizabeth Cohen, *A Consumers' Republic: The Politics of Mass Consumption in Postwar America* (Vintage Books, 2004), 8–9.

require less mental effort and appeal more directly to our emotions. So television, then the internet, replaces more-complex presentations in newspapers and journals. The more relentless this onslaught becomes, the more tenuous is our grasp of reality.

Social media exacerbates this process greatly, particularly by blurring the lines between information and entertainment until we cannot distinguish between them, and by sapping our willingness to engage with anything complicated. Soon we confound media influencers with experts and bible salesmen with scientists. The "bubble" effect of social media shrinks our mental horizons. Our ability to judge "truth" is constantly sapped. So too is our will to do so. We would rather feel good. We thus conspire, as it were, to make ourselves more stupid, for the simplistic world of dreams and illusions is much more manageable than the reality which often forces itself upon us. And the more pressurised the external world becomes, the lower are the critical standards we set ourselves. So we are still more readily manipulated, to the point at which the entire process looks extremely mechanical. And so we look from human to robot—to paraphrase the last lines of Orwell's *Animal Farm*—and cannot tell which is the more programmed.

The Ideology of Choice

This process would be still more frightening if it were not so elaborately disguised, and justified by a higher level of theory. A key to advertising's success is the ideology of "choice", which in turn is central to any defence of capitalism.[70] Freedom is such a successful modern concept because throughout history so few of us have actually enjoyed it, and because even the illusion of autonomy is pleasing. So we like being told that we are exercising our free will by buying things, and that this "choice" both proves we are free and in control, and makes us, as individuals, better people, because we are invested with moral responsibility. To Bauman "it is the urge of selection and the effort to make the choice publicly recognizable that constitutes the self-definition of the liquid modern individual."[71] Advertisers thus try to persuade us that their products empower us, give us choice, put us in command, make us positive and optimistic, and reveal how clever and powerful we are. All this is music to our ears, since it chimes with the feel-good narcissism and the endless desire for praise which

70. Its classic statement is Milton Friedman and Rose Friedman's *Free to Choose* (Secker and Warburg, 1979).

71. See Bauman, *Consuming Life*, 110: In the TV age the youth market, of course, also preceded the youth culture.

define the ethos of our times. If "Advertising as a whole celebrates capitalism", this is where the encouragement to spend wantonly is wedded to an ideology of infinite self-expansion, self-expression, and self-aggrandisement.[72]

Satisfying consumer choice is thus the chief form freedom assumes in late consumer society. Shortly after Roosevelt's famous Four Freedoms speech in 1941, a Hoover vacuum cleaner advertisement actually proclaimed that "the Fifth Freedom is Freedom of Choice".[73] *Free to Choose* was Milton Friedman's 1980 title announcing the new conservative revolution of Reagan and Thatcher. Few can doubt that "a personal right to consume . . . has become accepted by many as a marker of personal freedom, power and expression and as the flag bearer of democracy", as Kim Humphery puts it.[74] Based on a vague God-given "right to happiness", this ideology is omnipresent and quasi-omnipotent. Even semi-automatic rifles, with which dozens are murdered every year in the United States, are marketed as demonstrating "the sound of freedom".

The insistence that we freely will what we are obviously being manipulated to do still further proves our gullibility. And these ever-noisier cries for freedom are seemingly proportionate to the actual loss of power over our own lives most of us increasingly experience. We have ever-less control over our circumstances of work or our political rulers, so compensate by emphasising the joys of choosing one breakfast cereal from a hundred rivals. This bizarre dialectic may further drive our desire to possess. It is plausible that "Material possessions serve as pacifiers for the self-induced helplessness we have created".[75] And "Maybe luxuries are a substitute for spiritual, religious, or community-based needs."[76] We might readily concede that this relationship with things represents an inversion of belongingness, for here the possession of and domination over things gives us a sense of shared identity and space through extending our ego, rather than subordinating our ego to others, which relationships with people require. Vampire-like, it thus sucks our sociability away. With consumerism, Lasch and others agree, an inner discontent is inevitable.[77] Not only our needs but our very being is framed "according to the logic

72. Matthew P. McAllister, *The Commercialization of American Culture: New Advertising, Control and Democracy* (Sage, 1996), 59.

73. Richard Wightman and T. J. Jackson Lears, eds, *The Culture of Consumption: Critical Essays in American History* (Pantheon Books, 1983), ix. Roosevelt's freedoms were freedom of speech, freedom of worship, freedom from want, and freedom from fear.

74. Humphery, *Excess*, x.

75. Mihalyn Csikszentmihalyi and Eugene Rochberg-Halton, *The Meaning of Things: Domestic Symbols and the Self* (Cambridge University Press, 1981), 230.

76. Lunt and Livingstone, *Mass Consumption*, 150.

77. Christopher Lasch, *The Culture of Narcissism: American Life in an Age of Diminishing Expectations* (W. W. Norton, 1978), 11.

of commodity production rather than the logic of human development."[78] Through the haze of propaganda we sense that we have been fooled, and fooled again. Then along comes the next siren-symbol, and we forget once more.

Nonetheless, any narrative of Sisyphean disappointment has to be balanced against the fact that self-definition by commodities has some obvious advantages, or we would long since have abandoned it as a failed and rather pathetic experiment in vanity. On the side of real needs, we all recognise the comforts resulting from technological advancement. But even psychologically, great advantages have resulted from universalising consumerism. In this process most gain a sense of equality, and thus power, on a historically unparalleled scale. We identify with the objects we consume because they symbolise our advancement in rank and status and validate the effort required to attain them, and the idea that we are responsible for our own success. (We cover up the credit card bill at the same time.) Through clothing ourselves like the wealthy and famous, the most important example through the ages, we gain a sense of feeling like them, of sharing in their aura, of touching, like the saints' bones, their holiness, and thus, to all appearances, of being as worthy as them, or at least sharing a mite of their goodness and infinite wisdom. This process is intensified by a declining sense of identity as defined by fixed rank, family, class, place, nation, religion, ethnicity, and so on. Individuality in modern society strips us of, or reduces the impact of, many of these more conventional markers of self. But these are then offset by focusing ever more strongly on possessing things, which corresponds with the increasingly universal value of money to define ourselves.

The success of consumerism also results from our (increasing?) psychological weakness in dealing with other human beings. Consumption-centred identity can be more gratifying than the less stable, more tentative, and often fragile and conditional pleasures of social interaction. It is here that command over objects really shows its advantages over dealing with people. With people there is always competition over status, rank, power, ego. Objects, including pets, are mostly at our command, cats again excluded, being naturally already superior. At many levels, thus, interacting with things is much easier than sociability with people. Our unique power over objects, expressed through ownership and consumption, is virtually absolute, and includes the power of destruction. So we can master things in ways we never can people, especially our notional equals. As a result, our sense of self-recognition in consuming objects is by default more powerful than that achieved in personal relationships. Unlike love, unlike life, shopping rarely disappoints. Commodities don't talk back, challenge or criticise us, wound our feelings, or make excessive demands on

78. Slater, *Consumer Culture and Modernity*, 89, 125.

us. "People say love is a great experience. I prefer shopping", runs one advertising slogan.[79] In this sense consumerism probably weakens our willingness to make the sacrifices and compromises which genuine sociability requires, and to retain the necessary verbal and emotional skills. Where we possess fewer goods we will likely value the company of others more. Where things matter more, the reverse may occur. And so at a fundamental level our love of things dehumanises or desocialises us, by draining us not only of our sociability but of even the desire to interact with others. The apogee of this process is reached when we resent others for taking away our time from things. By this point materiality has virtually vanquished us.

Things Take Over

If consumerism is a disease, its terminal stage is defined by things taking over, and symbols taking over things. The more attached we become to objects, and what we think they mean, and the more identified we are with them, the more this process overwhelms us. Slaves to our appetites, our commodities, and their signifiers, we are unable to stem our relentless desire for novelty and self-affirmation through renewal. Our sociability suffers, and becomes more anxious, as it becomes filtered through our attachment to objects, as we have to prove ourselves to others by showing off our things. When new the objects we acquire empower us. As they age and their shine wears off they weaken us in turn, in the proportion in which we rank ourselves by their newness. Where social pressure dictates renewal, the poles are reversed: rather than goods empowering us, the necessity of renewing them weakens us. We become prisoners of their demands. Before long we do not consume goods, goods consume us. They come to own us: we belong to them. We don't have stuff; *stuff has us.*[80] So, says Ernest Dichter, "If you keep on buying new cars, new clothes, in order to impress your friends, it is the object that possesses you."[81] We become object-defined rather than personality- or character- or achievement-defined. Identifying ourselves with commodities means objectifying ourselves and becoming obsessed with what Stuart Ewen calls "beautiful thinghood", an ideal of our self which is so unrealistic that it "stirs painful feelings of inadequacy".[82] The inner,

79. Twitchell, *Living It Up*, 82.

80. Emerson famously wrote that "We do not ride upon the railroad; it rides upon us" (quoted in David Brooks, *Bobos in Paradise: The New Upper Class and How They Got There* (Simon and Schuster, 2000), 72).

81. Ernest Dichter, *The Strategy of Desire* (T. V. Boardman, 1960), 110.

82. Stuart Ewen, *All Consuming Images: The Politics of Style in Contemporary Culture* (Basic Books, 1988), 89.

fatal, ironic dialectic of commodifying the self is thus that the process of defining ourselves by possessing and consuming goods enhances our sense of self-mastery even as "our" objects in fact achieve mastery over us, and we become "their" human. And so the very rich must be extremely selfish and egotistical, for their sense of self is greatly inflated by the volume of things which has come to define them. The same holds true for corporate strategists who always seek to increase their profit margins. They cannot help themselves. This is what Orwell called the "money-god" demands.[83]

This phenomenon is often associated with the evolution of individualism, or the increasing differentiation of our sense of self from society.[84] In modernity a kind of "individualist materialism of the masses" takes hold, writes Roberta Sassatelli, where satisfying consumer desire becomes central to the social order, and the "sovereign consumer" gains priority over the sovereign citizen.[85] This individualism derives in part from secularisation, the erosion of the divine and the magical, and the decline of traditional forms of authority. It is most extreme in America because here traditional markers of status and identity are less fixed, self-definition is more fluid, and "community" alters more rapidly than elsewhere. As external determinants of behaviour decline, internal motivations come to the fore to take their place. The process parallels an intensified sense of our inner life and an introspective experience of reality through our inner world, which to Simmel made "psychologism" the "essence of modernity".[86] It is greatly reinforced by consumerist ideology, which, it has been suggested, persuades us that "personal experience alone—largely in the form of wants and desires—constitutes the highest authority."[87] The cult of this experience in turn assumes the form of what is usually called narcissism, the excessive admiration of our own self-image, which is in this sense the end and aim of all advertising.

Narcissism as the Consumerist Personality Type

The dominant personality type associated with consumerism is narcissism, the term being derived from the mythical Greek youth who fell in love with his own reflection. What is novel, thus, is not the type, but its universality and intensity. We all know the feeling: take a selfie or look in the mirror and admire

83. In *Keep the Aspidistra Flying* (1939), in Orwell, *Complete Novels*, 585, 604.

84. See Steven Lukes, *Individualism* (Basil Blackwell, 1973).

85. Sassatelli, *Consumer Culture*, 40.

86. Quoted in David Frisby, *Fragments of Modernity: Theories of Modernity in the Work of Simmel, Kracauer and Benjamin* (Polity, 1985), 46.

87. Karin M. Ekström and Helene Brembeck, eds, *Elusive Consumption* (Berg, 2004), 40.

what you see, the image you have elaborately constructed of yourself, the external appearance which reflects your inner essence. Then do it over and over again and begin to see yourself through the image, and to shape and judge your every move by it.

The actual ideals we model ourselves on have not changed much over the short centuries of humanity's history. Admiration of youth, strength, and beauty have always existed. What is new is the scale and magnitude, the proliferation of modes of expression of self-love, and their commodification, as the mirror gave way to the photograph, then to the moving picture. By the mid-twentieth century, in the prolonged period of growth following World War Two, a "Promethean generation" became fixated on what Henry Malcolm calls "the illusion of paradise", defined by a child-centred "environment of immediate gratification". This resulted in the "resurrection of narcissus".[88] An unparalleled self-absorption coupled with "meaningless consumption" and "compulsive shopping" came by the "Me Decade" of the 1970s to define the "Americanization of narcissism", in Elizabeth Lunbeck's phrase.[89] What Stuart Ewen terms "commodity defined *self-fetishization*" thus corresponds to a "culture of narcissism" or self-centredness, which in Christopher Lasch's well-known account is defined by an intense focus on "privatism", self-scrutiny, self-absorption, and "pseudo" self-awareness.[90] Originally associated with the "consciousness movement" of the 1960s, it is now much more pervasive and involves endless self-validation through flattery and constant references to happiness and well-being.[91] An obsession with increasing our feeling of happiness, and with having to appear to be constantly happy, and to smile to prove it, leads us constantly to ratchet up our apparent levels of satisfaction with hyperbole, by the law of self-breeding superlatives, by which the moderns infinitely hype adjectival responses to demonstrate their enthusiasm. Effusive expressiveness is now demanded as proof of our inner health, good nature, and outlook. So people are now "delighted" to respond to our "awesome" emails or solicit our attention when formerly they were only "happy" or "glad", or, once upon a time, "received with thanks".

To Richard Sennett this mentality leads us to become overwhelmingly interested only in how "'what this person, that event means to me'", at the expense

88. Henry Malcolm, *Generation of Narcissus* (Little, Brown, 1971), 4.

89. Elizabeth Lunbeck, *The Americanization of Narcissism* (Harvard University Press, 2014), 1–2, 268, 12.

90. Stuart Ewen, *Captains of Consciousness: Advertising and the Social Roots of Consumer Culture* (McGraw-Hill, 1976), 48.

91. Ewen, 48; quoted in McKendrick, Brewer, and Plumb, *Consumer Society*, 15; Lasch, *Culture of Narcissism*.

of any sense of what goes on outside our emotions, or how we should socially contextualise events.[92] An all-consuming search for inner self-gratification acts as a negative, "destructive gemeinschaft" which "treats intimate interchanges as a market of self-revelations. You interact with others according to how much you tell them about yourself; the more 'intimate' you become, the more confessions you have made."[93] A relentless process of self-justification and self-flattery encourages us to abandon guilt and "take pride" in every part of ourselves, including our own ignorance, by even more self-love, instead of acknowledging with humility the disaster we have wrought on our planet. (Pride, we recall, was for Thomas More what stands between us and utopia.)

As there is often not much to love in our mental accomplishments, it is the body which plays the key role here. Clothing is again central to the vanity associated with an extreme cult of self-image. But no aspect of appearance is left untouched. This process encourages an ever-more body-centred, perfectibilist idea of the self in which Botox, plastic surgery, extending our eyelashes, and having tattoos raise large questions about what is true and what is false in self-representation. Everywhere the image is king, or queen. Smartphone use and social media have expanded this process exponentially. Anxiety about our bodies is now more widespread than ever, and for ever younger groups. So-called Snapchat dysmorphia occurs when individuals become anxious because their real selves fail to live up to their retouched selfies, the value of the image here clearly trumping the reality. Indeed, a veritable epidemic of self-doubt occurs as the proliferation of potential personality types and forms of gratification overwhelms our capacity to understand the process, and leads to acute anxiety over "identity" in particular. Small wonder that our collective mental health is becoming so fragile.

The confluence of consumerism and narcissism also involves a kind of psychological regression, or a return to the Cockaigne-like version of the Golden Age of innocence, ludic imagination, and plenty.[94] As we have seen, the need to sell to the largest number brings down the mental level of the sales pitch to the lowest common denominator. But consumerism also makes adults more like children. The symbols which shape our desires and crowd our minds are simple, like a child's. The child's desire for immediate gratification finds a parallel in impulse buying, which perfectly expresses desire without consequences.

92. Richard Sennett, *The Fall of Public Man* (Cambridge University Press, 1977), 8.

93. Richard Sennett, "Destructive Gemeinschaft", in *Beyond the Crisis*, ed. Norman Birnbaum (Oxford University Press, 1977), 181.

94. This, however, presumes an unidentified or unscrutinised idea of the "adult" or mature personality, where thrift, saving, sacrifice, and other notionally "Protestant" virtues are ascendant.

We are told that we should be obsessed with happiness, and that this rarely involves self-denial. Most of us are happiest in childhood, before we are enslaved to work and by the responsibilities of adulthood, and so we cling to it. In adulthood we are happiest when entertained by simple thoughts which do not tax our reasoning capacity or demand critical reflection, and so allow us to celebrate our ignorance and prejudices.

So from the 1950s our culture has increasingly frozen our mental aspirations in adolescence, that great compromise between childhood and adulthood which purports to capture the best of both worlds. As the demand for self-renewal focused ever more on the young, teenagers became the model humans, and the target economic market. In the United States, consumerism shifted steadily towards the "youth market", then the "global teen market", increasingly a fluid amalgam of races, genders, and ethnicities, which is now the core of both consumer and popular culture. The "lifestyle identity" of the young has become more and more permeated with and defined by brands and their consumption, which are usually advertised by attractive media or sports "celebrities". The cult of youth, that eternal euchronia, also demands that the middle-aged and older look and at least pretend to feel younger as long as possible, until their bodies finally revolt at the effort. Ideals of appearance, outlook, and mental capacity get steadily younger, like the models who represent them, while contempt for the old and ageism grow steadily. In the mass media, again because of the imperative to maximise sales, an "infantilist ethos", in Benjamin Barber's words, inevitably downgrades the intellectual content of programmes and commercials alike to the lowest common average, thirteen years old for television content and blockbuster films. This depresses the mental horizons and aspirations of the entire society, and by extension the rest of the world.[95] Adulthood, both individually and as a species, is in retreat, along with literacy, the pursuit of complexity, subtlety, irony, and much else. "Adultescent" people, or "kidults", refuse to take responsibility as adults. A continuous decline in reading and writing skills, and a renewed focus on images, pushes us ever further into semi-literacy. Long words are too challenging: people will tune out. Reading at all becomes too demanding. YouTube videos are more amusing. They lead us to judge people solely on their appearance. Consumerism requires simple, pliable, easily manipulated consumers. We happily oblige by dumbing ourselves down. And so another paradox in our onward progress is revealed: humanity emerges from its childhood only to return to it in its dotage.

95. Benjamin R. Barber, *Consumed: How Markets Corrupt Children, Infantilize Adults, and Swallow Citizens Whole* (W. W. Norton, 2007), 3.

Consumerism and Identity: Summarising the Pros and Cons

Emulation, scientific and technological innovation, cultural materialism, hedonism, and individuation have driven consumerism inexorably onwards, until it has now reached its limits. But we will never supersede consumerism unless we understand its achievements. A consumer-centred identity has many advantages over more fixed definitions of self in more rigid social hierarchies in terms of mobility, flexibility, comfort, and rapidity of social advancement. The role of consumer feels superior to many others we might play or which are forced upon us. The ideology of choice which pervades advertising implies that we command the world around us, even as our power as citizens and workers is drained away constantly. Indeed, the former may compensate directly for the latter: the less control we really have, the more we project onto our possession of things. The power we feel in possessing objects encourages our desire to see ourselves as consumers first and foremost. Rather than being defined by our neighbourhood, class, religion, gender, and occupation, any or all of which may drag us down in self-esteem, fix us in predefined and inescapable expectations, and make many more demands on us, we can adopt new styles. Fluidity is the name of the game. We can remake ourselves simply by shifting the symbols—a new T-shirt, shoes, hairstyle, eyebrows, facelift, tattoo.[96] To Don Slater, we choose "self-identity from the shop-window of the pluralized social world; actions, experiences and objects are all reflexively encountered as part of the need to construct and maintain self-identity." Now "Consumerism simultaneously exploits mass identity crisis by proffering its goods as solutions to the problems of identity, and in the process intensifies it by offering ever more plural values and ways of being."[97]

The disadvantages of consumer-centred identity are equally tangible. Obsessively acquiring goods, and defining ourselves thereby, turns out not to rival, much less surpass, warm and intimate social relationships, but actually to undermine them, by making others jealous if nothing else. The greatest paradox of modern progress is that it creates a depressed and unfulfilled populace who are increasingly isolated from others. Recent studies of declining feelings of happiness and self-worth stress that "we get happiness primarily from people; it is their affection or dislike, their good or bad opinions of us, their acquiescence or rejection that most influence our moods." Commodities are "poor substitutes for friends", and those who are attracted to our

96. Gabriel and Lang, *Unmanageable Consumer*, 94, referring to Anthony Giddens.

97. Slater, *Consumer Culture and Modernity*, 85.

possessions are unlikely to value us for our human qualities.[98] The corollary here, which is vital to the argument of this book, is that we need objects less if we have warm, emotionally satisfying personal relationships. Consumerism presumes the self-validation gained by acquiring goods compensates for losing traditional forms of association—notably, religion and the family. It does not. In fact, when wedded to an increasingly urban life where community identity is weakened and even elementary social bonds diminish, our sense of self becomes depleted in proportion to our materialism. Things are soulless and bring us little if any emotional gratification. A sense of nearness to others, of human warmth, thus cannot be replaced by objects. This too the Covid pandemic has taught us.

But in the past twenty years, our increasing sense of malaise has been exacerbated by growing inequality, precarity, and declining living standards and expectations for the majority. At the outset of the 2020s, even before Covid exploded upon us, we were stressed, nervous, anxious, and depressed. Things were getting worse. And we have not yet entered the most dangerous period of our looming future, when these anxieties will be intensified a thousandfold.

Such a diagnosis suggests, however, that if we reject the supremacy of consumerism as a way of life we might also reverse this relationship, and to a substantial degree exchange material gratifications for psychological ones. We can give contact with nature, creative activities, and human relationships a higher value, and derive from them a greater satisfaction, than consuming goods. Donella Meadows puts this nicely in her book *Beyond the Limits*:

> People don't need enormous cars; they need respect. They don't need closets full of clothes; they need to feel attractive and they need excitement and variety and beauty. People don't need electronic equipment; they need something worthwhile to do with their lives. People need identity, community, challenge, acknowledgment, love, and joy. To try to fill these needs with material things is to set up an unquenchable appetite for false solutions to real and never-satisfied problems. The resulting psychological emptiness is one of the major forces behind the desire for material growth.[99]

98. Lane, *Loss of Happiness*, 8.

99. Donella Meadows, *Beyond the Limits: Global Collapse or a Sustainable Future* (Earthscan, 1992), 216.

This proposition will be teased out in the final section of this book. Beforehand, however, we must consider a series of challenges to the consumerist mentality, some of which imply that this process is less inevitable than may appear to be the case, and may be tamed. The "veritable cultural revolution" in "the material and spiritual situation of modern man" which consumerism represents could not, of course, have been contained within national boundaries.[100] Its great aim and ambition, its destiny indeed, was to conquer the world: the relentless logic of profit generation in capitalism could dictate no other end. But during the twentieth century, critics of capitalism constructed what initially appeared to be a highly successful alternative to it, and one which moreover hinted at a much less consumerist approach to commodities. This process began with the Russian Revolution of 1917.

Counter-ideals: The Soviet Response to Consumerism

We have seen that from the late eighteenth century, utopianism shifted towards a future-oriented image of the ideal society. The lure of the primitive, and of a more virtuous and innocent life on the land, if in retreat, still lingered. But traditional utopian goals of harmony and equality were now also increasingly wedded to cornucopian visions of unlimited needs, expanding luxury and consumption. As the twentieth century dawned, utopianism remained ambiguous about modern systems of mass production and consumption. This is nowhere more evident than in the then greatest and most ambitious experiment in social engineering ever undertaken, the Bolshevik Revolution. When it occurred, Russia was a largely peasant nation with little industrialisation or extensive commerce in most of the interior. In principle it should have been easier for a society aiming at much greater equality and justice and not yet addicted to consumerism to skip over this stage of development. Russia might have modernised without the destructive effects of excessive consumerism. And in part, for a time, it succeeded in so doing.

That this would prove a challenging task is, however, obvious today. In practice, modernisation was from the outset central to Bolshevism, and this soon implied a constantly rising standard of living, especially for the hitherto impoverished masses. Yet the USSR and later Soviet satellite regimes became far and away the most important single example of combining a high level of subsistence with an attempt to reject some central aspects of consumerism. Unlike Maoist China after 1949, where for some thirty years equality and

100. Timo Vihavainen and Elena Bogdanova, eds, *Communism and Consumerism: The Soviet Alternative to the Affluent Society* (Brill, 2016), xii.

raising the basic standard of living for the masses took priority, the Soviet model was intensively industrial from the outset. Renouncing technological modernity was never an option, for military reasons if nothing else, for Russia had been too often conquered, and most recently humiliated by Japan in 1905 and Germany in 1914–17. The new USSR which emerged from the Russian Empire thus aspired to be a western nation defined by science and technology, as well as by the supposed rule of an enlightened urban proletariat rather than by a larger but more backward rural peasantry.

This did not as such necessarily imply capitalist excesses of consumption. A centralised command economy combined with an anti-capitalist ethos to provide, in principle, the possibility of repressing the most individualistic, unequal, divisive, wasteful, self-indulgent, and class-based forms of consumerism. Respect for wealth as such dissipated rapidly, and was initially replaced by zeal for revolutionary goals. Consumption was to be rationalised and socialised and rendered more just and equal. So from early on advertising was suppressed, and consumer choice was restricted by prioritising basic needs. Abundant public facilities were created to satisfy demand, especially for food, efficiently and cheaply. In keeping, as we have seen, with a prominent trend in the socialist tradition, public luxuries were given precedence over private. The Moscow Metro was a powerful new symbol of this ideal. So were thousands of workplace canteens, new subsidised communal housing, hospitals and free medical care, more parks and gardens (in Moscow) than anywhere else, and, eventually, guaranteed holiday facilities and adequate pensions. Judged by rising longevity and literacy rates, this approach was enormously successful.

What does this vast experiment, lasting over seventy years, suggest about the prospects of superseding twenty-first-century consumerism? The Soviet example indicates exactly the same ambiguity with respect to consumerism which we earlier associated with Marx. The USSR and its Eastern European satellite states did not deviate substantially from the paradigm of expanding population, needs, and production which defined liberal political economy.[101] As with Marx, two ideals clashed here. On the one hand, Russia was on the road to capitalism in 1917. Her chief cities, at least, had witnessed a great expansion of retail consumption and culture in the late nineteenth century. Here, as elsewhere, cavernous palaces of opulence now touted the wares of the world to the well-to-do. In the court at Saint Petersburg, Paris set the tone in fashion, though Berlin, London, and even New York were not far behind. Though serfdom was abolished in 1861, the chasm between aristocratic and bourgeois opulence and the poverty of the masses was more gaping than in the west. So there

101. See, generally, Philip Hanson's *The Consumer in the Soviet Economy* (Macmillan, 1968).

was potentially vast untapped demand. On the other, Marxism indicated that another road might also be taken, and opulence defined very differently. The unknown factor in this equation remained competition with the west, and eventually with the American utopia.

When the Bolsheviks seized power in Russia in 1917, revolutionary mores were naturally associated with personal austerity and the spirit of sacrifice which a fifty-year struggle had demanded. Like France after 1789, when luxury in dress rapidly disappeared, Russians gave up their jewellery to support the cause and affected simpler styles. Luxury stood for the excesses of the tsarist court, and for oppression, exploitation, and the ostentatious extravagance of the old regime.[102] It implied idleness, frivolity, and waste on mere decoration, as well as a greater stress on wanton sexuality for women. By contrast, the new regime sought to encourage a worker- and productivist-centred ideal of humanity. Luxury meant the workers had to labour harder, which as we have seen was a staple argument against it in eighteenth- and nineteenth-century radicalism and utopianism.

From the outset the new regime thus opposed conspicuous consumption in principle. Initially the cult of the puritanical revolutionary promoted this. As in France after 1789, many early Bolsheviks adopted what Hans Jonas calls "a credo of public morality" and "a spirit of frugality, alien to capitalist society", commensurate with aiming to live "for the whole" and demanding asceticism and self-denial.[103] Yet some felt that such austerity was always temporary, required until the revolution achieved victory and plenty had been achieved. This frugality also had to coexist with a cult of technology within Bolshevism, and the promise of electrification and ever-expanding machinery. Heavy industry could provide railways, tractors, tanks, and eventually spaceships. But it could also be used to manufacture consumer goods if priorities shifted. And raising the standard of living for the poor majority was also an aim of the revolution.

There is no evidence, then, that the regime thought that growth might in principle be limited under socialism, and plenty for the reverse assumption. Lenin's famous slogan of 1920, that "Communism is Soviet power plus the electrification of the whole country", abundantly symbolises the centrality of modern technology to the new ideals.[104] The First Five Year Plan (1928–32) prioritised heavy industry, with light industry and food production following. The USSR still rightly feared war from both the east and the west. This factor alone drove it towards industrial modernisation.

102. Aileen Ribeiro, *Fashion in the French Revolution*, p. 60.

103. Jonas, *Imperative of Responsibility*, 147.

104. "Our Foreign and Domestic Position and the Tasks of the Party", in V. I. Lenin, *Collected Works*, 47 vols (Progress, 1964), 31: 419.

Such priorities were initially unproblematic and widely conceded by most. At the outset of the revolution, euphoria was common amongst its supporters, in whom a powerful sense of freedom and idealism had been awakened. The utopian vision of an egalitarian future where everyone was happy was captivating. Through the civil war (1918–23) an ethos of restraint and sacrifice was thus easy to maintain. Early enthusiastic visitors to the USSR, John Maynard Keynes commented in the 1930s, found that communism offered "an appeal to the ascetic in us", and saw no need to apologise for the shortages ordinary Soviet citizens endured.[105] A British communist, Maurice Dobb, observed that while lemons, tea, and bananas were scarce, "new Diesel engines are preferred in their stead."[106] Other socialists at the time, we recall, were also regarded as hostile to luxury.[107]

Outside of established communes and privileged Party outlets, however, regulating consumption was more difficult and less easily justified. Some shops were quickly transformed from retailing for the middle classes to selling to the workers. Others were simply closed. In 1920 the stalls at Moscow's most popular retail centre, the Sukharev Market, were demolished by the Cheka (political police), the crowd cheering them on. Abolishing middlemen between producers and consumers, it was hoped, would quickly lower prices. Instead, the number of retail outlets plummeted. Supplies dwindled, and rationing resulted. Food in particular was allocated on the basis of the difficulty of one's work, and housing also on political rank and connections. Soon, formerly expensive luxury goods had value only as barter for food.[108]

At least initially it was expected that the new Soviet Man and Woman would rise above any desire for superfluity. In the regime's early years the general outlook of "popular communism", Julie Hessler writes, was "radically egalitarian and hostile to anything considered a luxury."[109] Youth was regarded as extremely pliable, and the basis of the new personality Bolshevism hoped to foster. Formed in late 1918, the Komsomol, or Communist Youth League, was

105. Quoted in Mark Landsman, *Dictatorship and Demand: The Politics of Consumerism in East Germany* (Harvard University Press, 2005), 5.

106. Maurice Dobb, *Russia Today and Tomorrow* (Hogarth, 1930), 21.

107. E.g. J. P. Lockhart-Mummery, who claimed that "Luxury was condemned by the early Christians and by the Puritans on religious grounds, and it is opposed to-day by the modern Socialists" (*After Us; or, The World as It Might Be* (Stanley Paul, 1936), 157). The reference may be to Robert Blatchford in particular. But it would hold for Orwell too.

108. Marjorie L. Hilton, *Selling to the Masses: Retailing in Russia, 1880–1930* (University of Pittsburgh Press, 2012), 185, 187, 191.

109. Julie Hessler, *A Social History of Soviet Trade: Trade Policy, Retail Practices, and Consumption, 1917–1953* (Princeton University Press, 2004), 223.

to be the breeding ground and vanguard of the new type. Its ideal was rooted in self-sacrifice, and a constant struggle to make the revolution succeed, often couched with a Spartan, military emphasis on "strength, endurance, toughness and dexterity". Here, and in the various clubs and societies which formed the Proletkul't, a proletarian outlook was essential. Austerity in personal dress and habits was mandatory. Bourgeois respectability was suspect. Money-grubbing and conspicuous consumption were unthinkable attributes for a revolutionary. State officials and other authorities, Maurice Dobb observed, "take pains to be ordinary rather than distinctive in their dress."[110] Many middle-class Komsomols thought an "undisciplined, slovenly appearance", such as leather jackets, short hair for both sexes, and patched shoes, exemplified the purity of their revolutionary morality, and, by symbolic solidarity, their fusion with the workers. Learning to "speak Bolshevik" meant looking and acting accordingly. This might mean adopting pseudo-proletarian affectations, such as talking in street slang or wearing working-class overalls, even long hair and dark glasses, as the student radicals of the 1860s and 1870s had done, and those of the 1960s would do.[111] Early on, thus, more- and less-puritanical wings developed amongst young activists. At the libertarian, even libertine, end, the revolution seemed to offer a multitude of possibilities, including unparalleled sexual freedom. The more puritan, however, campaigned, for example, to stop swearing and smoking, and then were mocked for their efforts, which clearly clashed with efforts to mix with the working classes.[112]

The civil war changed this. In some areas, by the 1920s, a Komsomol uniform was adopted which consisted of a military jacket and breeches—and a pistol, a serious status symbol, and appropriately Spartan. A "macho-subculture" flourished in some areas which overlapped at points with the new asceticism.[113] For a time schoolchildren appeared at demonstrations fashionably barefoot and in "phys-cultural" sports clothes.[114] Amongst the Pioneers, the youth wing of the movement, drinking, smoking, and swearing were prohibited.[115] But elsewhere, drinking, "hooliganism", and rampant sexual aggression appeared in Komsomol cells. A survey of three hundred Leningrad youth in the mid 1920s discovered that common leisure activities included seeing films, drinking,

110. Quoted in Julian Huxley, *A Scientist among the Soviets* (Chatto and Windus, 1932), 6.

111. Anne E. Gorsuch, *Youth in Revolutionary Russia: Enthusiasts, Bohemians, Delinquents* (Indiana University Press, 2000), 16, 89–90.

112. Matthias Neumann, *The Communist Youth League and the Transformation of the Soviet Union, 1917–1932* (Routledge, 2011), 110–11.

113. Neumann, 88–90.

114. N. Ognyov, *The Diary of a Communist Schoolboy* (Victor Gollancz, 1938), 40.

115. Albert Rhys Williams, *The Russian Land* (Geoffrey Bles, 1929), 124.

pursuing love affairs, and fighting. Following the regime's insistence that dance halls were "gathering places for counter-revolutionaries", the Komsomols banned dancing as a "bourgeois prejudice". Now they sponsored their own dances.[116] Another assessment of Komsomols in 1926 found them pessimistic about the future, individualistic to the point of being petty bourgeois, and anti-Semitic to boot. In one small cell in Belyi, in the far west, eighteen out of thirty members were dismissed for hooliganism or drinking.[117]

The Komsomol experience was interwoven with the regime's desire to forge a more collectivist life, especially in the newly established communes, where revolutionary ideals had priority over traditional family blood ties. Here, what Andy Willimott calls "plain ascetic values and new comradely relationships" defined the new outlook.[118] By the late 1930s some 95 per cent of commune members, thirty thousand people in European Russia, were Komsomols. Their daily life was strictly regimented, sometimes down to specifying the hours when newspapers could be read.[119] Selfish interests were supposed to give way to the common good, whether by sharing food, drinking moderately, or not having sex outside marriage. Frugality was the order of the day. It did not pay to be too distinctive. Foreign observers noted that "no one dresses in anything but brown and black through fear of attracting attention."[120] Women were condemned for wearing make-up, gold jewellery, or anything which made "you stand out from the larger working masses."[121] Even "pretty light shades" and "decorations" in the bedroom, for those lucky enough to have one, were scorned.[122] Instead of dances there were political and scientific lectures and meetings, as many as fifty a month. We are not far off here, at points, from a revived Spartan model, now compounded with Bolshevik ideology and bureaucracy.

There was also pressure to feed, clothe, and house the masses. Lenin himself emulated the frugality of the French revolutionaries of 1789, condemning "slovenliness . . . carelessness, untidiness, unpunctuality" and "dissoluteness

116. Gorsuch, *Youth in Revolutionary Russia*, 61–62, 73, 121; Matthias Neumann. *The Communist Youth League*, 108–9. Over 90 per cent of those interviewed were Komsomol members.

117. Roger Pethybridge, *One Step Backwards, Two Steps Forward: Soviet Society and Politics in the New Economic Policy* (Clarendon, 1990), 352–53.

118. Andy Willimott, *Living the Revolution: Urban Communes & the Soviet Socialism, 1917–1932* (Oxford University Press, 2012), 30.

119. Gorsuch, *Youth in Revolutionary Russia*, 56, 89.

120. Elena Osokina, *Our Daily Bread: Socialist Distribution and the Art of Survival in Stalin's Russia, 1927–1941* (M. E. Sharpe, 2001), 92.

121. Gorsuch, *Youth in Revolutionary Russia*, 91.

122. Willimott, *Living the Revolution*, 101–2.

in sexual life".[123] But this model of a vanguard elite was not Spartan as far as mass consumption was concerned. Early on, Lenin acknowledged the need for "the equal distribution among the population of all consumer goods, so as really to distribute the burdens of the war equitably."[124] He recognised the "undoubtedly progressive phenomenon" of a "rising level of requirements for the entire population", including "clothing, housing, and so forth".[125] Indeed, one of capitalism's main problems, he thought, was that "the productive forces of society increase without a corresponding increase of consumption by the people". Under an "economy of associated producers" the surplus which resulted, he assumed, would serve to rectify this injustice.[126] This implied an unlimited expansion of both production and consumption. (But Lenin also told H. G. Wells he thought that "'The towns will get very much smaller".)[127]

A degree of revolutionary asceticism remained part of the regime's official ideology throughout its history. This was a genuinely anti-capitalist stance, though it also helped to mask or excuse shortages. Efforts were made to contain consumerist desires (*veshchizm*, or materialism), and to craft a society which was affluent without being consumerist. Ideally, cultural and social activities (*kul'turnost*, or "culturedness") could be substituted in part for consuming material goods. Two ideals were thus clearly at war from the outset: notionally "bourgeois" norms of rising affluence and communist austerity, wedded to various forms of public luxury. Cosmetics were criticised for a time and, indeed, became a kind of battleground between these competing world views: did beauty lie in unadorned natural simplicity or extensive but artificial self-decoration, which implied class distinction as well as a love of indolence and hedonism? Pre-revolutionary fashions were generally held in contempt. The image of the self-centred *meshchanstvo*, or philistine, petty-bourgeois egotist, often associated with city dwellers, was contrasted to that of the model Soviet citizen, who was supposed to be more

123. Quoted in Michael Walzer, *The Revolution of the Saints: A Study in the Origins of Radical Politics* (Weidenfeld and Nicolson, 1966), 314–15. See, generally, Bruce Mazlish's *The Revolutionary Ascetic: Evolution of a Political Type* (Basic Books, 1976), which tries to make the case for a "common personality" often defined by asceticism linking many revolutionaries, including Robespierre, Lenin, and Mao (3), and Melvin J. Lasky's *Utopia and Revolution* (Macmillan, 1977).

124. "The Impending Catastrophe and How to Combat It", in Lenin, *Collected Works*, 25: 348.

125. "On the So-Called Market Question", in Lenin, 1: 106–7.

126. "Reply to P. Nezhdanov", in Lenin, 4: 161, 165.

127. Quoted in W. W. Wagar, ed., *H. G. Wells: Journalism and Prophecy, 1893–1946* (Bodley Head, 1965), 229.

interested in intellectual and cultural self-development and in public service to socialism. Popular works portrayed the latter indeed as "neither indulgent hedonists nor calculating utilitarians, but fiery soldiers of their cause" who "very much looked like puritan moralists."[128] Mere pleasure-seeking was "bourgeois" (*burzhui*), and as such contemptible, especially for Party members and Young Communists. The regime also enjoined people not to evaluate others on the basis of material possessions. An ideology of "everyday asceticism" developed in the 1920s which treated goods as functional, rational, and practical rather than symbolic.[129] Debates about consumption also overlapped with a narrative which sought to reduce private life as such, and to merge what remained more closely with public activities. This is the ideal we associate with Stalin and the construction of communal flats, with common kitchens and toilets for several families, for much of the urban population.[130]

As portrayed in the classics of socialist realism, as early as Maxim Gorky's famous *Mother* (1906), social and moral, human goals, most notably solidarity, were thus supposed to have priority over material ones. A contempt for the wasteful, idle, parasitic bourgeoisie who killed millions solely "to keep the timber of their houses, secure their furniture, their silver, their gold, their worthless papers—all that cheap trash", all "for the sake of their possessions", dominated such assumptions.[131] Bolshevik literary utopianism echoed such themes. In Alexander Bogdanov's *Red Star* (1909), clothing was mass-produced with no embellishments, in a style which was "fundamentally the same" for both sexes.[132] And in the former Bolshevik Evgeny Zamyatin's famous satirical dystopia *We* (1921), a woman dressed in a "saffron-yellow dress of an ancient cut" is described as "a thousandfold more wicked than if she had had absolutely nothing on". Here similarity has eradicated envy, and there are plans even to equalise noses.[133]

Official antagonism towards excessive consumerism dominated the Soviet period. As late as 1960 the Party line was that "Communism excludes those narrow-minded people for whom the highest goal is to acquire every possible

128. Timo Vihavainen, "The Spirit of Consumerism in Russia and the West", in Vihavainen and Bogdanova, *Communism and Consumerism*, 15.

129. Olga Gurova, "The Ideology of Consumption in the Soviet Union", in Vihavainen and Bogdanova, *Communism and Consumerism*, 70–71.

130. Adele Marie Barker, ed., *Consuming Russia: Popular Culture, Sex, and Society since Gorbachev* (Duke University Press, 1999), 32–33.

131. Maxim Gorky, *Mother* (D. Appleton, 1907), 190.

132. Alexander Bogdanov, *Red Star* (1909; Indiana University Press, 1984), 69.

133. Zamyatin, *We*, 31, 83.

luxurious object."[134] Nonetheless, this was to some degree a futile aspiration. Demand for consumer goods surged after the deprivation of the war years. Variety, creativity, and innovation also came more to the fore, and some modernist and utopian trends promoted the proliferation of colour and of aesthetically driven rather than merely functional designs. During the relatively liberal NEP (New Economic Policy) period (1921–28), free market policies were introduced. Foreign experts were invited to the USSR, and, with state encouragement, merchants returned in large numbers by 1922. An affluent new class of wholesalers and retailers emerged. By 1928 they may have been responsible for as much as half the total volume of retail trade.[135] Shop windows were now full of formerly unobtainable goods like pastry, meat, and cheese. Small luxuries brightened everyday lives blighted by war and scarcity, even if many could not afford them. By 1929, Maurice Dobb observed, "the Soviet typist probably thought more of silk stockings and less of being a new woman than she did four years before".[136] But the principle remained that "socialist trade should guarantee everyone basic necessities before supplying anyone with 'luxury' items."[137]

The NEP brought greater disparities to Soviet society. Diehard Bolsheviks now wondered whether the old inequalities were being reborn. As early as 1922 "NEPification" (*Oneprivanie*) was a subject of contemptuous but worrying discussions amongst Communist Party activists, with the new luxury-oriented consumerism being called "the disgusting offspring of NEP".[138] A popular novel of the day, *The Diary of a Communist Undergraduate* (1929), denounced those who danced the foxtrot as "fake Communists", and insisted that as a "stupid and anti-revolutionary pastime . . . This nonsense ought to be stopped."[139] But plenty *were* dancing the foxtrot, and a culture war inevitably resulted. Youth wearing leather jackets and flamboyant NEPmen and their risqué escorts were both regarded with suspicion if not hostility. The urban young in particular became fascinated by foreign trends, like the flappers and American jazz, interests which by 1926 might attract the charge of hooliganism against the new "Soviet Bohemians".[140]

134. Quoted in David Crowley and Susan E. Reid, eds, *Pleasures in Socialism: Leisure and Luxury in the Eastern Bloc* (Northwestern University Press, 2010), 23.

135. Alan M. Ball, *Russia's Last Capitalists: The Nepmen, 1921–1929* (University of California Press, 1987), 164.

136. Dobb, *Russia Today and Tomorrow*, 47.

137. Hessler, *Social History*, 223.

138. Willimott, *Living the Revolution*, 85.

139. N. Ognyov, *The Diary of a Communist Undergraduate* (Victor Gollancz, 1929), 122.

140. Gorsuch, *Youth in Revolutionary Russia*, 119.

Restoring something like freedom of consumption would subsequently be seen as a "great retreat" from the regime's original revolutionary ideals.[141] The emergence of a new technical and political elite who had to be motivated by larger salaries and greater privileges also did much to stimulate demand from the 1920s onwards. The problem of providing sufficient consumer goods was always pressing. Those who could afford to shop at GUM, the state department store in Moscow established in 1922, embraced a new model of socialist consumption. Here, the concept of a just and egalitarian society was retailed as much as coats or cooking pots. "Everything for Everybody" was GUM's motto.[142] It boasted "the latest styles" and a "Colossal Assortment of High-Quality Merchandise."[143] Like other retailers at the time, it consciously sought to expand its horizons to a much wider public, the working classes. Socialism had to prove its superiority to capitalism, and providing the masses with material goods for the first time was an obvious way of doing so.

From the outset, thus, Marjorie Hilton writes, "In theory, a socialist society was supposed to lift the worker's standard of living, which in practice meant an increase in personal consumption."[144] Shortages of goods merely highlighted this norm and made its attainment even more desirable. A dearth of foodstuffs occurred again in the late 1920s when land collectivisation began in earnest and *kulaks*, "rich" peasants, began to hide or destroy their produce rather than let the state seize it. Food remained scarce, indeed, even after official rationing on bread and meat was abolished in 1935–36. Even the privileged elite received as little as 4 kilograms of meat per month in 1932, and the only decent harvest of the decade was 1937.[145] Yet by the 1930s those who could were acquiring luxury goods in state stores, rather than privately. This marked a key turning point in the regime's attitude towards luxury and consumption. Now, it can be claimed, a "survival-oriented material culture had begun to give way to a new interest in fashion and individuation through consumption on the part of a growing consuming elite."[146]

141. Nicholas Timasheff, *The Great Retreat: The Growth and Decline of Communism in Russia* (E. P. Dutton, 1946), 144–49.

142. Hilton, *Selling to the Masses*, 170. Hence the Soviet joke: "What is Communism? It is when anybody can buy anything. Like in the days of Nicholas II" (Anna Krylova, "Saying 'Lenin' and Meaning 'Party': Subversion and Laughter in Soviet and Post-Soviet Society", in Barker, *Consuming Russia*, 259).

143. Hilton, *Selling to the Masses*, 211.

144. Hilton, 221.

145. Jukka Gronow, *Caviar with Champagne: Common Luxury and the Ideals of the Good Life in Stalin's Russia* (Berg, 2003), 126.

146. Hessler, *Social History*, 245–46.

The NEP's values clearly contradicted those of the Bolshevik "moralists", and revealed a fundamental underlying tension within Marxism-Leninism itself. Though flaunting one's wealth could be risky, the new red bourgeoisie and NEPmen unsurprisingly soon acquired the same tastes as the pre-revolutionary elite. To them the foxtrot and tango, caviar and roast-beef, were the order of the day. For the few who could afford them, a trickle of luxuries soon became a flood, and not in foodstuffs alone. The NEP equivalent of the Roaring Twenties brought high-stakes casinos, race tracks, nightclubs, hotels, and brothels.[147] "No Komsomol would think of spending an evening there or be seen" in such places of entertainment.[148] This was the first great age of cinema, and American trends in particular were impossible to suppress. A veritable dance mania broke out in the early 1920s and soon spread even to the countryside, though some party functionaries protested that "democratic" folk dances were preferable. The latest dances required the latest fashions, too, and by 1924 flapper styles were appearing in leading magazines. In the late 1920s, lipstick, earlier condemned but now produced by a government monopoly, was in great demand. Privately produced clothes appeared, and some factory girls "literally starved because they spent all of their wages on silk stockings, makeup, and manicures." Thus "Western films, fashion magazines, and travelling artists and performers helped structure a consumer mentality similar to that of Western Europe and the United States." The demand for novelty which fashion embodied now fundamentally challenged socialist values, which were "organized around stability, fear of change, predictability, and eternity."[149]

This "alternative youth culture" inevitably clashed with official Bolshevik ideals. By the late 1920s the regime began to express anxiety about the morals of the young. A prolonged period of "cultural revolution", in Sheila Fitzpatrick's phrase, followed, defined by militant Communist youth resistance to the old intelligentsia and the new NEPmen. American films and touring jazz bands were caught in the crossfire of class war and banned.[150] (Jazz would be rehabilitated only in 1957.) Severe repression of NEPmen led to their virtual disappearance by 1929, taxed out of existence if not "liquidated" physically, adding a new meaning to the traditional term for going out of business. Now,

147. Ball, *Russia's Last Capitalists*, 41.

148. Maurice Hindus, *Under Moscow Skies* (Victor Gollancz, 1936), 271.

149. Djurdja Bartlett, *Fashion East: The Spectre That Haunted Communism* (MIT Press, 2010), x. The subtitle is telling.

150. Gorsuch, *Youth in Revolutionary Russia*, 131, 137; Sheila Fitzpatrick, ed., *Cultural Revolution in Russia, 1928–1931* (Indiana University Press, 1978), introd., 1.

even in Moscow, many shops were shuttered, their windows broken, their contents confiscated.[151]

But yet again, when supply faltered demand accelerated. As wage differentials in the USSR developed in the 1920s, the new elite had cash to spend, and a black market and system of *blat*, or informal exchange, developed which never entirely disappeared. For the masses, however, many goods were almost permanently in short supply. In 1932–33, admittedly a very difficult period, the average worker's family had access to one pair of shoes per year, two or three shirts' or two dresses' worth of cotton, and one piece of soap. Even in Moscow, industrial workers' daily rations included only 30–40 grams of meat or fish, and a glass of milk per week. Attempts to extort greater output from farmers backfired. In the Ukraine, the Holodomor, Stalin's state-induced and sustained famine, killed millions.[152]

During the 1920s a massive surge in providing consumer goods occurred. The production of cosmetics and perfume expanded greatly, as well as gramophones, bicycles, sewing machines, watches, radios, cameras, and, in the 1930s, the first refrigerators, which most food shops lacked at the time. A consumer crisis in 1929–30 followed the widespread failure of land collectivisation and resulted in shortages of nearly all necessities, and lengthy queues of hours or even days. In 1931 the Party commenced a campaign for "Soviet trade" aimed at "better satisfying the demand of the general public of the cities and countryside in relation to mass consumer goods".[153] If consumerism as such had been condemned as "bourgeois" decadence in the 1920s, the regime now tried to assuage demand and promoted an ideal of "cosiness, beauty and comfort" which formed a kind of middle ground between bourgeois luxury and revolutionary asceticism.[154]

When Stalin famously proclaimed in 1936 that "life has become better, life has become more joyous", this was what people thought, or were supposed to think, he meant: abundance was imminent. Admittedly, there was not much variety in what was produced, only one type of bicycle, for example. When things broke, as they often did, spare parts were scarce or non-existent. But a loose classification based on a distinction between wants and needs had been established, and the validity of the former conceded. The mid 1930s saw the expansion of symbolically important champagne production. This specifically

151. E. M. Newman, *Seeing Russia* (Funk and Wagnalls, 1928), 203.

152. Osokina, *Our Daily Bread*, 91.

153. Amy E. Randall, *The Soviet Dream World of Retail Trade and Consumption* (Palgrave Macmillan, 2008), 23.

154. Gurova, "Ideology of Consumption", 75.

aimed to demonstrate publicly that "the ordinary Soviet worker had access to a standard of living that was earlier restricted to members of the nobility or rich bourgeoisie".[155] It became one of Stalin's "great inventions" symbolising the "good life", which only the churlish might satirise as "champagne socialism".[156] It also indicated that universalising luxury was an implied aim of the system as such.

Providing food and supplying luxury goods were two quite different issues, however. Some official hostility to fashion remained, and to the idea of discarding useful goods simply because fashion deemed them obsolete. Lists of luxury goods were compiled, including things made from silk, velvet, or suede, many furs, imported goods generally, works of art, perfumes, cars, and cosmetics. Chandeliers, silver cutlery, linen cloth, and wine were regarded as inappropriate to a working-class lifestyle. Priority instead was given to goods required to satisfy everyday needs, including ready-made clothing, shoes, furniture, and linens.[157]

Unsurprisingly, once production began to pick up "the public's interest in material goods was immense".[158] Ironically, perhaps, the regime did much to fuel this. After 1935 "Stakhanovites", exceedingly productive "shock workers", were offered luxuries like cars as inducements to labour. They were rarely available otherwise for anything other than official use, and were highly prized as status symbols, despite the great expense of maintaining them.[159] (Leonid Brezhnev's collection included a Rolls Royce, a Mercedes, and a Cadillac.) Workers' congresses also offered costly clothing as incentives. Display was clearly becoming a priority. One woman who earned nine times the average wage vowed to spend her surplus entirely on clothing, including ivory-coloured shoes and a crêpe de Chine dress.[160] To some degree, there was a homogeneous "Soviet" look. The mass production of clothes accompanied a mandate for ethnic groups like the Uzbeks to dress like "cultured" Muscovites, without national costumes, as A. I. Mikoian put it in 1936.[161] Even Trotsky, known as a "popular" communist, suggested that Mikoian's description of Soviet women now desiring "fine perfumes" meant that, as Julie Hessler puts it, "material progress [should] be measured in relation to the consumption of luxury goods."[162]

155. Gronow, *Caviar with Champagne*, 33.

156. Gronow, 17–30.

157. Hilton, *Selling to the Masses*, 223–24.

158. Vihavainen, "Spirit of Consumerism", 4.

159. As late as 1980, only 15 per cent of Soviet citizens owned cars.

160. Bartlett, *Fashion East*, 68.

161. Quoted in Randall, *Soviet Dream World*, 42–43.

162. Quoted in Hessler, *Social History*, 226.

So communism and luxury were not incompatible; to the contrary, a proletarian ideal was now supplanted by a bourgeois one. Now Soviet papers proclaimed: "We endorse beauty, smart clothes, chic coiffures, manicures . . . Girls should be attractive. Perfume and make-up belong to the 'must' of a good Comsomol girl . . . Clean shaving is mandatory for a Comsomol boy." A magazine named *Fashion* appeared. Dancing again became all the rage. Then came a backlash against "promiscuity", and arrests of young women for creating an "immoral appearance" by flaunting the new fashions.[163] The more puritanical and the more libertine aspects of the revolution once again danced together in a dialectic. Critics observed that Soviet Russia now seemed as devoted to "an ever-expanding market for mass production" as the United States.[164] But many recognised that a distinctive shift in emphasis on the regime's part had taken place. As Elena Osokina describes it:

> A new image of a model Soviet citizen emerged. Things that Soviet propaganda had earlier despised as bourgeois luxuries became desirable and even necessary: jewelry, cosmetics, pretty clothes, permanent waves, manicures, and patent-leather shoes. Only a few years earlier a Komsomolka wearing lipstick would have provoked anger and horror and would have been excluded from the Komsomol for moral degradation. Now things changed.[165]

But access to the new commodities was always limited. Champagne was a rarity in the 1930s and remained so until the 1960s. Caviar, perfume, silk stockings, and chocolate were still scarce. Such advertising as existed often portrayed goods, like cheese and sausages, which only the elite consumed. Shortages remained common, including basic necessities like indoor toilets, central heating, telephones, light bulbs, scissors, coat hangers, soap, toothbrushes, salt, and many foodstuffs.[166] In 1950, a peasant had to work sixty days to buy a kilogram of butter, and more than a year to acquire a suit. Queues for goods were often lengthy. A service culture never developed successfully, and customers were as often treated with contempt as solicitude. Privilege alone permitted one to jump the queues and acquire scarce goods from Party or specialist Torgsin luxury shops, where only gold or foreign currency was accepted (but holding either could get you arrested), or from black market suppliers.[167] The obvious existence of a privileged *nomenklatura* who enjoyed a higher

163. Timasheff, *Great Retreat*, 316, 320.

164. Richard Gregg, *The Value of Voluntary Simplicity* (Pendle Hill, 1936), 5.

165. Osokina, *Our Daily Bread*, 133.

166. In 1971 63 per cent of the population lacked running water at home, 75 per cent had no shower or bathroom, and about 90 per cent no telephone: Vihavainen, "Soviet Project", 41.

167. Malcolm Muggeridge, *Winter in Moscow* (Eyre and Spottiswoode, 1934), 150.

standard of living, including, at the top, access to imported foreign luxuries, greatly undermined the system. When people realised that better quality products were available abroad, "communism was doomed."[168] For those aware of the higher standard of living in the west (many were not), the American utopia thus helped undermine its Soviet counterpart.

The retreat from asceticism was largely a Cold War phenomenon. By the 1950s socialist fashion began to emerge, though a taste for all things American, especially cars, music, and dance styles, also intensified. (Snack bars were called *amerikanki*, and ice-cream bars were also regarded as "American".)[169] This was notably evident during the 1957 Festival of Youth and Students, when some thirty thousand foreigners descended on Moscow. Many must have gone home clad like peasants, for a roaring trade developed in their apparel, many of the buyers being Young Communists or *Komsomols*.[170] The resulting gap between promise and reality meant that "the inability to fulfil the needs of consumers would become a major factor in destroying the Soviet regime, perhaps even the chief one."[171]

Though the Party line was still "modesty in personal appearance", by the late 1950s the USSR was beginning to imitate western goods more widely. In June 1959 Christian Dior was invited to present its latest lines in Moscow. Clearly signalling the importance of the issue, Khrushchev even attended a Dior show personally in Paris the following year. It proved very difficult to mass-produce new western styles, however, and prices remained high.[172] In 1961 the USSR claimed that its production and consumption would surpass that of the United States within a decade, and moreover that all commodities would be free, with money no longer being required. Even so, the 22nd Party Congress that year insisted that only "reasonable needs" could be fulfilled, implying that the entire society could not be supplied with the latest fashions. The official line thus still stressed "rational" consumption and a desire to "manage" needs, even while acknowledging constantly growing demand.[173]

168. Randall, *Soviet Dream World*, 1.

169. Gronow, *Caviar with Champagne*, 116.

170. Larissa Zakharova, "How and What to Consume: Patterns of Soviet Clothing Consumption in the 1950s and 1960s", in Vihavainen and Bogdanova, *Communism and Consumerism*, 104–5.

171. Vihavainen and Bogdanova, *Communism and Consumerism*, xi. We would today add the arms race, the Afghan war, and the Chernobyl disaster to this assessment.

172. Bartlett, *Fashion East*, ix.

173. Paulina Bren and Mary Neuburger, introduction to *Communism Unwrapped: Consumption in Cold War Eastern Europe*, eds Bren and Neuburger (Oxford University Press, 2012), 9.

Along with the space race and the arms race the USSR now faced a consumer race, and one moreover fuelled by a burning desire for foreign fashions. In the 1960s, commercial advertising began to increase markedly.[174] By the 1970s, the underground market in western music and clothing was huge, and local imitators when tolerated did a booming business. At this point the struggle against consumerism was essentially lost. As elsewhere, ordinary people imitated the styles of those wealthier than themselves, and foreigners in particular. But the system's failure lay more in its inability to provide basic consumer goods than in any promise of luxury for all. Yet, in the absence of verifiable comparisons, most Soviet citizens in 1976 evidently still believed that their standard of living was higher than that in the United States. In some areas, such as healthcare for workers and the poor, it probably was.[175] Those who had travelled abroad knew how great the gap was in consumer goods. But making invidious comparisons about the USSR invited arrest, and earlier, at least, could have landed one in the Gulag.

Eastern Europe

A somewhat different pattern developed in the Eastern European countries forcibly integrated into the Soviet bloc after 1945. Here shortages of many goods, wartime damage, practical necessity, and then communist norms initially dictated "an austere and simple style of dress".[176] In the north, reverting to a lower standard of living was particularly difficult, for pre-war levels of consumption had been high, and reductions were more obvious than in the poorer south-east. Ideologically, as we have seen, self-restraint was an integral aspect of socialist culture. Practically, consumer demand was soon on the march. At least in East Germany, socialist conceptions of fulfilling workers' needs as opposed to capitalists' profits actually hinted at "a sort of consumer paradise". Again, tension and ambiguity in the struggle between two quite different concepts of the new socialist society was evident. One assumed that "true" needs could be separated from mere "wishes" and "desires" and aspired to integrate cultural development into this concept. The other simply represented a long-awaited plenitude for the working masses, leaving them to define what this meant.[177]

174. Philip Hanson, *Advertising and Socialism* (Macmillan, 1974), 17.

175. Vihavainen, "Soviet Project", 41.

176. Bartlett, *Fashion East*, 1.

177. Judd Stitziel, *Fashioning Socialism: Clothing, Politics, and Consumer Culture in East Germany* (Berg, 2005), 13–14.

In the case of Germany both trends also had pre-war precedents. Under the Nazis, some reaction occurred in the 1930s against the supposed flamboyance and decadence of Weimar clothing, such as against suggestively "masculine" or androgynous styles for women, like wearing trousers in public. Peasant and rural dress also enjoyed renewed attention. As in fascist Italy, some of the potential demand for expanded fashion in clothing was taken up by the increasing prevalence of uniforms.[178] In the Hitler Youth Movement, in which membership was mandatory after 1939, uniform dress, short, military-style hair for boys, and long braids or rolls for girls were prescribed.[179] Nazi women's organisations discouraged cosmetics, alcohol, cigarettes, and conspicuous jewellery. (An exception was made for a small part of the Bund Deutscher Mädel, or League of German Girls, where stylishness was linked to cultivating a eugenically pure Aryan model for advertising purposes.)[180] Some attempt was made to shift "material" consumer demand towards tourism, sport, and other forms of experience. This occurred partly because of ideological opposition to American-style mass consumption and "the awakening of wholly unnecessary 'demands' for 'luxuries'." Hitler declaimed against the "fashion mania" and urged physical training to harden the population.[181] He warned that the nation had "no right to a life of material luxury and ease" unless it recognised the obligations incumbent on racial superiority, which mostly meant learning to enslave and despoil as much of the rest of the world as possible. Some forms of consumption, like smoking, were actively discouraged. But as the production of the Volkswagen, or "people's car", showed, mass demand had to be met, and often was.[182] It proved difficult to ban Hollywood films, though greater success was achieved with American jazz. Food supply was also a tough issue, since Germans proved reluctant to forego delicacies like imported tropical fruits.[183] During the war rationing curtailed demand, and the clothing ration card was described by the government as promoting a "reasonable frugality".[184] But until the end most foodstuffs were still widely available.

In the post-war communist German Democratic Republic, constant comparisons with standards in the west were unavoidable. Housing was scarce, and

178. Eugenia Paulicelli, *Fashion under Fascism: Beyond the Black Shirt* (Berg, 2004), 77.

179. Michael H. Kater, *Hitler Youth* (Harvard University Press, 2004), 28.

180. Kater, 93–95.

181. Adolf Hitler, *Mein Kampf* (Hurst and Blackett, 1939), 345.

182. Shelley Baranowski, *Strength through Joy: Consumerism and Mass Tourism in the Third Reich* (Cambridge University Press, 2004), 38–39.

183. See Hartmut Berghoff, "Enticement and Deprivation: The Regulation of Consumption in Pre-war Nazi Germany", in Daunton and Hilton, *Politics of Consumption*, 165–84.

184. Irene Guenther, *Nazi Chic? Fashioning Women in the Third Reich* (Berg, 2004), 217.

since 80 per cent of intact pre-war industrial plant was removed to the USSR, heavy industry often had priority over consumer goods. Rationing continued; in East Germany meat, milk, fat, sugar, and eggs were restricted until 1958.[185] The need to rearm against the threat of a "fascist" west and the United States provided some ideological justification for austerity. But few endorsed it with any enthusiasm. *Butter statt Kanonen* (Butter instead of cannons) was a slogan in the 1953 East Berlin uprising (*Kanonen statt Butter* had been painted on wartime trains). Short of butter themselves, the Soviets sent tanks instead. Prices on consumer goods were nonetheless quickly lowered.[186] But shortages then increased. Consumption of clothing, textiles, and many semi-luxuries remained below 1936 levels.[187] An insistence that the barrage of propaganda about the superior life of the west was false consciousness, and that "Diese amerikanische Lebensweise ist in Wirklichkeit nichts anderes als das Luxusleben einer kleinen Minderheit auf Kosten der Mehrheit" (the American standard of living is that of a minority based on exploiting the majority) was not always convincing.[188] In East Germany, a quarter of the population, over three million people, left before the Wall was constructed in August 1961. The lure of a higher standard of living was doubtless a key motive in this migration.

Riots were one way to determine consumer demand. But where everyday goods were concerned, Mark Landsman observes, it was difficult to establish exactly "where 'needs' ended and 'taste' began".[189] Amidst rationing and shortages, basic needs were hard enough to satisfy, never mind the aesthetic challenges of beauty, variety, colour, and style. Here too there was idealism, as the quest to construct a new, more egalitarian society inspired the will to rise above "bourgeois" norms. Nonetheless, the desire for luxuries across the next forty years would prove the undoing of socialism, as the gap between East and West widened despite efforts to reduce it. The West German *Wirtschaftswunder* (economic recovery) of the 1950s and 1960s exacerbated differences rapidly, and in the west observers noted, as in eighteenth-century Britain, that the old restraints of class now seemed superseded: "a great restraining threshold has fallen away. Who, now, can say, 'That is not befitting our station?'"[190] Yet again the liberating effects of mass production were evident.

185. Landsman, *Dictatorship and Demand*, 77.

186. Landsman, 118, 124.

187. Landsman, 147.

188. Ina Merkel, *Utopie und Bedürfnis: Die Geschichte der Konsumkultur in der DDR* (Böhlau Verlag, 1999), 152.

189. Landsman, *Dictatorship and Demand*, 96.

190. Landsman, 135.

As in West Germany, consumerism in the East thus denoted a new form of egalitarianism. Beginning in 1958, Berlin Fashion Week showcased new East German styles, albeit interspersed with vigorous lectures aimed at raising political consciousness.[191] In the early 1960s, the East German government inaugurated a new brand of "consumer socialism", which included state advertising agencies, more colourful packaging, modern furniture, magazines, and the like, with a particular emphasis on housing. In 1963, the regime asserted that "the goal of communist production" was the "constant progress of society [and] material and cultural goods for each member of society according to his/her growing needs, individual demands, and inclinations." Within certain limits, it was conceded that this even included various luxuries, a desire for which some now saw as "totally natural". "In fact", writes Judd Stitziel, "official propaganda promised that 'luxury' items that currently only a small proportion of the population could afford would become 'standard' items available to all in the future."[192] By the 1960s, thus, Mary Neuberger observes, "the very legitimacy of socialism was riding on its ability to provide luxury."[193] In the 1970s the Party also came to accept private life as a legitimate sphere, even acknowledging that retreating to small country cottages or garden plots was not a subversive effort to dodge public duties.[194] But here, and elsewhere in the Eastern bloc, the goalposts constantly moved: when food supplies increased, the demand for cars and televisions surged. In such circumstances marketing socialism as an ethos of restraint was well-nigh impossible.

Nonetheless, most luxuries remained in short supply, except for privileged higher Party members, who enjoyed hunting expeditions, villas with swimming pools, and imported foreign goods. Although prices of necessities were often set artificially low, shortages of bread, meat, and milk were common. A few with West German marks had access to the hard-currency Intershops opened in 1973. Here, scarce commodities like sausages, ham, coffee, chocolate, and honey were available.[195] For the rest, the black market was the only alternative to the red market.

Drab scarcity was never really popular or widely embraced as morally superior to capitalist consumerism, and even less since the latter model could occasionally be glimpsed tantalisingly on the horizon, just over the western

191. Stitziel, *Fashioning Socialism*, 74.

192. Stitziel, *Fashioning Socialism*, 15.

193. Mary Neuburger, "Inhaling Luxury: Smoking and Anti-smoking in Socialist Bulgaria, 1947–1989", in Crowley and Reid, *Pleasures in Socialism*, 241.

194. Paul Betts, *Within Walls: Private Life in the German Democratic Republic* (Oxford University Press, 2010), 131, 141.

195. Merkel, *Utopie und Bedürfnis*, 270–71.

borders, through the airwaves or via magazines and films. Since all goods were in short supply, there was no incentive in countries like Poland to create anything elegant.[196] This merely made western goods more attractive. But by the 1960s, throughout the Eastern bloc, a western lifestyle became increasingly desirable. A Hungarian commentator lamented in 1961, "Isn't the mentality of the petty bourgeois being reproduced in this television-automobile-weekend house-motorcycle lifestyle?"[197] In Hungary the chief response by the 1970s was "goulash socialism", which involved a greater concentration on consumer goods. Then came rising expectations for still more: for cars, refrigerators, radios, washing machines, vacuum cleaners, imported clothes, nylons, and electronics. In 1976, a reader of the Czech daily newspaper noted that "people chase after things" in both east and west and saw "no substantial difference" between the two.[198] Indeed, it was "members of the socialist middle classes who gradually turned into a new bourgeoisie" in promoting these styles.[199] They had access to western trends, hard-currency shops, and scarce imports. They were also keen to prove their status, and perhaps just to be different. In East Germany in the 1980s, this induced "an increase in social distinction, as well as in its disturbing visibility."[200] But there were long waits for very expensive goods. A colour television in the German Democratic Republic cost six months' wages for an average worker in 1989.[201] Everyone wanted one anyway.

As ever, clothing played a major role in debates over socialist consumerism. Early on, the new regimes usually officially condemned older bourgeois styles. A Hungarian fashion magazine, for instance, announcing "We are protesting against the waste fashion", described Dior "new look" clothes as "class struggle dresses" suitable only for the idle rich. "Elegance" was rejected in favour of "pleasant, good, smart, tasteful", attributes suitable to the "socialist clothing" of the future. Many past styles were suspect. A Czech fashion commentator insisted that "Clothing must be free from ornamentation originating in a different historical period". Much attention was given to functional clothing suitable to the greater physical strength and proportions of women now engaged in more manual labour. Delicate bodies and evening dress were thus

196. Anne Applebaum, *Iron Curtain: The Crushing of Eastern Europe, 1945–56* (Allen Lane, 2012), 374.

197. Quoted in Crowley and Reid, *Pleasures in Socialism*, 23.

198. Bren and Neuburger, *Communism Unwrapped*, 23.

199. Bartlett, *Fashion East*, 11.

200. Paul Freedman, "Luxury Dining in the Later Years of the German Democratic Republic", in *Becoming East German: Social Structures and Sensibilities after Hitler*, eds Mary Fulbrook and Andrew L. Port (Berghahn, 2013), 181.

201. Crowley and Reid, *Pleasures in Socialism*, 220.

particularly suspect as denoting the moral superiority of idleness. By the 1950s, however, these prejudices were generally in retreat, as western trends simply overwhelmed the "utopian", proletarian, and ascetic stress of the early years. A "back to femininity" campaign in the late 1950s promoted hats, gloves, and a new "feminine" ideal of elegance.[202] Just as American and West European teens began flaunting blue jeans, the traditional dress of workers, communists seemingly embraced high bourgeois fashion.

But communism could not lag far behind western trends. In the east, the proliferation of youth styles defining the new chic and cool from the late 1950s created a market for ever more niche goods like rock music records and blue jeans. Here, too, brand was king. When someone brought a pair of handmade blue jeans to the USSR, where they were very scarce and a great status symbol, he was crushed by the query, "where is the label?"[203] So Latvian hippies made their own bell-bottomed jeans from blue-dyed canvas and then put American labels on them.[204] These jeans were, Daniel Miller writes, "thoroughly semiotic": the label was vastly more important than the thing itself.[205] Miniskirts and brighter colours came in as well. The lure of life in the west, and of participating in a global counterculture in particular, now pervasively accessible through radio, film, and television, made the survival of any distinctive socialist fashion difficult.

As communication with the west became easier, the American utopia now became almost irresistible. Music was central to its appeal. In the 1950s, jazz and chewing gum were all the rage amongst the elite of Soviet youth, swiftly followed by rock and roll. In 1964 commenced a wave of Beatlemania. Leninism proved no match for Lennonism, though a Beatles' album cost half a month's worker's income on the black market and both fans and musicians risked official suppression and police-enforced haircuts.[206] In Moscow the *Stiliagi*, or style hunters, drank cocktails and listened to jazz. In Warsaw their Polish equivalents, the *Bikiniarze*, or Bikini boys, named after the atomic testing ground, smoked Camels, followed jazz, and danced the jitterbug when they could. They were condemned as hooligans by the Party, which insisted that the private life of youth should be "completely subordinated to the interests

202. Bartlett, *Fashion East*, 102.

203. J. Brooks, *Showing Off in America*, 4.

204. Mark Allen Svede, "All You Need Is Lovebeads: Latvia's Hippies Undress for Success", in *Style and Socialism: Modernity and Material Culture in Post-war Eastern Europe*, eds Susan Reid and David Crowley (Berg, 2000), 197.

205. Daniel Miller, *Consumption and Its Consequences* (Polity, 2012), 98.

206. Timothy W. Ryback, *Rock around the Bloc: A History of Rock Music in Eastern Europe and the Soviet Union* (Oxford University Press, 1990), 4–5.

of class warfare".[207] Prague teenagers in the early 1950s actually pinned American cigarette packets to their ties as a style statement.[208] In Bulgaria, cigarette producers retailed their goods with labels containing words from western languages, to make smoking more glamorous. (In the early 1980s half the population were heavy smokers.)[209] Goods labelled as "imported" were regarded as luxuries.

In 1989 it was a shortage of bananas in East Germany which symbolised its distance from West Germany. But other red lines indicated a perceived impossibility of convergence. Rock and roll "dancing apart" was prohibited as early as 1958. In 1959, to combat "capitalist decadence" in the shape of Elvis Presley, the East German regime actually invented its own asexual version of the jitterbug, called the "Lipsi", which was publicly endorsed by Walter Ulbricht, and which flopped completely amongst the young.[210] Repression continued throughout the 1960s, including attempts to regulate lyrics, clothing, and sound levels. But the airwaves could never be completely controlled, and neither ultimately could youth. A 1967 concert by the Rolling Stones in Warsaw provoked a riot. It was a metaphor of the times, for political dissidence and rock music were soon synonymous.

In such circumstances constructing a viable communist alternative to bourgeois consumerism was impossible. A vision of limitless abundance underpinned all modern theories of progress, and a convergence of styles seemed natural. Outside of China, Cambodia, North Korea, and a few other places, communist asceticism invariably had a limited and short-lived appeal. The Khmer Rouge revolution in Cambodia in 1975 induced a complete reaction against modernity, and the rejection for nearly all the population of luxury as such. Luxury goods were regarded as symbols of western imperialism, and piles of refrigerators, televisions, and even furniture were burnt when Phnom Penh was invaded. Nowhere else in the modern period has a reversion to primitivism been embraced so fervently, and with such brutally destructive results.[211] Cuba, in contrast, gambled on public luxury in health provision and social equality and has achieved remarkable advances in literacy and longevity. In the USSR a sense of dutiful austerity by fervent young Party members was often more than

207. David Crowley, "Warsaw's Shops, Stalinism and the Thaw", in *Style and Socialism: Modernity and Material Culture in Post-war Eastern Europe,* eds Crowley and Susan E. Reid (Berg, 2000), 29.

208. Susan E. Reid and David Crowley, eds, *Style and Socialism: Modernity and Material Culture in Post-war Eastern Europe* (Berg, 2000), 15.

209. Neuburger, "Inhaling Luxury", 244, 249–50.

210. Ryback, *Rock around the Bloc,* 29.

211. See my *Dystopia,* 219–35.

compensated for by their devotion to luxury consumption when they became senior members of the privileged elite. Eventually communism capitulated to consumerism and promised, if anything, even more consumer goods than capitalism. Yet many Eastern bloc countries also retained some nostalgia for pre-war, and peasant, ways of life. Some of these ideals would outlast the communist period and even enjoy a revival of sorts. But as the USSR was collapsing; the first McDonald's in Moscow opened in January 1990, marking the symbolic victory of western consumerism over communism. Thousands queued for their first taste of its symbolically exquisite capitalist delights. The intellectual battle against modern luxury had finally been lost for good, or so it seemed.

A Note on China

When Mao Zedong emerged victorious in the 1949 revolution, China was a desperately poor country. Both the Communist Party and the People's Liberation Army had survived for decades only by frugality, and not burdening the peasants with unnecessary demands. Until the reforms begun by Deng Xiaoping in 1978, which wedded a "free" market to communist political control, economic policy had concentrated on locating the rural population in self-sufficient communes, and ensuring adequate food supply for the cities. Outside of a period of large-scale economic catastrophe (1958–62), this ensured a low but sufficient standard of living for most, until the early twenty-first century. Maoism to some degree consciously presented itself as a frugal, Spartan alternative to mainstream Stalinism. Mao urged party members to "always keep to the style of plain living and hard struggle" when the revolution succeeded.[212] The American journalist Edgar Snow, who spent time with Mao in the 1930s, reported that even "After ten years of leadership of the Reds, after hundreds of confiscations of property of landlords, officials, and tax collectors, he owned only his blankets and a few personal belongings, including two cotton uniforms." His only luxury, indeed, seemed to be a mosquito net.[213] Mao took pains to sustain this image. On a visit to the USSR in 1957 he was housed in a palace but insisted on using a chamber pot brought from China, doubtless to show up his Soviet hosts, who had long since reconciled themselves to palatial living. The "Mao suit" became his trademark attire. Maoists elsewhere emulated this asceticism. In Italy they gave up all luxury goods to the Party, including

212. "Always Keep to the Style of Plain Living and Hard Struggle" (1949), in Mao Tsetung, *Selected Works of Mao Tsetung*, 5 vols (Foreign Languages Press, 1954–77), 5: 23.

213. Edgar Snow, *Red Star over China* (1938; Victor Gollancz, 1968), 93, 70.

record players, hairdryers, and toasters, to raise funds for the struggle. In Tanzania, teenagers assailed "Playboy and the Beatles, tight trousers and miniskirts, cosmetics and beauty contests" in Mao's name, and gathered in labour camps in the countryside where they were lectured "on the need for 'dedicated exertion, the suspicion of . . . consumption . . . [and] the virtues of frugality and self-denial.'"[214]

In China all this began changing in the late 1990s. Thereafter the proliferation of a desire for luxury goods became very marked, as a new affluent class emerged to champion the "Chinese Dream". The party rapidly became a privileged elite in the 2000s, and the dramatic expansion of a new middle class was marked by an obsession with luxury goods. But excessive flaunting of the wealth of the nouveau riche has occasioned some official hostility. Under President Xi, for instance, an internet video showing young people surrounded by luxury brands led to several participants being sent for "re-education" courses. During the various phases of rapid industrialisation, China's environmental record, like that of the USSR, was particularly bad.

Twentieth-Century Literary Utopianism: Green Shoots

The bourgeois utopia of indefinite progress for all was never realised, and indeed in principle never could be. In the later twentieth century, however, it seemed epitomised by post-war American prosperity, and alternative utopian literary visions declined markedly. As a utopia, nonetheless, its effects would ultimately prove more destructive than any other. This was not yet understood in environmental terms. As the debate over the consequences of decadence accelerated in the early twentieth century, the capitalist west witnessed some resurrection of older reservations about excessive luxury consumption. With increasing economic and geopolitical competition, bourgeois complacency about the effects of luxury began to be challenged on the old grounds of undermining military capacity. As early as the Boer War (1899–1902) it was evident that the factory system had a similar effect on workers, many of whom lacked the strength and stamina to serve as soldiers. By World War One we begin to see a revival of classical republican civic and military objections to luxury as inducing "softness" and physical weakness, now also often linked to racial competition and imperialism. H. H. Munro's *When William Came* (1914) warns that in the old society

> world-commerce brings great luxury, and luxury brings softness. They had everything to warn them, things happening in their own time and before

214. Julia Lovell, *Maoism: A Global History* (Bodley Head, 2019), 201–2.

> their eyes, and they would not be warned. They had seen, in one generation, the rise of the military and naval power of the Japanese, a brown-skinned race living in some island rice fields in a tropical sea, a people one thought of in connection with paper fans and flowers and pretty tea-gardens, who suddenly marched and sailed into the world's gaze as a Great Power.[215]

With a few years, a neo-Spartan attitude came to be associated with militarism and extreme authoritarianism. Walter Rathenau's *The Days to Come* (1921) noted that one outcome of the "mechanised spirit" was that "Within the brief space of a single war, the Spartan spirit of the armed nation, with all its merits of self-sacrifice and love of honour, has been diffused throughout the country."[216] The "Spartan precedent" was still to the fore in Alexander Moszkowski's *The Isles of Wisdom* (1924), where it was linked to ending "the thraldom of yesterday's fashion".[217] In Stefan Żeromski's *The Coming Spring* (1924), the narrator says "In the war I learned to live a Spartan life. I have no special needs", and laments that "Our money, our costly and comfortable furniture, our expensive plates and the fine meals served upon them were seasoned and thoroughly saturated with wrongs done to others."[218] By the mid-1930s, Sparta was increasingly linked to Nazi ideals in works like Storm Jameson's *In the Second Year* (1936). Here the English fascist leader Frank Hillier proclaims, "We shall become a self-sufficing and self-contained people again . . . The world was going rotten with greed and looseness, everyone trying to become rich, and cheating and lying, as though money and trade were the be-all and end-all of life on earth." He adds:

> It had to stop. We are stopping it. We have retreated into ourselves. We intend to have no debts to other nations and to rely only on ourselves. If it means that we are poor for a time, so much the better. We shall be all the sweeter for it. Some of the muck and rottenness will drop off. Sparta, not Athens, is going to be our model in the future.

And this meant—bearing in mind this is a satire—Little Englandism with a vengeance: "England for the English. No more foreigners allowed, except as envious visitors. No French, Boches, Eyetalians, or Scythians. The women shall spin English wool and the men wear it, and they shall eat English mutton

215. H. H. Munro, *When William Came: A Story of London under the Hohenzollerns* (Bodley Head, 1914), 200.

216. Walther Rathenau, *In Days to Come* (George Allen and Unwin, 1921), 143, 32.

217. Alexander Moszkowski, *The Isles of Wisdom* (George Routledge and Sons, 1924), 126.

218. Stefan Żeromski, *The Coming Spring* (1924; Central European University Press, 2007), 161, 52.

and cabbage, and keep early hours."[219] The parallels with present-day populist nationalism are uncomfortably close, if instructive.

Support for a simpler life was also reinforced by the psychological malaise of much of the inter-war period, and the realisation that modernity had failed to live up to its promises, and instead induced a spiritual emptiness which undermined its material affluence. Bourgeois opulence came to be seen increasingly as shallow, superficial, and unsatisfying. Sir Philip Gibbs's *The Day after To-morrow* (1928) hoped that "The simplification of life with less needs and less luxuries may bring back happiness which seems to have fled from many centres of our present civilisation."[220] The United States was often associated with vulgar hedonism. Orwell's 1939 portrait of a new-fangled milk bar is resonant:

> Everything slick and shiny and streamlined; mirrors, enamel, and chromium plate whichever direction you look in. Everything spent on the decorations and nothing on the food. No real food at all. Just lists of stuff with American names, sort of phantom stuff that you can't taste and can hardly believe in the existence of. A sort of propaganda floating round, mixed up with the noise of the radio, to the effect that food doesn't matter, comfort doesn't matter, nothing matters except slickness and shininess and streamlining.[221]

The futurological French writer Georges Duhamel anticipated many of Orwell's objections to American influence in complaining that:

> I belong to a community of peasants who for centuries have lovingly cultivated fifty different varieties of plum, and who find in each a taste deliciously unlike that of any of the others.
>
> Well, no one in America concerns himself about such delicate riches. The beings who today people the American ant-hills have no desire for these unsubstantial viands. They demand palpable, incontestable wealth, recommended, or, preferably, prescribed, by the national divinities. They yearn desperately for phonographs, radios, illustrated magazines, "movies", elevators, electric refrigerators, and automobiles, automobiles, and, once again, automobiles. They want to own at the earliest possible moment all the articles mentioned, which are so wonderfully convenient, and of which, by an odd reversal of things, they immediately become the anxious slaves.

219. Storm Jameson, *In the Second Year* (Cassell, 1936), 40–41.

220. Sir Philip Gibbs, *The Day after To-morrow* (Hutchinson, 1928), 130.

221. Orwell, *Complete Novels*, 442–43.

He added that "The whole philosophy of this industrial dictatorship leads to this unrighteous scheme: to impose appetites and needs on man."[222]

Aldous Huxley

Utopian and dystopian satires on consumerism also began to proliferate in the early twentieth century. The most influential of these critics was Aldous Huxley (1894–1963), who recognised how all-powerful advertising had become and lambasted it mercilessly in *Brave New World* (1932).[223] Here, in A. F. (After Ford) 632 (2495 CE), the norm is enforced consumption in order to guarantee industrial output and solve the problem of overproduction and underconsumption, which Huxley saw as endemic to capitalism. Huxley described his politics in 1926 as "Fabian and mildly labourite".[224] In 1932 he called "the deliberate planning of our social life in all its aspects" "an ideal which must, it seems to me, appeal to every reasonable man."[225] *Brave New World*'s famous phrase, "ending is better than mending", encapsulated the early Depression mentality, which suggested that maximum stimulation to consumption was needed to provide markets for goods.[226] The human costs of mass consumerism worried him constantly. Huxley was fascinated by the prevalence of mass advertising in America, which he first visited in 1926. These techniques are freely and centrally adapted in his hedonistic utopian satire, where we find many echoes of the "superlatively Good Time of modern California", and especially the Rabelaisian "Joy City", Los Angeles. In America, he found, "Most things in this modern land are provisional, made to last only till something better, or at any rate something newer, shall appear to take their place."[227]

222. Georges Duhamel, *America the Menace: Scenes from the Life of the Future* (George Allen and Unwin, 1931), 201–2, 200.

223. Huxley also developed these themes in *Now More than Ever*, a play written in the same period (eds David Bradshaw and James Sexton (University of Texas Press, 2000)).

224. Aldous Huxley, *Jesting Pilate: The Diary of a Journey* (Chatto and Windus, 1930), 259.

225. A. Huxley, *Hearst Essays*, 116.

226. A key source was Fred Henderson's *The Economic Consequences of Power Production* (George Allen and Unwin, 1931), which argued that while inequality meant that only a fraction of productive power could be used and the masses remained at "the poverty level" (10), an increase in "effective demand" could increase production tenfold (67).

227. A. Huxley, *Jesting Pilate*, 268, 281. He added: "How Rabelais would have adored it! For a week, at any rate. After that, I am afraid, he would have begun to miss the conversation and the learning, which serve in his Abbey of Thelema as the accompaniment and justification of pleasure" (269).

Brave New World shows how a fully planned society might fuel mass consumption. "In the nurseries" of the new society, we are informed, when

> the Elementary Class Consciousness lesson was over, the voices were adapting future demand to future industrial supply. "I do love flying," they whispered. "I do love flying, I do love having new clothes, I do love . . . But old clothes are beastly," continued the untiring whisper. "We always throw away old clothes. Ending is better than mending, ending is better than mending, ending is better . . ."[228]

The refrain is all too familiar today. So, too, we are told that "The more stitches, the less riches": why repair, which does not bolster employment, when you can buy new clothes, which does? In the future world, thus, "Every man, woman and child [is] compelled to consume so much a year. In the interests of industry", provoking one rebellious character to say, "Anything not to consume. Back to nature." The new world is defined by a mentality hostile to nature, solitude, and self-reflection. Constant titillation, mechanical distraction, and mind-numbing repetition are the norm. A key character in the novel, Bernard Marx, who has many qualms about the system, takes his lover Lenina on a flight over the ocean, and asks her to look down at the scenery. Her response:

> "But it's horrible," said Lenina, shrinking back from the window. She was appalled by the rushing emptiness of the night, by the black foam-flecked water heaving beneath them, by the pale face of the moon, so haggard and distracted among the hastening clouds. "Let's turn on the radio. Quick!" She reached for the dialling knob on the dashboard and turned it at random.
>
> ". . . skies are blue inside of you," sang sixteen tremoloing falsettos, "the weather's always . . ."[229]

It then becomes clear that one of the leading movements during the Nine Years' War, when the choice was between "World Control and destruction", was a "Simple Life" movement, for eight hundred "Simple Lifers" are described as being "mowed down by machine-guns at Golders Green."[230] So we might surmise that it is the descendants of Thoreau and Carpenter who have been excluded from this new world, and who stand for what Huxley himself sympathised with.[231]

228. Aldous Huxley, *Brave New World* (1932; Vintage Books, 2007), 41–42. Unless otherwise noted, subsequent references are to this edition.

229. A. Huxley, 78.

230. A. Huxley, 42–43.

231. Isaiah Berlin mentions Carpenter as amongst the influences on Huxley's generation: in Julian Huxley, ed. *Aldous Huxley, 1894–1963: A Memorial Volume* (Chatto and Windus, 1965), 144.

Yet Huxley was no "simple lifer" either. In the novel it is the Savage Reservation and the character of John the "Savage" who represent this retrograde, now defeated tendency. But in 1927 Huxley wrote that:

> I am old-fashioned enough to believe in higher and lower things, and can see no point in material progress except in so far as it subserves thought. I like labor-saving devices, because they economize time and energy which may be devoted to mental labor . . . Discomfort handicaps thought; it is difficult when the body is cold and aching to use the mind. Comfort is a means to an end.[232]

Nonetheless, Huxley was sceptical about the idea of progress, writing in 1928 that

> the colossal material expansion of recent years is destined, in all probability, to be a temporary and transient phenomenon. We are rich because we are living on our capital. The coal, the oil, the niter, the phosphates which we are so recklessly using can never be replaced. When the supplies are exhausted, men will have to do without. Our prosperity has been achieved at the expense of our children.[233]

He added in 1932 that

> Tolstoyans and Gandhi-ites tell us that we must "return to Nature"—in other words, abandon science altogether and live like primitives or, at best, in the style of our mediaeval ancestors. The trouble with this advice is that it cannot be followed—or rather, that it can only be followed if we are prepared to sacrifice at least eight or nine hundred million human lives.[234]

With the passage of time, Huxley became increasingly pessimistic about the possibility of humanity surviving modernity. In 1940 he wrote that "I see no hope except in a reversal of existing trends and a deliberate return to a more decentralized form of society with a wider distribution of land and other property. But the probability of such reversal taking place seems almost infinitely small."[235] In 1941 he worried that "Most political idealists have no doubt at all; liquidate the people who don't agree with you, and you will have Utopia."[236] In 1949 he condemned the progress of capitalism as meaning that "Today every

232. "Comfort", in Aldous Huxley, *Proper Studies* (Chatto and Windus, 1927), 298–99.

233. "Progress", in Aldous Huxley, *Complete Essays*, 6 vols (Ivan R. Dee, 2000–2002), 2: 293.

234. Aldous Huxley, *The Hidden Huxley: Contempt and Compassion for the Masses*, ed. David Bradshaw (Faber and Faber, 1994), 106. The world's population then was around 2 billion.

235. Aldous Huxley, *Letters of Aldous Huxley* (Chatto and Windus, 1969), 451.

236. Aldous Huxley, *Grey Eminence: A Study in Religion and Politics* (Chatto and Windus, 1941), 113.

efficient office, every up-to-date factory is a panoptical prison, in which the worker suffers . . . from the consciousness of being inside a machine."[237] In 1958, *Brave New World Revisited* stated that "The prophecies made in 1931 are coming true much sooner than I thought they would", and that the "nightmare of total organization, which I had situated in the seventh century After Ford, has emerged from the safe, remote future and is now awaiting us, just around the next corner".[238]

The most powerful mechanisms assuring this destination were, as in 1931, the techniques of mass manipulation associated with advertising. The mass media, Huxley thought, were controlled by the Power Elite, which was "not something of which a Jeffersonian democrat could possibly approve." One of this elite's weapons was simply distracting the population by constant exposure to "the irrelevant other worlds of sport and soap opera, of mythology and metaphysical fantasy".[239] Another was consumerism, which required "the services of expert salesmen versed in all the arts (including the more insidious arts) of persuasion". Given "the mass producer's chronically desperate need for mass consumption", there were no limits to its exercise. Huxley understood well how symbols were manipulated, and how far "Simple-minded people tend to equate the symbol with what it stands for, to attribute to things and events some of the qualities expressed by the words in terms of which the propagandist has chosen, for his own purposes, to talk about them."[240] Thus, he claimed, quoting a cosmetics manufacturer, that its purveyors "are not selling lanolin, they are selling hope." So too "'We no longer buy oranges, we buy vitality. We do not buy just an auto, we buy prestige' . . . And so with all the rest. In toothpaste, for example, we buy, not a mere cleanser and antiseptic, but release from the fear of being sexually repulsive." For all this Huxley had boundless contempt. But he was willing to go so far as to advocate "the decentralization of economic power and the widespread distribution of property", and to applaud "a new kind of community living on the village and small-town level" of the type proposed by the Bellamyite and New Deal engineer Arthur E. Morgan during World War Two, in the belief that "life in a huge modern city is anonymous, atomic, less than fully human".[241]

It is sometimes assumed that Huxley's final work, *Island* (1962), offered a utopian antidote to *Brave New World*'s anxiety-ridden modernity. Here the

237. A. Huxley, *Prisons*, 15.

238. Aldous Huxley, *Brave New World Revisited* (Harper and Row, 1958), 1, 5.

239. A. Huxley, 35, 37.

240. A. Huxley, 48, 50.

241. A. Huxley, 54, 114–15.

relatively primitive, peaceful, and co-operative island paradise Pala appears as a respite from the idea of progress devouring the rest of humanity. Mass consumerism is a key target, and the "Tree of Consumer Goods" represents a kind of original sin. The Palans are too preoccupied with sex to have time for excessive consumption: they value human contact above all else, and save the world the fun way.[242] But to David Bradshaw *Island* is in fact "perhaps Huxley's most pessimistic book", for Pala ultimately succumbs to its enemies.[243] It does seemingly embody the values Huxley cherished most in 1946:

> economics would be decentralist and Henry-Georgian, politics Kropotkinesque and cooperative. Science and technology would be used as though, like the Sabbath, they had been made for man, not (as at present and still more so in the Brave New World) as though man were to be adapted and enslaved to them.[244]

Pala reverses the poles of the Soviet formula. Its slogan is: "Electricity minus heavy industry plus birth control equals democracy and plenty. Electricity plus heavy industry minus birth control equals misery, totalitarianism and war."[245] The problem was that all this was more unachievable than ever. Yet Huxley, initially dismissive of Gandhi, came to see him as "not only an idealist and a man of principle, but also an intensely practical politician", and persisted in thinking that non-violent resistance to tyranny might yet effect change.[246] The popularity of Huxley's later vision was nowhere more evident than in the cult-like reception achieved by E. F. Schumacher's *Small is Beautiful* (1973), which championed a "Buddhist economics" of "right livelihood" as a "Middle Way between materialist heedlessness and traditionalist immobility".[247]

The revolt against consumerism of the later twentieth century took place in the west rather than the east. As we have seen, proto-ecological ideas appeared in utopian form in the late nineteenth century, with Morris in particular. After World War Two, most utopias acknowledge the need to confront waste.

242. Aldous Huxley, *Island* (Penguin Books, 1964), 139–40.

243. David Bradshaw, in Aldous Huxley, *Island*, ed. Bradshaw (Flamingo Books, 1994), xvi.

244. Aldous Huxley, *Brave New World* (Penguin Books, 1955), 8.

245. A. Huxley, *Island* (Penguin), 150.

246. Aldous Huxley, *Science, Liberty and Peace* (Chatto and Windus, 1950), 7–8. Huxley even considered writing the script for a film of Gandhi's life (*Letters*, 638).

247. E. F. Schumacher, *Small Is Beautiful: A Study of Economics as if People Mattered* (Abacus, 1974), 51.

B. F. Skinner's *Walden Two* (1948), for instance, specifically targets "conspicuous consumption". The utopian community projected here aims "to avoid the waste which is imposed by changing styles, but we don't want to be wholly out of fashion. So we simply change styles more slowly, just slowly enough so we needn't throw away clothing which is still in good condition." At the relevant point, people

> simply chose the kind of clothes which suffer the slowest change—suits, sweaters and skirts, or blouses and skirts, and so on. You won't find half a dozen "party dresses" among us—and those aren't from the community supply. Yet each of us has something that would be in good taste except at very formal functions.[248]

In Orwell's great dystopia *Nineteen Eighty-Four* (1949), however, consumerism, which Orwell loathed, is recognised as having a revolutionary function. Here, the state is described as engaging in constant war in order "to use up the products of the machine without raising the general standard of living." This was because:

> an all-round increase in wealth threatened the destruction—indeed, in some sense was the destruction—of a hierarchical society. In a world in which everyone worked short hours, had enough to eat, lived in a house with a bathroom and a refrigerator, and possessed a motor-car or even an aeroplane, the most obvious and perhaps the most important form of inequality would already have disappeared . . . if leisure and security were enjoyed by all alike, the great mass of human beings who are normally stupefied by poverty would become literate and would learn to think for themselves; and when once they had done this, they would sooner or later realize that the privileged minority had no function, and they would sweep it away.[249]

Ernest Callenbach

The most impressive utopian literary confrontation with environmentalism in the later twentieth century came with *Ecotopia* (1975) by Ernest Callenbach (1929–2012). This portrays a secessionist state in the US north-west whose capital is San Francisco. It is organised along explicitly environmentalist lines. International air travel is banned "on grounds of air and noise pollution", but an efficient rail system and electric cars exist, with a far lower environmental

248. B. F. Skinner, *Walden Two* (1948; Macmillan, 1962), 35.

249. Orwell, *Complete Novels*, 855.

cost, and all public transport is free. Recycling is mandatory, and streets are smaller and lined by many thousands of trees. Cities have been redesigned to reduce their size and improve their convenience. Workers co-own and co-manage enterprises. Ethnic minorities manage their own city-states. Energy is conserved wherever possible. Production and consumption are organised around the principle of a "stable-state ecological system". Pesticides have been eliminated. There is universal healthcare. The working week has been reduced to twenty hours, and those made unemployed by the disruption of the transition period have found new positions in the expanding transport system and elsewhere. The population is slowly declining. The Ecotopians' costumes are somewhat flamboyant, like the hippies', but clothing is expensive, being made locally and from recyclable materials. So synthetic fibres are out, as are many forms of glass and pottery, while plastics degrade after only a month's exposure to sunlight. The variety of goods available has been reduced "to curb industrial proliferation", while many basic necessities are standardised. The general attitude is that the sacrifice of present consumption for future survival is justified in principle.[250]

With the new ecological outlook come new manners. The Ecotopians are very active physically, walking and cycling wherever possible. They have a touchy-feely, Californian sociability and hug and hold hands regularly. Indeed "The propensity of Ecotopians to touch one another is remarkable", partly because they use marijuana frequently. They are free and open sexually, and "family" is a fluid social concept. Women play a much larger role than in the earlier society. Ecotopians criticise shortcomings in restaurant food or service. We are told that they have "lost the sense of anonymity which enables us to live together in large numbers". This brings a much greater sense of familiarity and egalitarianism. They expect this as a right and demand it when necessary. They refuse to be treated as servants, or cogs in the machine. Thus, on buying a train ticket, the narrator discovers that:

> The Ecotopian at the train ticket window simply wouldn't tolerate being spoken to in my usual way—he asked me what I thought he was, a ticket-dispensing machine? In fact, he won't give you the ticket unless you deal with him as a real person, and he insists on dealing with you—asking questions, making remarks to which he expects a sincere reaction, and shouting if he doesn't get it.[251]

250. Ernest Callenbach, *Ecotopia: A Novel about Ecology, People and Politics in 1999* (Pluto, 1978), 61, 10.

251. Callenbach, 10. Callenbach has also been accused of ecofascism, however. See David Pepper, *The Roots of Modern Environmentalism* (Routledge, 1984), 206–7.

(Shouting does seem an odd response amongst such a laid-back people.)

In a sequel, Callenbach outlined the chief principles on which a transition should hinge. These were:

> *No extinction of other species.*
> *No nuclear weapons or nuclear plants.*
> *No manufacturing of carcinogenic (cancer-causing) or mutagenic (mutation-causing) substances.*
> *No adulterants in foods.*
> *No discrimination by reason of sex, race, age, religion, or ethnic origin.*
> *No private cars.*
> *No advertiser-controlled or broadcast television.*
> *No limited-liability corporations.*
> *No absentee ownership or control—one employee, one vote.*
> *No growth in population.*[252]

Many variations on these themes have appeared since the 1970s, often intermingled with science fiction, which makes their strict demarcation from utopias difficult, as well as with dystopian narratives. Writers of note in this period include Ursula K. Le Guin, Marge Piercy, and Kim Stanley Robinson. A number of women writers have also developed themes related to simplicity or a reversion to more primitive societies in feminist directions.[253] The all-female world of Sally Miller Gearhart's *The Wanderground* (1979), where "simplicity" is the creed of the "hill women" who live in modest dwellings sharing their property, is a good case in point.[254] As Lisa Garforth notes, however, "there have been few full-blown green utopian novels" since the early 1990s.[255] The magnitude of the problem has helped to render both the small-scale solutions to environmental issues and the science-fiction orientation of Piercy and Le Guin less relevant now than earlier, though the later works of Margaret Atwood are less far-fetched. By the early 2000s, a subgenre now known as climate fiction, or cli-fi, had emerged in which many possible scenarios were played out. But most future-oriented fiction from here on is anxiously and

252. Ernest Callenbach, *Ecotopia Emerging* (Banyan Tree Books, 1981), 34–35.

253. A good overview of these trends is Nan Bowman Albinski's *Women's Utopias*.

254. Sally Miller Gearhart, *The Wanderground: Stories of the Hill Women* (1979; Women's Press, 1985), 2.

255. Lisa Garforth, *Green Utopias: Environmental Hope before and after Nature* (Polity, 2018), 94.

consciously apocalyptic rather than utopian, and we can speak of a "dystopian turn" occurring from the 1990s onwards. Floods, fires, and pandemics often figure in the newer titles.[256]

Post-apocalyptic fiction by definition is generally, thus, dystopian rather than utopian, even if the reader's hope of being amongst the few intrepid survivors is encouraged.[257] Rather than confronting the prospect of a worldwide catastrophic environmental breakdown, and analysing its causes in detail, the disaster is often simply taken to be a fait accompli. Grappling with the causes seems beyond our imagination. Even more so on screen, bleak scenarios like that of the *Hunger Games* are typical, and the need to provide lucrative visual entertainment discourages practical engagement with real-world issues. Few such works offer a realistic assessment of what we today regard as a plausible breakdown scenario, or explanations for their systematic causes. Few address the transition problem as such—how we might get to utopia—or appreciate the imminent nature of the coming catastrophe. (Some more-practical utopian schemes are the exception here.)[258] Even fewer attack the wealthy, or consumerism, or the American way of life. Nor do the many computer games which feature post-apocalyptic narratives.[259] In offering a detailed model of how a sustainable society might work, the Ecotopian model, we will see, can still serve as a point of departure insofar as hints at detailed solutions are provided.

Amongst the more recent works to take up this challenge is Kim Stanley Robinson's thoughtful and engaging near-future dystopia/utopia *The Ministry of the Future* (2020), a graphic portrayal of a much hotter world in the 2030s, when CO_2 is at 447 ppm (it rises to 475 before beginning to decline) and the Arctic ice has gone in 2032. Here people steal air conditioners at gunpoint, and millions die in heatwaves. A clear narrative of failed efforts to deal with warming from the Paris Agreement onwards is offered, with the United Nations'

256. See, e.g., Clare Morrall, *When the Floods Came* (Hodder and Stoughton, 2016); Stephen Baxter, *Flood* (Gollancz, 2008); Marcus Sedgwick, *Floodland* (Orion, 2001). An excellent pandemic-centred dystopia is Emily St. John Mandel's *Station Eleven* (Picador, 2014). See, generally, my *Dystopia*, chap. 8.

257. For an updated version of these texts, see the online database version of Lyman Tower Sargent's *British and American Utopian Literature, 1516–1985* (Garland, 1988): Sargent, *Utopian Literature in English: An Annotated Bibliography from 1516 to the Present* (database; Penn State Libraries Open Publishing, 2016 and continuing), doi:10.18113/P8WC77.

258. See, e.g., Karin Bradley and Johan Hedrén, eds, *Green Utopianism: Perspectives, Politics and Micro-practices* (Routledge, 2014). An overview of recent developments is offered in Mark Stephen Jendrysik's *Utopia* (Polity, 2020, 108–16).

259. That is, they stretch the bounds of our credibility unduly, following my distinction in *Dystopia* (284–90). See, generally, Lisa Garforth's *Green Utopias* (72–95).

Ministry for the Future being the body set up in 2025 to address failures of implementation, with a pathetically small budget of $60 billion a year. Oil corporations are named and shamed. When serious environmental change begins, social change is led by India, which remedies the effects of the caste system and defeats Hindu nationalism, as part of "the Great Turn". But the wealthier countries take care of themselves first, ignoring pleas for greater equality, which Robinson defends, though since he asserts that "morality is a question of ideology" and laments that "We have never been rational", a more robust attack on capitalism is ruled out.[260] There is little "science fiction" here, but plenty of science; various geo-engineering strategies are discussed, amid much near-future projection of current trends in 2020. Hints at a new "Earth religion" where we "are all family" are floated.[261] Mondragon-style cooperation expands, and is touted. The use of violence and sabotage to attain ecological ends, notably by attacking fossil-fuel burning aircraft, container ships, and power plants, and killing cows, is discussed, and the potential efficacy of such methods suggested. Davos is hijacked by eco-propagandists. Socialism is mentioned, along with Green-New-Deal-style plans, and the resistance of the wealthy to reform. Personal annual income is capped at ten times universal basic income. "Red Plenty", central planning, and progressive taxation are approved, as is nationalisation of water resources. The suggestion that we might retreat from technology is rejected, though other options are not excluded, since the "orienting principle" here is that "we should be doing everything needed to avoid a mass extinction event".[262] Robinson does not fight shy of some difficulties, however: a hundred million climate refugees; bankers dominated by self-interest; corporations demanding compensation; humanity's general lethargy and inability to understand, much less confront, the problem; police and political repression of the movement. His rescue scenario is greatly helped by some fortuitous coincidences and a fair amount of sheer good luck. Saudi Arabia's renunciation of oil production for burning, an environmental revolt in Brazil, the decline of the Russian oligarchs, a popular uprising in China, and student strikes play an important role, as does a massive depression. Schemes for the prevention of further glacial melting are successful. Animal populations return. Climate migration is solved by creating global citizenship, enabling refugees to live anywhere. There is much that is intended to be plausible here. To Robinson the most important goal is gaining control over the world's financial institutions, and an ingenious plan is offered for doing so.

260. Kim Stanley Robinson, *The Ministry for the Future* (Orbit Books, 2020), 74, 88.

261. K. Robinson, 539.

262. K. Robinson, 166.

Some major problems, like the possibility of a large-scale fascist resistance by the wealthy, nuclear war, and endemic racism, are largely avoided here, and the success of eco-terrorism remains unexplained. Overpopulation is dealt with coyly; we are told that the population is dropping, but not why, and some doubt is suggested as to whether this is a good thing, though an "optimum number" of two to four billion humans is discussed. COP (Conference of the Parties) UN climate-change meetings are described as ignoring the issue as "too hot to handle". So we are given no clue as to how to get there, though the rewilding projects described, and all the gains in energy conservation, are contingent on resolving it. Consumerism as such is not addressed.[263] Robinson shies away from portraying his projected end state as a "utopia". It is a moot point what the literary form of his argument contributes to the social theory. But he does not permit the fictional genre to detract from his confrontation with real problems, and this is a transition scheme which merits close study and discussion. Much of the programme it suggests is discussed at greater length below.

Such transformations, however, require a cultural basis, and a shift of feelings as well as intellect. The counterculture of the 1960s, which we must now briefly examine, provides one way forward for thinking how this might happen.

263. K. Robinson, 502, 477. The word "utopian" is mentioned only twice in the book: in relation to gatherings which are "spaces of hope" and with respect to Charles Fourier.

10

Counterculture and Consumerism

THE 1960S

Prelude

From the 1930s onwards, utopianism seemingly began to pass out of fashion. Critics began identifying Stalin's failings with socialism as such, into which most modern utopian enthusiasm had been invested. They now depicted it as a self-defeating search for perfection. To Ludwig von Mises, writing in 1936, socialism was obliged

> to construct for its Utopia a type of human being totally different from the race which now walks the earth, one to whom labour is not toil and pain, but joy and pleasure. Because such a calculus is out of the question, the Utopian socialist is obliged to make demands on men which are diametrically opposed to nature.

Socialism had only begun to turn towards asceticism, he thought, to excuse its low economic productivity.[1] In 1944, Friedrich von Hayek insisted that "the Road to Freedom" promised by the "great utopia", socialism, was actually "the High Road to Servitude". All attempts at central planning eventuated in totalitarianism, and Stalin's was worse than Hitler's.[2] In 1945, Karl Popper contrasted "piecemeal social engineering" to "utopian social engineering", and described the latter's pursuit of perfection as inevitably producing "an intolerable increase in human suffering" and the "strong centralized rule of a few", which was "likely to lead to a dictatorship."[3]

Ignoring the existence of two communist regimes, the USSR and China; the success of Roosevelt's New Deal; the election of a post-war Labour government

1. Mises, *Socialism*, 452.

2. Friedrich Hayek, *The Road to Serfdom* (University of Chicago Press, 1945), 27.

3. Popper, *Open Society*, 1: 22, 160.

in Britain, which created the National Health Service; the widespread success of the trade union movement in raising workers' wages; and the triumph of social democracy and the social market economy in many countries elsewhere, the refrain was often heard by the mid-1950s that utopia was dead—meaning dead amongst western intellectuals. Liberal and conservative thinkers alike celebrated the supposed triumph of capitalist democracy over communism. Judith Shklar's *After Utopia: The Decline of Political Faith* (1957) declared belief in a "messianic age", the "classless society", or any other version of "life after death" impossible. Such false hopes were steeply discounted—the market was glutted. What remained was mediocre organisation at best.[4] (An essay from the 1980s, "What Is the Use of Utopia?", called this a premature judgment.)[5] In 1958, Ralph Dahrendorf described utopias as "societies from which change is absent", and promoted "a universal consensus on prevailing values and institutional arrangements", including a recognition of the value of pluralism.[6] Daniel Bell's *The End of Ideology* (1960) celebrated "the exhaustion of utopia", adding that "Few serious minds believe any longer that one can set down 'blueprints' and through 'social engineering' bring about a new utopia of social harmony."[7] (And, smarting from Marx's objection to "blueprints", many on the left have agreed.) Isaiah Berlin conceded that "Utopias have their value—nothing so wonderfully expands the imaginative horizons of human potentialities". But he added that

> as guides to conduct they can prove literally fatal . . . the very notion of a final solution is not only impracticable but, if I am right, and some values cannot but clash, incoherent also . . . For if one really believes that such a solution is possible, then surely no cost would be too high to obtain it: to make mankind just and happy and creative and harmonious for ever—what could be too high a price to pay for that?

In "The Decline of Utopian Ideas in the West" (1978) Berlin insisted that all utopias erred fatally in endorsing "a static perfection in which human nature is finally fully realised, and all is still and immutable and eternal".[8] Conservatives like Michael Oakeshott similarly identified utopia with the "politics of perfection", from which, curiously instancing Godwin, "springs the politics of uniformity; a scheme which does not recognize circumstance can have no place for variety."[9]

4. Shklar, *After Utopia*, 111, 149.

5. Shklar, *Political Thoughts*, 175–90 (essay undated otherwise).

6. Ralph Dahrendorf, *Essays in the Theory of Society* (Routledge and Kegan Paul, 1968), 107.

7. Daniel Bell, *The End of Ideology: On the Exhaustion of Political Ideas in the Fifties* (Free Press, 1960), 16, 373.

8. Isaiah Berlin, *The Crooked Timber of Humanity* (1959; John Murray, 1990), 14–15, 21.

9. Michael Oakeshott, *Rationalism in Politics and Other Essays* (1962; Liberty, 1991), 10.

As we saw above, such criticisms are misplaced insofar as they depict utopianism as a whole. They evidence a confusion between millenarianism and utopia, and are accurate only insofar as this is also shared with some utopian thinkers. To its critics, however, utopia had become a dirty word, a symbol of hubris, synonymous with impossible perfectionism and an unwillingness to concede human fallibility. This limited society's horizons to technocratic tinkering, and extending the American way of life to a grateful humanity. Utopia and perfectionism were inexorably wedded to violence, extreme collectivism, and an all-powerful state. If the utopian baby went out with the totalitarian bathwater, so much the better.

Such assessments turned out to be short-sighted and unimaginative. In fact, the quest for utopia, having faltered temporarily in the west, was soon vigorously renewed. Critiques of capitalist plutocracy from a social democratic perspective gave rise to the Scandinavian utopia, and as post-war Europe recovered, viable regimes which included substantial welfare systems and large state sectors emerged, without risk of totalitarianism. Despite its prosperity, the American utopia was itself beginning to show cracks by the later 1950s. A new-found sense of malaise regarding the vapidity of popular consumer culture accompanied mounting fears of the growing power of large, machine-like corporations. In the late 1940s, existentialists probed the spiritual void left when meaning was dissolved into materialistic hedonism. Many Americans had found redemption in the "promised land", even though it was already occupied, and saw this as "progress". But by the late 1950s others lamented that abundance seemed overrated, dull, and shallow. There now arose a new spiritual rebellion, pressing its claims in the name of "liberation" and "freedom". The spirit of Rousseau and romantic authenticity burst forth once again. The term "counterculture" was coined by 1951 to herald a new way of life.

The Counterculture: A New Model of Sociability

The youth counterculture of the 1960s marks what Frank Musgrove calls "a major discontinuity in the history of our times".[10] A generational rupture occurred, with consciousness of an age-defined cohort uniting a new oppositional

10. Frank Musgrove, *Ecstasy and Holiness: The Counter Culture and the Open Society* (Methuen, 1974), 20. A good general introduction to the subject is J. Milton Yinger's *Countercultures: The Promise and Peril of a World Turned Upside Down* (Free Press, 1982).

group. For a time, at least, some thought this trumped identities of race, gender, nation, and class. Originally composed mostly of affluent middle-class students, the counterculture came to epitomise the idealism of the epoch. It demanded a fundamental redefinition of values, including new attitudes towards commodities and consumerism. Sweeping claims were made about the superior moral ideals of the "Age of Aquarius". "It promises", that great bible of the counterculture Charles Reich's *The Greening of America* (1971) asserted,

> a higher reason, a more human community and a new and liberated individual. Its ultimate creation will be a new and enduring wholeness and beauty—a renewed relationship of man to himself, to other men, to society, to nature and to the land . . . Their protest and rebellion, their culture, clothes, music, drugs, ways of thought and liberated life-style are not a passing fad or a form of dissent and refusal, nor are they in any sense irrational . . . it promises a life that is more liberated and more beautiful than any man has known.[11]

Such claims may seem quaint and overblown today. But we still enjoy, to some degree, their legacy. Their relevance to twenty-first-century anti-consumerist culture will quickly become obvious.

Though our focus in this chapter is the United States, many other nations shared in the sense of the epoch as a moment of global liminality, cultural catharsis, and emotional purification. The year 1968 in particular represented the coincidence of many movements and events which were experienced in very diverse ways throughout the world, but which collectively, to the young at least, gave the sense of a worldwide movement. A sequence of events defined the moment: from January to August, the Prague Spring and its Soviet suppression; in February, the Tet Offensive in Vietnam; in March, the massacres at My Lai, which exploded into world public awareness in late 1969, and an anti-Semitic campaign in Poland; in April, Martin Luther King's assassination; in May, the student riots ("*les événements*") in Paris, which commenced as a protest that men were not allowed to stay over in women's college dormitories, and the largest series of strikes in French history; in June, the assassination of Bobby Kennedy, who symbolised his brother's legacy. Throughout the second half of the 1960s there were student movements; feminism; the anti-war resistance; black liberation and anti-imperial struggles; the continuing Maoist Cultural Revolution; a radical intellectual revival centred on the young Marx's theory of alienation; and the progress of sexual liberation, the growth of toleration, and the discovery of "cosmic" consciousness. This was an age of testing limits, and then exceeding them.

11. Charles Reich, *The Greening of America* (Allen Lane, 1971), 1.

In the United States and Western Europe, at least, these developments seemingly represented a generalised revolt against a monoculture centred on older white, male, middle-class experience. (Parallels with the 2020s should be obvious here too.) They often invoked one ideal, equality, and attacked hierarchies, technocratic, political, or otherwise, in the name of "freedom" and "liberation".[12] In Britain, youth rebellion commenced with the emergence of the "mods" and the "rockers" in the late 1950s and progressed stylistically through the skinheads in the late 1960s. (But the latter and the Teddy boys were often racist and politically conservative, and a world away from the New Left.)[13] In France, the counterculture was fundamentally, Alain Touraine writes, "more social movement than political action."[14] It privileged spontaneity over planning, the improvisation of rock music over the orchestra. Even where vaguely political, it was often anarchistic and anti-bureaucratic. It rejected traditional partisan politics as intrinsically corrupt and emphasised decentralisation and democracy from below. How could one organise ecstasy and the celebration of disorder?

In the United States, the cultural changes of the 1960s assumed two main forms: the hippie movement, which erupted in the "Summer of Love" in San Francisco in 1967, when "Diggers" handed out free food and thousands converged on the Haight-Ashbury district, and the Monterrey Pop Festival in June showcased the new musical and personal styles; and the more political explosions which rocked the 1968 American presidential campaign, which built on the momentum established by the free-speech, anti-war, and civil rights movements. Our concern here is mainly with how the first of these promoted a new form of sociability, indeed possibly, as Richard Neville, the founder of a leading countercultural magazine, *Oz*, put it, "a complete rethink of Western Civilization".[15] This was clearly a liminal utopian moment in terms of the definitions adopted here—indeed, a form of prolonged festival which hints at a superior permanent state of being. Though at one level a revolt of feeling against reason, it also had solid intellectual backing. Its many disparate elements included, as Theodore Roszak's seminal study *The Making of a Counter Culture* puts it,

12. A good later overview of these developments is Vladimir Tismaneanu's *Promises of 1968: Crisis, Illusion, and Utopia* (Central European University Press, 2011).

13. Kenneth Leech, *Youthquake: The Growth of a Counter-culture through Two Decades* (Sheldon, 1973), 3–6.

14. Alain Touraine, *The May Movement: Revolt and Reform; May 1968—the Student Rebellion and Workers' Strikes—the Birth of a Social Movement* (Random House, 1971), 26.

15. Quoted in Boyle, *Authenticity*, 122.

> a continuum of thought and experience among the young which links together the New Left sociology of Mills, the Freudian Marxism of Herbert Marcuse, the Gestalt-therapy anarchism of Paul Goodman, the apocalyptic body mysticism of Norman Brown, the Zen-based psychotherapy of Alan Watts, and finally Timothy Leary's impenetrably occult narcissism, wherein the world and its woes may shrink at last to the size of a mote in one's private psychedelic void.[16]

Fissures were soon evident between the counterculture's "alternative lifestyles" and the relative Puritanism of the coat-and-tie bureaucrats of the Old Left, as well as working-class union organisers, both of whom had more in common than either had with scruffy, young, dope-smoking longhairs. The New Left attempted to bridge this gap. But some thought cultural and political rebellion were essentially incompatible. "Tune in—turn on—drop out" was not a slogan commensurate with organising campaigns, or leafleting, or mass demonstrations, the work of the traditional Left. Psychedelic drugs certainly helped drive a wedge between the two camps, for the New Left's other-worldly mentality seemingly defied traditional efforts at organisation. Tom Wolfe, author of *The Electric Kool-Aid Acid Test* (1968), an experiment in stream-of-consciousness journalism, said LSD was destroying the New Left on the US West Coast and fostering a mentality which rejected the old "political games".[17] The acid guru Timothy Leary, founder of the League for Spiritual Discovery, if now derided as "the great charlatan of the sacramental underground", categorically dismissed political action "unless it's based on expanded consciousness". The Berkeley Free Speech Movement, he thought, was "playing right onto the game boards of the administration and the police".[18] The satirical mockery of the counterculture also jarred against the seriousness of the Left. The Yippies (Youth International Party), who nominated a pig for president in 1968, were sometimes described as "Marxists in the tradition of Groucho, Harpo, Chico and Karl".[19] Few traditional Marxists were amused. Revolution was no joking matter, and Marx epitomised a serious engagement with social and political ideas, and with history, which the cosmic babble of the acid-heads manifestly did not.

16. Theodore Roszak, *The Making of a Counter Culture* (Faber and Faber, 1970), 64. Few of these authors are household names today.

17. Tom Wolfe, *The Electric Kool-Aid Acid Test* (Farrar Straus and Giroux, 1968), 357.

18. Quoted in David Caute, *The Year of the Barricades: A Journey through 1968* (Harper and Row, 1988), 57.

19. Yinger, *Countercultures*, 91.

Central to this rift was the counterculture's exposure of crucial fault lines in ideas of human progress. The age-old problem as to whether radical change should commence individually, with consciousness, or collectively, with action, was now posed anew. Its communal ethos notwithstanding, the counterculture in many respects centred on individuality and a "politics of authenticity". Indeed, insisted Marshall Berman, "being oneself" was the aim of community as such, even if this implied radically redesigning society.[20] Many thought that individual and cultural change and moral improvement had to precede organised rebellion. The "revolution of the new generation", Charles Reich promised, "will originate with the individual and with culture, and it will change the political structure only as its final act."[21] But the drama never progressed so far. Its focus was also much more on pleasure than rebellion. Janis Joplin is famously supposed to have said that her music "isn't supposed to make you riot, it's supposed to make you fuck", while Jim Morrison has been quoted as proclaiming, "I ain't talking about revolution, I'm talking about having a good time".

Our concern here is chiefly with the parallels between these developments and the rapidly emerging and largely still-latent social movements of the 2020s. To assess these, we must consider the countercultural currents which flowed through the 1960s, then the political movements, and finally the tensions between them.

Origins

The counterculture emerged out of and extended the same broadly liberal humanist movement which championed the prison reform, anti-slavery, female suffrage, and other agitations of the preceding century. Now it produced demands to abolish the death penalty; to protect the environment; to loosen divorce, abortion, and contraception laws; to extend the rights of women, indigenous peoples, and other oppressed groups; and to support Third World independence movements. It was also defined by the sexual revolution brought about by the birth-control pill, the availability of recreational drugs, the popularity of eastern philosophies like Buddhism, the ecstatic trend in popular music, and the quest for more enduring and authentic life-ideals valid for the younger generation. This rights-centred fusion of humanism and hedonism was partly an attempt to universalise the privileges of the most affluent generation in history. The process also continued that long moment of America-as-utopia

20. Marshall Berman, *The Politics of Authenticity: Radical Individualism and the Emergence of Modern Society* (Atheneum, 1970), ix, xv.

21. Reich, *Greening of America*, 1.

which began with the conquest of the "New World" and now defined the United States as the model of the ideal life, symbolised above all by California, or at least Hollywood's rose-tinted projection of it, the last destination of the infinite road westwards towards the American dream. It would end in our own day with the collapse of that model as a focus of universal aspiration, as California burns in extreme temperatures and respect for brand America sinks dramatically as a result of right-wing nationalism and racist populism.

The counterculture also had its roots in earlier Romantic and bohemian rebellions against conformity; in artistic and "folk" revolts against bourgeois conventionalism and commercialism, associated with Rousseau; and in a culture of self-confession, with appeals to nature against artifice, and to feelings over reason. One of its central themes was resistance to the idea of progress as the increasing conquest and exploitation of nature, as opposed to harmonious coexistence with it.

This built on various earlier traditions. The US national park movement began by according Yellowstone protected status in 1872. Yosemite followed in 1890, largely as a result of the efforts of John Muir, the Scottish-born founder of the Sierra Club. Awareness of the damage caused by climate change, especially the extreme aridity of the Midwest Dust Bowl, was heightened by various New Deal rural programmes. A small revival of communitarianism had also occurred in the 1930s. Influential here was Ralph Borsodi (1886–1977), a Manhattan-raised radical and early opponent of "high pressure marketing" who began plotting an escape to the land in his twenties.[22] Borsodi's works include *Flight from the City: An Experiment in Creative Living on the Land* (1933) and *This Ugly Civilization* (1929), the latter a wholesale attack on the factory system which condemned the "noise, smoke, smells, and crowds" of modern civilisation. He insisted that "*the idea that mankind's comfort is dependent upon an unending increase in production is a fallacy*."[23] Encouraged by Franklin Delano Roosevelt's proclamation that "the continuance of migration from country to the city" might be reversed, he urged groups of "artists, craftsmen and teachers" to flee to small family farms, where they could avoid the stress, congestion, and "human stupidity and ugliness" of large cities. An admirer of Gandhi's village communities, Borsodi hoped these farms could promote more closely knit families through baking, singing, dancing, and working together in the fields. Homesteading might also alleviate unemployment, for Borsodi thought merely paying the unemployed was degrading

22. Ralph Borsodi, *The Distribution Age: A Study of the Economy of Modern Distribution* (D. Appleton, 1927), 286.

23. Ralph Borsodi, *This Ugly Civilization* (Simon and Schuster, 1929), 1, 16.

and counterproductive.[24] Founded in 1938 at Suffern, New York, his "School of Living" still exists, and some thirteen communities were established to apply its ideas.[25] Promoted by writers like Lewis Mumford, Borsodi's ideas became "part of the hippie counterculture", and especially of the drop-out, back-to-the-land movements of the 1960s.[26]

The counterculture also emerged more directly out of the 1950s beatnik generation of Jack Kerouac and Allen Ginsberg, who were convinced, as Paul Goodman put it, "that society is a Rat Race". Being "hip" ("hipsters" appeared before "hippies") was an alternative which "costs very little and gives livelier satisfaction", largely because its small-group culture was "handmade, not canned", and was "communally improvised".[27] To Lawrence Lipton, the beats were "holy barbarians" rebelling against the "square", homogeneous, mass-manufactured, conformist, corporate "Organization Men".[28] Beat men—it was a male-centred movement—were bearded and sandaled, bongo-drumming, poetic, and intoxicated whenever possible, affecting, at least, the lifestyle of the artistic drop-out or ne'er-do-well. Beats endorsed the romantic values of unconventionality, excitement, individuality, spontaneity, and immediacy. They challenged the complacency and superficiality of the 1950s, demanding something deeper, more meaningful, more adventurous, ecstatic: anything but monotonous and mechanical. Some saw the uniformity and efficiency of machines, and the demand that humans imitate them, as the key enemy. They embraced the idea of the stream of consciousness—think of the roll of paper on which the entire text of Jack Kerouac's *On the Road* (1957), the beat bible, was written. They celebrated the erotic before the sexual revolution proper had begun, and taboos began to fall, as Henry Miller, D. H. Lawrence, and William S. Burroughs were published. They relished drug taking both to induce expanded consciousness and as a "*social ritual*" which defined the intimacy of groups of furtive outlaws.[29] Living in the shadow of the Bomb, the beats promoted peace before a generation would embrace the ideal. Some were political. But some identified Marx with Stalinist repression and saw radical politics as too concerned with "the manufacture of things", and too ignorant of "inner gratification and lasting value".[30] This was a different kind of revolution.

24. Ralph Borsodi, *Flight from the City* (1933; Harper Colophon Books, 1972), xxv, xxix, 129.

25. Sutton, *Secular Communities*, 112–13.

26. Paul Goodman, writing in Borsodi, *Flight from the City*, xi.

27. Paul Goodman, *Growing Up Absurd: Problems of Youth in the Organized Society* (1956; Victor Gollancz, 1961), 64–65. "Hip" dates from the early 1900s.

28. Lawrence Lipton, *The Holy Barbarians* (W. H. Allen, 1960) (in reference to William Whyte's *The Organization Man* (1957)).

29. Lipton, *Holy Barbarians*, 171; Leech, *Youthquake*, 39.

30. Lipton, *Holy Barbarians*, 283.

The beats also rebelled explicitly against the waste of resources, both human and natural, which American opulence epitomised. Though largely unaware of the environmental dangers of mass consumerism, they hinted at the ideal of the "open green city", following the tradition of Veblen, Patrick Geddes, and Ebenezer Howard.[31] Their emphasis was more upon the demoralising and dehumanising aspects of post-war urban congestion, as cities were abruptly carved up by massive freeways and white flight created suburban sprawl and the decay of city centres and of neighbourhood-centred life. But there was already awareness about the corrosive effects of commodification. "One cannot help distrusting the goods, thinking they are only packages and brand names", wrote Paul Goodman, a leading figure in this generation of rebels, in 1956.[32]

The new hedonism also had intellectual precursors, amongst the most influential being Herbert Marcuse, who wedded Freud to Marx in *Eros and Civilization: A Philosophical Enquiry into Freud* (1955). This established a key doctrine for the 1960s, that repressing erotic instincts produces guilt and anxiety, and was intertwined with capitalism insofar as workers subordinated sexual desire to the demands of the workplace. *One-Dimensional Man* (1964) extended the critique of the mentality of industrial society, opening the way for that engagement with Marx's theory of alienation which fascinated many intellectuals in 1968. "Alienation" indeed proved to be the great theme linking individual and social aspirations. *An Essay on Liberation* (1969) cemented the idea that rebellion was a multidimensional process involving personal, social, cultural, and political emancipation simultaneously.[33] Here, the ethos of self-realisation was placed in a collective context which an older generation, at least on the left, could recognise. Many others would echo such themes. The radical psychiatrist R. D. Laing, for example, proclaimed that "We are born into a world where alienation awaits us. We are potentially men, but are in an alienated state, and this state is not simply a natural system."[34]

The 1960s

Too ethereal for most, and unsuitable to those who had not yet tasted and spat out the American Dream, beatnik culture was a niche movement which never became mainstream. But rebellion was in the air, fuelled by James Dean, the sanitising of black culture by white musicians, and emergence of the mass rock

31. Paul Goodman, *Growing Up Absurd*, 218.

32. Paul Goodman, 91.

33. Touraine comments that "This revolt matched the analyses of Herbert Marcuse more than it was influenced by them" (*May Movement*, 265).

34. R. D. Laing, *The Politics of Experience and The Bird of Paradise* (Penguin Books, 1967), 12.

music festival, with its capacity for fusing larger groups than ever before, though its parallels with earlier utopian expressions of mass euphoria are clear. The first signs of pent-up emotional overflowing—"rocking"—came with Bill Haley and the Comets in 1956, when mass dancing (still partly the foxtrot) in the aisles of music venues commenced, along with antagonism towards police interference. Then came Presley and Beatlemania, when huge frenzied crowds revealed pent-up desires for emotional release from repression and conformity. And then a great distance was travelled from the Beatles to Hendrix.

An age of ecstasy commenced as the urge to ditch the restraints of social convention seized millions. The 1960s broke from respectability precisely when the Beatles and the Rolling Stones abandoned coat and tie for tie-dyed psychedelia, the hair grew, and Clapton became God. A drug-centred consciousness marked this break, which we can date with some precision to 1966–67. So did the use of the amplified electric rather than the acoustic guitar: the new instrument was intoxicatingly loud, and designed to overwhelm thought with feeling. (Recall the controversy which erupted when Bob Dylan first went electric at the Newport Folk Festival in 1965.) Now, in music, clothing, language, and hairstyles, disaffected youth, especially the most privileged, attained a distinctive, instantly recognisable identity defined by appearance as much as values. In came paisley, the miniskirt, bell-bottomed trousers, bra burning, tie-dyed T-shirts, dayglo fluorescent colours, leather and sheepskin jackets, long hair for men and women alike, and abundant decoration to boot. All of this aimed at celebrating difference, distinction, and membership in the new group of Youth. Many would experience the era as "liberating". One later recalled it as "one long Indian summer of shining brightness, long hair, short dresses, long legs . . . an experience of metaphysical joy and utopian sharing."[35] Sex was at the centre of it all, the core of the rebellion against bourgeois constraint. As the hemlines, the symbolic marker of convention, rose, so did the expectations of the novelties and delights of freedoms to come. The emergence of a new stage in feminism also marked a revolt against a machismo central to western culture, though still embedded in the counterculture.

Appearances notwithstanding, these developments were not a revolt against bourgeois society as such. Admittedly, countercultural participation was often correlated with a widespread and growing sense of anomie or loneliness, heralded by many sociologists in the period and suggesting the need for groups of a type not nurtured by modern society.[36] The new styles and the groups they defined also indicated another phase in consumerism's progress,

35. Sara Maitland, ed., *Very Heaven: Looking Back at the 1960s* (Virago Books, 1988), 166.

36. Musgrove, *Ecstasy and Holiness*, 11.

and an extension, even a universalisation, of hedonism, in powerful demands for instant gratification. Anything which impeded the increasingly all-consuming desire for pleasure had to go. "We want everything" was one Paris slogan of 1968. Another was "total freedom, total experience, total love, peace and mutual affection". Still another was "total orgasm". One Berlin student leader insisted that "All the established institutions must be destroyed".[37] ("Overthrow everything" was a Maoist slogan during the 1966 Cultural Revolution.) To Zygmunt Bauman, consumer society "rests its case on the promise to satisfy human desires in a way no other society in the past could do or dream of doing".[38] The 1960s unleashed the prospect of unlimited desire. It may have offered an "alternative" way of life. But the counterculture's search for a "real self" in some respects merely echoed the demand for constant renewal which characterised the commodified self. Its utopia did not so much negate as epitomise consumer desire, albeit with a greater emphasis on being over having, and on expressing over possessing.[39]

Nonetheless, there were many clear demarcations between culture and counterculture. Middle-class culture in the 1950s was defined by the suburban cocktail or Tupperware party, the golf club, the corporate ethos, and the family sitting in front of the television devouring the inedible, processed "TV dinner" and the equally bland lily-white platitudes of 1950s popular culture. This was all increasingly derided as inauthentic, false, insincere—in a word, "plastic", the term which epitomised this culture, the artificial again confronting the natural. The contrasting ideal was less formal, more relaxed, emotional rather than rational, sincere, more egalitarian, and less manipulative. "Real", "authentic", "pure", and "sincere", were "natural", like wood, stone, wool, and non-synthetic materials. For similar reasons, the idea of the noble savage enjoyed a revival, for the "primitive" too had a strong claim to be more "authentic". To Baudrillard, societies of hunter-gatherers possessed "transparent and reciprocal human relationships", which were more rewarding than those of later stages.[40] Native Americans were romanticised as never before, and began to reclaim their own identity. Their actual treatment on tribal lands did not improve.

Authentic sociability was thus the hallmark of the new movement. Its ethos of togetherness, sharing, and belonging was later epitomised, for the "Woodstock generation", by the great 1969 festival. It was heralded by the words of a song from the Youngbloods, released in 1967, entitled "Get Together", whose

37. V. Turner, *Dramas, Fields, and Metaphors,* 261–63; Lasky, *Utopia and Revolution*, 469.

38. See Bauman, *Consuming Life*, 113.

39. Zygmunt Bauman, *Liquid Life* (Polity, 2005), 80.

40. Baudrillard, *Consumer Society*, 11.

key refrain is: "Come on people now / Smile on your brother / Everybody get together / Try to love one another right now". This appeared on the title page of Charles Reich's *Greening of America*.[41] The new generation, thought Reich, recognised that "Friendship has been coated over with a layer of impenetrable artificiality as men strive to live roles designed for them." It acknowledged people's "need for each other", for honesty with others, and for rejecting "relationships of authority and subservience". The new consciousness "seen in the smiles on the streets" had "begun to transform and humanize the landscape". It was being realised in some types of communes, thousands of which were created at the time, where the "bourgeois" nuclear family was often rejected and "free love" prevailed.[42]

But we should not exaggerate the extent or magnitude of this communalism either. In opposing conformity, the counterculture promoted an ethos of "doing your own thing", which was not so distant from old-fashioned American individualism. "Liberation" and "emancipation", for example, implied release from sexual repression. But they could also mean that everyone, but especially men, might have more sex with less responsibility. Other forces pushed in the same direction. The search for "higher states of consciousness" through "Eastern" wisdom implied a retreat to interiority, and away from the social. Eastern mysticism represented the ascetic, want-suppressing side of the counterculture, which stressed simplicity of life and identities not based on ownership, domination, and consumption. It revived what Roszak called "the essential religious impulse . . . exiled from our culture".[43] Its regimen was tough and demanding, and for many it was ultimately a journey of self-discovery. But reduced to jargon, "consciousness raising" could easily be interpreted as being more focused in whatever one did. Here, "self-awareness" and narcissism were no great distance apart.[44] Soon there was no conflict in proposing yoga retreats for business executives, and in a few years "mindfulness" training would be introduced by human resources departments who spent the rest of their time ratcheting up stress levels and workloads, bullying, and promoting ever-greater inequality in the workplace.

41. Reich, *Greening of America*. Here "the" counterculture refers specifically to the American model of the 1950s and 1960s. Clearly, many countries, notably Eastern European and other dictatorships, possessed subcultures which in local political terms were "counter" to the mainstream orthodox communist cultures, but emerged from quite different sources and had their own trajectories and developments.

42. Reich, 5, 167–68, 290. See T. Miller, *60s Communes*.

43. Theodore Roszak, *Where the Wasteland Ends* (Faber and Faber, 1973), xx.

44. See Edwin Schur, *The Awareness Trap: Self-Absorption instead of Social Change* (Quadrangle, 1976).

The philosophy of drug use also exemplifies the uneasy tension between the social and the individual in the counterculture. "Recreational" drugs expressed "the baby boomers' boundless faith in science", and perfectly symbolised the mentality of instant gratification.[45] From about 1966, drugs marked the fault line between the old culture and the new, demarcating the colourful, stoned, turned-on, hip, progressive "heads" or "freaks" from the "juicers", "boozers", or "squares". The majority went no further than cannabis, or marijuana, which had a gentle, laid-back ethos associated with it, since unlike alcohol it depresses aggression. The more adventurous framed a philosophy of life from experiencing psychedelic drugs like LSD, psilocybin, and mescaline. Their avant-garde, inspired by Huxley's *The Doors of Perception* (1954), was led by Timothy Leary, Ken Kesey, and Ram Dass (Richard Alpert). They experienced a philosophy of time which prioritised being in the moment over racing towards the future. They supposed that LSD in particular, by dramatically altering, slowing down, and intensifying and magnifying perceptions of space and time, could break down barriers between individuals and promote a rapturous sense of oneness with both nature and other people. Here, none of the discipline of yoga or Zen was required. Personal perfection was instead a kind of Kool Aid, add-water-and-stir, instant enlightenment, whose "born-again" qualities indeed had parallels with deep-rooted American Christian beliefs as well as consumer society generally.

These voyages to inner space by pioneering psychonauts were, however, largely individual experiences, not social events.[46] At the extreme mystical fringe of the movement, communion with the shamans, as commended by another guru of the period, Carlos Castaneda, was a means of re-enchanting the universe and returning to the age of magic. This too was always intensely personal. If it implied solidarity, it was of the few mystical elect or adepts, not of the many. This suggests that the drug subculture was ill-equipped to produce a philosophy of life capable of rising above solipsistic navel-gazing. It often reflected the dominant individualism of an evangelised Christian consumer culture, rather than projecting an alternative to it. It demonstrated just how liquid identity had become, to use a term Bauman agreeably adapts from Marx's claim that in capitalism "all that is solid melts into air".[47] But that liquidity was still contained within the receptacle of consumer culture.

45. Gerard Degroot, *The 60s Unplugged: A Kaleidoscope History of a Disorderly Decade* (Macmillan, 2008), 211.

46. The term "internaut" is now used as a parallel description regarding the internet.

47. "Manifesto of the Communist Party", in Marx and Engels, *Collected Works*, 6: 487.

The proximity of culture to counterculture was also evident in the latter's approach to technology. In an age when humanity was more dominated by machines than ever before, writers like Jacques Ellul, Jean Meynaud, and Lewis Mumford challenged the automation of the mind and body alike. US technology was regarded as the core of its industrial and military might. Roszak saw Bacon's vision of scientific domination in *New Atlantis* as epitomised in the RAND corporation, which was deeply implicated in the computer-centred analysis and number-crunching, body-counting approach to the Vietnam War.[48] And so that conflict became in many respects symbolically a David and Goliath war of machines, and corporations as machines, against Spartan peasant virtues and a desire for self-determination.[49] Culturally, television dominated the age, and its blatantly commercial ethos was widely rejected by the young. But increasingly sophisticated stereos took their place, and other machines were welcomed. The electric guitar and synthesiser were the products of an electronic age, as were portable music devices like cassette recorders. Good technology was always welcome.

Legacies and Relevance

The years 1967–69, from the Summer of Love to Woodstock, marked the apogee of the counterculture. Initially resistant, by 1970 the wider society had begun embracing the more adaptable elements in the new music, clothes, hairstyles, and drugs. The styles were easily co-opted. Selling youth rebellion was lucrative. Youth was now more than ever the ideal condition of life. It now also became the focus and "financial centre" of American culture.[50] Then the mood soured, with the Manson murders (1969), which were linked to a "Hippie" philosophy; with the deaths at the Altamont rock concert (1969) at the hands of Hells Angels; and with the Kent State killings of anti-war demonstrators (1970). As the decade ended, peace and love were in increasingly short supply, and the previous half-dozen years were increasingly looking like a stoned haze rather than a mission to remake the world in the image of brother- and sisterhood.

Such bonds of solidarity as had been forged often dissolved quickly. Intensely individualistic and narcissistic, the "Me decade", where everyone was

48. Roszak, *Where the Wasteland Ends*, 148.

49. Numerous participants linked the American conquest of the west with the battle against the new "Indians", the Vietnamese.

50. Twitchell, *Lead Us into Temptation*, 222. On the process of co-optation, see Joseph Heath and Andrew Potter's *The Rebel Sell: Why the Culture Can't Be Jammed* (Capstone, 2005).

encouraged to "do their own thing", was easily transformed, Sheila Rowbotham notes, into a justification for living completely for yourself.[51] Gutted of any vestige of political radicalism, it was soon reduced to style alone, while its intellectual, or anti-intellectual, core, "liberation", degenerated into the psychobabble of New Age spirituality and self-help mumbo jumbo. Soon it was also discovered that the socially liberal ethos of self-celebration and "lifestyle choice" was easily married to a libertarian, economically conservative, minimal-state, low-taxation politics which perfectly suited the corporate ethos and, especially, by the late 1990s, the up-and-coming hi-tech world of Silicon Valley. Hippies distrusted the state; so did corporations. To a later generation, in the laid-back, libertarian, corporatist world view of Mountain View, Silicon Valley, the entirety of the counterculture could be reduced to a New-Agey egotism of libertarian "self-awareness". Now this was harnessed to a revitalised corporate culture even greedier and less socially responsible than those which preceded it, and far more dedicated to a machine-centred and machine-dominated future. In a grotesque parody, the spirit of the 1960s was indeed zombified. Thus did the language of liberation help to facilitate new forms of oppression.

But worse was yet to come, as the world turned sharply in a different direction. From the viewpoint of 1968 the road from the 1980s to the present is marked by the milestones of failure. One participant, the British radical Tariq Ali, observes that "During the last two decades of the twentieth century . . . Utopia was erased from the map of the world. In its place there emerged a Washington Consensus, embodying a neo-liberal dystopia." Writing in 2005, Ali adds that the politics of 1968 often "tends to be ignored" amidst the sentiment that "The music, the new sexual permissiveness etc. . . . and a form of happy mysticism", represent "the meaning of 1968".[52] Another veteran of the epoch reflected that "we were entirely wrong in a belief that if we smoked enough dope and screwed enough people the world would be transformed, the revolution brought in, and Eden replanted".[53] It became fashionable to mock the pretensions of the counterculture. At its most extreme, as a kind of Disneyland writ large, it has been condemned as a regression to infantilism which grew from the overheated fantasies of the spoiled children of the bourgeoisie, chiefly Americans, for whom Ken and Barbie, the ubiquitous macho- and bimbo-style toy dolls of the period, were the new Adam and Eve.[54] Daniel

51. Sheila Rowbotham, *Promise of a Dream: Remembering the Sixties* (Allen Lane, 2000), 198.

52. Tariq Ali, *Street Fighting Years: An Autobiography of the Sixties* (Verso, 2005), 7, 198.

53. Maitland, *Very Heaven*, 13.

54. Roszak estimated in 1972 that the United States, then some 7.5 per cent of the world's population, consumed 30 per cent of the world's non-renewable resources, including 37 per cent

Bell lambasted the epoch's "anti-cognitive, anti-intellectual mood", fixated upon the Dionysian and apocalyptic and exulting in a new sensibility defined by genuineness of feeling. To Bell, this was all a "conceit", and little more than "longing for the lost gratifications of an idealized childhood", "a children's crusade that sought to eliminate the line between fantasy and reality and act out in life its impulses under a banner of liberation". Indeed, Bell thought its rejection of "technocratic society" was "an attack on reason itself".[55] As Anglo-American politics shifted rightwards in the 1980s, many were stunned into silent submission by the virile counter-counterculture of Christian zealots who took heart at the triumph of Reagan, who had mocked the hippie as someone "who dresses like Tarzan, has hair like Jane, smells like Cheetah", and sought to return to sanctimonious puritanical rigidity. "The door of idealism opened briefly and was then slammed shut", comments a study of the era.[56]

But this is an unduly harsh judgement. The counterculture permitted a general relaxation of morals and mores, which over the past half-century has greatly facilitated everyday sociability between different ages, genders, races, classes, and nationalities. Our present culture wars notwithstanding, we are much more tolerant, informal, and relaxed in many ways than our forebears. We dress alike much more, and more casually, and speak, socialise, and marry with an ease and lack of awkwardness unthinkable a century ago. We are less repressed sexually. We have much greater respect for diversity generally, and for racial, class, ethnic, and other forms of difference. This egalitarian tolerance, though now outpaced by the spectre of grave anxiety and of much greater economic inequality, is all to the good. But the future requires much more than this of a new counterculture.

The utopianism of the 1960s remains relevant today in at least six areas. Firstly, it represents a surge in democratic demand against entrenched and largely authoritarian structures, including universities, the military, political "machines", capitalist corporations, and Stalinist bureaucracies. Remaking the world in the image of the hyper-bureaucratic corporation aimed only at revenue extraction seemed repulsive. Populist and other forms of democratic assertion now again indicate that restrictions on sovereignty and political expression are key themes in the early twenty-first century. Egalitarianism is on the march again; whither, however, remains unclear. Participatory democracy

---

of energy, 25 per cent of steel, 28 per cent of tin, and 33 per cent of synthetic rubber (*Where the Wasteland Ends*, 441). The figure is now about 25 per cent.

55. Daniel Bell, *The Cultural Contradictions of Capitalism* (Heinemann, 1976), 81, 122, 133, 144.

56. Degroot, *60s Unplugged*, 2.

was a key demand of the New Left. We begin to hear these echoes again today, when plutocracy is so much stronger than it was then. But they are intermixed with a growing backlash of right-wing populism, vicious nationalism, and downright fascism, as the victims of globalisation strike out in anger at their deprivation.

Secondly, much of what made the 1960s as a whole such a distinctive epoch was the youthful character of its revolt. Once again we see hints of generational resentment coming to the fore. A sense of worldwide consciousness based on an age cohort occurred for the first time during this period. It heralded, though it did not yet mark, the creation of a single, homogeneous world culture. The young again, now, feel betrayed, and rightly. Youth is meant to be enjoyed, and abundantly, not sacrificed. More than any group in human history, they are being asked to pay the price, in perhaps twenty years of austerity and rationing, for the excesses of past generations. Today they are again beginning to revolt against their supposed fate. An explosive liminal moment which like the 1960s juxtaposes a brilliant vibrant idealism with a dark impending sense of catastrophe may yet give them a powerful common identity. But we must compensate them for their losses in some areas with pleasures in others.

Thirdly, a rights- and identity-based politics grew out of the 1960s which has not yet reached fruition. Its relentless egalitarianism constitutes a utopian project of its own in the demand to reshape institutions in order to give reality to rights agendas.[57] This is particularly true for claims based on gender and race. Feminism, in particular, was sidelined by and in the counterculture, and as the Me Too movement has made clear, newly "liberated" women often found that male demand was the chief result of their "liberation". (Recall that James Bond and Hugh Hefner were mainstream male role models throughout this period.) Abortion rights and gender identities which vary from the mainstream are not universally acknowledged. Racism remains as grave an issue as ever. In the 1960s, blacks and other ethnic minorities were often at the margins, rather than the centre, of any movement other than their own, and were underreported. (They occupied one building at Columbia in 1968.) The Yippie proclamation that "Long hair makes us the new niggers" sounds very hollow indeed today—it did nothing of the sort. The counterculture was often white, male, and middle class in its most public expressions. So was its political counterpart. The civil rights movement still has a great distance to travel, as Black Lives Matter has abundantly evidenced.

57. See Samuel Moyn, *The Last Utopia: Human Rights in History* (Harvard University Press, 2010).

Fourthly, students of utopia recognise that the counterculture sought new forms of community defined by friendship and solidarity. Such moments, liminal in their expectation of dramatic alterations in human behaviour, have often been associated with millenarian anticipation, or the promise of salvation from without. Both the countercultural and the political sides of the 1960s bore out some of this expectation. A noble ideal of common humanity still shines forth from this period, which rejects nationalism and great-power chauvinism in particular. The counterculture was a form of *Gemeinschaft* whose conscious target was the *Gesellschaft* defined by a selfish materialism. It pitted an agora against a marketplace which was now further defined as a mall, consumerism's most obvious symbol of utopia as abundance.[58] It challenged the idea that the apogee of civilisation was creating the autonomous, sovereign individual, by indicating the centrality of sociability to our humanity. Its "failure" does not negate the effort as such.

Fifthly, while it instigated new forms of consumerism, the countercultural ethos at root rejected modernity defined as the commodification of everyday life, the subordination of being to having, and the eternally nagging insistence on instant gratification and conspicuous consumption designed to make our neighbours envious rather than to satisfy our true needs. It recognised that communitas could be fatally overturned by consumer market forces.[59] It condemned the elevation of shopping into a religion. It rejected the soul-destroying nature of work under capitalism. These goals, we will shortly see, coincide with our own: the consumerist world view has now reached its expiry date.

Sixthly, the countercultural rejection of consumerism fuelled an attitude towards nature which is indispensable to our forward movement today. Following the path of environmentalist activism charted by William Vogt's *The Road to Survival* (1947) and Rachel Carson's *Silent Spring* (1962), the hippies condemned pesticides and pollution and endorsed vegetarianism and organic food, bringing about a "radical consumerism" in the demand for healthy produce.[60] The famous *Whole Earth Catalog* (1968–72) showed the counterculture at its greenest, striving to fulfil an ethos of self-sufficiency, harmony with nature, and alternative education.[61] Paul Ehrlich's *The Population Bomb* (1968)

58. This contrast is from Zygmunt Bauman's *Collateral Damage: Social Inequalities in a Global Age* (Polity, 2011, 10–26). See David M. Potter, *People of Plenty: Economic Abundance and the American Character* (University of Chicago Press, 1968), 175.

59. Bauman, *Liquid Love*, 74.

60. Warren J. Belasco, *Appetite for Change: How the Counterculture Took On the Food Industry, 1966–1968* (Pantheon Books, 1989).

61. See Andrew G. Kirk, *Counterculture Green: The Whole Earth Catalog and American Environmentalism* (University Press of Kansas, 2007).

warned of the dangers of overpopulation. When "acid rain" and "forest-dying" became an issue in Europe in the 1980s, a new green movement arose to oppose it. The Chernobyl disaster in 1986 provoked a similar reaction and suggested the limits of nuclear power. Green parties are now integral to the politics of many nations. All these issues are more important than ever today, when the destruction of nature has progressed much further. Everywhere environmentalist rebels, the heirs of the 1960s, are leading the campaign to halt the process. These works and phenomena laid the foundation of anti-consumerism, and of the very mentality, we will shortly see, which might, indeed must, define our own future.

11

Life after Consumerism

UTOPIANISM IN THE AGE OF SUFFICIENCY

WE NOW RETURN to applied utopianism, or utopistics. We need to decide how far the various forms of restraint on consumption outlined earlier, both utopian and non-utopian, have contemporary utility, and what, if anything, might be adapted from historical experience. We can start with an incontrovertible premise: affluenza *is* curable—of course it is! We existed quite happily before it, and we will survive its disappearance. The question is whether we can stay the course required for the cure. The medicine may taste bitter at first, but the treatment will not kill us. It will, however, be much more pleasant if we sweeten the pill, and here utopia can help greatly. But if this knowledge inspires hope for the future, this hope must be grounded realistically. It cannot be a ritualised formula of official optimism—the reassuring feel-good narrative of "hopium"—or hint at secular salvation, or reviving the old religions. It must be intermingled with sufficient fear to motivate us to act without paralysing our will to do so. This is an enormous challenge. As we have seen, many who have envisioned transitioning to a post-luxury society have also deemed it impossible to actually achieve the task, once the process of degeneration has advanced sufficiently. Which modern Savonarola will defy these predictions and awaken the millions? Which Fénelon or Mercier will sketch this transition? Which Fourier, Owen, or Marx will make it practical?

The Spectre of Extinction

Let us recall the urgency of our predicament, for this forces us to think and do things which would ordinarily seem impossible. At the outset here we noted that increasing carbon emissions by as much as 50 per cent from around 2020 to 2050 will mean that temperature rises of 3–4°C may occur by 2040, or even earlier. This will bring catastrophic wildfires, floods, desertification, crop failures, extreme biodiversity loss, mass human deaths, rising sea levels, vast and rapid population movements, and much more. As all this is happening with a warming rate of only 1.25°C, these processes will likely accelerate as the world gets hotter. A notional target of 1.5°C is therefore much too high, and much more radical solutions are accordingly required. We have to stop thinking about compromises, and about appeasing "the market", and realise that our survival is at stake.

There is an unparalleled *immediacy* about this scenario. As early as 2030 the environmental destruction now taking place may become *irreversible.*

Think on that last word. It should have the effect of an immensely loud thunderclap coming from a deep black cloud above your head. If we continue on our present course, ignoring the signs all around us, clinging to the old mantras of "normality", "growth", and "business as usual", thinking "prosperity" can somehow return, most if not all of us are doomed. Warnings about civilisational breakdown have been around for decades.[1] They have usually been dismissed as hysterical. Now they are realistic. We cannot wait until 2050, or vainly hope for technological fixes to our problems. We will not invent our way out of this mess. Slowly, fighting a deep-rooted instinct of denial, we are awakening to the implications of these changes. But we need to act now. You may, of course, having reached this point feel that we've already lost the battle, and thus the war. But we have not. Before despairing, please read on. Utopia beckons.

But which parts of this rich tradition remain useable? This book has argued that utopianism has functioned throughout the ages to hold up a critical mirror to deepening social inequality and oppression by the rich, as well as the folly of obsessively pursuing wealth and luxury. The concept of utopia gained meaning in reaction to greedy landlords dispossessing the poor of their land. Utopia long remained the ghostly nostalgic memory of the imagined original equality of a lost paradise or Golden Age, before the wealthy ruined everything. From the sixteenth century onwards, some social revolts took such images seriously and projected a return to this equality. Others glimpsed a possible alternative sociability in the customs of the indigenous peoples of the "New World", or pointed admiringly to Christian monastic and communitarian practices. From Thomas More onwards, the literary genre of imaginary ideal societies offered the dream of such equality being practised on a considerable scale. A society not corroded by vanity, egotism, and greed but rooted in human solidarity, mutual assistance, and respect might not only have once existed but, better still, might again be possible. The long gone "once was" became the looming "not yet", the "might be", still suspended in imaginary

1. See, e.g., Fairfield Osborn's *Our Plundered Planet* (Faber and Faber, 1948), which discusses land degradation, population growth, and other factors.

time and space. By the late eighteenth century, an increasingly confident humanity began shifting this image towards an imminent future, and away from a now increasingly irrelevant past. It became "us" and "here", and "soon" or even "now", rather than nowhere or the hereafter. In the modern period, utopian aspiration then became equally invested in specific spaces, notably France, America, and the USSR, as well as in scientific and technologically based ideals of the future.

Equality remains the quintessential utopian social theme from Thomas More onwards. Early modern utopians realised that emulating the display of luxury by the wealthy had to be stemmed. A closer-knit sense of community and mutual assistance required mastering desires and restraining needs. Spartan solutions to this problem long remained appealing to some. But by the nineteenth century, utopian writers and actors alike began to argue that greater public luxury and a rich and rewarding public life might compensate for reduced private accumulation and emulation. This chapter will consider, amongst other things, how far such compensatory sociability, and some new forms of pleasure, might play a central role in easing ourselves into a post-consumer mentality and mitigating the anxieties we will necessarily suffer in so doing.

We should start, however, on a cautionary note. Sceptics might well contend at this point that many of the utopian solutions to the problems discussed so far here, even up to the early twentieth century, were designed for circumstances vastly different to those we now confront. Let's play devil's advocate for a moment. What has been described here so far might make an interesting or amusing, even morally instructive, historical survey, or an exercise in applied moral philosophy, but that's all it is. Even our most recent countercultural movements were more culture than counter. We are wedded to a commodity-centred life, and would rather extinguish ourselves than relinquish it. It has made us vain, lazy, and spoiled, and that suits us fine. We don't care about tomorrow; we want gratification now.

This is an all-too-plausible assumption. So how much of the ethos of restraint and frugality, in particular, we have examined here is still relevant today, in a vastly different cultural, social, political, and economic context? Six objections might be seen as undermining much of the argument suggested so far.

Firstly, the earth's population is now mostly urban. In the United Kingdom and United States, about 83 per cent live in cities. Globally, this figure, now at about 56 per cent, will reach 70 per cent by 2100. Unless or until many existing cities become uninhabitable because of the heat, rising sea levels, and other threats, this probably removes the possibility of utopian agricultural communes being a viable alternative to city life for more than a very few, never mind tropical islands to escape to. Rural communitarianism was the option preferred by religious sectarians as well as most pre-Marxian socialists, though not the three greatest utopian modernists, Marx, Bellamy, and Wells.[2] Until the stress of population growth eases, urban concentration similarly eliminates garden-city proposals, which entail substantial new green land use. Green space instead must be devoted to rewilding, though brown or wasteland sites are a different matter. Though some intentional eco-communities have been and will be formed, we do not have time to drop out en masse and return to the land, even if we wanted to, which most do not.[3] Farming is very hard work. And then the internet connection is terrible, transport connections poor to non-existent, and the social life dull to dire. This is not to say that the more ecologically conscious contemporary communities cannot serve to inspire us. They can, and doubtless more will be formed.[4] It is only to acknowledge that, for the vast majority, their ideals of sustainability will need to be adapted in cities.

Secondly, powerful religious arguments against luxury and favouring asceticism have mostly disappeared. Christian guilt and civic humanist as well as socialist restraint alike have been replaced by near-universal hedonism. We no longer associate luxury with sin, and rarely go to church to have the point driven home. Vanity is rife in later modernity, and we are forever encouraging one another to be prouder. As with campaigns for sexual abstinence before marriage, voluntary renunciation of luxury inspired by religious principle is therefore an unlikely way forward for most. The cause of virtue as such has an uphill struggle in hedonistic societies. Nor would it be wise to fuel any revival of religious enthusiasm, which is more often a source of division than of unity. Religion also encourages "faith" in general, meaning belief without empirical, verifiable evidence. This fosters scepticism about science, whose damaging results have become all too evident in the Covid-19 pandemic. Challenging empirical forms of proof makes people more generally susceptible to propaganda and misinformation. We will be lucky to escape the wrath excited by the prophets of doom who will doubtless arise in the coming decades urging spiritual revival and declaring that God's vengeance alone will expiate our sins. What works for the Amish and Hutterites, then, for example, is probably not adaptable to the rest of us, though secular lessons can be drawn from their experience.

2. A general survey of contemporary trends is given in Ben-Rafael, Oved, and Topel's *Communal Idea*.

3. On such communities, see David Pepper's *Communes and the Green Vision: Counterculture, Lifestyle and the New Age* (Green Print, 1991) and, more recently, Lucy Sargisson's *Fool's Gold?* (129–45), and Anitra Nelson's *Small Is Necessary: Shared Living on a Shared Planet* (Pluto, 2018).

4. See Graham Meltzer, "Contemporary Communalism at a Time of Crisis", in Ben-Rafael, Oved, and Topel, *Communal Idea*, 73–90.

Thirdly, modern individualism inhibits the power of collective restraint on behaviour. The rigidly ordered hierarchies of many early modern utopias, so useful for regulating behaviour, are repugnant to us now. Patriarchy is everywhere in retreat. More people live alone, fewer are marrying or staying married, more live in the surreal worlds of gaming or YouTube videos. Though powerful countervailing trends towards conformity exist, collective pressure is weaker just as we need it to help limit consumption. Overwhelmed by neoliberal ideology, we generally feel little obligation to others. Thousands complain even when asked to wear masks to protect others, and themselves, in a pandemic. We demand our rights, without presuming that commensurate duties follow such claims. We cannot abandon rights language, however. Indeed, protecting rights will be central in the difficult decisions of the coming decades. But our collective self-preservation is the final arbiter here. Nature's limits must define our rights, not vice versa. So the rights of nature trump human rights. We must free ourselves from the tyranny of anthropocentrism to admit this. And the rights of humanity are more pressing than the rights of property. We cannot permit the selfishness of the 1 per cent to destroy the world. Our obligation to future generations, and to the planet, must be foremost.

Fourthly, the relentless pursuit of novelty, which was confined to relatively small groups in a few societies before the twentieth century, is now virtually a universal phenomenon affecting billions of people. Wanting things to remain the same, and freeing ourselves from an obsessive desire for change-as-self-renewal, is now extremely difficult. Appeals to the superior virtues of simplicity fall on deaf ears in the early twenty-first century. The few remaining indigenous peoples attract our sympathy, but not our admiration. They have no television, no smartphones, no internet: what kind of life is this? They live in the Stone Age, not the Golden Age. They have nothing we prize.

Fifthly, economic inequality in most developed countries has become much more extreme since the 1980s, especially since 2008, and has been greatly exacerbated by the Covid-19 pandemic. In 2019, 1 per cent of the US population owned as much wealth as the entire middle and upper-middle classes, who in the 1980s had had a third of the nation's wealth. The richest 1 per cent have received 95 per cent of all economic gains since 2009. In September 2020, following the massive accession of wealth to a few, three people owned more than the bottom 50 per cent of Americans, and by early 2021 two owned more than the bottom 40 per cent. Globally, 1 per cent now own more than the other 99 per cent, and eight people, six of them Americans, are as wealthy as the poorest half of the planet's entire population. In the United Kingdom, the combined wealth of the richest thousand people was £256 billion in 2009, and £724 billion in 2018. By 2021 the richest 1 per cent had nearly a quarter of the nation's wealth. CEO pay has outdistanced that of the average worker massively in the past few decades. In the United Kingdom, CEOs earn on average

129 times the average wage of their employees. In the United States it is 271 times (it was 299 times in 2014). In 1965 in the United States it was about 15 times, and in 1980, about 42 times. This disastrous growth in inequality has in turn driven an increased desire for luxury goods, including private jets and yachts, some of the latter costing $500 million. This process has propelled us rapidly in exactly the opposite direction to that in which we should be moving. The extravagance it fuels sets a terrible example for the rest of humanity, and encourages more greed and corruption. And it facilitates the control of the very wealthy, who are almost invariably right wing, over information, politics, government, and environmental policy.

Finally, our public morals have shifted discernibly in recent years. The public sector has in many countries become so privatised that distinctions between public and private interest are difficult to discern. Across the world extreme corruption in government is regarded as normal. Politicians accept contributions to promote policies which favour particular corporations or interests. Government ministers profit personally from policies they introduce themselves. Jobs and contracts are given to cronies, who kick back party contributions. Large-scale corporate lobbying and constant movements from public to corporate life and back are the norm. We expect great corporations to profit from government and turn a blind eye when privatisation further reduces public services. We know that many who routinely declaim most violently against socialism will demand immense government bailouts when their assets are threatened, and then resist throwing a few scraps to the poor. ("Socialism for the rich, capitalism for the poor".)[5] We are generally less concerned with, and less capable of identifying, the public interest than ever. At elections we demand that politicians serve our interests rather than those of the public as such. We rarely challenge them when, in power, they seek short-term gain and are unconcerned by long-term planetary interests. Instead, we are distracted by their false patriotism, pompous xenophobia, and arrogant nationalism. We are more pliable, more gullible, more susceptible to manipulation than ever before. The sober maturity and calm calculation of what humanity's long-term future requires, which the changes proposed here demand, largely elude us.

These realities suggest that few previous utopian visions suit our current plight. Consumerism is triumphant. It blinds us to the possibility of any alternative. We are all Sybarites today, and we know full well that Sybarites cannot

5. The phrase dates to at least the late 1950s, according to Michael Harrington's *The Other America: Poverty in the United States* (Macmillan, 1962, 58).

become Spartans. Many of us, indeed, would probably rather not live than endure Spartan restraint and the curbing of pride and ostentation prescribed by some earlier utopians. Pride, indeed, practically defines the later moderns. So there is no return to the lost "home". This *is* the Golden Age, and we will not have our desires frustrated. We want what we want, and to hell with the consequences.

This is a childish attitude, and if anything kills us, it will be this state of mind. Making a sustainable utopia attractive will clearly be extremely difficult as a result. Some of its more compelling, dramatic, and emotionally satisfying incarnations have lost their allure. Secular millenarianism is, frankly, a bad idea: surely the bloody twentieth century teaches us this much. We should no longer seek complete human regeneration, the attainment of "grace", the cult of the free spirit, or "complete", "total", or radically "new" states of being. We want no popes of hope, messianic millenarians, pious prophets, or sanctimonious saints, no "salvation" or "redemption", secular or otherwise, no demand to end depravity or sin, or to cure forever our existential woes. Alienation will never be abolished. All ideals polluted by theology must be rejected. But so too must we surpass the mainstream "solutions", the two grand ideas of the modern period: progress defined as blind faith in indefinite material improvement, and the vision of the classless society inaugurated by the revolutionary communist proletariat seeking to universalise the benefits of progress. The millennium, liberal progress, and traditional Marxism cannot help us.

But a secular and sustainable version of utopia remains viable, as the successor concept to the old idea of progress. Some components of both liberalism and Marxism will have to be integrated into it. A sustainable lifestyle cannot abandon the benefits of science and technology. But we must offer more than this. Relinquishing unsustainable consumption can succeed only if the majority are offered what Paul Wachtel calls "an alternative that is not just a bitter necessity but holds out promise of a genuinely better life." People must be persuaded that while their consumption must be reduced, they can nonetheless be happier. Only then can we

> think beyond material goods as the defining essence of life and . . . focus instead on the quality of our relations with others; on the clarity and intensity of our experiences; on intimacy, sensual and aesthetic experiences, and emotional freedom; and on the ethical, spiritual, and communal dimensions that give the entire enterprise meaning.[6]

6. Paul L. Wachtel, "Alternatives to the Consumer Society", in Crocker and Linden, *Ethics of Consumption*, 199.

But this "better life" cannot be left vague. It must have a specific content and aim, to reduce the stress and anxiety of everyday life by offering economic security for the majority, and opportunities for enhanced sociability similar to those we have explored here. This utopia consists in strong and rewarding individual relationships; a supportive social network; a powerful sense of belongingness; good education, employment, and health opportunities; reasonable life expectancy; a sense of being free, equal, and rewarded for one's labour; sufficient and increasing leisure; and a safe and sustainable natural environment.

Much of this agenda will be familiar to readers. This book has given unusual priority to the value of enhanced sociability and belongingness in this formula, as not only valuable in themselves, but more specifically as a means of offsetting our desire for superfluous consumption. If we can gain self-esteem and pleasure through interacting with others, and through sustainable experiences and consumption, we will make do with fewer unsustainable experiences and commodities. It has been conceded, however, that we cannot mandate loving strangers, or even our neighbours. We cannot even demand kindness. Here religion has failed, as did twentieth-century totalitarian forms of contrived association, or "compulsory solidarity".[7] We cannot even make people *like* one another: "like" is a *feeling*, not a form of behaviour, and voluntariness is the essence of the feeling. But liking is based on knowing, and we can provide circumstances which facilitate sociability and show how kindness and consideration not only help others but make us feel better too. If successful this would be a fine step forwards. We can make it much easier to identify with others, to respect them, and to benefit from their company, even in brief encounters. This is utopia's compromise with perfectionism, the rescuing of the good from the perfect. How might it work?

Compensatory Sociability in the Twenty-First Century: Some Hindrances

The sociability component in this formula assumes reduced work time and introducing a universal sustainable income, which are discussed in more detail below. We also want to make work as pleasant as possible, and to ensure that the workplace reinforces solidarity and association. We must also rise to the challenge of greater leisure, which cannot translate into increased demand for unsustainable goods and services. We might well heed Aldous Huxley's warning, discussing Tolstoy, that extending leisure might produce "an enormous

7. Claeys. *Dystopia*, 8.

increase in the demand for such time-killers and substitutes for thought as newspapers, films, fiction, cheap means of communication and wireless telephones", resulting in large numbers being "afflicted by ennui, depression and universal dissatisfaction."[8] So free education for all to the highest possible level is crucial too.

Further objections exist regarding the sociability hypothesis presented here. Sceptics may query whether a real reciprocity exists in potentially exchanging unsustainable consumption for the sustainable enjoyment of both goods and experiences, greater sociability, and the increased sense of belonging it will bring. It has been contended here that warm personal relationships are superior to consuming excessively. But other factors may reduce our desire to associate with people. The more individualistic we become, the less prone we are to compromising with others. This may help explain why friendship as such is declining. In 1985, 80 per cent of Americans had more than one close friend, by 2004, only 57 per cent did. Women have more friends than men (in 1981, 4.7 as opposed to 3.2).[9] Some suppose they possess a greater capacity for making friends in the first place. Female intimacy also certainly needs to be understood in different ways from its male counterpart.[10] But the main problem for the population as a whole is that loneliness is increasing steadily, to virtually epidemic proportions, and not just because of the rising number of people living alone. A few may live a satisfactory and rewarding life which involves considerable solitude. (Hermit-like academics will recognise this model immediately.)[11] But many more lament and suffer from enforced isolation, as the pandemic has sharply reminded us. Some even blame the rise of populism on this loneliness. Isolation is not, however, markedly worse in cities. Close proximity to large numbers of other people brings many compensations, particularly in the form of subcultures of specialised small groups, and the greater tolerance and liberality which results from consorting with such a variety of often more liberal people.

8. Aldous Huxley, *Along the Road: Notes and Essays of a Tourist* (Chatto and Windus, 1930), 240. He later hinted as much: "The whole of Ireland was put on to the four-hour day. What was the result? Unrest and a large increase in the consumption of soma; that's all" (Huxley, *Brave New World* (1955), 176).

9. R. Bell, *Worlds of Friendship*, 63.

10. See Elizabeth Susan Wahl, *Invisible Relations: Representations of Female Intimacy in the Age of Enlightenment* (Stanford University Press, 1999). See further Lillian Faderman's *Surpassing the Love of Men: Romantic Friendship and Love between Women from the Renaissance to the Present* (William Morrow, 1981) and Nina Auerbach's *Communities of Women: An Idea in Fiction* (Harvard University Press, 1998).

11. Lane, "Road Not Taken", 238–39.

Then there is the problem introduced early on here, that even if we had more time to form friendships, our capacity for affection is limited. Gradations of intimacy must be recognised. Not everyone can be accorded the honour of the personal *Du* or *Tu* (the now-archaic thou in English), for this devalues the status, and those who breach the convention by being overfamiliar are regarded as condescending if not deranged. Tarde noted the "serious objection" to cosmopolitan ideas of equality that friendship is "a circle which deforms itself in *stretching out too far.*" For "the social circle to widen itself out to the limits of humankind", this objection has to be met.[12] There is the challenge, noted by Bertrand Russell, that in modern cities we no longer need to know our neighbours because we no longer depend on them.[13] In at least two other areas where relationships are key to subjective well-being, marriage and religious affiliation, general social trends are moving against both. Increasing anxiety in the face of environmental breakdown and economic uncertainty also makes people more rather than less competitive.

There is also the objection that national traditions of sociability vary greatly. Pandemics aside, Mediterranean peoples like the French, Spanish, Portuguese, and Italians hug and embrace friends and acquaintances readily and frequently. In France, anywhere from one to four kisses, mostly in the air, may be the norm. Some cultures rub noses or bump foreheads. Touching releases a hormone, oxytocin, which makes us feel good, and which to Robin Dunbar is the chemical basis of trust.[14] So for many, as Covid has reminded us, hugging, embracing, and cuddling are about the best things in the world. They suggest empathy, co-operation and non-aggression, sharing and, again, belonging. These peoples are better at recognising this, at *feeling* and *expressing*, we might say. Other nationalities, however, are squeamish about physical closeness, and guard their personal space much more jealously. In more disengaged, standoffish cultures, touch is rarer. Shaking hands may already appear sufficiently intimate. Tipping hats may have once sufficed, or a cap, if class is at issue. A wave or salute may occur, or a palm placed on the heart, or two palms pressed together. Bowing at a suitable distance is common. The English are renowned for such reticence; Graham Wallas noted of interaction on public transport in the early twentieth century that "no event short of a fatal accident is held to justify a passenger who speaks to his neighbour."[15]

12. Tarde, *Laws of Imitation*, 350.

13. Claude S. Fischer, *To Dwell among Friends: Personal Networks in Town and City* (University of Chicago Press, 1982), 103.

14. Dunbar, *How Many Friends*, 64.

15. Graham Wallas, *Human Nature in Politics* (Constable, 1919), 49.

Another range of problems emerges from the growing role technology plays in our lives. Once heralded as a new "utopian communitarianism", the internet now seems more like a litter-strewn playground dominated by bad-mannered, foul-mouthed bullies who exploit every opportunity to vent their ill-concealed frustrations, as well as by giant corporations more concerned to monetise our data, encourage right-wing opinions, and sell us stuff than to inform us responsibly. Hounding and trolling, political manipulation, and the unleashing of deep-seated anxieties and prejudices seemingly dominate its politics.[16] A key problem here is anonymity, which in the absence of personal shame some use to cloak all manner of rudeness and insult. New technologies also breed a constantly increasing love of stimulation, especially through smartphones. These are physical addictions: as shopping heightens the intensity of our sensations, withdrawal from smartphone clicking, swiping, scrolling, and checking, ping-anticipation, generates nervous anxiety, impatience, an inability to concentrate, and the ubiquitous FOMO—fear of missing out.[17] There is a pronounced and for some very alarming tendency for phones to take over: we all know how common it is to see groups of friends sitting together silently at a table, each passively staring at his or her loved one, the device. Recent research even suggests that some regard their phones as a kind of "home", in the sense of a safe place where they are in control and surrounded by reassuring images. There is no doubt that the internet expands our range of contacts. But does digital communication compensate for a decline in other forms of interaction? Are phone or online contacts qualitatively of the same value as face-to-face encounters? Does dependency on digital media weaken direct personal interaction? Bauman thinks so: "In fact, we grow shy of face to face contacts. We tend to reach for our mobiles . . ."[18] So does Turkle, who adds that it may be easier for shy people "to make friends because they have fewer inhibitions when they can hide behind a screen."[19] So the art of conversation is declining, as we come to prefer texting to talking. But everyone recognises that even Zoom, Teams, or Facetime, which the Covid-19 pandemic forced on millions, provide pale, detached, and inferior forms of human interaction, and fail to inspire us with the energy, motivation, and

16. Robert D. Putnam, *Bowling Alone: The Collapse and Revival of American Community* (Simon and Schuster, 2000), 170.

17. A surprising proportion of people recently deceased have asked to be buried with their mobile phones, so far have they become an extension of our personalities: Graaf, Wann, and Naylor, *Affluenza*, 58.

18. Zygmunt Bauman, *Wasted Lives: Modernity and Its Outcasts* (Polity, 2004), 130.

19. Sherry Turkle, *Alone Together: Why We Expect More from Technology and Less from Each Other* (Basic Books, 2011), 259.

warmth which personal contact provides. Here too our social skills have been weakened.

Yet these platforms also facilitate interactions which were previously impossible or much more limited. The positive side of digital communication, like online chatting, some think, is that it allows us to form groups and make contacts which manifest "the closest, strongest, most involving and generally most intense form of social bond", which "does not necessarily require face-to-face contact or even sustained mediated connectedness."[20] This may help expand our personalities and allow us to share experiences with others, and thus increase our pleasure, especially through images. Games may perform the same function. A 2001 survey indicated that some 20 per cent of virtual-world players regarded the virtual world as their true home.[21] Studies show that social media augment communication by permitting more frequent interactions (Skype, Snapchat, Facebook, WhatsApp, Twitter, etc.), and by encouraging the sharing of pictures, experiences, and feelings. This has made us obsess about the number of "friends" and "likes" we accumulate, as if these reflected our real personalities. But Facebook "friendships" are not "real" friendships, and Twitter "likes" may reflect an obsessive need for self-reinforcement. And what is the value of friends who can be purchased for 20 cents on uSocial.net?[22] (They may be better than nothing.) What is missing by way of depth is compensated for by the multitude of sources of stimulation. But substantial emotional sustenance is lacking in most of these encounters. Digital worlds are also often defined by fantasy, falsehood, illusion, and manipulation to an extraordinary degree. Fake news, conspiracy theories, and disinformation abound as never before. Mistrust and a tendency to think the worst of people accompany digital confrontation. But if we insult each other more freely, online "deception may . . . serve as a kind of social lubricant that actually makes it easier for people to get along and thus help to sustain connectedness", much as the artificiality of conventional politeness or wearing masks at a carnival may facilitate sociability by permitting anonymity.[23]

Let us leave these queries about sociability to one side temporarily, and turn to confront the problem of how a post-consumer society might be structured. Sociability, we might say, is the cream on top. This is the cake.

20. Mary Chayko, *Connecting: How We Form Social Bonds and Communities in the Internet Age* (State University of New York Press, 2002), 118–19. See also Sherry Turkle's *Life on the Screen: Identity in the Age of the Internet* (Weidenfeld and Nicolson, 1996).

21. Edward Castronova, *Exodus to the Virtual World: How Online Fun Is Changing Reality* (Palgrave Macmillan, 2007), 13.

22. Harry Blatterer, Pauline Johnson, and Maria M. Markus, eds, *Modern Privacy: Shifting Boundaries, New Forms* (Palgrave Macmillan, 2010), 94.

23. Chayko, *Connecting*, 125–26.

Neither Sybaris nor Sparta: Envisioning a Post-consumerist Society

Can we imagine a post-consumerist society which has broken the cycle of emulation, decommodified our selves, and reined in consumption substantially? *Yes.* But . . . how realistic a prospect is it that we will realise this vision in time? The brief answer, to be blunt, is *not very*. By 2000, the prospect of rescuing utopia from the all-engulfing vortex of consumer society seemed well-nigh exhausted. The mood of the epoch was rapidly becoming dystopian. The massive increase in global inequality since 2008 has meant that billionaires have ever-greater control over the media, politics, and the judiciary. Big money favours environmental exploitation and business as usual. Vast sums have been spent—as much as £2.8 trillion, if not £8 trillion, by banks just since 2016—on promoting the most destructive forms of agriculture, and on disseminating climate disinformation and denial ($2 billion 2000–2016, ten times more than environmental groups spent).[24] Coal still generates a third of the world's electricity, while producing twice the CO_2 of natural gas. Some 2000 new coal power stations—one a week is being built—are planned in the coming decades in China alone, the world's largest coal consumer and CO_2 emitter (27 per cent of the global total). Australia continues to deny that coal production is harmful. Indonesia has become the world's leading coal exporter. Britain may open a new coal mine, which would annually produce the carbon footprint of 650,000 people.[25] Poland and Germany drag their feet. Canada continues its oil-sand and coal development. America fracks. (This alone releases as much methane annually as Florida's entire contribution to global warming.) Oil extraction is slated to rise by some 35 per cent from 2020 to 2030. The Paris Agreement, already weak enough, seems doomed by such folly, despite newly announced (April 2021) Biden-initiated goals of 50 per cent reduction in US emissions (11 per cent of the global total) by 2030, and the notionally improved targets trumpeted at COP26.[26] The goals announced, however, simply do not

24. Rich, *Losing Earth*, 6. Fossil fuel is currently subsidized at the rate of $11 million a minute.

25. Mike Berners-Lee, *How Bad Are Bananas? The Carbon Footprint of Everything*, 2nd ed. (Profile Books, 2020), 163. This book is a fantastic source of information as to how to reduce personal consumption.

26. COP26 proposed new targets in methane emissions, coal mining, and forest preservation in particular. Many leading producers did not sign up to the coal commitments. Delays in implementing the forest provisions would do little to stem the loss of the Amazon rainforest, while Indonesia insisted that developmental priorities would continue vis-à-vis palm-oil plantations.

match the measures actually being taken to combat emissions, or to avoid disaster. Evasiveness, bad faith, NIMBYism, and the usual greed prevent real movement from occurring. The challenge of pulling a utopian white rabbit out of the hat in the face of the dystopian spectre of catastrophic environmental breakdown is enormous.

But however slender the odds, we have no option but to make the effort. Our task here is to sketch out a new paradigm which is neither capitalist nor Marxist, but defined by utopian sustainability. Capitalism as such, defined as the modern form of exploiting human labour, is no longer the key problem facing us. Because exploitation, racism, the oppression of women, and poverty fuel inequality, they must still be minimised. But the primary goal is to preserve the natural world. As a rationale for constructing a new system, this supersedes Marxism's emphasis on capitalism's exploitation of labour and rejects both capitalism and Marxism's emphasis on the indefinite expansion of needs, production, and population. Such a society might be called post-capitalist, or state capitalist, or socialist.[27] But the name is not important. What is vital is sustainability.

We face three critical periods this century if we are to achieve this goal. In the present decade, to about 2030, urgent efforts are required to stabilise and then reduce warming by reaching zero carbon emissions. In the second phase we will need to cope with the effects of existing warming, which will last for many years yet, and ensure minimal temperature rises in the interim (*c.*2030–50). The third phase will be, on the best case scenario, stabilisation and recovery (*c.*2050–2100). The greatest disruption will therefore occur during the first thirty years or so of this transition. Gen L, as we might call it, the Lost Generation, will suffer the most and will require the greatest compensation. But every sacrifice made now will avoid far greater difficulties thereafter.

The most obvious goal for the first period is to relinquish carbon-based energy, especially coal, oil, and natural gas, by 2030 at the latest. The year 2050, still bandied about widely at present relative to zero emissions, is much too late. Hence the slogan #keepitintheground: all exploitation of these resources must cease rapidly to keep global warming to 1°C or, better still, less—not 1.5°, much less 2°. This requires reducing emissions to zero, not just stabilising them. It must occur in the face of a rise in energy demand of some 50 per cent in the coming decades. But we have seen immense advances in renewable technologies, especially wind and solar power, which have become financially

27. See the discussion in Wolfgang Streeck's *How Will Capitalism End? Essays on a Failing System* (Verso, 2016). On the disinformation process, see Michael Mann's *New Climate War.*

preferable far more rapidly than was supposed possible only a few years ago. Renewables are now often cheaper than fossil fuels, and many countries are within sight of being able to abandon coal, oil, and gas. Renewables provided 25 per cent of global energy in 2019, a percentage projected to rise to 30 per cent by 2030. We could easily make that 100 per cent by 2030. They exceeded coal as a source of new power in 2020. Wind power alone could give us more than eighteen times the electricity required in 2019. So could solar power. All that is lacking is the will to make it happen.

Beyond these immediate goals lies the need to create a post-consumerist mentality, and to restrain demand for non-renewable energy and resources. This is a harder nut to crack, harder than shifting to sustainable energy, and we will need to muster a supreme amount of ingenuity and goodwill to achieve it. We have seen that two forms of affluenza result from consumerism, one focused on commodities and an addiction to shopping, the other on the people we use consumption to try to excel. No "real" utopia can exist without solving these problems, and all utopian proposals hitherto conceived which have not confronted them are of little help. Merely demanding "socialism" or social, class, racial, and gender justice is no longer sufficient. "Human rights" are not the supreme goal, even where they implicitly demand greater social equality through institutional change. We must reduce our desire for excessive consumption, or all the rest will be unachievable—indeed, is meaningless.

This does not mean renouncing the benefits of labour-saving devices like refrigerators, microwaves, vacuum cleaners, air conditioning, and central heating, or the amusements offered by radio, film, television, and the internet. It means, rather, using and adapting, and when required limiting, such devices, according to the criterion of sustainability. Such inventions and discoveries were conceived to be harmful luxuries only when they generated excessive inequality. They are more harmful as sources of planetary destruction. So we must make these technologies sustainable, especially those which protect us from heat.

How, then, do we move forward? Like Thomas More, we must start with the central utopian principle, equality. No utopia is possible where great inequality exists. The proposals which follow are not set in stone and are intended to contribute to an ongoing debate. They indicate the direction we need to take, rather than the precise course required to get there. But they are underpinned by great urgency. If we do not act dramatically, and quickly, we may not survive.

No More Billionaires: The Rationale for Equality

All societies have some gradations, ranks, or means of distinguishing functions or achievements. The problem is not status differentiation, which is inevitable. A title, or ribbon, or a feather in a cap, applause and public approbation, can indicate accomplishments and claims for respect. The problem is rather inequality based on wealth. Besides making people feel inferior and unvalued, extreme economic inequality distorts social relationships and makes people want to consume well out of proportion to sustainability, or even their own needs, or what they can afford. Reducing such inequality will reduce unhappiness. More-equal societies have lower crime rates and less mental illness, anxiety, violence, and other problems. If they are affluent, they also have universal healthcare, long life expectancy, decent holidays and pensions, considerable social and intellectual freedom, low population growth rates, and good friendship networks. Compared with less sympathetic societies like Britain and the United States in particular, they waste little time on blaming the poor for being poor and invest more in raising them from poverty and sparing them from misery. This is why relatively equal nations like Denmark, Finland, Norway, Iceland, Switzerland, and the Netherlands come at the top of international happiness rankings year after year. (If we measure happiness by the amount of time people spend laughing and smiling, and their tranquillity, less well developed societies do much better.)

As Richard Wilkinson and Kate Pickett show, then, materialistic societies like the United States and Britain make us more unhappy, and being unhappy makes us more materialistic.[28] In the United States, the number of those describing themselves as "very satisfied" peaked in the 1950s. The far greater range of commodities available today has not increased satisfaction.[29] So there is a clear correlation between excessive consumerism and social unhappiness. Above the poverty level, Paul Wachtel wrote nearly forty years ago, "the relationship between income and happiness is remarkably small."[30] When we have little money, getting more can make us happier. But when we have

28. Richard Wilkinson and Kate Pickett, *The Inner Level: How More Equal Societies Reduce Stress, Restore Sanity and Improve Everyone's Well-Being* (Allen Lane, 2018), 106. See also Wilkinson, *The Impact of Inequality: How to Make Sick Societies Healthier* (Routledge, 2005), and Wilkinson and Pickett, *The Spirit Level: Why More Equal Societies Almost Always Do Better* (Allen Lane, 2009). On different measurements of happiness, see Daniel M. Haybron's *Happiness: A Very Short Introduction* (Oxford University Press, 2013).

29. Humphery, *Excess*, 43.

30. Paul L. Wachtel, *The Poverty of Affluence: A Psychological Portrait of the American Way of Life* (Free Press, 1983), 39.

enough, more does not make us happier: this is one paradox of affluence.[31] In the United States, incomes above $75,000 do not markedly increase personal happiness, and certainly not emotional well-being. And having more money beyond a reasonable income may actually decrease our sense of subjective well-being by raising our expectations ever more unrealistically and making us still more competitive.[32]

There is also a clear and dramatic correlation between inequality and unsustainability. More-equal societies consume less per capita than more-unequal ones. Inequality is a major driver of excess consumption. Globally, the richest 1 per cent produce double the greenhouse gas emissions of the poorest 50 per cent of the world's population, with 10 per cent generating as much as the other 90 per cent. Billionaires are estimated to individually produce some 8000 tons of emissions per annum, mainly through private yachts and planes.[33] A quarter of global emissions are in the United States.[34] This is the clearest indicator that inequality drives compensatory consumption, as we struggle to keep up. Making societies more equal will reduce the interpersonal competition which drives overconsumption.[35] Thus, Wilkinson and Pickett write, "we move from a society that maximizes consumption and status, to a society that uses each increase in productivity to gain more leisure and reduce the demands of work."[36] Where leaders and trendsetters live in modest comfort, this process will be accelerated. Ceasing to praise or admire extreme wealth and greed, and, within limits, displaying contempt for both, is also very important. Such proposals do not challenge the fundamental psychological tendencies revealed by Mandeville, Veblen, and Tarde. Emulation will inevitably continue. But display and distinction can avoid taking the form of wasteful conspicuous consumption.

Reducing inequality will require two policies with respect to wealth and consumption which are familiar to us: restrictions on luxury, through sumptuary laws, and on property ownership, through "agrarian" laws, or wealth taxes.

31. Lane, *Loss of Happiness*, 6, 8, 59.

32. Robert H. Frank, *Luxury Fever: Money and Happiness in an Age of Excess* (Princeton University Press, 1999), 72; Haybron, *Happiness*, 119. Those describing themselves as "very happy" are never more than 40 per cent of the total.

33. Dario Kenner, *Carbon Inequality: The Role of the Richest in Climate Change* (Earthscan, 2019), 13. This translates into 55 metric tons of emissions for the top 1 per cent in 2013, as opposed to 3.6 metric tons for the poorest 10 per cent, and the gap is widening. The formula is thus simple: the higher the disposable income, the larger the carbon footprint.

34. Kenner, 27.

35. A starting point here is Lunt and Livingstone's *Mass Consumption*.

36. Wilkinson and Pickett, *Inner Level*, 262.

Amongst many others, we recall, Fénelon saw such measures as essential for the transition from a luxury-centred to a post-luxury society. Reviving them today, with suitable adjustments, is not preposterous. Making goods which endure and abolishing planned obsolescence might be combined with certain forms of rationing, sumptuary laws, and luxury taxes which restrict the ownership and display of luxury goods. Some are described below.

The relevant analogy here is wartime, or a pandemic. Rationing of goods is endured more or less willingly during severe crises, which parallel what we face in the coming decades. Again, equality is the key to success. We will endure reduced consumption only if the sacrifice is perceived to be shared by all, and the privileged do not find ways of exempting themselves from every burden and shortage. Rationing of food, clothing, and other consumer goods existed in both world wars in many countries.[37] It may be less manageable with declining agricultural production and proportionately larger urban populations than in the twentieth century, when farmers were less affected by rationing and shortages were greater in cities.[38] Rationing "growth" will be difficult if we assess every country by its population and judge its carbon consumption and emissions accordingly, for, as George Monbiot points out, this would allow less developed nations to increase their emissions.[39] But what were reasonable proposals a decade or so ago are now inadequate. Initially, also, goods must be rationed, not because of absolute scarcity but because of their unsustainability, their carbon cost, or footprint. Within a few decades, however, food will likely be in much shorter supply and "digging for victory", converting wasteland into agricultural land, in both towns and the countryside, will become necessary as dearth and famine occur. Not now widely rationed, water will become increasingly scarce, and its regulation will have to be co-ordinated with flood control, by saving vastly more rainfall to offset shortages as temperatures rise, and by huge irrigation projects to move it from wetter to drier regions. This will help end the cycles of drought and flood which many countries already experience. During the transition away from petrol and diesel vehicles, fuel consumption, engine and vehicle size, and speed limits will need to be reduced. Domestically produced goods should be prioritised over imports. Common stores might provide low-cost or free second-hand clothing and furniture. Recycling must be compulsory. Manufacturers will need to offer cheap repair options.[40] Common

37. For the example of Britain, see Ina Zweiniger-Bargielowska's *Austerity in Britain: Rationing, Controls, and Consumption, 1939–1945* (Oxford University Press, 2000).

38. *Food Rationing and Supply 1943/44* (League of Nations, 1944), 52.

39. Monbiot, *Heat*, 45.

40. Limited right-to-repair legislation was introduced in the EU in 2021, requiring products to last ten years.

kitchens might furnish free or very-low-cost food both to reduce waste and to feed the hungry. In some areas, such as beef and lamb imports, outright bans might be considered, or quotas imposed. (Beef production alone is about 4 per cent of all global gas emissions.) As the oceans warm, fish will become increasingly scarce, and industrial fishing, very high in CO_2 emissions and extremely destructive of both fish stocks and the sea bed, will have to be sharply curtailed and in some cases cease altogether. Cheese, milk, cereals, sugar, rice, coffee, and tea will also be affected. Synthetic or laboratory-grown substitutes will become increasingly widespread.

To equalise the burden, regulating consumption will require assessing our individual global carbon footprints, and establishing a per capita maximum, or personal carbon budget. So if you travel more unsustainably, you consume fewer unsustainable commodities. As transportation becomes less polluting, some of these restrictions may ease. They will be offset, however, by the effects of temperature increases. These will also affect power consumption. But if the transition to renewables is rapid enough, by around 2030, we should be able to cool ourselves in the very hot summers of the 2030s and beyond. But the problem is not only energy use and consumption. Warming will cause shortages of many kinds, which is another reason why we must change our attitudes towards consumption as such. Rationing will obviously be unpopular until the general population realises its necessity—this is where we most require leadership, and the repeated message that the end justifies the means. As in World War Two, when "the affluent classes felt the greatest culture shock", the wealthiest 15 per cent or so, the world's middle classes, will have to give up more, relatively, and the wealthiest the most. But they are unlikely to want to do so, and well placed to evade any sacrifice.[41] Environmental necessity, however, dictates that the high-carbon lifestyles of the "polluter elite", as Dario Kenner terms them, must end.[42] A real risk remains that the wealthy will try to ensure, as in wartime black markets, a steady supply of scarce or prohibited goods to themselves, regardless of the burdens borne by others. But if we lose the battle for equality, we will lose the war on consumerism. Only by being seen to be fair can any such system of restrictions succeed. So consumption must be as transparent as possible.

Voluntary renunciation can work, then, only if others are seen to comply. Thankfully, twenty-first-century technology provides us with a degree of public transparency Thomas More would have lauded. Campaigns against the symbols of excess wealth and conspicuous consumption will be necessary. (Shades

41. Katherine Knight, *Rationing in the Second World War* (Tempus, 2007), 20.

42. Kenner, *Carbon Inequality*, 27.

of *Nineteen Eighty-Four* . . . or at least Savonarola.) In creating the new mentality which can alone make this transition possible, public opinion will be at least as effective as legal regulation or technological surveillance. Utopians across the ages have realised that all the wealth in the world counts for nought if no one respects its possessor. Public opinion is our most powerful ally. The cold shoulder is its unspoken expression. The green glare will provoke discomfort. Shunning and ostracism will work much better than sumptuary legislation. Soon we will regard those who display luxury purely for the sake of showing how rich they are with the same contempt we now feel for trophy hunters who kill lions or giraffes. Like the children in More's Utopia, who laugh at foreign ambassadors clad in jewels, pearls, and golden chains, we will mock the ostentatious. We will unfollow them on social media, and shun their presence. We will stare silently but intently at their cars as they drive by, or perhaps jest and tell jokes. By cultivating a sense of luxophobia and ceasing to esteem people who flaunt luxury goods we can show a fine contempt for such excesses.

Such acts may seem rude, but they are proportionate: destroying the planet is vastly worse. We need not boo or make obscene gestures at the drivers of expensive cars to make this point, or, like Diogenes the Cynic, spit in the faces of the rich. Acts of symbolic violence are to be avoided. A Godwinian disdain is sufficient. (Laughing or pointing at conspicuous consumers, or giving them a thumbs-down gesture, or holding our noses—the stench!—might be acceptable.) We should not forget that the cause of virtue and violence against the vicious have often been associated, in imperialism, for example, and in revolution. But moral condemnation of excess will play a vital part in reversing the emulation cycle, and a thumbs-down gesture, or turning our backs, might be quite appropriate.[43] This much moral suasion and symbolic disapproval we will simply have to accept, and it is certainly compatible with the liberal tradition, which recognises many limits on tolerating antisocial behaviour.[44] "Flight shame" (*flygskam*), the anti-flying movement which began in 2018 in Sweden, offers an instructive lesson here. But some symbols of excess will simply need to be outlawed: Maseratis and private jets, corporate helicopters, yachts, and cruise ships are all hugely polluting. As with restrictions to limit infections in a pandemic, this is not a private rights matter but a public health issue. But there are further implications. Private collections of rare and expensive art and antiquities, too, ought to be housed in public galleries or made otherwise accessible as the

43. As related by R. Frank in *Luxury Fever* (203).

44. On justifications for non-coercive pressure, see my "Mill, Moral Suasion and Coercion", in *Ethical Citizenship: British Idealism and the Politics of Recognition*, ed. Thom Brooks (Palgrave-Macmillan, 2014), 79–104.

common patrimony of humanity, and to set the standard of public luxury first, private luxury second. (So thought the liberal John Stuart Mill too.)

Then there is the problem of a new "agrarian" law, now not so "agrarian", but broadly a wealth tax. A social model dominated by billionaires is the most destructive in history. Billionaires think that because they are billionaires they shouldn't have to pay much if any tax: they, the modern saints living in a state of grace above mortal constraints, are too special. So we need to make it impossible for individuals to accumulate great wealth. The rich tell us that limiting inequality limits wealth production. We will tell them that limiting excessive consumption requires reducing inequality of wealth and the concentration of ownership. While reducing extreme inequality, however, we also need to abolish poverty. Providing a universal *middle-class* income for all, not a "basic income", which is not a "real utopia",[45] but a universal sustainable income (USI), would clearly be an important step in stemming this inequality, and in addressing the problem of looming widespread unemployment.[46] Such proposals have been discussed for many years. They now need to be fully developed: their time has come.[47] They must be accompanied by schemes for outlawing very large fortunes by an environmental wealth tax. This is what Thomas More's utopian republicanism might look like in the twenty-first century. These proposals are fleshed out in greater detail below.

I Am Not Your Servant

Relative equality and greater happiness will not be achieved merely by abolishing billionaires. We also need to modify greatly the way we treat other people, to become more alert to their need for dignity and humanity, and to extend them more respect than is customary in countries where rigid class structures exist. This will give most people a stronger sense and feeling of self-respect, a crucial component in our sense of well-being. The greater equality sustainable societies require will also diminish the cycle of emulation which drives much excessive consumerism. Seeing others as equals involves not treating them as competitors, or as a means to advance our own interests, and more as simply fellow human beings trying to make the best of life. We must resist status differentiation based on wealth. But this requires a tremendous exertion, for power relations everywhere interfere with such efforts.

45. See, generally, Rutger Bregman's *Utopia for Realists* (Bloomsbury, 2017).

46. E. Wright, *Envisioning Real Utopias*, 5.

47. See Boris Frankel's *The Post-industrial Utopians* (Polity, 1987, 73–86) for a survey of such proposals.

Here utopian sociability confronts its greatest test, and utopia is most explicitly both a method and an end. Maximising friendly interaction will extend that sociability or friendliness commonly associated with "real" utopian or heterotopian spaces to everyday life, not necessarily with the same intensity, but on the same principle. (In a show of unity we might hug strangers at a festival, for instance, but can't expect this will occur on the street.) This involves an investment of time, and slowing life down, in order to create meaningful encounters, which come mostly from shared experience. We need a great deal more small talk with neighbours, bank tellers, shop assistants, waiters, cleaners, colleagues, and so on—in fact, everyone we encounter. We need to realise that just because other people serve us does not mean they are our servants or should be treated as servile. Putting ourselves in a service role routinely is the most effective way of learning this. We must also actively endeavour to be kinder, not only enquiring about other people's lives but trying to assist them. Try one gratuitous act of friendliness a day. Every such gesture builds up a reservoir of meaningful sociability which encourages further such acts. Acts of amiability do not necessarily lead to friendship. But they make everyone feel better.

Practical or realistic utopianism begins with just these everyday acknowledgements that people are not our servants, but merit our respect, for in fact we should all ideally serve each other in turn. This is the most basic form of expressing community. Treat people with kindness if possible, and politeness and openness if not, and expect no gain in return. This requires no great effort, and the rewards are tangible and immediate. You can say "hello" or "good morning" to people if you are out for a walk. One word can break a bad mood and dispel fear, anxiety, and paranoia. Every greeting demarcates inclusivity. A grin of acknowledgement can be a subversive act, a wedge inserted into the system of exploitation and alienation. One smile can make someone's day and lift the gloom of despair, and of withdrawal into our shells, where we stew in our anxieties. Then look into people's eyes: we don't want the routinely insincere, fake "have a nice day". We want the heartfelt, authentic "thank you very much for helping me—I really appreciate it". There's a challenge to the heart! And some chatter, and banter, about books, politics, sports, the weather . . . this is how everyday utopianism commences. We all know how differently we feel when we are treated this way: we feel recognised, accepted, worthy, relaxed, fully human.

Every such encounter with others proves this principle. A sense of equality empowers us, and feeling acknowledged makes us happier. Not having to imitate the wealthy will also make us much more likely to form our own characters, to cultivate our own individuality, in the Millian sense, and choose our own paths in life, however varied and diverse they may be.[48] This process

48. J. S. Mill, *On Liberty*, chap. 3.

would be greatly facilitated by ensuring that everyone engages at some point in more difficult menial or serving activities, and in public service of some kind. This principle, recognised by Marx and many other socialists, is the best means of ensuring that the division of labour does not create fixed categories of rank or class. It is treated in greater detail below. But first let's consider the economic measures which must necessarily accompany the transition to sustainability.

The Great Change: The Sustainability Paradigm

Completely unparalleled in human history, the situation we face demands great sacrifices from all of us. Rich and poor, of every country, class, and race, we must adjust our expectations of life and concede that without a major alteration in our lives we will have no planet to live on, and no one alive to inhabit it. The sustainability paradigm involves reassessing our desires and recognising that survival involves reducing wants. The infantile insistence that every desire can and must be satisfied must be superseded by greater self-command, and a rational approach to needs.[49] The new paradigm owes to capitalism an emphasis on maximising technical efficiency, and an element of technological utopianism. In its dedication to equality and antipathy to exploitation it is indebted to socialism. But in the centrality it gives to sustainability it differs greatly from both the two main modern paradigms. It is uniquely utopian.

Central to this paradigm is its emphasis on sufficiency, or "enoughness". To James Nash, this involves "moderation, temperance, thrift, cost-effectiveness, efficient usage, and a satisfaction with material sufficiency." Respecting possessions, and consumption, it assumes a degree of austerity, or asceticism, and simplicity, rural or otherwise. It also represents a different concept of abundance, "a quest for *being* more rather than *having* more—that is, a qualitative rather than quantitative enrichment".[50] We do not need immense choice, expensive packaging, scarce and out-of-season imported delicacies. To survive we can make do with less, while enjoying what we have just as much, and even increasing pleasurable experiences so long as they remain sustainable. Regardless of what we want, we do not "need" any product which destroys the environment; our need to preserve nature is much more fundamental. This implies

49. See, generally, Regenia Gagnier's *The Insatiability of Human Wants: Economics and Aesthetics in Market Society* (University of Chicago Press, 2000).

50. James A. Nash, "On the Subversive Virtue: Frugality", in Crocker and Linden, *Ethics of Consumption*, 421, 427.

giving up things the wealthy regard as theirs by right.[51] "Sovereign citizens" and "sovereign consumers" may baulk at proposals to limit their "right" to consume anything they can afford. To them, freedom of consumer choice *is* freedom as such.[52] But there is no freedom in collective self-destruction. We must relinquish the sense of power and command, of self-validation and self-worth, which such choice evidently offers. We now revel in our ability to bring resources to our doorstep with a few phone clicks. We must learn to consider how little the delivery person is paid, or what the environmental cost is, then legislate to increase the former and diminish the latter. "Freedom to choose" cannot include anthropocide or ecocide, our species' and planetary destruction, any more than it does the right to infect others in a pandemic.[53] There is such a thing as social liberty, keeping societies free for all even at the expense of the individual liberties of the few. Now is the time to support it.

So the orgy of consumption which defines the past two centuries must now end. Hints at how to halt the process have long been present. The first international Buy Nothing Day was in 1992. Black Friday can become Green Friday. Watch No TV Week has been floated—No Smartphone Day might raise a rebellion today, though. But the principle applies to all forms of energy and resource consumption. That is not to say that the transition process will not occasionally be shocking and dramatic, and not only for the wealthy few. As early as 1975, Robert Heilbroner described a period of "harsh adjustment" to an ethos of no growth.[54] But forged consciously and democratically, the Great Change need not be a Great Leap Backward.

Voluntary Simplicity

So this scenario is not all sackcloth and ashes. Green may be lean, but it does not have to be mean. A shift away from private luxury does not imply asceticism for all, but rather greater public luxury mixed with some private restraint. Obviously, the more voluntary this is, and consensual, the less painful and coercive it will be. So a luxury debate comparable with that of the eighteenth century, but with even greater import, must clearly commence once again. Pleas for a "simple life", we have seen, have been recurrent since the Industrial

51. Rudolf Bahro suggested that our standard of living might have to be reduced by as much as nine-tenths: *Avoiding Social and Ecological Disaster: The Politics of World Transformation* (Gateway Books, 1994), vii.

52. Zygmunt Bauman, *Freedom* (Open University Press, 1988), 59.

53. William Ophuls, *Ecology and the Politics of Scarcity: Prologue to a Political Theory of the Steady State* (W. H. Freeman, 1977), 151–52.

54. Robert Heilbroner, *An Inquiry into the Human Prospect* (Calder and Boyars, 1975), 136.

Revolution: in the 1790s, the 1840s, the 1890s, the 1930s, and again in the 1950s and 1960s. The term "voluntary simplicity" (or VS) was popularised by a student of Gandhi, Richard Gregg, whose *The Value of Voluntary Simplicity* (1936) urged "singleness of purpose, sincerity and honesty within, as well as avoidance of external clutter, of many possessions irrelevant to the chief purpose of life."[55] Revived in the 1960s, the ideal has been defined as embracing "frugality of consumption, a strong sense of environmental urgency, a desire to return to living and working environments which are of a more human scale, and an intention to realize our higher human potential—both psychological and spiritual—in community with others."[56] Duane Elgin promoted the term in his 1981 book *Voluntary Simplicity: Toward a Way of Life That Is Outwardly Simple, Inwardly Rich,* which wedded the concept to "ecological consciousness", and emphasised "the importance of relationships" in human well-being, and of "connectedness and community".[57]

VS is now often associated with the "slow" movement, which began in Italy in the 1980s in response to the arrival of McDonalds in Rome. Besides championing healthy eating, it combines hostility to conspicuous consumption, an ethical contempt for unnecessary luxury, an emphasis on qualitative rather than quantitative consumption, and the desire for ecological sustainability. It is sometimes linked to "de-modernisation", insofar as this implies accommodating ourselves with nature rather than merely exploiting it.[58] It involves weaning ourselves from that constant lust for novelty which keeps the moderns' adrenaline pumping. It implies a modest stoicism now in short supply. It will be very good for our collective blood pressure, and our waistlines.

VS is also linked to the desire to reduce work time, while ensuring that the resulting leisure does not mean increased unsustainable consumption. It implies more sociability with family and friends, more public entertainment, and a willingness to exchange some luxuries for more free time. It does not entail a uniform and all-encompassing anti-hedonist asceticism. To the contrary, as we will see, it can be explicitly hedonist, as long as our pleasures are sustainable, and implies not merely more but more varied and complex amusements.

55. Gregg, *The Value of Voluntary Simplicity*, 4.

56. Duane Elgin and Arnold Mitchell, "A Movement Emerges", in *Voluntary Simplicity: Responding to Consumer Culture*, eds Daniel Doherty and Amitai Etzioni (Rowman and Littlefield, 2003), 146.

57. Duane Elgin, *Voluntary Simplicity: Toward a Way of Life That Is Outwardly Simple, Inwardly Rich* (William Morrow, 1981), 49, 64, 40.

58. See Peter L. Berger, Brigitte Berger, and Hansfried Kellner's *The Homeless Mind: Modernization and Consciousness* (Pelican Books, 1974), 180–204, for a discussion of the countercultural approach to this issue.

Greater personal simplicity is not even hostile to "growth" or progress as such, provided these are sustainable, and construed in more qualitative than quantitative terms, once we—and especially the poor—have reached a certain standard of living. But, as Amitai Etzioni stresses, VS insists on satisfying basic needs for all before comforts and luxuries. It also addresses the problem of a recurring desire for status differentiation by using a range of "lower cost but visible consumer goods" to indicate the choice of a sustainable lifestyle.[59]

VS thus implies desiring fewer material goods and unsustainable commodities or activities. This would appear to make the Spartan or primitivist traditions in utopianism useful precedents. But the technologically obsessed spirit of our own times, and the need for a massive expansion of sustainable technologies, makes this unlikely. In fact, recalling two great late nineteenth-century examples, Bellamy's *Looking Backward* will probably prove more relevant for the twenty-first century than the romantic medieval nostalgia of Morris's *News from Nowhere*. This does not accord well with the temper of much literary utopian scholarship today, where Morris is usually favoured. But most people today will probably not be willing to renounce labour-saving or entertainment devices, though they may accept sustainable use limits, like not upgrading them so frequently. And the VS vision needs to be portrayed in an urban and communitarian context, as we will see shortly. If "Individual efforts to live more simply are more likely to succeed in a supportive community", this requires building stronger communities within great cities, especially at the neighbourhood level.[60] Local initiatives will prove immensely important to building the future.

Political Implications

What does the process described so far imply for politics?[61] This is a delicate question, and the more so as we recognise just how much more fragile democracy is in the 2020s than we imagined for the past half-century and more, how disillusioned many are with the electoral process, and how willing the mostly corporate opponents of sustainability and/or the victims of globalisation are to resort to authoritarianism. Then there is the problem of introducing the measures necessary for sustainability to be achieved. It is true, we will see, that for

59. Daniel Doherty and Amitai Etzioni, eds, *Voluntary Simplicity: Responding to Consumer Culture* (Rowman and Littlefield, 2003), 17–18.

60. Jackson, *Prosperity without Growth*, 150.

61. An account of a potential utopian citizenship is given in Rhiannon Firth's *Utopian Politics: Citizenship and Practice* (Routledge, 2012).

the 1 per cent some restraints and downright prohibitions will have to be imposed. The fast lane will be closed, and speed restrictions imposed on a high-octane lifestyle. To the degree to which we can forge a consensus around sustainability, however, coercion can be minimised. Creating this consensus will be the primary responsibility of governments worldwide for the next decade or so.

But no one imagines this will be easy. Disseminating accurate information about the dangers we face is the starting point. Then comes the task of persuading the public with regard to the structural changes required. The programme sketched below, we will see, does involve substantial wealth transferral from the very rich. This is the only way we can pay to save the planet. But no one really loses by such a transfer, or return, of wealth to the common treasury, since billionaires too, fantasies of Mars colonisation or cocooning in vast bunkers notwithstanding, will not survive an environmental apocalypse, or the nuclear wars and nuclear winter which would inevitably accompany it. If a few did, improbably, exiting to their boltholes like first-class passengers escaping in the *Titanic*'s lifeboats, their quality of life would be pathetic. But none will—and the wealthy need to learn this as soon as possible. We will all die together or we will all live together.

There are other implications, too. The Great Change will require a much more interventionist state and international order than has been common during the past half-century. The main arguments against central planning in particular, since Hayek, now fall to the wayside. In 1943 Hayek wrote that:

> The "social goal," or "common purpose," for which society is to be organized is usually vaguely described as the "common good," the "general welfare," or the "general interest." This is perhaps the most dangerous and muddle-headed argument mounted in modern politics. It does not need much reflection to see that these terms have no sufficiently definite meaning to determine a particular course of action.[62]

This was always a ridiculous proposition. But now, more than ever, it is abundantly clear that preserving the planet is absolutely in the "general interest", and that "the market", shorthand for large-scale economic interests, will not do the job. There is such a thing as "society": it is our common interest, and our collective self. Neoliberalism, and the insistence on the need for an extremely small state with little obligation to assist the majority, is dead. The sustainable state, with a mighty mission before it, will take its place.

As William Ophuls already stressed in the late 1970s, we thus require "political institutions that preserve the ecological common good from destruction

62. Hayek, *Road to Serfdom*, 57.

by unrestrained human acts." Democracy based on extreme individualism, and on defining self-interest as maximising consumption and portraying this as the epitome of freedom, "cannot conceivably survive" this process of adaptation.[63] Analogies with the War Communism phase of early Bolshevism are sometimes made in describing a green planned economy.[64] But popular consent is still vital—indeed, more so than ever. Everyone must feel themselves to be a part of the transformation, and to be saving the world *they* and their family and friends live in, and which their descendants will inherit. This means that disseminating accurate information about climate crises and suppressing disinformation are essential. It also requires electoral reform, including encouraging voter participation, and shortening electoral cycles. The 2020 US election cost $14 billion, money we can now ill afford to spend this way. We would be better off with a maximum three-month campaign, with limited advertising, and each major party spending the same amount, out of public funds rather than campaign contributions. The stranglehold which the wealthy have over the processes which facilitate democracy must be broken.

Eliminating corruption in our current electoral and governmental systems requires equally great efforts. The link between the major fossil fuel corporations and the political process must be severed. The sustainable state and the planning process must be much more transparent than any preceding political order. The financial transactions of all public servants must be subject to full public scrutiny. Politicians must be barred from accepting favours from lobbyists. To avoid conflicts of interest all should have only their one, public income, hold no "offshore" accounts, and ensure their tax returns and major financial transactions are completely public.[65] Electoral processes must be purged of the corruption of lobbying, electronic interference, and the proliferation of "fake news". Everything possible must be done to distance the democratic process from the influence of billionaires, as well as malevolent internal and external actors. Public subsidies to ensure the expression of a wide range of views are necessary; there is no "free press" where a few billionaires dictate opinion. Requiring the press and social media to give equal space to a range of viable candidates would reduce the influence of the wealthy. Breaking up press and social media monopolies to ensure a wider spectrum of opinion is also essential. Public media must be tasked with selling anti-consumerist lifestyles and encouraging civic virtue. Regulating disinformation in social media is vital. The process

63. Ophuls, *Ecology*, 151–52.

64. See, e.g., Andreas Malm's *Corona, Climate, Chronic Emergency: War Communism in the Twenty-First Century* (Verso Books, 2020).

65. In Norway all tax returns are publicly available.

of political deliberation must also become more popular, in order to eliminate that political alienation which is now so common. Assemblies drawn at random from the citizenry, guided by appropriate scientific expertise, have a policy role to play here. This is a task the many must be involved in, and which cannot as such be trusted to governments alone, much less "the market".

Clearly we also need a renewed sense of political engagement and civic responsibility. Decades of praising selfishness and denigrating the public sphere have left us with an anaemic sense of public duty. Now a stronger spirit of dedication to the common good must come to the fore, and be instilled from childhood onwards. Some of the identity we have lost to consumerism must be recaptured as citizenship. There is nothing sinister in such pleas, which are in everyone's interest. It is not totalitarian indoctrination to teach children the value of recycling, or respect for nature, or the history of environmental degradation, or how to repair things, or why plutocracy is destructive, or how democracy is supposed to work, or why corruption is corrosive. Laissez-faire, the "invisible hand", and "economic man" have now reached their expiry date. The sustainable citizen must be our new human type.

The international order must be similarly transformed. Currently, more or less hostile large national blocs—notably, the United States, Russia, and China—seek through every means short of direct war to undermine one another. Size really matters here. As ever, great states have bloated egos proportionate to their imagined importance and continue to tyrannise over the smaller and foster the most oppressive forms of nationalism, nursing their historic grudges and grievances like bruised children in the playground. We would be better off breaking up the larger into less malignant groups, while retaining a federative structure. But every country is threatened by climate disaster, and only through international unity and close co-operation can disaster be averted. This is the most global problem humanity has ever faced. We cannot shove it off on our neighbours or cheat by making others assume greater burdens, though everyone will try to do so. It can be solved only with greater collaboration than has ever previously been possible. Whether such international co-ordination will amount to something approaching world government, as Wells suggested, remains to be seen, but is probable. Certainly, the era of petty nationalism must finally end, and petty rivalries over territory will have to be suspended, while accommodating necessary population movements. A grand rapprochement in international relations must take place, with a cessation of all hostile acts towards other nations. Much greater regional and global co-operation will also involve levelling out "development". Debt forgiveness in exchange for sustainable policies, and raising the standard of living of poorer countries, should be a norm. The facile populist nationalism which has emerged in the wake of the failures of globalisation only hinders our capacity to guarantee future sustainability.

(De)globalisation is an issue which must also be addressed. Compensation will have to be offered to those nations which have profited from entering global markets but who will suffer when long-distance transport and tourism are temporarily reduced and local self-sufficiency encouraged, as well as those who are suffering already from dramatic climate change. Clearly it would be much preferable to transfer wealth to the global south and poorer areas to mitigate environmental damage than to spend the same sums on military forces and constructing barriers to movement. The impetus for such mutual aid, admittedly, is more likely to come from catastrophe than from voluntary planning. A sudden shock, like evidence of an imminent tipping point, is most likely to lead us to realise what is required. Disasters foster such an ethos, which, Rebecca Solnit notes, "is always coming into being in response to trouble and suffering".[66] But by the time the worst shocks take place it may be too late. So we have to anticipate them.

Obviously, some of the measures proposed here involve interference with the liberty of a few to assist that of the many. Politically, however, the risk of extremist, right-wing, authoritarian regimes born out of fear, especially of vast environmentally driven refugee migration, estimated by the UN to be around 250 million by 2050, is much more likely than eco-fascism. We must accept that the imperatives facing us do threaten to undermine our liberties, and that some freedoms must be curtailed in order to ensure others, as well as our survival. But the much more equal society sketched here would also extend the liberties of the majority well beyond what they now enjoy. We must also acknowledge the growing risk that we will not act in time, and the international economic and political order will collapse into anarchy, barbarism, and war, which will also encourage corporatist totalitarianism. A reversion to primitivism could occur if civilisation collapses into the kind of post-apocalyptic scenario beloved of modern dystopian writers. This risk is sufficiently great to justify measures which are considerably more radical than those usually included under the Green New Deal rubric. But let us first consider what the latter is usually supposed to consist of.

The Green New Deal

Mooted since at least 2008, the idea of a Green New Deal derives from Franklin Delano Roosevelt's recovery programme during the Depression, when, amongst other things, large-scale state intervention and investment led to the public employment of about 5 per cent of the male US population, the

66. Rebecca Solnit, *A Paradise Built in Hell: The Extraordinary Communities That Arise in Disaster* (Viking, 2009), 313.

planting of some two billion trees to stem topsoil erosion, a vast dam-building programme, and many other public works projects.[67] The parallel is not exact, however. Our problems are global, vastly more is at stake, and proposals involving only one nation and without very substantial modifications of capitalism, even its abolition, are senseless. So far the phrase is chiefly a slogan, with little sense of urgency attached: a proposed €1 trillion EU Green New Deal measure passed in December 2019, for instance, gives 2050 as a goal for zero carbon emissions, which is much too late.[68] So comfortably distant as to be meaningless, this date also figures in many other current discussions of emissions reductions. It is a sop to the green lobby by those who do not even want to contemplate the gravity of our situation. It symbolises apparent action but disguises real inaction.

Certain assumptions unite most plans of this type. The British Labour Party's initiatives and Alexandria Ocasio-Cortez's proposals of 2019, or EU plans in 2020 to reduce carbon emissions by 55 per cent by 2030, assume massive state intervention to promote sustainability and limit environmental damage.[69] These schemes are often bundled with infrastructure repair or development plans and welfare and job guarantees, and, in the United States, a long-overdue universal healthcare programme.[70] Some versions envision reducing military spending—the US military alone is the forty-seventh-largest greenhouse gas emitter on the planet. The aspirational norms for reducing global warming are generally more ambitious than in previous targets (100 per cent renewable energy by 2030, for instance), and a 2°C ceiling on warming is often regarded as too high. Costs have been estimated at up to $93 trillion in the United States alone, with the price of carbon removal estimated at about $5.1 trillion per year, leading critics to claim such plans are simply unaffordable, as if such calculations are relevant.[71] (An analogy: we let the patient die

67. A recent statement of economic issues is given in Ann Pettifor's *The Case for the Green New Deal* (Verso, 2019), which dates these discussions to 2007 (2-3). Pettifor argues for a "collaborative" approach rather than the implementation of a plan by a single global authority (68). She defines a Green New Deal according to seven principles: a steady-state economy; limited needs, not limitless wants; self-sufficiency; a mixed-market economy; a labour-intensive economy; monetary and fiscal co-ordination for a steady-state economy; and abandoning principles of infinite expansion (93-108).

68. Of this, about two-thirds is slated to come from the European Union's budget, the rest from member nations. A 50–55 per cent cut in emissions is proposed by 2030. The overall scheme is still described as a "growth" strategy in which "prosperity" will not be affected.

69. On "green growth", see Daniel J. Fiorino's *Good Life*.

70. The most substantial part of Biden's early 2021 initiative addressed such repairs.

71. Bill Gates, *How to Avoid a Climate Disaster: The Solutions We Have and the Breakthroughs We Need* (Allen Lane, 2021), 63. The investment bank Morgan Stanley has estimated that $50

because he or she does not have insurance or a credit card.) A temperature rise of 3.7° would cost an estimated $551 trillion. (The world's gross domestic product (GDP) was *c.* $80 trillion in 2017.)[72] In the United States the case has been broadly stated in Naomi Klein's *The (Burning) Case for a Green New Deal* (2019), which indicates the nature of the crisis we face. Klein's six-point programme corresponds with many of the proposals outlined below, including reviving the public sphere; reintroducing planning; reining in corporations; relocalising production; ending the cult of shopping; and taxing the rich, and especially oil corporations.[73] More detail is offered here to flesh out the "egalitarian" and "communitarian" elements Klein mentions. Some delicate issues she ignores—notably, population growth, or does not flesh out, like and income and wealth taxation—are also tackled here.

Most contributors to this debate recognise that "green growth" is also partly "de-growth", involving a massive shift in industrial emphasis and adopting steady-state strategies. When the "limits to growth" hypothesis entered common currency during the energy crisis of the 1970s, it was suggested that the earth would exhaust its resources in about a century.[74] Pleas for a "sustainable society" appeared which now look strikingly prescient, if often ominously over-optimistic.[75] Nonetheless, we have not yet broken free of the growth paradigm or conceded that future progress must be qualitative rather than quantitative in all areas except sustainability technologies. As recently as 2009, green economists were writing that "There is no case to abandon growth universally", because "there is a strong case for the developed nations to make room for growth in the poorer countries."[76] This is now true only for sustainable growth, which must of course also raise the standard of living in poorer countries. To Marius de Geus, there is no possible "sustainable development"

---

trillion will be required to achieve 1.5°C warming, including $20 trillion for hydrogen power development, $14 trillion for wind, solar and hydro power, and $11 trillion for electric vehicles.

72. Wallace-Wells, *Uninhabitable Earth*, 166.

73. Naomi Klein, *The (Burning) Case for a Green New Deal* (Simon and Schuster, 2019), 81–90.

74. *Limits to Growth: A Report for the Club of Rome's Project on the Predicament of Mankind* (Potomac Associates, 1972), 23.

75. E.g., Lester W. Milbrath's *Envisioning a Sustainable Society: Learning Our Way Out* (State University of New York Press, 1989), which puts world population at at least a billion less in 2020 than it was (13).

76. Jackson, *Prosperity without Growth*, 41.

beyond the parameters of a Green New Deal.[77] The earth is simply degrading too quickly. So, writes Hans Jonas, "the watchword will have to be contraction rather than growth, and this will be harder to adopt for the preachers of utopia than for the ideologically unencumbered pragmatists."[78] But utopians, too, can concede that production, consumption, distribution, and demand in many areas must be reduced.

The transition to sustainability obviously risks an unprecedented global depression just as we recover from the damage inflicted by Covid-19. Entire sectors of the economy employing millions may simply disappear. Avoiding disastrous economic consequences will require careful planning of alternative employment, providing substantial retraining of displaced workers, and ensuring a safety net for all affected by this momentous transformation. Every Green New Deal implies massive expansion in both investment and production of sustainable technologies and infrastructure, dwarfing, James Lovelock suggests, our previous space and military expenditure combined, both of which will have to be reduced.[79] But this process must also involve a contraction in the overall volume of production, to reduce emissions and preserve resources. Large-scale investment is required in solar, tidal, and wind power; desalination; insulating buildings and replacing billions of windows, roofs, and walls; public transport; sustainably fuelled aircraft; fossil-fuel decommissioning; rewilding, species renewal, and reforestation; fire-fighting equipment; relocating environmental refugees; building vast underground reservoirs and walkways under cities; replacing roads, electrical wiring, and other infrastructure destroyed by higher temperatures and extreme weather events; constructing flood defences and tidal walls; creating overground shelters and underground residential facilities in the hottest areas; and much more. Every technology involving carbon capture and greenhouse gas reduction must be rapidly developed. All this involves a green cost—namely, rising short-term emissions in order to reduce long-term emissions.

These measures do not imply that there is a technical "fix" for environmental destruction. We may envision solar panels, wind farms, electric cars, efficient air conditioning, recycling schemes, carbon capture, organic food, reforestation, and desalination as solutions to our problems. Clearly intensive research into and development of every technology which might aid such

77. Marius de Geus, *Ecological Utopias: Envisioning the Sustainable Society* (International Books, 1999), 13. See also Geus, *The End of Over-Consumption: Towards a Lifestyle of Moderation and Self-Restraint* (International Books, 2003).

78. Jonas, *Imperative of Responsibility*, 161.

79. James Lovelock, *The Revenge of Gaia* (Allen Lane, 2006), 12.

efforts will be needed. But there is still no substitute for an essential change in our attitudes towards the earth and its resources. No magic wand will remove greenhouse gases from the atmosphere, or acidity from the oceans, or reduce the earth's temperature. Billionaires like Bill Gates think their own lifestyles, and their private jets, need not be affected by a dramatic reduction of CO_2 emissions, and expect both wealth and population growth to continue to expand, with all the attendant problems being solved by technological innovation.[80] This may comfort his wealthy readers. But it is a wildly over-optimistic scenario. We cannot allow billionaires to jet about while the rest of us have our travel restricted, or fail to alter the more destructive aspects of our lifestyles.

For similar reasons there is no market "fix", either. The "market" is just rich people who control vast amounts of money. And it has so far mostly assisted the process of environmental destruction. This "market", here chiefly a few large banks, has invested some $1.9 trillion in fossil fuels since the Paris Agreement of 2015. Private companies can assist in producing labour- and energy-conserving devices. But stimulating their use, especially by providing immediate subsidies for their introduction, and encouraging waste reduction, requires massive state intervention. Thousands of factories will have to shift from producing harmful or non-essential goods to producing sustainable ones. This will be extremely expensive, and we cannot afford corporate profiteering on top. So corporations generally will have to accept much lower profit margins, and, in some areas, declining demand. The size of the corporate dividend can no longer be the aim of production. Companies cannot be allowed to blackmail us by insisting that they either profit from cleaning up the environment or they won't play along at all. To ensure this, both the structure and function of the corporation will need to change dramatically.

All this can be incorporated into a global planning process which ensures that the system of needs as understood in capitalism is transformed in order to prioritise sustainability. Critics who insist that twentieth-century socialist central planning was a manifest failure should note that environmental planning is not for the market or consumers as such, but rather for humanity generally. It can be more or less efficient, but cost as such is not the issue. There is no choice. We want to survive, not to ensure corporate profits while humanity dies. No price is too high to ensure life itself.

Many countries, including the most powerful, will try to thwart such proposals. Even now, most do not understand what is at stake and why special pleading must be rejected. Compliance must be assured by popular agitation,

80. Gates, *Climate Disaster*, 113. There is no discussion of consumerism here at all, or of taxation.

including mass non-violent campaigns of protest and the boycott of "climate outlaws", whether they be individuals, corporations, or nations. With sufficient consent, what used to be called a "dictatorship over needs" may be introduced democratically without generating profound social disruption.[81] International agreements will enact measures to transfer wealth to countries too poor to create sustainable infrastructures on their own. Countries extremely affected by climate degradation, already including Australia, India, Chile, Zimbabwe, and South Africa, will require special assistance. In some cases reforestation will have to be paid for by wealthier nations, as will compensation for reducing beef and palm oil production and fossil fuel extraction. To take one key example, the cost of recognising our common need for the Amazonian rain forest must be considerable assistance to Brazil and adjoining countries to begin reforestation, as President Biden has suggested—effectively buying the Amazonian jungle to protect it (and well before 2031, the date agreed at COP26, by which time, at the current deforestation rate of over 11,000 km^2 a year, there will be little left to protect). The same is true for Indonesia. Nations refusing to conform to these norms will be subject to boycotts and embargoes, which are crucial means of moving forward, and other types of economic and political pressure, short of war. Even the biggest players, like the United States and China, might be subjected to such shunning if they refuse to implement these programmes. There are no opt-out clauses in this scenario. There cannot be one rule for rich nations or individuals and another for the rest. We are all in this together, like it or not.

A Radical Green New Deal

Given the statistics offered here, most Green New Deal plans still fall far short of what is required, in their estimate of the severity of the crisis, their goals, and their timetable. They might have worked half a century ago, but now they offer too little, too late, and are too often accompanied by empty promises and green posturing rather than a robust and sincere commitment to action. But the remedies must be proportionate to the severity of our condition. Measures supposedly suitable to containing world temperature increases at 1.5°C will be insufficient to reduce warming to less than 1°C, which is what is required. So we must act proportionately, which means much more dramatically, much more rapidly, and with a much greater sense of urgency.

More radical versions of these proposals face even greater problems in their implementation, but are also far more likely to succeed in their ultimate goals.

81. See Ferenc Fehér, Agnes Heller, and György Márkus, *Dictatorship over Needs* (Basil Blackwell, 1983).

The immediate aim of radical environmental legislation will be to greatly reduce greenhouse gas emissions to slow global warming. Harmful emissions need to fall by 7 per cent a year for the next twenty years or so; they are currently rising at a rate of 2–4 per cent each year. In 1990, CO_2 emissions were about 6 billion tons (gigatons, or GT). They rose to 33.2 GT in 2011, and 37.1 GT by 2018—more than sixfold in thirty years. They are projected to reach over 43 GT by 2050. Methane gas is another matter entirely. It is thirty-four times more potent than CO_2, and has a global warming potential eighty to a hundred times greater. It accounts for about one-sixth of warming, and is increasing rapidly, from around 750 ppb (parts per billion) for the past ten thousand years to around 1850 ppb now, owing to fracking, flaring natural gas, beef production, and the melting of the Siberian tundra. An agreement at COP26 to reduce methane emissions by 30 per cent did not include many major emitters, and looks likely to fail to reduce total output. These developments alone are potentially fatal to the planet.

It will take about forty years for significant emissions to clear from the atmosphere, and up to two hundred for much of the rest. One goal thus dominates the coming decades: *warming must be limited to less than 1°C, since extreme damage is already evident at the current average of 1.25°. This means rapidly and dramatically reducing our carbon footprint; hence the slogan #onedegreeistoomuch.* The United Kingdom's current (2019) CO_2 footprint is about 13 metric tons of emissions per person, down from 15 in 2010.[82] The United States is at around 20, and Australia slightly more (owing to coal use), with the global average about 7. CO_2 concentration was about 280 ppm from 1780 to 1850, and only 316 in 1958. It reached 350 around 1987 and is now at 420. Keeping CO_2 levels at 350 ppm would require individual emissions, or our personal carbon budget (PCB), to average at most (in Mike Berners-Lee's calculations) 5 tons per person, assuming no growth in population. This would require eliminating a large percentage of carbon-emitting activities in the United Kingdom, and more in the United States. (For comparative purposes, a return flight from London to New York produces 0.986 tons, and London to Hong Kong 4.5 tons.[83] Using your mobile phone an hour daily for a year: 1.25 tons.) So one long-haul flight would take up virtually your entire annual PCB. Other estimates, however, insist on a need to reduce emissions by 2030 to a PCB of 2.3 tons (or even lower) to meet a 1.5°C target. Meanwhile, the wealthiest 1 per cent are now estimated to emit an astonishing 70 tons annually. Here we see where the gravest problem lies.

82. Berners-Lee, *How Bad Are Bananas?*, 11. Three-quarters of CO_2 released into the air dissolves in the ocean within several decades.

83. Berners-Lee, 12.

Phrased this way, such a project sounds impossible. But it is not. Renewable energy alone can reduce emissions by 70 per cent or more quite quickly, with little disturbance to most of our daily routine or individual lifestyles, except where high emissions are generated. Here, restraint is required, and rationing must be the main way forward until sustainable technologies are introduced. Installing solar panels alone can reduce your CO_2 emissions by 1 ton a year, as would insulating your loft.

To understand how we might achieve this we need a sense of where the most destructive human sources of fossil fuel emission originate.[84] (The natural sources are much larger, but the human sources have created our present crisis.) These can be portrayed in a table (overlapping activities result in >100 per cent outcomes):

Emissions by Category

Source of Human Emissions Overall	Breakdown by Category	Further Detail
Fossil fuels, 75–87% of all greenhouse gas emissions; 75% of CO^2 emissions	Coal, 43%[a] Oil, 36% Natural gas, 20%	Electricity and heat generation, 41% of total emissions
Transportation of people and goods, 22%	Road transport, 72% Marine shipping, 14% Aviation travel, 11%[b]	EU container traffic, *c.*25% of all automobile emissions (2019) Aviation (3% of total CO^2): International flights, 62% Domestic flights 38%
Industry, 20%		
Industrial processes (mostly chemical reactions), 4%		
Fashion, 10%		
Agriculture, 10–20% of total emissions, of which 60% is meat production		
Deforestation and land use, 9%		
Concrete, 4% of total emissions		

[a] For every ton of coal burned, 2.5 tons of CO_2 are produced.

[b] In the United Kingdom, CO_2 emissions are divided into approximately these ratios: 34% air travel, 27% gas, 16% electricity, 19% car use, 10% agriculture, 2% bus, 2% rail.

84. At present, fossil fuel companies spend some $200 million annually denying this narrative.

The principal means of coping with rising emissions is firstly transferring to renewable energy sources, then moving to a wider sustainability programme. A radical Green New Deal requires at least nine strategies which prioritise sustainability. All involve deglobalisation, or producing food and other commodities locally where possible, to reduce transportation costs, as well as downshifting, greater frugality, and prioritising public over private luxury. The initial focus, as the statistics above indicate, must be on immediate and dramatic reductions in fossil fuel consumption, by keeping coal, oil, and natural gas in the ground and massively expanding sustainable energy. This is relatively easy and straightforward. Reforestation is the second key priority and is also not difficult. Reducing other forms of unsustainable consumption then follows. Time is not on our side. Reaching any number of possible tipping points—at least nine have been identified—could negate all our efforts to prevent greater warming and induce catastrophe.[85] Some, in 2021, are virtually imminent, like a spike in Siberian methane emissions. So some painful emergency restraints will need to be introduced, some of which will last decades. But there must also be compensation for this much sacrifice, in the form of greater opportunities for sociability and self-development.

All this will necessarily demand greater international co-ordination than has ever been achieved before, for any purpose. An environmental convention far more encompassing than the Paris Agreement must be the starting point for this process. It must form an international body or environmental commission (here called ECOCOM) devoted to the purpose, stipulate its duties, and invest it with sweeping powers and a multi-trillion-dollar budget. Its edicts must be binding on every major nation, and it will also be tasked with supervision to ensure compliance. Its powers should transcend national sovereignty claims in environmental matters, including energy, industrial policy, land use, deforestation, and transportation. ECOCOM will need matching subordinate ministries in every nation, with the power to prioritise sustainability above other national goals. A vast amount depends on its efficiency, stamina, and clout as an organisation, and its abilities to police its policies, largely through satellite and drone surveillance. Its programme will be the subject of much discussion and negotiation, but should be integrated with the compensatory utopian measures outlined here.

85. They include: deforestation of the Amazon, disintegration of the Greenland ice sheet, the loss of the Siberian permafrost, breakdown of the Atlantic Ocean's circulation system, a shift in the boreal forest system, loss of the West Antarctic ice sheet, the vanishing of coral reefs, and shifts in the West African and Indian monsoon patterns.

1. *Energy*

A rapid transition to 100 per cent renewable energy sources everywhere is the first goal. In the United States, this would save some $320 billion annually. Renewables now generate about 25 per cent of electricity worldwide (38 per cent in the United Kingdom in 2019). They can be increased very rapidly and very economically and are already the cheapest source of new energy. Coal use must be ended immediately by a coal convention which terminates mining, transport, use, and funding, with retraining for displaced workers. Then oil and gas must be sharply reduced. All energy misleadingly categorised as "renewable" which involves logging or wood burning(!) (20 per cent in the European Union at the moment) needs to be reclassified and, as with fossil fuels, no longer subsidised, and reduced as soon as possible.[86] Transportation needs shifting where possible from automobile to bicycle and carbon-neutral free public transport. All new buildings will have to be carbon neutral, or indeed negative, generating surplus energy. Older buildings will have to be slowly retrofitted with heat pumps, better insulation and windows, and the like. Household electrical consumption can be reduced by turning lights off wherever possible; installing cheap, subsidised solar panels; reinforcing and subsidising home insulation; paying users to contribute to national grids; and making our entire building stock more energy efficient. Shops will need to close their doors to conserve energy, and empty buildings to turn their lights off. Physically, this implies a darker world, at least initially. There may be brownouts and energy shortages as coal, oil, and natural gas are phased out more quickly than demand can be reduced. Nuclear power is immensely expensive and inherently unsafe and should not be developed further.[87] But neither can it be decommissioned until cheaper sustainable energy sources are ready. Moreover, increased demand for air conditioning (now 10 per cent of electricity usage), as summer temperatures commonly rise to 35°C (95°F) or more, and over 50°C (122°F) in the hotter areas, for months on end, will require 40 per cent more power. A massive expansion of solar, wind, tidal, and other forms of sustainable power must occur to provide this. None of this can be accomplished without (inter) nationalising energy companies with minimal (or no) compensation to shareholders, but with reemployment assistance to employees. ECOCOM

86. The United Kingdom imports over 10 million tons for wood p.a. for burning, mostly from the southern United States.

87. Decommissioning costs in Britain alone are estimated at over £300 billion. A large number of nuclear waste sites are also located alarmingly close to coastlines, or near sea level.

can use their expertise. But the entire energy sector must be brought under public control quickly.

TRANSPORTATION: Let's begin with cars, which in the European Union emit about 12 per cent of CO_2, though there are great individual differences in emissions. Key policies here include providing free public transport, as in Luxemburg, which has the highest ratio of car use per capita in Europe, and as in Kansas City in the United States from 2019, and replacing private cars with a fleet of low-cost and eventually driverless public vehicles useable like taxis. Public mass transport will reduce pollution and energy consumption dramatically as well as costs for consumers, many of whom will no longer need private cars. Smaller and more efficient engines will aid fuel economy, until electric vehicles become standard around 2035. Most SUVs, five to six times more polluting than smaller cars, need to be banned immediately, beginning in cities. In the last decade they have become the second-biggest cause of increasing global CO_2 emissions.[88] Petrol and diesel cars must follow in a few years. But they cannot be replaced one-for-one by electric cars, which require too many scarce materials, like lithium, cobalt, and nickel. In any case, the same volume of car ownership is unnecessary and a waste of space and energy.[89] In cities, most cars sit unused 95 per cent of the time. So here, one public car for every thirty to forty people, bookable like a taxi, would probably be quite sufficient. Rush hours can be staggered to allow this.

Then air travel, which though now 3 per cent is forecast to generate about 20 per cent of all CO_2 emissions in the coming decades. Ninety per cent of people have never flown on a plane, so this is an issue for the wealthy 10 per cent. Flying produces about a billion tons of CO_2 annually (out of 33 billion total in 2018), with domestic flights outstripping foreign in terms of emissions, and has increased 70 per cent from 2005 to 2020 (32 per cent from 2013 to 2018 and 5 per cent in 2018–19 alone). Sustainably powered planes capable of carrying hundreds of passengers are perhaps twenty years off, though smaller models (up to 180 passengers) may be ready by 2030. So, in the short to medium term (ten to fifteen years), aviation travel must be rationed. If we aim for an average personal carbon budget of at most 5 metric tons of CO_2, but hopefully considerably less, recall that a return flight from London to New York takes up

88. In the United Kingdom, emissions from cars actually rose by 7 per cent between 2018 and 2019, owing to larger engine sizes. Emissions for long journeys by SUV are actually higher than flying.

89. There are currently about a billion cars in the world, but sufficient cobalt for at most only 375 million electric vehicles.

0.986 tons, a big chunk of our prospective annual PCB. Only 4 per cent of the world's population flew to a foreign country in 2018. But 1 per cent of us are responsible for 50 per cent of aviation emissions, flying an average of 35,000 miles a year.[90] If air fuel consumption were reduced by 2.5 per cent p.a., it has recently been estimated, emissions could be held steady, and then reduced.

Merely taxing flying will not disturb the 1 per cent, who can pay higher prices (and business users will pass the costs on to consumers). But it will affect many others. So in the interests of fairness, rationing must continue until environmentally neutral technologies and sustainable aviation fuels are introduced. A carbon tax on flying should be introduced immediately, with national plane travel over land, say anything up to 500 km (300 miles), abolished until sustainable flights are possible.[91] International air travel might be limited to a maximum of two trips per year per person, for business or pleasure, emergencies aside. (High-speed dirigibles flying at 135 kmph/85 mph might work on some routes.) First- and business-class seating, with a carbon expenditure three to four times higher than tourist class, should be eliminated, along with frequent-flyer programmes. All public subsidies on flying, such as for aviation fuel, should be removed. Air freight can also be reduced substantially; it has a hundred times the carbon footprint of sea freight. No food should be shipped by plane, though again dirigible substitution may be possible, and full carbon costing needs to be imposed on all types of shipment. Private jets using aviation fuel, one of the most potent symbols of the age of excess, should be banned outright; some eight thousand such aircraft are projected to be built in the next decade, each producing 10 times more CO_2 than standard passenger jet aircraft, and 150 times more than an equivalent rail journey.[92] Airport capacity should also be capped. To make this possible, the expansion of high-speed rail lines is necessary. Though very expensive, maglev trains can travel at up to 600 kilometres per hour (370 mph)—but normal high-speed trains (up to 500 kph/300 mph) are more sustainable.[93] Domestic train travel might well be free or very cheap—in 2021 Austria introduced a one-day €3 ticket.

90. *Guardian*, 18 November 2020, 20.

91. The first such move, announced in France in April 2021, prohibits domestic air travel where train times are two and a half hours or less. (The original proposal was four hours.) Half of all flights globally are less than 500 km.

92. Moreover, some 40 per cent of flights are "empty leg" journeys which return aircraft to more convenient locations. The global emissions output in the United Kingdom of private jets alone (8 per cent of air travel) is the same as 450,000 cars on its roads. Ironically, perhaps, the first carbon-neutral aircraft will be small and useable only for short-haul flights under 500 miles. There are plans for 800-mile capacity zero-emissions aircraft for around 2030.

93. Monbiot, *Heat*, 183.

International train travel should also be much cheaper, to encourage that interchange of peoples which does so much to make life interesting and pleasurable, to extend and enrich our personalities, and to promote peace. But here, too, business- and first-class travel should be eliminated.

Then sea travel. Shipping currently generates 2.2 per cent of global CO_2 emissions. Container cargo ships can be powered by solar or hydrogen engines in the near future, and a reversion to sails, perhaps augmented by airship transport, would ease the burden of international cargo shipments. Reducing speeds by half would also halve emissions. These will, however, also diminish as the principle of local production is introduced. It will take longer to acquire things we need from abroad, and we will have to learn to do without anything which is not sustainable.

2. *Reforestation, Water Management, and Species Protection*

We have cut down at least 3 trillion trees since civilisation began. Replanting them can reduce emissions by 10 per cent or more.[94] Britain alone needs to replace some 1.5 billion trees. This can be done in only a few years—a trained forester can plant 1000 trees a day. Under Roosevelt, the United States planted 3 billion trees in nine years; in 2019, Ethiopia achieved 350 million in one day. After rapid reductions in fossil fuel use, this will be the single greatest immediate step in environmental protection. So ECOCOM will need many foresters. And why not plant more trees than we have removed? Each tree absorbs some 4 tons of CO_2 over its lifetime. We cannot have too many: each person in the G7 nations is responsible for four trees per year being cut down.

Extreme heat and water shortages mean that water management will have a very high priority as alternatively drought and then flooding affect many regions in the globe. Vast underground reservoirs will be needed to retain excess water; we can use the earth removed to raise floodplains and reinforce shorelines. Increasing desertification will occur where this is impossible. In hotter countries, water control and preservation will be the highest priority in the coming decades. There are already severe shortages in India, southern Africa, Australia, and parts of the United States.

Rewilding will play a major role in reducing water pollution in urban areas, and in preserving biodiversity.[95] Britain has lost some 70 per cent of its animal

94. On some prospective limits to the benefits, see Mann's *New Climate War* (165–66).

95. For a current plan for Britain, see Dieter Helm's *Green and Prosperous Land: A Blueprint for Rescuing the British Countryside* (William Collins, 2019). Bird numbers in the United States have declined by nearly a third in 50 years.

and insect populations since the late twentieth century; James Lovelock suggests that as much as a third of the country, where 97 per cent of natural meadows have been lost, might be left unfarmed and unoccupied.[96] Eliminating harmful pesticides which poison our rivers and food and create dead zones in oceans, the issue with which the green movement began with Rachel Carson, is now more necessary than ever. The loss of bee stocks (30–60 per cent worldwide) alone makes this obvious. Much of this may be achieved only if the land is nationalised or socialised, so that public control over environmental initiatives is assured and given the highest priority. In England, at least 30 per cent of the land is still owned by the aristocracy or gentry.[97]

3. *Food*

Some 40 per cent of agricultural land worldwide is now degraded, and as much as half of existing farmland may be unusable by 2050, while demand for food will rise by 50 per cent. Prices will have to be subsidised to remain affordable, and both cultivation and supply possibly (inter)nationalised. As much as 80 per cent of fish stocks are overexploited, and by 2050 the oceans will contain more plastic than fish at current pollution rates (around 8 million tons a year).[98] Agriculture will once again need to be prioritised over manufacturing, and necessities above luxuries, to avoid food shortages if not widespread or even global famine, as the world's great breadbaskets (e.g., the US and Canadian Midwest, the Ukraine) suffer prolonged droughts, which have already commenced. The international division of labour will have to alter dramatically, as carbon pricing determines what is grown or made and where. ECOCOM will also avert famine by ensuring that food supplies can be shifted sustainably to areas in grave need. Laboratory-grown substitutes will doubtless help to fill the gap.

New policies here might thus involve moving from a global market to sourcing locally grown and entirely organic produce where possible, with a view to maximising local and regional self-sufficiency and reducing the carbon cost of food, so food may cost more when it is grown in a high-wage economy, but its carbon footprint is smaller; reducing food waste (nearly 40 per cent is thrown away in the developed world, and in the United States this waste

96. Lovelock, *Revenge of Gaia*, 133.

97. Guy Shrubsole, *Who Owns England? How We Lost Our Green and Pleasant Land & How to Take It Back* (William Collins, 2019), 267.

98. Kate Raworth, *Doughnut Economics: Seven Ways to Think like a 21st-Century Economist* (Business Books, 2017), 5. Six billion tons of plastic waste exists.

accounts for 25 per cent of water use); transforming fish farming to avoid the disasters we have seen with salmon fisheries; and shifting consumption from animal to vegetable produce, including sharply reducing if not eliminating eating red meat and especially beef entirely. (Each serving of beef produces around 3 kg (6.61 lb) of CO_2, compared with 0.57 kg (1.26 lb) for poultry and 0.4 kg (0.89 lb) for eggs. Curtailing beef consumption dramatically would release some 11 million square miles of land—equivalent to the United States, Canada, and China. In the United Kingdom, 51 per cent of land is devoted to raising meat.) A high carbon tax on beef could commence this process. Agricultural subsidies must be shifted dramatically in favour of sustainability. Only 1 per cent of some $700 billion annually is currently used this way, and the European Union's Common Agricultural Policy is one of the worst offenders. Food must be cherished and savoured. We also need to reduce dramatically the amount of fat in our diets—40 per cent of what Americans consume—as well as sugar and artificial additives.

4. *Avoiding Waste and Restraining Demand and Consumption*

About 2.12 billion tons of waste is produced globally each year, with some 99 per cent of everything bought being disposed of within six months. The average American throws away about 2 kilograms (4.4 lb) daily, about twice the global average of 1.18 kilograms (2.6 lb). Policies required here include: reducing variety where extreme wastefulness or energy inefficiency in production is evident, or where branding disguises the fact that similar commodities are being sold with different labels; minimising, shrinking the size of, and reusing packaging; and shifting to non-plastic, paper, and recyclable coverings. Plastic pollution, it goes without saying, needs to be curtailed dramatically: the use of plastic bags and bottles is actually increasing, with one trillion projected to be produced before 2025. Packaging can become more functional and less glittery. Wherever possible we must repair, rent, and borrow instead of buying, and use in common rather than acquiring individually. Libraries are a good model! Ending planned obsolescence, or the deliberate design of goods to have the shortest viable shelf life, is essential. All products should be made to last as long as possible, and where feasible should be repairable at publicly convenient shops or be bought back by their manufacturers. Our attitude must be, reversing Aldous Huxley's phrasing, that "mending is better than ending".[99] *Everything* will need to be recycled, and that does not mean exported to poorer countries. In 1942 it was estimated that half of America's

99. Paraphrasing Huxley in *Brave New World* (1955, 49).

production was wasted; the figure now is probably much higher.[100] Waste production has risen tenfold in the last century, and will double between 2013 and 2025. A very large proportion of military activity, in particular, is highly wasteful and produces a vast amount of carbon emissions. (But military emissions were excluded from COP26 discussions.) The peace dividend implied by a co-operative global sustainability programme may be immense.

It might be plausible for ECOCOM to begin non-profit retail distribution in competition with the private sector, by opening sustainable goods shops on a large scale, as discount shops now do, especially to supply basic commodities, including food, if need be below cost price. Here, anyone using USI cards could meet their basic needs, while economies of scale would ensure low prices and quality guarantees, and buy-back clauses and the like where applicable would help built a new clientele. As a transitional measure, dual economy proposals of this kind, with states competing with the private sector, have been present throughout the history of socialism. In most such schemes, public ownership of essential assets, including mines, water and power resources, and railways, is presumed. One shop selling necessaries could be located within every neighbourhood (see below), and linked to public kitchens and entertainment complexes.

In patterns of consumption, we want particularly to avoid the US model. Here there is far more waste, pollution, excessive consumption, lower quality produce (additives in meat and so on), and destruction of scarce resources, as well as less free time, more advertising, and greater poverty and social inequality than elsewhere. This extreme-consumption-centred model privileges the desire for and possession of money and commodities over other forms of social and cultural interaction and individual self-definition. It has acted as a universal paradigm from the early twentieth century, thanks to emigration, American global power, and the vast advertising apparatus fuelled by Hollywood. It is no longer a suitable model for the United States, or anyone else.

ADVERTISING: Artificially stimulating desire for things we don't need, or need fewer of, can be greatly reduced. This means ending not only planned obsolescence of quality but, even more, as the great pioneer in this field, Vance Packard, stressed, eliminating obsolescence of desirability as well—the will to want the new.[101] In a "functional society", a Veblen disciple suggested in 1942, advertising could easily be a mere 10 per cent of what it was in a society defined by "illth", where "nine-tenths and more of advertising is largely competitive

100. Chase, *Tragedy of Waste*, 270.

101. Vance Packard, *The Waste Makers* (Longmans, 1961), 69.

wrangling as to the relative merits of two undistinguished and often indistinguishable compounds—soaps, tooth powders, motor cars, tires, snappy suits, breakfast food, patent medicines, cigarettes."[102] Sweden and Norway have banned advertising to children under twelve. Abolishing images of attractive young men and women to sell anything has been proposed.[103] But why stop here? The point is to suppress unnecessary demand. Let us sell fewer products, and fewer fantasies: will our lives really be the poorer for it? We might eliminate advertising which associates youth, strength, happiness, vigour, and sexual appeal with commodities. The same is true for using abstract appealing images, like babies, cute animals, virgin forests, tropical beaches, which have no discernible relation to the product itself.[104] Famous people should not be allowed to appear in any other than publicly minded commercials (try wording the legislation for this . . .). Advertising luxuries and superfluities might well be eliminated or sharply reduced; it works with tobacco and alcohol, and emission-intensive goods are just as harmful. Music might be eliminated from commercials too. The UK Green Party has proposed banning the advertising of flying. The same might be done regarding exotic holidays and fancy cars. The puppeteers of the imagination may protest, but desires will be suppressed, as they must be, by not being artificially stimulated. We may well end up eliminating virtually all advertising of non-sustainable products—but not sustainable experiences—completely. Then there is the problem of our obsessions with the lifestyles of the rich and famous. Hello Hollywood! We need no more films or programmes about how great it is to wallow in luxury. Stop being a fantasy factory for the lifestyle of opulence and instead give us propaganda which accords with a desperate, wartime scenario, at least some of the time. Entertain *and* educate—remember when fascism was the enemy?

These steps will not release us from the tyranny of branding or end the emulation of harmful social ideal types. We will not cease to buy things we do not need or lust after what we cannot afford. But such policies would move us in the right direction. We will also have to curtail the creeping commercial branding of everyday activities, including education. Publicly minded advertising, however, might be strengthened. Instead of encouraging consumption, ads could depict lifestyles based on diminished consumption, as in wartime, and highlight the harm caused by overconsumption. Ads might compare, say, the effects on air quality of using a publicly owned electric car to owning a

102. Chase, *Tragedy of Waste*, 113.

103. Oliver James, *Affluenza: How to Be Successful and Stay Sane* (Vermilion, 2007), 333.

104. In the United Kingdom, one bank advertises its services by showing a herd of horses galloping on a beach.

private diesel one, to take a current example. Children also need to be educated from an early age in the psychology of advertising to make them less gullible and more resistant to temptation. (Finnish schools already lead the way here.) The aim of advertising has hitherto often been to create an ideal type of character: cool, hip, sexy, desirable. Emulation will need to shift towards sustainable and socially responsible character types. We might also reduce or eliminate incentives to buy more, such as instalment plans or "buying on time" using credit cards. But the poor cannot be more adversely affected by such measures. So rationing clothing purchases, for example, will be necessary on the grounds of fairness. We might also register the carbon footprint of purchases, introduce proportionate carbon taxes on all products, and set a digital maximum on consumption using a universal carbon budget card to track all acquisitions.

As with food, we need to reduce sharply distribution costs and factor in their carbon footprint. Clothing which comes from Bangladesh or Vietnam may actually be more expensive in carbon terms than what can be produced locally or regionally in Texas, Spain, or Lancashire. We also need to reduce our expectations that speed of delivery, rapidity of operation, and the size and appearance of the product are the ultimate goals in consumption. This process, the "McDonaldization" of society, places a premium on quantity over quality, and haste ("fast food") and instant gratification over sociability and delayed satisfaction.[105] It also encourages indebtedness and the downward spiral of shopping to compensate for the depression we feel from being indebted as a result of shopping too much. Again, the instant gratification mentality must be curbed. Slower is often better: the "Slow movement" now embraces a much wider range of activities than eating and specifically targets "the spell of status".[106] The moment must become more important; life is not a race to exhaust either ourselves or the planet. Too much speed, an overly rapid pace of life, is killing us, and eroding our capacity for enjoying many pleasures we might otherwise delight in.

FASHION: Fashion will clearly be affected by such measures. After fossil fuels, the clothing industry is the second-most polluting in the world (8 per cent of carbon emissions). "Fast fashion"—discarding clothes which have hardly been worn—is immensely wasteful. In the United States, clothing production

105. George Ritzer, *The McDonaldization of Society*, 9th ed. (Sage, 2019).

106. Cecile Andrews, *Slow Is Beautiful: New Visions of Community, Leisure and Joi de Vivre* (New Society, 2006), 43. See also Carl Honoré, *In Praise of Slowness: Challenging the Cult of Speed* (HarperOne, 2004).

doubled between 2000 and 2014, with the average consumer buying 60 per cent more items, while keeping them half as long. This is one of the least justifiable forms of carbon emission. In the United Kingdom alone, fashion generates about a million tons of waste per annum, and up to 35 per cent of microbead pollution in the sea. Garments are typically used for only 2.2 years. Before Covid-19, consumers bought thirty-eight million items every week, and eleven million went into landfill. A large proportion of unsold clothes are simply burnt. Beyond protecting us from the elements, much of the purpose of clothing is purely ornamental. But we can satisfy our vanity and need for decoration without being so destructive. "Uncooling consumption", in Kalle Lasn's useful phrase, involves buying fewer things, and rejecting fur and other animal products.[107] Does this mean the end of "style" or "fashion" as such? No: only the most wasteful and non-sustainable forms of self-expression. Self-distinction, self-decoration, and imitation are ingrained aspects of social existence. People will always seek to be different, to appear distinct, and to innovate. But variety in appearance need not involve massive waste or undue consumption, and personal differentiation can persist in many forms. Start by taking the pledge: no new clothing for a year. Then persuade your friends.

5. *Population Restraint*

Amongst the measures discussed here, controlling the birth rate is the last taboo, and discussions about it still arouse strong passions. But all other measures to protect our environment put together will be useless if we do not *end*, not reduce the rate of, population growth. Nationalists, particularly in countries with declining populations, will shriek. So will those who insist on their God-given right to procreate ad infinitum. A leading study of this subject in 2005 warned that the earth could not sustain a population of 7.5 billion, which it reached in 2016. It is now 7.8 billion.[108] Some suggest that 9 billion "need not signal catastrophe".[109] However, everyone agrees that the earth has a limit to the population it can support. So everyone concedes that population control will be necessary at some point. The crucial question is where we draw the line. Clearly, the more people there are, the more will be poor and suffering. The more people there are, the fewer plants, insects, and animals. If we could only

107. Lasn, *Culture Jam*, 169.

108. Donella Meadows, Jørgen Randers, and Dennis Meadows, *Limits to Growth: The 30-Year Update* (Earthscan, 2005), 240.

109. Martin Rees, *On the Future: Prospects for Humanity* (Princeton University Press, 2018), 23.

solicit their opinion! For if we do have a "creed of kinship" with animals, we owe them the obligation of not exterminating them. Who will record the death of the last soaring eagle, the last majestic elephant, the last lonely whale? Who will shed a tear? Animalism is as central to our future as humanism. Restraining population growth reduces demand for commodities, housing, roads, transport, and so on, and makes room for all the other species we are now killing off. Then, too, the more people there are, the less privacy and solitude. A more realistic goal would be 5 billion, which is often associated with sustainability at a reasonable standard of living, and an acknowledgment of the rights of other species. More extreme current projections, moreover, suggest 11–15 billion by 2100, though this does not allow for potential catastrophes like pandemics, or for a sudden and dramatic surge in warming.

Population restraint can no longer be seen as some type of capitalist or western plot foisted on the less developed countries, as it was in the 1980s and later.[110] It must also be seen centrally as a key feminist issue. Women's education and other factors are already reducing population growth rates, but not fast enough. Limits on family size will be necessary. But they must be applied equally, and not punitively; we must learn from previous experiments, and particularly China's. Proposals to trade in "birth vouchers" are unfair to the poor. The same norms must apply to all. Incentivising through taxation is insufficient: we need more positive incentives. Such restraints will be possible only if we promote a vibrant feminism which results in gender equality across society. Until girls' lives are accorded the same value as boys' throughout the world, this policy will not succeed. Women, who possess considerably more power than men in disposing of household budgets, need full choice over their reproductive rights. Taxation needs to be used to nudge families in the right direction. The successful transition to a sustainable society thus hinges on a much more powerful global feminist movement.

6. Work

Billions of people still engage in toilsome drudgery in the twenty-first century. The circumstances of work are, moreover, deteriorating for many. Precarity, low wages, constantly rising productivity norms, repressive bosses and managers, stress, intrusive supervision, and poor pensions and job security are widespread. But the dreams of the early socialists, some anarchists, and Marx with respect to making work human again, and vastly less oppressive, can now be realised. The "de-growth" process *must* indeed involve a dramatic alteration

110. E.g., Murray Bookchin, *Toward An Ecological Society* (Black Rose Books, 1980), 37.

in attitudes towards work. We can reverse the extension of the working day and year, which has created a culture of overwork, exhaustion, and unhappiness. The first step will be the four-day week, which is already being trialled in Spain in 2021, and then three days, which is probably a good work/life balance. Before the seventeenth century, English peasants worked an average of 1440–2300 hours per annum. British and American workers today labour for some 3150–3650 hours.[111] So much for progress! Long hours and stressful work shorten our lives and make us more miserable. They also lead us to consume more in our limited leisure time, wages permitting, and then to work more in order to pay for our excessive acquisitions.

A four-day working week would reduce carbon emissions by up to 16 per cent without lowering efficiency. It would greatly reduce daily stress. As the philosopher Bertrand Russell suggested, it would also tend to make us "more kindly and less persecuting and less inclined to view others with suspicion."[112] We can move towards Jonas's "earthly paradise of active leisure", where "Sundayness" increases.[113] A universal sustainable income (USI), as discussed above, needs to be paid throughout the world, thus ensuring that income is not used for excess conspicuous consumption by restricting purchases to sustainable commodities. Some form of green currency might also be adopted to the same end, preventing ordinary money from being used for hoarding. Profit-sharing, co-ownership, and co-operative measures also need to be introduced into the workplace to enhance our sense of well-being at work. The wages of public servants like nurses, policemen, firefighters, and teachers need to be raised. Workplaces might also introduce rotation schemes where workers occupy bosses' positions for a week a year, or more, and vice versa. In universities, for instance, all administrators might be required to teach and do productive research for part of each year.

This implies a dramatic change in the current dominant corporate model. The twentieth and, even more so, twenty-first centuries teach us that this model is relentlessly dishonest and will eradicate all competition via fraud, force, and corruption. Megacorporations not only know no democratic values, they resolutely oppose them. They resist the protection which trade unions offer workers. They recognise the value of socialism, but only for the rich, who expect public subsidies at every stage, from the location of and investment in

111. Juliet B. Schor, *The Overworked American: The Unexpected Decline of Leisure* (Basic Books, 1991), 51.

112. Bertrand Russell, *In Praise of Idleness and Other Essays* (George Allen and Unwin, 1936), 28–29.

113. Jonas, *Imperative of Responsibility*, 194–55.

new plants to stepping in to supply pensions for workers in bankrupted firms whose assets have been stripped by greedy managers. This model of capitalism—but it may well be the only model capitalism can produce—knows only profit, and that unto death. Collectively, corporations would prefer authoritarianism to renouncing business as usual. The corporate ethos now dominates international life and thought as never before. It is the chief barrier to the transformation proposed here, largely because of its defence of unbridled inequality and unlimited growth. So this model must be changed during the transformation.

This might be accomplished by the following transformational economic policies. ECOCOM could aim to restructure corporations and businesses in a more democratic and egalitarian manner. During the transitional phase, a newly introduced worldwide environmental renewal programme and an environmental reconstruction fund (ERF) should also focus on working conditions and aim at restructuring the economic order. No contracts should be issued by ERF administrators in ECOCOM to companies with a greater than 10:1 ratio of executive to average earnings or which engage in aggressive corporate tax avoidance, or are registered offshore, or hold offshore assets or accounts. Any bank which fails should be bailed out only on condition of co-ownership by ECOCOM—in other words, the public. Co-operative and profit-sharing measures will have to be in place to receive such contracts, with all companies receiving ERF funds ensuring decent wages and pensions, increasing workers' ownership, trade-union involvement, and workers' participation at all levels of corporate structure. We can call this the Fair Earnings model. Distributing funds to corporations built on existing models cannot be contemplated. Thus, a radical Green New Deal can lead the way in transforming the workplace and capitalism as such. We got over cannibalism and slavery, and feudalism and imperialism (mostly!): we can do this too.

7. Public Service

As noted above, this programme requires a renewed idea of public duty towards other people and the natural world, or "patriotism now directed to the planet itself", as Kim Stanley Robinson phrases it.[114] We need not, like Lycurgus, attempt to compel permanent civic responsibility, but we can certainly encourage it more. Some of this can be achieved through education, which can emphasise the practical obligations of citizenship and care for the

114. K. Robinson, *Ministry for the Future*, 358. But not an "explicit religion", as one of his characters suggests.

natural world. An ethos of civic virtue and dedication to the common good might also be created by introducing some scheme of civilian service.[115] A form of conscription, for a green service or civil conservation corps, lasting a short period in everyone's youth, say a year, or a few months a year over several years, or a day a week engaging in local maintenance, might be exclusively dedicated to environmental preservation, including water conservation, supply, and control; forestry; firefighting; rewilding; clearing litter; and painting whole cities white. Serving in this body needs to be portrayed as an honour, and few exemptions should be permitted. In exchange, the young should receive free higher education or technical training, and all should have guaranteed employment at a USI wage and the assurance of affordable housing. A sense of duty to the public and of devotion to the common good is an essential element in the good life, and it can be united with the need for such service. It can also be combined with neighbourhood obligations to maintain our immediate environment. Anything we can do to promote a sense of reciprocity and mutual duty is worthwhile.

8. Wealth and Inequality

Far and away the most radical changes necessary concern the worldwide distribution of wealth. We have seen that there is a clear relationship between growing inequality, luxury, excessive consumerism, and environmental degradation. Leaving issues of exploitation aside, a radical reinvestment of wealth is just and necessary, insofar as its extreme concentration has caused great environmental destruction in the first place. So this is only a restitution for wrongs done. As most of the wealthy are born that way, it also helps to rectify accidents of birth, and the iniquity of exploitation. Abolishing poverty will also create more happiness and less antisocial behaviour than any other course of action. Most importantly, this is what saving the earth will cost. But we need also to avoid the dystopian perils of extreme collectivist levelling, and putting anyone in personal physical danger. A peaceful revolution will forestall inevitable widespread violent disorder if dramatic changes in social structure do not occur.

This issue is twofold: the cost of the transition to sustainability, and the need to hamper the cycle of greed and emulation which results from and then fuels extreme inequality. The Great Change will be immensely expensive, at least $100 trillion and probably much more. But cost as such is not the issue.

115. A similar proposal has recently been put forward by Fredric Jameson et al., in *An American Utopia*, edited by Slavo Žižek (Verso, 2016).

We cannot put a price on humanity's survival, and there is no market on a dead planet. This sum will have to be raised by a progressive tax on wealth and income, most of which will fund the ERF and ECOCOM. Additional measures required include eliminating tax evasion; abolishing offshore tax havens; and introducing total public transparency for all large-scale financial transactions, including requiring identifiable individual names on all accounts and all property deeds. As secrecy assists crime, corruption, and despotism, so transparency aids democracy and accountability. Countries which refuse to cease being tax havens will simply face an international boycott, and economic and political blockade, including prohibiting all travel by their citizens to other countries, and by others to them, freezing their assets, and confiscating any which are transferred in an effort to avoid the ERF tax. The United Kingdom, which sustains havens like the British Virgin Islands, the Isle of Man, and Jersey, is the largest global "laundromat" of money. It should commence this policy, or itself face these consequences.

Income tax might commence with a progressive assessment on larger incomes, say 70 per cent on those above $500,000 rising to 100 per cent at $1 million per annum. A wealth tax might then reallocate to the ERF all individual property in excess of, say, $10 million per person, which should be the maximum anyone can inherit.[116] No one needs more, and this is probably too high. There are perhaps ten million people worldwide worth this much.[117] Should we let them destroy the planet and the lives of the rest of humanity by hoarding their wealth? Obviously not. The two thousand plus billionaires in the world are worth more than 4.6 billion other people, or 60 per cent of the world's population, or over $10 trillion (million million) in 2020. The current richest person in the world, Jeff Bezos, was worth $6.8 billion in 2009, and some $211 billion in July 2021. On one day in 2020 he made $13 billion. Yet he paid less than 1 per cent tax annually for 2014–18, and in 2011 even claimed a $4000 tax credit for his children. Twenty-four of the other richest Americans paid an average of 3.4 per cent, a ridiculously small proportion. He and his fellow billionaires will survive being reduced to multimillionaires. Their

116. There are precedents: In late 2020, Argentina proposed a one-off 3.5 per cent tax on wealth. In the United Kingdom, the Wealth Tax Commission has proposed a levy of 1 per cent on assets per household of over £1 million, with a view to raising some £260 billion. At least a dozen countries have taxes of this type. In the United States, Elizabeth Warren has recently proposed a tax of 2 per cent of net wealth above $50 million, and 6 per cent above $1 billion, to yield *c.* $2.6 trillion over ten years. Other current proposals include up to 8 per cent tax on holdings of over $10 billion.

117. This exceeds the limits suggested in John Stuart Mill's proposals to limit wealth to something like a "moderate independence": see my *Mill and Paternalism*, 76.

demotion does not merit our sympathy. The very wealthy are mainly obsessed with getting richer because others are still richer. Greed consumes them. The fate of the rest of us and of the earth itself concerns them very little, if at all. It is difficult to muster much sympathy for this compulsive psychological disability. Incentives for entrepreneurial economic activity will still exist for many people. Such a tax might raise $9 trillion or more.[118] This money will be essential in the fight to save the earth.

Those who retort that this looks like Marxist-Leninist levelling should be reminded that taxes on high incomes of up to 90 per cent were common in Britain, the United States, and other countries until the low-tax revolution of Reagan and Thatcher of the 1980s, which began the current trend towards extreme inequality.[119] A key argument for high, redistributive taxation is that so-called "trickle-down" tax cuts never benefit the majority, but instead "trickle up" to the wealthy few and end up as luxury-goods expenditure. (Hence the invisible hand exists, the joke runs, but it is usually found picking our pockets.) Corporate taxation has also fallen sharply in the past seventy years.[120] It should be a reasonable percentage of profits in every country where products are made or services rendered, with the same tax rate levied everywhere, and as high as personal income tax.[121] Corporate tax avoidance has risen sharply in the same period, and now costs at least another £395 billion every year; again, the United Kingdom is a substantial enabler of such avoidance.[122] A tax on all stock market transactions would also be useful.

Redistributing excess wealth to this degree may seem like a revolutionary proposal. But the extreme concentration of wealth is a key cause of our current problems. The richer you are, the easier it is to get richer, and also to disguise your wealth. The ultra-rich, worth $45 million or more, are ten times more likely to hide income, and save about 30 per cent in tax thereby. Some $21–32

118. We cannot, of course, sell the helicopters, yachts, etc. of the rich, though some can be deployed for public service. The 400 richest Americans are reported to be worth around $3 trillion in 2019. In 2018, for the first time, they paid a lower proportion of tax than any other group: 23 per cent, as opposed to 70 per cent in 1950.

119. Tax rates in the United States peaked at 91 per cent in 1955, and were still at 70 per cent in the early 1980s.

120. From 32.1 per cent of US revenue in 1952 to 6.6 per cent in 1983, and 7 per cent in 2018. In the US the Federal corporation tax was 50 per cent in 1950, and 13 per cent in 2020.

121. The G7 2021 initiative proposes a 15 per cent taxation rate, which is much too low, and allows loopholes which make evasion easy.

122. About 40 per cent of international corporate profits are moved to tax havens annually, or around $600 billion. An effort to avoid greater tax scrutiny played a major role on the pro-Brexit side of the referendum campaign of 2016.

trillion is held in or passes through some fifty million accounts in tax havens, a third of the world's GDP. A one-off immediate tax of 50 per cent on these accounts, prior to their dissolution, is quite justifiable. (There are good grounds for confiscating the lot.) No one will go hungry as a result. Here, the international finance system will need to be frozen temporarily to prohibit capital flight, while a universal law on levels of taxation and transparency of assets which clearly names all owners of property and capital is passed to ensure that no shell companies or other shadowy devices for tax avoidance can survive the transition period.

This combination of income and wealth taxes would give the ERF and ECOCOM around $20 trillion to commence at the outset. This sum has to be set against the estimated cost of global environmental damage of some $54 trillion between now and 2040, based on an estimate of 1.5°C global warming, which is too low. The cost of any Green New Deal will occur over many decades, perhaps half a century, and this investment will reduce the costs of environmental damage. There will be considerable savings in energy costs, but also great expenditure in retraining and new plant. So we are still well short of what we require, which is at least $100 trillion. This reinforces the argument for confiscating offshore wealth entirely. It also implies that about 10 per cent of GDP per annum will have to go towards funding ECOCOM. The longer we dither, the more costly the process will become. A worst-case scenario, such as letting the entire problem slide for another decade, will make even greater sums necessary. This also implies the need for immediate reductions in other areas, such as arms expenditure. Halving these would save us around $850 billion annually. So the sooner we act the fewer confiscatory measures will be needed, and the longer we dally and dither the more drastic the remedies required will be.

These are admittedly drastic measures. But without a substantial reduction of inequality and a huge investment strategy, no programme of the kind described here is possible. The entire planet must pull together in a spirit of equal sacrifice and produce climate justice, or no one will survive. To do this a change in economic system is required.

9. Urban Renewal

The pain of transition to a sustainable economy, which involves renouncing or restricting a number of things we, or at least the wealthy, now do, needs to be offset by improving the lives of the majority and preventing the degradation of their standard of living. This is the quid pro quo which will make any future utopia viable. To a substantial degree it involves making cities survivable in much hotter temperatures, and increasing our comforts and opportunities for enjoying ourselves in them. But we need to resolve a number of paradoxes in order to do this.

The history of utopianism demonstrates the distinct advantages of small-scale life for nurturing those warm relationships so essential to our sense of well-being. In the realistic sense adopted here, as an optimal balance between order and warm personal relationships, utopia is clearly achievable with small numbers of devotees living in relative isolation united by a common cause, with a bit of good luck and nature on their side. Back-to-the-land movements have often resulted from adverse reactions to modernity, and a hankering for authentic ways of life. But the personal familiarity which appears to be the most appealing aspect of rural and small-town life is also its greatest drawback: it can be downright stifling, and boring. The allure of cities is correspondingly great, and that is also where most of the work now is. This combination of push and pull has greatly reduced the appeal of rural life. So we now rarely desire to flee the city, or seek solace in solitude, for longer than a holiday. We no longer feel that the solidarity we crave can be found only in small towns. Indeed, we have no choice but to reject this claim. Most of us live in, want to stay in, or aspire to inhabit cities. But this doesn't mean that we like our cities the way they are. Many are well past their prime, or have been "developed" badly and too quickly. Many have evolved as places to work, rather than being planned as places to live. Business, not life, dominates their ethos. We endure much of our lives here, while yearning for something better.

So any "utopian city" must replicate the solidarity of small-town life on a much larger, urban scale.[123] This was Bellamy's great aim, and one we can now take up anew. The sense of belonging, we have seen, is central here: more than anything else it defines utopian aspiration. "If alienation is the point on which our crises converge", the environmentalist George Monbiot suggests, "belonging is the means by which we can address them."[124] People care more for what is close to them, and for places they identify with. This local patriotism and activism makes us attentive to our immediate environment. When we are in foreign cities we do not usually pick up rubbish on the streets, or tell people off for littering or spitting on the pavement, or assist elderly people with their shopping, or give them our seat on the train or bus, or help water the flowers planted in public spaces. But we might do all these in our own neighbourhood, because here we feel a sense of co-ownership and belonging, as members of a group. Here we can still feel personal responsibility: we are co-owners of this space. We

123. A starting point here is Malcolm Miles's *Urban Utopias: The Built and Social Architectures of Alternative Settlements* (Routledge, 2008).

124. George Monbiot, *Out of the Wreckage: A New Politics for an Age of Crisis* (Verso, 2017), 71.

suffer from a kind of spiritual agoraphobia when drifting untethered amongst large masses of people. We need a sense of place, or *Heimat*, in which to anchor our identity in a feeling of belongingness. The vast possibilities opened up by Covid-related closures give us many opportunities to help create this.

The Neighbourhood Model

The key to a human-centred city is the feeling of neighbourhood, and the vivacity, familiarity, and energy of street life for young and old alike. As the primary urban unit of the future, the neighbourhood or urban village must be small enough to foster a sense of group membership, participation, and place. It must be defined by a communal ethos, institutions, and identity, not merely a place name. The streets must be safe, for women in particular. Neighbourhoods have size limits, as urban theorists in the 1950s like William Whyte discovered: cohesiveness and a "happy group" can be correlated. They are partly defined by the amount of communal space available, and how it is configured.[125] Think of recreating, in twenty-first-century terms, what Raymond Williams saw as the model modern village: a "parish council, a reading-room, a gymnasium, council cottages, a women's institute", and concentrating these in each neighbourhood.[126]

This does not imply that neighbourhoods need be identical. To the contrary, their differences, even uniqueness, should be a strength. Their requirements can be reshaped to meet their inhabitants' needs, wants, and opinions. Specifically lonely or isolated groups, like students, young mothers, prisoners, carers, immigrants, the disabled, and pensioners, will have to be catered for.[127] Some may feel more comfortable in neighbourhoods where their own group, however defined, predominates. Political affiliations might bring people together in anarchist or other communes. Heterotopian enclaves could permit specific rules to prevail, so long as the criteria of voluntariness and no harm to others or coercion are observed. (Polygamy, including polyandry, and other forms of group marriage or cohabitation may well increase in future, for example.) Each neighbourhood should be a cohesive unit, with a name and a meeting centre, with a maximum size of perhaps five thousand people, the number of faces we can remember. A sense of communal identity can then be constructed on a number of levels, and space can be "moralised" in the sense

125. Whyte, *Organization Man*, 348.

126. R. Williams, *Country and the City*, 195.

127. Robert S. Weiss, *Loneliness: The Experience of Emotional and Social Isolation* (MIT Press, 1973), 79–89.

of creating reciprocal obligations on the basis of a shared feeling of ownership. All newcomers should be made welcome to the neighbourhood and invited to block meetings or street parties and enrolled in the residents' internet group. People with special needs, like the elderly, or those who live alone, should be enrolled in support networks designed to ease everyday life, all the while heeding rights of privacy and individuality. "Fast friend" meetings and neighbourhood internet groups can be formalised so that we can readily find people with similar interests. Every effort must be made to combat loneliness, the great enemy of utopia. Building belongingness requires more than a reliance on chance acquaintance. We must make opportunities to meet others, not wait for them to arise. And we must make the process as relaxed, informal, and fun as we can.

"Urban villages" will prioritise low-cost living, housing, and easy interaction.[128] The "village effect", in Susan Pinker's phrase, will promote face-to-face contact and define "social capital" as "the knowledge and mutual trust captured in our relationships".[129] The idea of the "sociable city" was long ago identified with the Garden City movement.[130] It now needs to be expanded through the "neighbourhood unit" concept, while still permitting that degree of anonymity which so many cherish in the city.[131] Neighbourhood-style urban villages will still be places to see and be seen, to exhibit, demonstrate, and encounter. They will also have sustainability at their heart. But, as Sharon Zukin insists, we must avoid mere gentrification, where bistros, bodegas, and cocktail bars sprout for upmarket custom and traditional customers are driven away.[132] Rents must be kept low to keep sociability affordable. Most of what people need should be locally provided, as we saw Bellamy suggested. The mayor of Paris, Anne Hidalgo, champions the idea of the "fifteen minute" city, to express the maximum time required to get most of what we require. This can do much to enhance our sense of neighbourliness. The pandemic-driven necessity of working from home, for those who can, and accessing local services much more, has helped to drive this point home.

128. See Sharon Zukin's *Naked City: The Death and Life of Authentic Spaces* (Oxford University Press, 2010), referring to New York City.

129. Susan Pinker, *The Village Effect: Why Face-to-Face Contact Matters* (Atlantic Books, 2014), 62.

130. Peter Hall and Colin Ward, *Sociable Cities: The Legacy of Ebenezer Howard* (John Wiley and Sons, 1998).

131. On the origins of the neighbourhood unit concept, developed for post-war London, see Roger Gill's "In England's Green and Pleasant Land", in *Utopias*, edited by Peter Alexander and Gill, 109–18.

132. Zukin, *Naked City*, 4.

Other strategies can assist in building what the renowned twentieth-century urban theorist Jane Jacobs calls an ethos of "togetherness":

> Under this system, it is possible in a city-street neighbourhood to know all kinds of people without unwelcome entanglements, without boredom, necessity for excuses, explanations, fears of giving offence, embarrassments respecting impositions or commitments, and all such paraphernalia of obligations which can accompany less limited relationships. It is possible to be on excellent sidewalk terms with people who are very different from oneself, and even, as time passes, on familiar public terms with them.[133]

To establish such "sidewalk terms", we can do worse than to start with Jacobs's four conditions for urban renewal:

1. The district, and indeed as many of its internal parts as possible, must serve more than one primary function; preferably more than two. These must ensure the presence of people who go outdoors on different schedules and are in the place for different purposes, but who are able to use many facilities in common.
2. Most blocks must be short; that is, streets and opportunities to turn corners must be frequent.
3. The district must mingle buildings that vary in age and condition, including a good proportion of old ones so that they vary in the economic yield they must produce. This mingling must be fairly close-grained.
4. There must be a sufficiently dense concentration of people, for whatever purposes they may be there. This includes dense concentration in the case of people who are there because of residence.[134]

To these, other proposals can be added. There will be many more elderly people as population declines and we live longer. Fearing the noise and bustle which the young find endearing, but not wanting exile from it either, their need for peace will have to be respected. Urban communes or co-housing initiatives where the elderly share resources, larger versions of care homes, can help here. So can assistance with shopping, medical appointments, and social excursions. Given the declining number of blood relatives we might consider introducing something like Pala's "Mutual Adoption Clubs" in Huxley's *Island*, where a larger "voluntary family" of twenty houses extends one's social circle,

133. Jacobs, *Life and Death*, 72.

134. Jacobs, 162–63.

ideally to the limit of 150 suggested by Robin Dunbar.[135] Every person should have access to a basic social network or virtual neighbourhood of this type, without any compulsion to participate in it or to surrender any privacy to it—it would be available merely to provide support. Other types of groups could also be created. Women may want safe havens where men's movements are restricted. Free child- and healthcare are essential. Promoting a sense of common ownership is also important. More co-operatively owned and managed buildings, even whole neighbourhoods, for instance, would create public spaces for common deliberation as well as relaxation. New apartment buildings should have a communal area like a hotel lobby where people can stop, rest, and chat, and where refreshments are available to encourage common mixing. A sense of democracy and local involvement will assist growing local identity. What Richard Rogers and Anne Power call "cohesive cities" will also require local economic stimulation, giving residents control over local development, offering equal public services, and prioritising environmental care.[136] Educational facilities are also vital. Cheap or free workshops, colleges, theatres, cinemas, and recording studios will ensure that creative outlets exist for all. Self-development is as important as, and intimately interwoven with, sociability.

The specifics of new urban design remain to be fleshed out. Reimagining urban space will involve conceiving our cities as communities where people live rather than merely places where they work and buy. Cities should be places we want to go, on free public transport, to meet others. To create them we must reverse the trend of privatising public space.[137] The air will be cleaner, so we can sit outside more when the heat does not drive us indoors or underground. (Thousands of lives will be saved by improving air quality alone.) Streets must be made much more attractive, and people-centred. The "streets without joy" long ago condemned by the French architect Le Corbusier will need to be replaced by "streets of repose . . . tucked in among the foliage of the trees."[138] By narrowing roads, restricting traffic to public electric vehicles, creating squares and open spaces, and widening pavements or making pedestrian-only zones, we can provide meeting places everywhere. Copenhagen, which has successfully reduced traffic and created a vibrant street culture,

135. A. Huxley, *Island* (Penguin), 93.

136. Richard Rogers and Anne Power, *Cities for a Small Country* (Faber and Faber, 2000), 285–87.

137. See Margaret Kohn, *Brave New Neighborhoods: The Privatization of Public Space* (Routledge, 2004).

138. Quoted in Robert Fishman, *Urban Utopias in the Twentieth Century* (Basic Books, 1977), 208–9, 238.

is an excellent model here. Open and closed group meeting spaces will be greatly enlarged, and become closer to the Greek *agora* or the Renaissance plaza than the modern mall. Plazas are "one of the most sociable of places", in William Whyte's words.[139] Here people can rest, watch and chat, play games like chess or basketball, and give life to a city, rather than being always in movement. Many more underground spaces will be created, for this is where we, and especially the elderly, will increasingly flee during the scorching summers. Buildings might be redesigned according to Le Corbusier's L'Unité concept, where workshops, meeting rooms, cafes, restaurants and shops, gyms, sports grounds and swimming pools, day-care centres, schools, and laundries are integrated within a single unit.

Greening urban space is essential and crucial to our future happiness. Parks will be everywhere: studies show visiting one can give an emotional boost equivalent to Christmas. A sense of the common ownership of common space is crucial here to cultivating belongingness. Trees and gardens with benches, vertical gardens, outdoor art, playgrounds, and public toilets will be plentiful. Roofs and walls will be lush and green with plants, side by side with solar panels. Flowers will everywhere provide invigorating colour, calming texture, and alluring smells. Larger parks will permit some isolation, for we need also to retain the possibility of solitude, and fleeing the mass. There will be plenty of places to rest, and no spikes on ledges or dividers on benches to discourage the homeless. For there won't be any.

Architects and urban planners will play a more important role in crafting this future than they have ever before had. There is plentiful scope for creativity here. We do not want all these entities to look alike, like Thomas More's cities: they should reflect our diversity, our geographical and historical traditions, our imaginations. And they must be as beautiful as we can make them: beauty inspires, delights, humbles, and refreshes us. Ugliness makes us more brutish. Where are the budding Le Corbusiers, Howards, and Lloyd Wrights who will rise to this challenge and create these cities of the future? There are many . . .

Urban Sociability: Towards Neo-Fourierism

These developments will have to be calibrated against the specific psychological needs of the coming decades. Nervousness and a profound depression of a type and on a scale never before seen will soon seize and demoralise millions, if measures to counteract them are not taken. Rising mass subliminal anxiety

139. William H. Whyte, *The Social Life of Small Urban Spaces* (Conservation Foundation, 1980), 53. And also more extensively in Whyte's *City: Rediscovering the Center* (Doubleday, 1988).

about the climate was already evident well before Covid-19 appeared, as images of forest fires, melting glaciers, rising seas, massive tropical storms, and dead coral reefs began to merge into a darkly threatening portrait of an impending and unstoppable dystopian future. Increasingly, this anxiety is coming to define the everyday life of millions. For others, absolute denial, a refusal even to acknowledge these problems, and a resentment of those who protest against them, is a common reaction. But panic, hysteria, and anger too will commence when many more people realise, and even more begin to feel, the colossal tragedy we face. People will cease to reproduce. The suicide rate will skyrocket. The physical stress of coping with the summer heat will tax even the sturdiest. Taut with fearful, nervous apprehension, some will become crazed and erratic. Even the traditional language of the apocalypse, which after all had meaning insofar as salvation for some lay at the end of it, will fail in the face of the prospect of a meaningless extermination of humanity. We will not understand what is happening, only that it is happening. Many will seek scapegoats, blaming anyone but themselves. Eventually, widespread convulsion will result unless a sense of mutual assistance reinforces our collective will to avoid this fate.

We can offset this mounting despair by entertaining ourselves as much as we can, and by caring for each other as much as possible, while still solving our problems. This is where a new hedonism must compensate for the new asceticism, maximising some pleasures in the face of being deprived of others, and offsetting the risk of a repressive new puritanism. We need bread, as well as the means of self-fulfilment and of attaining the "higher" pleasures of culture and education. But there must also be circuses, and the promise that we will be compensated to some degree for our sacrifices. The principles of public luxury and satisfying our passions which have been elaborated, most notably in Fourierism, must now come to the fore. Here we may have novelty galore, for our desire for it will not disappear. Some will embrace epicureanism, taking pleasure in more limited desires, but most probably will not. Mass amusement in the face of scarcity will be one of the most appealing aspects of the new sustainable world view. Without it, indeed, the concept is not saleable and will be, rightly or wrongly, rejected by large numbers. People will not want a future which demands continuous self-sacrifice. But sacrifice is necessary. So to mitigate its impact, we have stressed repeatedly here, we must provide as much and as varied pleasure as possible. We must party to forestall the apocalypse.

Public celebrations will play a vital role here. Carnivals of the type now enjoyed in Rio, Cologne, or New Orleans might be universalised as week-long affairs. Christmas might be extended to last a month. Great festivals like the Saturnalia might be revived, and their egalitarian tendencies drawn out. Local fairs are also important. Perhaps quarterly, marking the seasons, all should be invited at public expense to a neighbourhood festival. Every opportunity should be taken to promote ideals associated with sustainability, so at least one festival should be Earth Day or Week. Once a year each neighbourhood might host alternatively one or other adjoining neighbourhoods, with games, sport, music, and feasts, extending a kind of guest-friendship by establishing ritual reciprocity of service and hospitality, and cultivating a sense of mutual obligation as well as civic identity. Other festivals might pay specific tribute to public servants like doctors, nurses, sanitation workers, police, and fire officials, not politicians, who should be given civic awards and public recognition of their work and sacrifice. Marks of distinction, like ribbons, might be given to them to wear for a year. A few times a year, too, but only for those who wished it, a common neighbourhood uniform, or merely a similar colour of clothing, or a neighbourhood armband, or emblem, like a coat of arms, might be displayed for a day-long grand party to cement our association with those who live around us and to acknowledge our common humanity. Common dining might also be encouraged, perhaps by weekly subsidised public feasts. All such measures will heighten our sense of belonging and extend our sociability.

Then there are means of expanding everyday pleasures in our increasing leisure time. Unlike many earlier, small-scale communities, where fear of God and sin dominated, later modern cities embrace sensual gratification. Here, a sybaritic balance can offset the Spartan or stoic demands of sustainability, or something like the "alternative hedonism" Kate Soper defends, which "dwells on the pleasures to be gained by adopting a less high-speed, consumption-oriented way of living".[140] This is the cherry on top of the cream on top of the cake. Here, community will consist not just in shared experience but specifically in shared pleasure. The ethos of the future will be closer to the universal hedonism of Huxley's *Brave New World* than the drab self-renunciation of a Shaker village. Vast theme parks might enable us to travel in time and space. Music and art festivals, public feasts, cafes, restaurants, sports facilities, theatres, and virtual-reality sensoriums will proliferate, to exercise, entertain, divert, amuse, and promote sociability. Alcohol, cannabis, and other popular stimulants will be available to excite and/or dull the senses as required, subject to cultural constraints. Heterotopian spaces will permit the free expression of non-harmful interactions which define and validate the self-identities of participants. We need this closeness, especially touching and sensory gratification. Where more relaxed attitudes towards sensuality prevail, sexual encounters can be encouraged, and catered for, with free birth control, for we need more pleasure but fewer babies. Great palaces of gratification will cater to

140. Kate Soper, *Post-growth Living: For an Alternative Hedonism* (Verso, 2020), 50.

every taste; we could call these Fourier centres in honour of the father of pleasure theory. And we can play at much else. In Rousseau forests we can cast ourselves back to nascent society. Nostalgic wonderworlds will allow us to go back to any era we please. Here will be Utopia stadiums, Soma cafés, Dionysian halls, and Cockaigne cocktail bars. Bloch Hobby emporia could cater to a thousand interests. Virtual sci-fi futureworlds or Houses of Enchantment will be all the rage. There will be new art, new music, new styles in everything: an unprecedented golden age of culture may result. And we might dance the apocalypse away: it is an eminently sustainable activity.

This sybaritic sociability or sustainable hedonism can balance the stoicism we will require elsewhere, and dull the pain of our greater private asceticism. Luxury of public sustainable subsistence can trump luxury of private ornamentation. This represents a turning point in humanity's history. We have worked for millennia; now is the time for revelry. Indulging our desires could be crucial to attaining environmental stability. As one member of The Farm in Tennessee, one of the most successful mid-twentieth-century American communes, Albert Bates, put it, whatever we do, "*it had better be fun or no-one will want to do it*".[141] To which we can add: it had better be sustainable or there is no point in doing it.

This is what a twenty-first-century utopia might look like, and none will be possible without many of these measures being introduced. Much more than hope, utopia symbolises possibility beyond everyday horizons. Utopia as blueprint is back with a vengeance, for all this is possible, and is more blueprint than dream. We need not abolish class entirely, nor the market, though neither can retain its present form. This vision concedes the case, made so often over the past three centuries, that nations which have once embraced luxury cannot renounce it. We are sybarites, and few will want to be Spartans. But it also acknowledges that public luxury will satisfy much that private luxury cannot, and in sustainable form help to avoid an otherwise desperate fate. This vision also remains loyal to the Morean utopian-republican paradigm insofar as far greater social equality and more virtuous citizenship are recognised as the only basis for creating happier societies. But the modern adaptation of More must be more joyous and carefree, while aiming all the while at sustainability. It must give us the main components of happiness, defined by Daniel Haybron as security, outlook, autonomy, relationships, and skilled and meaningful

141. Quoted in Metcalf, *Shared Visions, Shared Lives*, 192.

activity.[142] But it must do so sustainably. Its success will be contingent upon the fact that, unlike the Soviet Union, or the kibbutz, no competing materialist utopia will be able to undermine a sustainable regime once it has been universalised. But if this does not occur, its failure is equally guaranteed.

Such arrangements will help allay fears that any future utopia will be changeless, uniform, hostile to individuality, or despotic. But it is obvious that no political party could get elected today on the basis of the programme outlined here, except perhaps its hedonistic aspects. For it is as dramatic and revolutionary as anything Marx or Lenin envisioned. We could call this programme green socialism, insofar as it promotes sociability, equality, and sustainability, or just sustainabilism for short. But what is defended here is not egalitarianism for its own sake. It is the inhibition of the excessive desire for emulative consumption which underpins consumer society. What is proposed here is not a classless society, or communism, much less salvation, paradise, heaven, perfection, a state of grace, the "total abolition" of alienation, or ensuring "complete harmony" amongst human beings, a "totality" of transformation, a "Novum", or any other variant on secular millenarianism. We cannot ask utopia to answer every religious and metaphysical conundrum thrown up by human experience. It cannot. To be human is to live in a state of incurable existential anxiety which terminates only in death. We will continue to suffer, to feel anguish, fear, and pain, to fail at times to curb our arrogance and our destructive and aggressive instincts, to be deluded, to find attaining pleasure elusive and temporary, and ultimately to die. Utopia will not make the great problems go away, stop us from being fickle, obtuse, spiteful, petty, and stubborn, or cure selfishness, malice, delusion, or stupidity. But by making utopia less ambitious we also make it more practical and attainable. The goals sketched here are achievable. They do not demand remaking humanity, or dramatic improvements in behaviour. They require no revolution or dictatorship. They imply upsetting at most a few million people, a very small proportion of the world's population. They offer a model of society in which the everyday life of the majority will be, subject to environmental constraints, very considerably improved. So making utopia turns out not to be utopian after all.

These principles, or something like them, can guide us into an uncertain future. To implement them three things are required, besides a programme: leadership, from intellectuals, as well as public figures and elected representatives, to reshape public opinion; a sea change in corporate behaviour; and a modest alteration in individual behaviour. Leaders will emerge spontaneously as people are inspired to take charge. They will likely be Stoic or ascetic in

142. Haybron, *Happiness*, 54.

outlook and, as a group, like Wells's Samurai, must set an example of personal restraint in consumption. For them, and possibly many others, the movement will also be a way of life. But radical chic will find new styles too. Changes in corporate and state behaviour will be essential to avoiding the worst-case scenarios. Corporations need to be held accountable, socially and ethically, as publicly responsible entities. The time has come to make them pay for the damage many have caused. Sometimes only public ownership and/or management will effect this end. And then there is our behaviour as individuals. Sacrifice will be required from each of us, and everyone can and must do their part. Let us briefly look at what this might entail.

What You Can Do

We can do all this, with sufficient will. *We* means first and foremost the hundred corporations who produce 71 per cent of global greenhouse gas emissions. But *we* is also you and me. As individuals we must learn to make do with less and take pleasure in greater simplicity, without feeling unhappier at same time. This requires asking some basic questions about who we are, and how much our sense of identity depends on consuming and possessing an inordinate amount of complex, scarce, and unsustainable and expensive things and experiences: this is what we want to diminish. The momentous changes in corporate behaviour required must be driven in part by consumers acting responsibly, and not demanding what the planet cannot afford. We need not like Thomas More suffer a hair shirt to accentuate our sacrifices. We can alter our behaviour without reducing our pleasures unduly. And the price to be paid in terms of individual behaviour is extraordinarily small given the reward reaped: our survival.

Individually, anyone can, and should, reduce meat, dairy, fat, and sugar consumption; choose slow over fast food; eat leftovers and what is in season; reject plastic packaging and single-use plastic like coffee cups, plates, and cutlery; try to fix and repair things, and eliminate waste; reduce clothing acquisition, mend clothes, and wear them longer; choose generic over luxury brands; resist brand tyranny; shun those who display wealth ostentatiously; reduce water use and employ eco-friendly cleaning products; turn down the heat and turn up the air conditioning; turn the lights off; get solar panels and storage batteries; cease using halogen bulbs; replace gas- and oil-fired boilers with heat pumps; increase household insulation; cultivate houseplants and balcony gardens; resist advertising and complain about commercialisation and product placement; use public transport frequently; drive and fly less; visit local rather than distant tourist destinations; write to our political representatives; strike and resist, agitate and sign petitions, go on demonstrations, join and subsidise green organisations, and encourage friends, neighbours, and colleagues to do the same. Agitate—agitate—agitate: this is the greatest and most momentous movement in the history of our species. It will take all our efforts to succeed, and so far people's willingness to engage in behaviour change has been minimal. But, as Dale Jamieson writes, our willingness to connect our own actions to environmental destruction, to overcome the mentality of "it doesn't matter what I do", is central to any forward movement,[143] and indeed to the wider question of living a meaningful life in the face of the gravest threats. Activists are now our greatest need: "O brave new world that has such people in it." With enough of this type, we can prevail. All of this can make a difference—you too! Renouncing beef and flying much less would be a great start. We should not be distracted by micro-consumerist issues, however, nor be consumed with guilt if we fall short sometimes in our achievements. The big issues are corporate more than individual. But endeavouring to alter our own behaviour can make us feel members of a group who are engaged in this greater change, and energise us to pursue sustainability at all levels and to place pressure on others to do so. This sense of common identity is vital to the transition before us. All this will help to reduce demand, and thus the consumption of energy, while not unduly hampering our quality of life. So you can make a difference.

Pitfalls and Paradoxes

There are obvious paradoxes in the programme offered here. No violent revolution is supposed, although the odds of the extremely wealthy voluntarily allowing their assets to be used for the common good are poor. Some, perhaps most, will prefer fascism. Others will build gigantic bunkers or construct fortress-microstates in the futile hope of saving themselves. The merely well-to-do but luxury-addicted middle classes are being asked to forsake some of their most-cherished pleasures, like frequent long-haul holidays. Millions more are being asked to stop aspiring after upper-middle-class lifestyles and lusting after novelty for its own sake. Sumptuary laws, like banning SUVs and private jets, are proposed, but will be resisted. Massive investment in green cities and green technologies is required at the same time as we want production, consumption, and the extraction of raw materials generally to decline, and other industries, like fashion, to shrink. Increased agricultural cultivation is required, but fewer people than ever will want to return to the land for this purpose. (And food and trees will be competing for space.) We also want such

143. D. Jamieson, *Reason in a Dark Time*, 179.

investment to occur in various countries, notably China, which have specialised in manufacture for export, which must necessarily be reduced. We imagine an appreciation of enhanced sociability just as direct human sociability unmediated by technology seems to be declining, and family size is shrinking, while also asking for greater concern for the public good. We invite a sanctimonious condemnation of conspicuous consumption while asking for greater sociability and friendliness in public. We want, even demand, utopia, in effect, just as the world slides into dystopia.

But these are not insurmountable challenges. We can unlearn what it is to be later moderns, without demanding that we become angels or create a City of God on earth. The measures proposed here can facilitate large, real, or everyday sustainable utopias; that is, spaces in which enhanced sociability and belongingness can flourish without costing the earth. Suitably designed and adapted, they can permit us to enjoy the more restrained and simple standard of living which the coming decades will demand, balanced by the plethora of delights we can still offer ourselves. This schema can provide us with the means of human flourishing dreamt of by the greatest reformers throughout the ages. Far more importantly, it can permit us to survive the catastrophe looming over us. But so far this is all still a dream. So let us now, finally, consider just how all of this can be commenced.

Conclusion

The Great Change

CREATING ENHANCED SIMPLICITY

To live Luxuriously is but a Custom: If it was broke off, no Body would miss it, and evidently it would be of infinite advantage to the Society that it were so.

[*AN ESSAY CONCERNING ADEPTS*, 1698][1]

a vicious people can never return to virtue . . . One can never go back to the time of innocence and equality, once one has departed from it.

—JEAN-JACQUES ROUSSEAU, 1752[2]

Cyrus having asked Croesus, how he could most effectually enslave the Lydians, who had revolted from him: he advised him to debauch their manners, and to encourage intemperance among them: for the chains of luxury are easily borne, and the hardest to break of any in the world. Accordingly Cyrus commanded their arms to be taken from them, erected taverns, gaming houses and stews; enjoined them to wear vests and buskins, and to teach their sons to sing, to play on the harp, and to frequent the public houses. And thus that nation which before was remarkable for their industry and valour, now dissolved in ease and luxury, soon became a prey to their own sloth and laziness.

—JAMES BURGH, 1764[3]

"I do love flying," they whispered. "I do love flying, I do love having new clothes, I do love . . . But old clothes are beastly," continued the untiring whisper. "We always throw away old clothes. Ending is better than mending, ending is better than mending, ending is better . . ."

—ALDOUS HUXLEY, *BRAVE NEW WORLD*, 1932[4]

1. Quoted in Claeys, *Restoration and Augustan*, 227.

2. In the preface to *Narcisse*: Cranston, *Jean-Jacques*, 243.

3. [James Burgh], *An Account of the First Settlement, Laws, Form of Government, and Police, of the Cessares, a People of South America* (J. Payne, 1764), 123–24.

4. A. Huxley, *Brave New World* (2007), 41–42.

READERS WHO HAVE gotten this far will probably be wondering, quite legitimately, what hope we have in confronting the reality of the 2020s, much less the 2030s or 2040s. One of the most striking aspects of much of the environmentalist literature produced between the late twentieth century and just a few years ago is that it now seems shockingly outdated. Few books written before about 2016 confronted the prospect of real, imminent catastrophe, and that in the short and medium term rather than the distant future. Now, quite suddenly and dramatically, the normality of many decades, even of the human condition itself, has disappeared. Slowly it is dawning on us that this normality will never return. Everything has altered and is changing swiftly on a daily basis, whether we choose to acknowledge it or not. The worst-case scenarios of only a few years ago are now commonplace predictions, as yesterday's nightmare becomes tomorrow's horror. A catastrophe without parallel in history is unfolding before us, and we stand nearly helpless before it, shocked, dazed, awed, overwhelmed, and usually in denial. Most of what concerns us in our daily lives, intellectuals especially, is now largely irrelevant except where it addresses this crisis.

The novelty of this scenario also makes most past utopias outdated, and many of the definitions associated with them. Modern utopian enthusiasts are fond of identifying their aspirations with Bloch's principle of hope. But most of what gave Bloch himself hope has disappeared. Now we are as near to being in a condition of absolute hopelessness as humanity has ever been. A creeping sense of despair increasingly defines our era. We have long since lost any belief in a past golden age. Now faith in a better future, the old "progress", is gone as well. But merely demanding hope, and fetishising the concept, in the name of offering solace, is obviously no answer. As Greta Thunberg puts it, "I don't want your hope. I don't want you to be hopeful. I want you to panic . . . and act as if the house was on fire."[5] Another environmentalist, David Wallace-Wells, adds, "As someone who was awakened from complacency into environmental advocacy through alarm, I see real value in fear."[6] Fear can produce doomism, but it can also lead to action. Hope-mongering which fails to offer solutions is thus grossly irresponsible—mere "hopium". But we can still avoid negativity, depression, and despair, as well as downright panic, in the coming decades. There is every reason to believe that this crisis will generate other visions of alternative futures, like that proposed here. Every past topia, Gustav Landauer noted, has been followed by a utopia, "a combination of individual

5. Speaking at the World Economic Forum, Davos, as recorded on Twitter, 25 January 2019.

6. Quoted in Mann, *New Climate War*, 212.

and heterogeneous manifestations of will that unite and organize in a moment of crisis to form a passionate demand for a new social form: a topia without ills and injustices."[7] Once we begin to pull together, a great momentum for change will carry us forward.

Intellectual conviction as to how to proceed is still the first step. A major premise of this book has been that both the great nineteenth- and twentieth-century paradigms of modernity, economic (not social) liberalism and Marxism, now need to be discarded. The new paradigm which will replace them is defined by egalitarianism, voluntary simplicity, sustainable consumption, and as much joyous and celebratory sociability and belongingness as we can muster and afford. There will never be a return to the "normality" of "growth". That epoch—call it the age of excess—has passed. Ours is an era of profound and liminal transition, and a future defined by sufficiency is upon us.

This future, if we can create it, will resemble the plans of some previous utopian writers, insofar as they plotted a sustainable society which placed preserving the planet at the centre of human activity. It will owe something to More, Marx, Wells, Morris, Bellamy, and Callenbach, as well as Fourier and Huxley. It will likely commence with a long liminal moment of eco-awakening, accompanied by the emergence of a great new movement which will embrace greater simplicity as a cultural ideal while adapting and improving technology as far as is feasible, especially with the aim of increasing sustainability as well as our capacity for pleasure. Like previous revolutions, and indeed the birth of the utopian concept itself, this liminality will involve a consciousness of transition to a morally superior future. A negation of past egoism will form part of its expression. In the past, scarcity has often weakened solidarity by encouraging a mentality of "every woman for herself".[8] What may come to pass will be something like a cultural revolution, though differing from any which has gone before. As in the 1960s, it may coincide with a generational revolt, and a huge surge in idealism. It will probably be led by the young, and by more women than men, at least initially—they are in the forefront of current environmental protests like Extinction Rebellion and the Fridays for Future School Strike movement. Soon it may become the mother of all social movements, the largest conversion to a secular idea in history, and a tsunami of force to save Mother Earth. It will spawn new political parties, or rejuvenate the best of the old, and will foster new ways of life.

In the shadow of the apocalypse, this movement will sometimes have millenarian overtones. Cults will form where communes would be more useful, and salvation will be sought instead of utopia. Sanctimonious prophets will clad

7. Gustav Landauer, *Revolution and Other Political Writings* (PM Press, 2010), 113.

8. B. Turner and Rojek, *Society & Culture*, 29.

themselves in the mantle of the old religions, and distract us with denunciations of sin and promises of redemption. Many people will blame themselves for our earth's fate—and angry propagandists will egg them on. But powerful antagonisms will also arise to the old consumerist way of life, and to the symbols of luxurious excess. Vast numbers will come to want the change and not to feel it is being forced on them—the growth in vegetarianism and veganism offers a useful parallel here. Millions will come to realise that the destruction of nature is a price too high to pay for "progress". This movement will generate massive protests, widespread civil disobedience, hugely successful boycotts of key polluters, and much more. It will be met by much resistance from those who profit from destroying the earth, and from those who are simply afraid of change. Eventually it may prevail—it must. But there are many risks along the way.

The Great Change would be best signalled by a well-defined moment of emotional and moral shift shared by a large part of the population. A liminal point, call it the Utopian Month, might be chosen when millions will commence lifestyles based on sustainability. Symbolically, at least, we might have Savonarola-like bonfires of the modern vanities, like fast cars and other luxury goods, to herald our retreat from consumerism, before we put the things themselves into museums of the age of excess. A grand worldwide Saturnalia-style festival might inaugurate the new mentality and cement the association of the well-being of the many with the abolition of extreme inequality. This might be the biggest party the world has ever known. This must be a moment of emotional conquest, and of a palpable sense of catharsis, for if we do not win with the emotions, we have won nothing. Bring tears to millions of eyes and you can conquer the world. Utopia will need new symbols, like the Extinction Rebellion hourglass, and new costumes to mark its identity. It will need a new soundtrack, a sustainability anthem; the "Internationale" was the last one. This might be the moment when some of the more enlightened among the very wealthy renounce their excesses and humble themselves before the rest of humanity in a spirit of self-sacrifice. Then we will need assemblies everywhere to plot our course. A great international convention will found and establish the goals of ECOCOM, and replace the Paris Agreement and the still-too-limited goals and timetable of COP26 with a truly sustainable programme. During this period, a new currency might be introduced worldwide which would invalidate all old stored wealth in a specified time, with a maximum of $10 million per person being convertible.[9] A grand jubilee of debt forgiveness might also occur.

All the while it will be getting hotter.

9. When East Germany collapsed, the maximum savings permitted for conversion into West German marks was 4000 East German marks.

Who will set this process in motion? Students? Scientists and intellectuals? Social media influencers? Celebrities? Governments? The poor and dispossessed? A core of selfless and intrepid advocates of simplicity, dressed in homespun clothing? What groups even in combination can possibly defy and then defeat the greatest concentration of propaganda power and information control ever amassed outside of a totalitarian regime? To term this an uphill battle invites an alpine comparison. Many readers may regard the agenda for rescuing us described so far as improbable, to say the least, and much too radical. To embrace sustainability in an age of widespread self-indulgence seems to fly in the face of most of our experience over the last three centuries. For those who enjoy its benefits, opulence is more than satisfying. It epitomises who we are, and what we desire to be. We have long since abandoned any association of luxury with vice. Reversing this development seems well-nigh impossible. We recall Rousseau's sad reflection: "a vicious people can never return to virtue". The well-to-do are all "vicious" today, in the sense of corrupt, spoiled, vain, selfish, and greedy. Moreover, attempts at centralised planning and social reorganisation of the type suggested here are redolent of the failed socialist regimes of the twentieth century. This history we clearly do not want to repeat.

Our way forward lies in sustainability. But simplifying our lives by reducing excessive consumption does not require Spartan austerity or primitivism. We cannot return to Rousseau's ideal state, or to noble savagery. There was no Golden Age, and no paradise: these were dreams, satisfying but whimsical. Now is the time to shed such illusions, for they hinder our progression. We began with the Stone Age, and life pretty much remains short, nasty, and brutish for those who still live in such conditions. The notion that this was somehow a more harmonious or authentic stage of existence has long since disappeared, and with it much of the appeal of the primitive.[10] So we must imagine a new future which preserves the best of the modern. Our world, like Thomas More's, is "nothing but a conspiracy of the rich, whose objective is to increase their own wealth while the government they control claims to be a commonwealth concerned with the common welfare."[11] But this can change. Our survival depends on wedding a much stronger sense of community to an ethos of sufficiency, and substituting an atheism of sustainability for a religion of consumerism. Can we agree that "it is possible to build a society based on not *more* but *better*, not selfishness but sharing, not competition but community"?[12] We can. If we

10. For a summary of the argument, see Robert B. Edgerton's *Sick Societies: Challenging the Myth of Primitive Harmony* (Free Press, 1992).

11. More, *Utopia*, 157.

12. Graaf, Wann, and Naylor, *Affluenza*, ix.

concede with Marius de Geus that "the reduction of levels of material consumption is an essential strategy in the abatement of environmental pollution and the realisation of sustainability", then there really is no choice in the matter.[13]

So can affluenza be cured with utopian medicine? The answer has to be yes. *Has* to be; as Alan Thein Durning writes, "The future of life on earth depends on whether we among the richest fifth of the world's people, having fully met our material needs, can turn to nonmaterial sources of fulfilment."[14] So let us leave the last word to Mandeville, who insisted that "the fewer Desires a Man has and the less he covets, the more easy he is to himself".[15] This is an attainable ideal. The alternative to not suppressing our excessive wants and not creating a sustainable world is far worse. Our candle is flickering. Let us not snuff it out. So go out there—fight back. We can still do this, and we must. We have everything to lose, and everything to gain.

13. Geus, *End of Over-Consumption*, 16.

14. Alan Thein Durning, *How Much Is Enough? The Consumer Society and the Future of the Earth* (Earthscan, 1992), 143.

15. Mandeville, *Fable of the Bees*, 1: 355.

Afterword

Covid-19 and Sociability

THE SOCIAL DISTANCING and self-isolation demanded by the Covid-19 pandemic have had an obvious bearing on a number of the central themes of this book. A larger number of people than ever before were removed from public spaces and gatherings, often into cramped domestic conditions. Boredom, loneliness, and irritation soon became significant challenges. Social interaction rapidly shifted to online contacts, as families and friends attempted to maintain their connections, and businesses, schools, and universities to ensure continuity. Social distancing reduced our sense of touch and feeling of warmth and support from others. With the progress of the virus, the more vulnerable in particular, forced to shield, became increasingly fearful of one another, as if strangers were lepers. Some, often the victims of social media disinformation and propaganda campaigns, proved resolute sceptics, about the disease, masks, social distancing, and vaccines, impeding our ability to fight this menace. They proved surprisingly vicious in their assaults on the medical profession and scientists associated with the pandemic.

One lesson here is that our incapacity for critical analysis is extremely widespread, and susceptibility to propaganda remains great, indeed greater than at any time in many decades. Though the disease will linger for a lengthy period, and will likely be succeeded by other pandemics, its antisocial results will likely be fairly quickly forgotten, as was the case after the 1918–19 outbreak. Yet, beyond the obvious public health issues, there is much to be learned from the experience. We can legislate to make it mandatory to reveal who funds news programmes and advertisements, especially where denial of public health risks is expressed—the parallel with climate-change denial is obvious. So-called think tanks, in particular, must reveal their financial sources on every occasion their views are represented as "impartial". As elsewhere, the operative principle here is transparency. Wherever a collision between apparent public opinion and "the science" appears, we need to know why people adopt the opinions they do, and ensure that "the science" gets a fair hearing. This is of supreme importance as far as environmental education is concerned.

The Covid crisis also invoked our capacity to act collectively when faced with a deadly challenge. Covid proved that a rapid and effective response to a global emergency was possible. This response was admittedly extremely uneven, with some countries acting much more rapidly, and thus saving many more lives, whilst others, like the United States and United Kingdom, dithered, and suffered vastly more deaths. Nations where habitual compliance and mask-wearing were widespread, like Taiwan and China, and a zero-tolerance approach to the disease was adopted, tended to do much better than those who objected at apparent violations of individual rights, like Britain and the United States, where a lack of concern for the more vulnerable was pronounced. Yet once collective action began, even under conservative governments, it was astonishing how quickly large-scale and often quasi-socialist proposals became the norm in dealing with the crisis. As in the 2008 economic crisis, vast sums of money appeared merely as a result of government policy, though corrupt profiteering on procurement contracts was widespread in the United Kingdom. Above all, Covid revealed the fundamental systemic weakness in the American social model, where so many lacked health insurance, and where hundreds of thousands of unnecessary deaths occurred as a result. But even here, government intervention by President Biden ensured a speedy vaccination programme. Enforced sociability in locked-in households of course had its disadvantages, in an increase in domestic violence. In the United Kingdom, it became apparent that there was no need for homelessness. It was abolished, for a time. Worldwide, the need for centralised planning of and fairness in vaccination distribution became clear by early 2021, as wealthy countries assured themselves of supplies while leaving the poorer to fend for themselves.

Covid also proved, as a leading UK newspaper editor, Alan Rusbridger, expressed it in an article entitled "Amid Our Fears, We're Rediscovering the Utopian Hopes of Our Connected World", the utopian potential of human sociability. Not only did thousands of people volunteer to help the less able, and come out of retirement to assist the medical and other emergency services, and to feed the hungry—more than this, wrote Rusbridger, "In our isolation we are rediscovering community", as a host of concepts like "generosity, community, participation, sharing, openness, cooperation, sociability, learning, assembling, imagination, creativity, innovation, experimentation, fairness, equality, publicness, citizenship, mutuality, common resource, information, respect, conversation" came to the fore.[1] These qualities, so often discarded or downplayed in

1. *Observer*, 29 March 2020, 37.

neoliberal discourse, now challenge everyday ways of doing things as they have not done in historic memory. Whether they can continue to do so is the great question. Such events will become more likely as infrastructures weaken worldwide. We must learn from what has often been handled very poorly to date. Everything we learned in 2020–21 tends to make us trust science more and faith less. And it proves what our collective will is capable of once our minds are made up.

BIBLIOGRAPHY OF WORKS CITED

Primary Sources

Periodicals

Alarm Bell; or, Herald of the Spirit of Truth. c.1842.

Associate. 1829.

Economist. 1821.

New Moral World. 1834–44.

Register for the First Society of Adherents to Divine Revelation at Orbiston. 1827. Cited as the Orbiston *Register*

Union. 1842.

Books

Abbott, Leonard. *The Society of the Future*. J. A. Wayland, 1898.

Andreae, Johann Valentine. *Christianopolis*. Edited by Edward H. Thompson. Kluwer Academic, 1999.

Aristotle. *Nicomachean Ethics*. Bobbs-Merrill, 1962.

———. *The Politics of Aristotle*. Edited by Ernest Barker. Oxford University Press, 1962.

Bacon, Francis. *The Essayes; or, Counsels Civill and Morall* [1597]. J. M. Dent, 1906.

Bacon, Francis, and Tomasso Campanella. *The New Atlantis and The City of the Sun: Two Classic Utopias*. Introduction by G. Claeys. Dover, 2018.

Bakunin, Michael. *Bakunin on Anarchy*. Edited by Sam Dolgoff. George Allen and Unwin, 1973.

———. *From Out of the Dustbin: Bakunin's Basic Writings, 1869–1871*. Ardis, 1985.

Banks, Joseph. *The Endeavour Journal of Joseph Banks, 1768–1771*. 2 vols. Public Library of New South Wales, 1962.

Barbon, Nicholas. *A Discourse of Trade* [1690]. Johns Hopkins Press, 1905.

Baxter, Stephen. *Flood*. Gollancz, 2008.

Bebel, August. *Women under Socialism* [1883]. Schocken Books, 1971.

Bellamy, Edward. *Edward Bellamy Speaks Again*. Peerage, 1937.

———. *Equality* [1897]. D. Appleton Century, 1937.

———. *Looking Backward 2000–1887* [1888]. Edited by Matthew Beaumont. Oxford University Press, 2007.

———. *Selected Writings on Religion and Society*. Edited by Joseph Schiffman. Liberal Arts, 1955.

———. *Talks on Nationalism*. Books for Libraries, 1938.

Bentham, Jeremy. *The Works of Jeremy Bentham*. 11 vols. Simpkin, Marshall, 1843.

Bestor, A. E., ed. *Education and Reform at New Harmony: Correspondence of William Maclure and Marie Duclos Fretageot, 1820–1833* [1948]. Augustus M. Kelley, 1973.

Bird, Arthur. *Looking Forward*. L. C. Childs and Son, 1899.

Blatchford, Robert. *Britain for the British*. Clarion, 1902.

———. *Merrie England*. Clarion, 1908.

———. *The Sorcery Shop: An Impossible Romance*. Clarion, 1907.

———. *What's All This?* Labour Book Service, 1940.

Bloch, Ernst. *Atheism in Christianity*. Verso, 2009.

———. *Das Prinzip Hoffnung*. 2 vols. Aufbau Verlag, 1955.

———. *Heritage of Our Times*. Polity, 1991.

———. *Literary Essays*. Stanford University Press, 1998.

———. *Man On His Own: Essays in the Philosophy of Religion*. Herder and Herder, 1970.

———. *Natural Law and Human Dignity*. MIT Press, 1986.

———. *A Philosophy of the Future*. Herder and Herder, 1963.

———. *The Principle of Hope*. 3 vols. Basil Blackwell, 1986.

———. *The Spirit of Utopia* [1918]. Stanford University Press, 2000.

———. *Thomas Münzer Als Theologe der Revolution*. Kurt Wolff Verlag, 1921.

———. *The Utopian Function of Art and Literature*. MIT Press, 1988.

Blount, Godfrey [pseud. Hans Breitman]. *For Our Country's Sake*. A. C. Fifield, 1905.

Blount, Godfrey. *The Gospel of Simplicity: A Plea for Country Life & Handicrafts*. Simple Life, 1903.

———. *A New Crusade: An Appeal*. Simple Life, 1903.

Bogdanov, Alexander. *Red Star* [1909]. Indiana University Press, 1984.

Borsodi, Ralph. *The Distribution Age: A Study of the Economy of Modern Distribution*. D. Appleton, 1927.

———. *Flight from the City* [1933]. Harper Colophon Books, 1972.

———. *This Ugly Civilization*. Simon and Schuster, 1929.

Bougainville, Louis-Antoine de. *The Pacific Journal of Louis-Antoine de Bougainville, 1767–1768*. Hakluyt Society, 2002.

[Bower, Samuel]. *A Brief Account of the First Concordium*. 1843.

Bradbury, Ray. *Fahrenheit 451*. Ballantine Books, 1953.

Bray, John Francis. *Labour's Wrongs and Labour's Remedy*. David Green, 1839.

Brinsmade, Herman Hine. *Utopia Achieved*. Broadway, 1912.

Brisbane, Albert. *Social Destiny of Man; or, Association and Reorganization of Industry*. C. F. Stollmeyer, 1840.

Brown, John. *An Estimate of the Manners and Principles of the Times* [1757–58]. Liberty Fund, 2019.

Brown, Paul. *Twelve Months in New Harmony*. C. H. Woodward, 1827.

Buckingham, James Silk. *National Evils and National Remedies*. Peter Jackson, 1849.

Bulwer-Lytton, Edward. *The Coming Race*. George Routledge and Sons, 1874.

Buonarroti, Philippe. *Buonarroti's History of Babeuf's Conspiracy for Equality*. Translated by Bronterre O'Brien. H. Hetherington, 1836.

[Burgh, James]. *An Account of the First Settlement, Laws, Form of Government, and Police, of the Cessares, a People of South America*. J. Payne, 1764.

———. *Political Disquisitions*. 3 vols. E. and C. Dilly, 1774–75.
Burton, Captain Sir Richard F. *Personal Narrative of a Pilgrimage to Al-Madinah & Meccah* [1853]. 2 vols. Tylston and Edwards, 1893.
Cabet, Etienne. *Travels in Icaria*. Translated by Leslie J. Roberts. Introduction by Robert Sutton. Syracuse University Press, 2003.
———. *Voyage en Icarie* [1840]. Paris, 1848.
Callaway, Joseph Sevier. *Sybaris*. Johns Hopkins Press, 1930.
Callenbach, Ernest. *Ecotopia: A Novel about Ecology, People and Politics in 1999*. Pluto, 1978.
———. *Ecotopia Emerging*. Banyan Tree Books, 1981.
Campanella, Tommaso. *The City of the Sun*. Translated by Daniel J. Donno. University of California Press, 1981.
Camus, Albert. *The Rebel*. Alfred A. Knopf, 1954.
Carlyle, Thomas. *Works*. Vol. 10. Chapman and Hall, 1899.
Carpenter, Edward. *Civilization: Its Cause and Cure*. George Allen and Unwin, 1921.
———. *England's Ideal and Other Papers on Social Subjects*. Swan Sonnenschein, 1902.
———. *My Days and Dreams: Being Autobiographical Notes*. George Allen and Unwin, 1916.
———. *Sex-Love, and Its Place in a Free Society*. Labour Press Society, 1894.
Chastellux, Jean François, Marquis de. *An Essay on Public Happiness*. 2 vols. T. Cadell, 1774.
Chaucer, Daniel [Ford Maddox Ford]. *The Simple Life Limited*. Bodley Head, 1911.
Chesterton, G. K. *Twelve Types*. Arthur L. Humphreys, 1910.
Child, William Stanley. *The Legal Revolution of 1902*. Charles H. Kerr, 1898.
Cicero. *Agrarian Speeches*. Oxford University Press, 2018.
———. *The Offices, Essays on Friendship and Old Age and Select Letters*. J. M. Dent, 1909.
Claeys, Gregory, ed. *Late Victorian Utopias*. 6 vols. Pickering and Chatto, 2009.
———, ed. *Modern British Utopias*. 8 vols. Pickering and Chatto, 1997.
———, ed. *Owenite Socialism: Pamphlets and Correspondence*. 10 vols. Routledge, 2005.
———, ed. *Political Writings of the 1790s*. 8 vols. Pickering and Chatto, 1995.
———, ed. *The Politics of English Jacobinism: Writings of John Thelwall*. Penn State University Press, 1995.
———, ed. *Restoration and Augustan British Utopias*. Syracuse University Press, 2000.
———, ed. *Utopias of the British Enlightenment*. Cambridge University Press, 1994.
Claeys, Gregory, and Lyman Tower Sargent, eds. *The Utopia Reader*. 2nd ed. New York University Press, 2016.
Clark, Henry C., ed. *Commerce, Culture, & Liberty: Readings on Capitalism before Adam Smith*. Liberty Fund, 2003.
Cobbett, William. *Advice to Young Men* [1829]. Henry Froude, 1906.
Codman, John Thomas. *Brook Farm: Historic and Personal Memoirs*. Arena, 1894.
Combe, Abram. *The Sphere of Joint-Stock Companies*. G. Mudie, 1825.
Comte, Auguste. *System of Positive Polity*. 4 vols. Longmans, Green, 1875–77.
[Condorcet, Marquis de]. *The Life of M. Turgot*. J. Johnson, 1787.
———. *Sketch for a Historical Picture of the Progress of the Human Mind* [1794]. Greenwood, 1979.
Conway, Moncure. *Autobiography, Memories and Experiences*. 2 vols. Cassell, 1904.
Cook, James. *Captain Cook's Journal during His First Voyage round the World*. Elliot Stock, 1893.

Cook, William Wallace. *A Round Trip to the Year 2000*. Street and Smith, 1903.
[Coste, Frank Perry]. *Towards Utopia (Being Speculations in Social Evolution)*. Swan Sonnenschein, 1894.
Craig, E. T. *An Irish Commune: The Experiment at Ralahine, County Clare, 1831–1833* [1920]. Irish Academic, 1983.
Davenant, Charles. *The Political and Commercial Works*. 5 vols. 1771.
Dawson, W. J. *The Quest of the Simple Life*. Hodder and Stoughton, 1903.
———. *Savonarola: A Drama*. Grant Richards, 1900.
Defoe, Daniel. *Robinson Crusoe* [1719]. Penguin Books, 1965.
Devinne, Paul. *The Day of Prosperity*. G. W. Dillingham, 1902.
Diderot, Denis. *Political Writings*. Edited by John Hope Mason and Robert Wokler. Cambridge University Press, 1992.
———. *Rameau's Nephew and Other Works*. Anchor Books, 1956.
———. *Selected Writings*. Macmillan, 1966.
———. *Supplément au voyage de Bougainville*. Librarie E. Droz, 1935.
The Doctrine of Saint-Simon: An Exposition; First Year, 1828–1829. Edited by Georg Iggers. Schocken Books, 1972.
Donnelly, Ignatius. *Atlantis*. Harper and Bros, 1885.
———. *Caesar's Column*. Sampson, Low, Marston, 1891.
Duhamel, Georges. *America the Menace: Scenes from the Life of the Future*. George Allen and Unwin, 1931.
Durkheim, Emile. *Sociology and Philosophy*. Translated by D. F. Pocock. Cohen and West, 1953.
———. *Suicide*. Translated by John A. Spaulding and George Simpson. Routledge and Kegan Paul, 1952.
Emerson, Ralph Waldo. *The Conduct of Life and Society and Solitude*. Macmillan, 1892.
Epictetus. *The Discourses and Manual*. 2 vols. Clarendon, 1916.
Erasmus. *In Praise of Folly*. Edited by Hoyt Hopewell Hudson. Princeton University Press, 1941.
Etzler, John Adolphus. *The Collected Works of John Adolphus Etzler*. Edited by Joel Nydahl. Scholars' Facsimiles and Reprints, 1977.
Evans, Frederick William. *Autobiography of a Shaker*. United, 1888.
Fénelon, François de Salignac de la Mothe-. *The Adventures of Telemachus*. Edited by O. M. Brack. University of Georgia Press, 1997.
———. *Les aventures de Télémaque* [1699]. 2 vols. Librairie Hachette, 1920.
———. *Letters*. Edited by John McEwen. Harvill, 1964.
Ferguson, Adam. *An Essay on the History of Civil Society* [1767]. Edinburgh University Press, 1966.
Fichte, Johann Gottlieb. *The Popular Works of Johann Gottlieb Fichte*. 2 vols. Trübner, 1889.
Fletcher, Andrew. *Select Political Writings and Speeches*. Edited by David Daiches. Academic, 1979.
Foigny, Gabriel de. *The Southern Land Known* [1676]. Edited by David Fausett. Syracuse University Press, 1993.
Food Rationing and Supply 1943/44. League of Nations, 1944.
Fourier, Charles. *Design for Utopia: Selected Writings of Charles Fourier*. Schocken, 1971.
———. *Le nouveau monde amoureaux: Oeuvres complètes de Charles Fourier*. 12 vols. Éditions Anthropos, 1966–68.

———. *The Passions of the Human Soul*. 2 vols. Hippolyte Bailliere, 1851.
———. *Selections from the Works of Fourier*. Introduction by Charles Gide. Swan Sonnenschein, 1901.
———. *The Theory of the Four Movements* [1808]. Edited by Gareth Stedman Jones and Ian Paterson. Cambridge University Press, 1996.
———. *The Utopian Vision of Charles Fourier*. Edited by Jonathan Beecher and Richard Bienvenu. Beacon, 1971.
The Free State of Noland. J. Whitlock, 1696.
Frost, Thomas. *Forty Years' Recollections*. Sampson Low, Marston, Searle, and Rivington, 1880.
Fuller, Alvarado M. *A.D. 2000*. Laird and Lee, 1890.
Gandhi, M. K. *An Autobiography*. Phoenix, 1949.
———. *Hind Swaraj and Other Writings*. Cambridge University Press, 1997.
———. *The Moral and Political Writings of Mahatma Gandhi*. Edited by Raghavan Iyer. 3 vols. Clarendon, 1986.
Gauguin, Paul. *The Letters of Paul Gauguin to Georges Daniel de Monfreid*. William Heinemann, 1923.
———. *Letters to His Wife and Friends*. World Publishing, 1949.
———. *Noa Noa*. Nicholas L. Brown, 1919.
Gearhart, Sally Miller. *The Wanderground: Stories of the Hill Women* [1979]. Women's Press, 1985.
George, Henry. *Progress and Poverty*. Wm M. Hinton, 1879.
Gibbon, Edward. *The Decline and Fall of the Roman Empire*. 3 vols. Modern Library, 1932.
Gibbs, Sir Philip. *The Day after To-morrow*. Hutchinson, 1928.
Gillette, King. *The Human Drift*. New Era, 1894.
Godwin, Parke. *A Popular View of the Doctrines of Charles Fourier*. J. S. Redfield, 1844.
Godwin, William. *The Enquirer*. G. C. and J. Robinson, 1797.
———. *Enquiry Concerning Political Justice*. 2 vols. G.G.J and J. Robinson, 1793.
———. *Enquiry Concerning Political Justice*. 3rd ed. [1798]. Penguin Books, 1976.
Gorky, Maxim. *Mother*. D. Appleton, 1907.
Gott, Samuel. *Nova Solyma: The Ideal City; or, Jerusalem Regained* [1648]. 2 vols. John Murray, 1902.
Gray, John. *An Efficient Remedy for the Distress of Nations*. Longman, Brown, Green, and Longmans, 1842.
———. *A Lecture on Human Happiness*. Sentry, 1826.
Greaves, James Pierrepont. *Letters and Extracts from the MS. Writings of James Pierrepont Greaves*. Vol. 1. Ham Common, 1843.
Griffith, George. *The Angel of the Revolution*. Tower, 1895.
Gronlund, Laurence. *Our Destiny*. Swan Sonnenschein, 1890.
Günzburg, Johann Eberlein von. *Wolfaria*. 1521.
Hale, Edward Everett. *Sybaris and Other Homes*. Fields, Osgood, 1869.
Hall, Charles. *The Effects of Civilization on the People in European States*. 1805.
Hall, James Norman. *Lost Island*. Little, Brown, 1944.
Harrington, James. *The Political Works of James Harrington*. Edited by J.G.A. Pocock. Cambridge University Press, 1977.
Hawthorne, Nathaniel. *The Blithedale Romance*. 2 vols. Chapman and Hall, 1852.

Hay, W. D. *Three Hundred Years Hence; or, A Voice from Posterity*. Newman, 1881.
Helvétius, Claude. *De l'Esprit; or, Essays on the Mind* [1758]. 1810.
———. *A Treatise on Man*. 2 vols. Albion, 1810.
[Hennell, Mary]. *An Outline of the Various Social Systems & Communities Which Have Been Founded on the Principle of Co-operation*. Longman, Brown, Green and Longmans, 1844.
Hertzka, Theodor. *Freeland: A Social Anticipation*. Chatto and Windus, 1891.
Hess, Moses. *The Holy History of Mankind and Other Writings*. Edited by Shlomo Avineri. Cambridge University Press, 2004.
Hillquit, Morris. *History of Socialism in the United States* [1903]. 5th ed. Dover, 1971.
History and Constitution of the Icarian Community [1917]. AMS, 1975.
Hitler, Adolf. *Mein Kampf*. Hurst and Blackett, 1939.
[Holbach, Baron d']. *System of Nature* [1770]. B. D. Cousins, 1841.
[Holberg, Ludwig]. *Journey to the World Under Ground* [1755]. In *Popular Romances*, edited by Henry Weber, 115–200. John Murray, 1812.
Horton, Harry Howells. *Community the Only Salvation for Man: A Lecture*. A. Heywood, 1838.
Howard, Ebenezer. *Garden Cities of Tomorrow* [1898]. Faber and Faber, 1945.
Howell, John. *The Life and Adventures of Alexander Selkirk*. Oliver and Boyd, 1829.
Howells, William Dean. *Through the Eye of the Needle*. Harper and Bros, 1907.
———. *A Traveller from Altruria*. David Douglas, 1894.
How I Became a Socialist: A Series of Biographical Sketches. Twentieth Century, 1896.
Hume, David. *Enquiries Concerning Human Understanding and Concerning the Principles of Morals* [1748, 1751]. Clarendon, 1975.
———. *Essays Moral, Political and Literary* [1758]. 2 vols. Longman, Green, 1882.
———. *The Letters of David Hume*. Edited by J.Y.T. Greig. 2 vols. Clarendon, 1932.
———. *Philosophical Works*. 4 vols. Adam and Charles Black, 1854.
Huxley, Aldous. *Aldous Huxley's Hearst Essays*. Garland, 1994.
———. *Along the Road: Notes and Essays of a Tourist*. Chatto and Windus, 1930.
———. *Brave New World* [1932]. Penguin Books, 1955.
———. *Brave New World* [1932]. Vintage Books, 2007.
———. *Brave New World Revisited*. Harper and Row, 1958.
———. *Complete Essays*. 6 vols. Ivan R. Dee, 2000–2002.
———. *Grey Eminence: A Study in Religion and Politics*. Chatto and Windus, 1941.
———. *The Hidden Huxley: Contempt and Compassion for the Masses*. Edited by David Bradshaw. Faber and Faber, 1994.
———. *Island*. Penguin Books, 1964.
———. *Island*. Edited by David Bradshaw. Flamingo Books, 1994.
———. *Jesting Pilate: The Diary of a Journey*. Chatto and Windus, 1930.
———. *Letters of Aldous Huxley*. Chatto and Windus, 1969.
———. *Now More than Ever*. Edited by David Bradshaw and James Sexton. University of Texas Press, 2000.
———. *Prisons*. Trianon, 1949.
———. *Proper Studies*. Chatto and Windus, 1927.
———. *Science, Liberty and Peace*. Chatto and Windus, 1950.
Huxley, Julian, ed. *Aldous Huxley, 1894–1963: A Memorial Volume*. Chatto and Windus, 1965.

———. *A Scientist among the Soviets*. Chatto and Windus, 1932.
Jameson, Storm. *In the Second Year*. Cassell, 1936.
Kant, Immanuel. *On History*. Bobbs-Merrill, 1963.
Kropotkin, Peter. *The Conquest of Bread*. Chapman and Hall, 1906.
———. *Mutual Aid*. William Heinemann, 1903.
La Bretonne, Restif de. *The Corrupted Ones* [1776]. Neville Spearman, 1967.
———. *Les Nuits de Paris; or, The Nocturnal Spectator* [1739]. Random House, 1964.
———. *My Revolution: Promenades in Paris, 1789–1794*. Edited by Alex Karmel. McGraw-Hill, 1970.
Lactantius. *Divine Institutes*. Liverpool University Press, 2003.
Lafitau, Joseph François. *Customs of the American Indians Compared with the Customs of Primitive Times*. 2 vols. Champlain Society, 1974.
La Roche, Sophie von. *Sophie in London in 1786*. Jonathan Cape, 1933.
Lee, Charles. *Memoirs of the Life of the late Charles Lee*. J. S. Jordan, 1792.
Leibniz, G. W. *Discourse on Metaphysics and Other Essays*. Hackett, 1991.
Lenin, V. I. *Collected Works*. 47 vols. Progress, 1964.
Levy, J. H. *An Individualist's Utopia*. Lawrence Nelson, 1912.
[Lithgow, John]. *Equality: A Political Romance; Without either Kings, Ghosts, or Enchanted Castles*. 1802.
Lloyd, Henry Demerest. *Wealth against Commonwealth*. Harper and Bros, 1894.
Locke, John. *Two Treatises of Government*. Edited by Peter Laslett. 2nd ed. Cambridge University Press, 1970.
Lockhart-Mummery, J. P. *After Us; or, The World as It Might Be*. Stanley Paul, 1936.
Lucian. *The Works of Lucian of Samosata*. 4 vols. Clarendon, 1905.
Lukács, Georg. *Tactics and Ethics: Political Writings, 1919–1929*. Verso, 2014.
Macaulay, Catherine. *Loose Remarks on Certain Positions to be found in Mr. Hobb's Philosophical Rudiments of Government and Society: With a Short Sketch of a Democratical Form of Government*. T. Davies, 1767.
Machiavelli, Niccolò. *The Prince and the Discourses*. Modern Library, 1950.
Mackay, Charles. *Memoirs of Extraordinary Popular Delusions*. 3 vols. Richard Bentley, 1841.
Macrobius. *Saturnalia*. 3 vols. Harvard University Press, 1977.
Madden, Samuel. *Memoirs of the Twentieth Century: Being Original Letters of State under George the Sixth*. Osborn, Longman, Davis, and Batley, 1733.
Mandel, Emily St. John. *Station Eleven*. Picador, 2014.
Mandeville, Bernard. *Enquiry into the Origin of Honour*. J. Brotherton, 1732.
———. *The Fable of the Bees* [1723–28]. 2 vols. Clarendon, 1924.
Manuel, Frank E., and Fritzie P. Manuel, eds. *French Utopias: An Anthology of Ideal Societies*. Schocken Books, 1966.
Mao Tsetung. *Selected Works of Mao Tsetung*. 5 vols. Foreign Languages Press, 1954–77.
Marriott, Joseph. *Community: A Drama*. A. Heywood, 1838.
Martyr, Peter. *De Orbe Novo: The Eight Decades of Peter Martyr D'Anghera* [1530]. 2 vols. G. P. Putnam's Sons, 1912.
Marx, Karl. *The First International and After: Political Writings*. Vol. 3. Edited by David Fernbach. Penguin Books, 1974.

Marx, Karl, and Frederick Engels. *Collected Works*. 50 vols. Lawrence and Wishart, 1975–2005.
———. *Werke*. 43 vols. Dietz Verlag, 1964–90.
Melon, Jean-François. *A Political Essay upon Commerce* [1734]. 1738.
Memoirs concerning the Life and Manners of Captain Mackheath. A. Moore, 1728.
Memoirs of the Court of Lilliput. 2nd ed. J. Roberts, 1727.
Mercier, Louis-Sébastien. *De J. J. Rousseau considéré comme l'un des premiers auteurs de la Révolution*. 2 vols. Buisson, 1791.
———. *Fragments on Politics and History*. 2 vols. H. Murray, 1795.
———. *L'an 2440: Rêve s'il en fut jamais*. Introduction and notes by Christophe Cave and Christine Marcandier-Colard. La Decouverte, 1999.
[———]. *Memoirs of the Year Two Thousand Five Hundred*. 2 vols. G. Robinson, 1782.
———. *The Picture of Paris: Before & after the Revolution* [1788]. George Routledge and Sons, 1929.
Michelet, Jules. *Historical View of the French Revolution*. G. Bell and Sons, 1896.
Michener, James. *Return to Paradise*. Secker and Warburg, 1951.
Michener, James. *Tales of the South Pacific*. Fawcett Crest, 1947.
Mill, James. *The History of British India*. 6 vols. Baldwin, Craddock, and Joy, 1820.
Mill, John Stuart. *Collected Works*. 33 vols. Routledge and Kegan Paul, 1963–91.
———. *Dissertations and Discussions*. 2nd ed. 2 vols. 1862.
———. *On Liberty*. John W. Parker and Son, 1859.
———. *Principles of Political Economy*. 2 vols. John Parker, 1848.
———. *The Spirit of the Age*. University of Chicago Press, 1942.
Millar, John. *The Origin of the Distinction of Ranks* [1771]. 4th ed. William Blackwood, 1806.
Mirandola, Pico della. *Oration on the Dignity of Man*. Edited by Francesco Borghese, Michael Papio, and Massimo Riva. Cambridge University Press, 2012.
Montaigne, Michel de. *The Complete Works of Montaigne*. Edited by Donald M. Frame. Hamish Hamilton, 1958.
Montesquieu. *The Spirit of the Laws* [1748]. Hafner, 1949.
More, Thomas. *The Complete Works of Thomas More*. Vol. 4. Edited by Edward Surtz and J. H. Hexter. Yale University Press, 1965.
———. *The Complete Works of Thomas More*. Vol. 8. Edited by Louis A. Schuster, Richard C. Marius, James P. Lusardi, and Richard J. Schoeck. Yale University Press, 1973.
———. *A Dialogue of Comfort against Tribulation*. Yale University Press, 1977.
———. *Utopia* [1516]. Edited by Edward Surtz. Yale University Press, 1964.
———. *Utopia* [1516]. Edited by David Wootton. Hackett, 1999.
Morelly. *Code de la nature* [1755]. Edited by Gilbert Chinard. 1950.
Morgan, John Minter. *Hampden in the Nineteenth Century*. 2 vols. Edward Moxon, 1834.
———. *The Revolt of the Bees* [1826]. 3rd ed. 1839.
Morrall, Clare. *When the Floods Came*. Hodder and Stoughton, 2016.
Morris, William. *The Collected Letters of William Morris*. Princeton University Press, 1987.
———. *A Dream of John Ball; and, A King's Lesson*. Longmans, Green, 1912.
———. *News from Nowhere* [1890]. Longmans, Green, 1899.
———. *The Political Writings of William Morris*. Edited by A. L. Morton. Lawrence and Wishart, 1984.

———. *Signs of Change*. Reeves and Turner, 1888.

Moszkowski, Alexander. *The Isles of Wisdom*. 1924. George Routledge and Sons.

Munro, H. H. *When William Came: A Story of London under the Hohenzollerns*. Bodley Head, 1914.

Müntzer, Thomas. *The Works of Thomas Müntzer*. Edited by T. Matheson and T. Clark, 1988. T. and T. Clark.

Murphy, G. Read. *Beyond the Ice*. Sampson Low, Marston, 1894.

Nedham, Marchamont. *The Case of the Commonwealth of England, Stated* [1650]. University Press of Virginia, 1969.

———. *The Excellencie of a Free-State* [1656]. Liberty Fund, 2011.

The New Political Economy of the Honey Bee. W. C. Featherstone, 1823.

News from New Cythera: A Report of Bougainville's Voyage, 1766–1769. University of Minnesota Press, 1970.

Noto, Cosimo. *The Ideal City*. 1903.

Noyes, George Wallingford, ed. *Free Love in Utopia: John Humphrey Noyes and the Origin of the Oneida Community*. University of Illinois Press, 1971.

———, ed. *Religious Experience of John Humphrey Noyes*. Macmillan, 1923.

Noyes, John Humphrey. "Bible Communism". In *John Humphrey Noyes: The Putney Community*, edited by George Wallingford Noyes, 116–22. Oneida, 1931.

———. *Strange Cults and Utopias of 19th-Century America*. Dover, 1966. Originally published as *History of American Socialisms*, 1870.

Noyes, Pierrepont. *My Father's House: An Oneida Boyhood*. Rinehart, 1937.

Ogilvie, William. *An Essay on the Right of Property in Land*. J. Walter, 1781.

Ognyov, N. *The Diary of a Communist Schoolboy*. Victor Gollancz, 1938.

———. *The Diary of a Communist Undergraduate*. Victor Gollancz, 1929.

Olerich, Henry. *A Countryless and Cityless World*. Gilmore and Olerich, 1893.

Orwell, George. *Homage to Catalonia*. Introduction by Helen Graham. Macmillan, 2021.

———. *I Have Tried to Tell the Truth: 1943–1944*. Secker and Warburg, 1998.

———. *The Penguin Complete Novels of George Orwell*. Penguin Books, 1967.

———. *Smothered under Journalism: 1946*. Secker and Warburg, 1998.

Owen, Robert. *A New View of Society and Other Writings*. Edited by G. Claeys. Penguin Books, 1991.

———. *Report of the Proceedings at the Several Public Meetings Held in Dublin*. 1823.

———. *Selected Works of Robert Owen*. Edited by G. Claeys. 4 vols. Pickering and Chatto, 1993.

Paine, Thomas. *Agrarian Justice*. J. Adlard, 1797.

———. *Complete Writings*. Edited by Philip S. Foner. 2 vols. Citadel, 1945.

———. *Rights of Man* [1791–92]. Edited by G. Claeys. Hackett Books, 1992.

Paley, William. *The Principles of Moral and Political Philosophy* [1786]. Baldwyn, 1819.

"A Paradox: Proving the Inhabitants of the Island, Called *Madagascar*, or *St. Lawrence* (in Things Temporal) to Be the Happiest People in the World". *Harleian Miscellany* 1 (1744): 256–62.

Pears, Thomas, and Sarah Pears. *New Harmony: An Adventure in Happiness; Papers of Thomas & Sarah Pears* [1933]. Augustus M. Kelley, 1973.

Peck, Bradford. *The World a Department Store: A Story of Life under a Cooperative System*. Bradford Peck, 1900.

Plato. *The Collected Dialogues of Plato*. Edited by Edith Hamilton and Huntington Cairns. Translated by Hugh Tredennick. Pantheon Books, 1961.

Plockhoy, Peter. *A Way Propounded to Make the Poor Happy*. 1659.

Plutarch. *The Lives of the Noble Grecians and Romans*. Modern Library, n.d.

Price, Richard. *Observations on Reversionary Payments*. 7th ed. 2 vols. 1812.

———. *Richard Price and the Ethical Foundations of the American Revolution* [1785]. Edited by Bernard Peach. Duke University Press, 1979.

Priestley, Joseph. *Lectures on History and General Policy* [1788]. Thomas Tegg, 1826.

———. *Political Writings*. Cambridge University Press, 1993.

Proudhon, Pierre-Joseph. *What Is Property?* Humboldt, *c.*1890.

Pufendorf, Samuel. *On the Duty of Man and Citizen according to Natural Law* [1673]. Edited by James Tully. Cambridge University Press, 1991.

———. *On the Natural State of Man* [1678]. Edwin Mellen, 1990.

Pure, Michel de. *Épigone, histoire du siècle future* [1659]. Edited by Lise Leibacher-Ouvrard and Daniel Maher. Presses de l'Université Laval, 2005.

Rathenau, Walther. *In Days to Come*. George Allen and Unwin, 1921.

Raynal, Guillaume Thomas. *A Philosophical and Political History of the Settlements and Trade of the Europeans in the East and West Indies*. 3 vols. J. and J. Robinson, 1811.

Reid, Thomas. *Essays on the Active Powers of the Human Mind* [1788]. MIT Press, 1969.

Reybaud, Louis. *Études sur les réformateurs ou socialistes modernes*. 2nd ed. 2 vols. Société Belge de Librairie, 1844.

Ricardo, David. *The Works of David Ricardo*. John Murray, 1846.

Richter, Eugene. *Pictures of the Socialistic Future* [1893]. George Allen, 1912.

Robbins, Caroline, ed. *Two English Republican Tracts*. Cambridge University Press, 1969.

Roberts, J. W. *Looking Within*. A. S. Barnes, 1893.

Robertson, Constance Noyes, ed. *Oneida Community: An Autobiography, 1851–1876*. Syracuse University Press, 1970.

Rousseau, Jean-Jacques. *Discourses on the Sciences and Arts and Polemics* [1754–55]. Collected Writings, vol. 2. University Press of New England, 1992.

———. *Emile* [1762]. J. M. Dent, 1911.

———. *Emile; or, On Education* [1762]. Collected Writings, vol. 13. Dartmouth College Press, 2010.

———. *Julie; or, The New Heloise* [1761]. Collected Writings, vol. 6. University Press of New England, 1997.

———. *Oeuvres complètes*. 5 vols. Gallimard, 1959–95.

———. *The Plan for Perpetual Peace, On the Government of Poland, and Other Writings on History and Politics*. Collected Writings, vol. 11. University Press of New England, 2002.

———. *Political Writings*. Edited by Frederick Watkins. Nelson, 1953.

———. *Social Contract* [1762]. Collected Writings, vol. 4. University Press of New England, 1994.

———. *The Social Contract and Discourses*. E. P. Dent, 1973.

Royce, Josiah. *The Philosophy of Loyalty*. Macmillan, 1908.

Ruskin, John. *Unto This Last*. Smith, Elder, 1862.

———. *The Works of John Ruskin*. 11 vols. George Allen, 1882.

Saint-Simon, Henri de. *The Political Thought of Saint-Simon*. Edited by Ghita Ionescu. Oxford University Press, 1976.

———. *Selected Writings on Science, Industry and Social Organization*. Edited by Keith Taylor. Holmes and Meier, 1975.

Sandford, Mrs Henry, ed. *Thomas Poole and His Friends*. 2 vols. Macmillan, 1888.

Satterlee, W. W. *Looking Backward, and What I Saw*. Harrison and Smith, 1890.

Saxe-Weimar Eisenach, Bernhard, Duke of. *Travels through North America, during the Years 1825 and 1826*. 2 vols. Carey, Lea, and Carey, 1828.

Say, Jean-Baptiste. *An Economist in Troubled Times*. Writings selected and translated by R. R. Palmer. Princeton University Press, 1947.

———. *Olbie; ou, Essai sur les moyens de réformer les moeurs d'une nation* [1802]. 1830 edition translated by Roy Arthur Swanson, in *Utopian Studies* 12 (2001): 79–107.

———. *A Treatise on Political Economy* [1803]. Claxton, Remsen and Haffelfinger, 1880.

Scott, Sarah. *Millenium Hall* [1762]. Virago Books, 1986.

Sears, John Van der Zee. *My Friends at Brook Farm*. Desmond Fitzgerald, 1912.

Sedgwick, Marcus. *Floodland*. Orion, 2001.

Seneca. *Letters from a Stoic*. Penguin Books, 2014.

Shaftesbury, Anthony, Earl of. *The Life, Unpublished Letters, and Philosophical Regimen*. Swan Sonnenschein, 1900.

Shaw, Albert. *Icaria: A Chapter in the History of Communism*. Knickerbocker, 1884.

Sidney, Algernon. *Court Maxims*. Cambridge University Press, 1996.

———. *Discourses concerning Government*. 2 vols. G. Hamilton and J. Balfour, 1750.

Sismondi, J.-C.-.L. Simonde de. *New Principles of Political Economy* [1819]. Transaction, 1991.

———. *Political Economy and the Philosophy of Government*. Chapman, 1847.

Skinner, B. F. *Walden Two* [1948]. Macmillan, 1962.

Sledge, E. B. *With the Old Breed: At Peleliu and Okinawa*. Presidio, 1990.

Sloan, Alfred P. *My Years with General Motors*. Sidgwick and Jackson, 1966.

Smith, Adam. *An Inquiry into the Nature and Causes of the Wealth of Nations* [1776]. 2 vols. Clarendon, 1869.

———. *The Theory of Moral Sentiments* [1759]. Henry Bohn, 1853.

Snow, Edgar. *Red Star over China* [1938]. Victor Gollancz, 1968.

Southey, Robert. *Sir Thomas More; or, Colloquies on the Progress and Prospects of Society*. 2 vols. John Murray, 1829.

Spence, Thomas. *Pig's Meat: The Selected Writings of Thomas Spence*. Edited by G. I. Gallop. Spokesman, 1982.

———. *The Political Writings of Thomas Spence*. Edited by H. T. Dickinson. Avero, 1982.

Spence, Thomas, and Charles Hall. "Four Letters between Thomas Spence and Charles Hall". *Notes and Queries* 28 (1981): 317–21.

Spence, William. *Tracts on Political Economy*. Longman, Hurst, Rees, Orme, and Brown, 1822.

Steere, C. A. *When Things Were Doing*. Charles H. Kerr, 1907.

Steuart, Sir James. *An Inquiry into the Principles of Political Economy* [1767]. 4 vols. Pickering and Chatto, 1998.

[Symmes, John Cleve?]. *Symzonia: A Voyage of Discovery* [1820]. Scholars' Facsimiles and Reprints, 1965.

Tarde, Gabriel. *Underground Man*. Duckworth, 1905.

Thomas, Chauncey. *The Crystal Button; or, Adventures of Paul Prognosis in the Forty-Ninth Century*. Houghton, Mifflin, 1891.

Thompson, William. *An Inquiry into the Principles of the Distribution of Wealth*. Longman, Hurst, Rees, Orme, Brown, and Green, 1824.

Thoreau, Henry David. *Walden; or, Life in the Woods*. Walter Scott, 1886.

Tolstoy, Leo. *A Confession and Other Religious Writings*. Penguin Books, 1987.

———. *The Russian Revolution*. Free Age, 1907.

———. *What Shall We Do?*. Free Age, 1910.

Trenchard, John, and Thomas Gordon. *Cato's Letters; or, Essays on Liberty, Civil and Religious, and Other Important Subjects* [1720]. 2 vols. Liberty Fund, 1995.

Trotsky, Leon. *Literature and Revolution* [1925]. University of Michigan Press, 1975.

Tuckwell, Reverend W. *The New Utopia; or, England in 1985*. Hudson and Son, 1885.

Turgot, Anne Robert Jacques. *Turgot on Progress, Sociology and Economics*. Edited by Ronald L. Meek. Cambridge University Press, 1963.

Urwick, E. J. *Luxury and Waste of Life*. J. M. Dent, 1908.

Veiras, Denis. *The History of the Severambians*. Edited by John Christian Laursen and Cyrus Masoori. State University of New York Press, 2006.

Volney. *The Ruins; or, A Survey of the Revolutions of Empires* [1791]. H. D. Symonds, 1801.

Voltaire. *Candide; or, Optimism* [1759]. Penguin Books, 2005.

———. *Oeuvres complètes*. Voltaire Foundation, 2003.

Wagar, W. W., ed. *H. G. Wells: Journalism and Prophecy, 1893–1946*. Bodley Head, 1965.

Wagner, Charles. *The Simple Life* [1897]. Ibister, 1903.

Wallace, Robert. *Various Prospects of Mankind, Nature and Providence*. A. Millar, 1758.

Warren, Josiah. *Practical Applications of the Elementary Principles of "True Civilization"*. 1873.

Warville, J. P. Brissot de. *New Travels in the United States of America, Performed in 1788* [1792]. Woodstock Books, 2000.

Webster, Charles, ed. *Samuel Hartlib and the Advancement of Learning*. Cambridge University Press, 1970.

Wells, H. G. *After Democracy*. Watts, 1932.

———. *The Autocracy of Mr. Parham: His Remarkable Adventures in This Changing World*. William Heinemann, 1930.

———. *The Bulpington of Blup*. Hutchinson, 1933.

———. *An Englishman Looks at the World*. Cassell, 1914.

———. *The Fate of Homo Sapiens*. Secker and Warburg, 1939.

———. *The Future in America*. Chapman and Hall, 1906.

———. *God the Invisible King*. Cassell, 1917.

———. *Mankind in the Making*. Chapman and Hall, 1911.

———. *Men like Gods*. Cassell, 1923.

———. *A Modern Utopia*. Chapman and Hall, 1905.

———. *The New America*. Cresset, 1935.

———. *The Salvaging of Civilisation*. Cassell, 1921.

———. *Science and the World Mind*. New Europe, 1942.

———. *The Shape of Things to Come*. Hutchinson, 1933.

———. *Socialism and the Family*. A. C. Fifield, 1906.
———. *This Misery of Boots*. Fabian Society, 1907.
———. *When the Sleeper Wakes*. Harper and Bros, 1899.
Wheeler, David Hilton. *Our Industrial Utopia and Its Unhappy Citizens*. A. C. McClurg, 1895.
Winstanley, Gerrard. *The Complete Works of Gerrard Winstanley*. 2 vols. Oxford University Press, 2009.
Xenophon. *Xenophon's Spartan Constitution*. Edited by Michael Lipka. Walter de Gruyter, 2002.
Zamyatin, Evgeny. *We* [1924]. Translated by Bernard Gilbert Guerney. Jonathan Cape, 1970.

Secondary Sources

Abrahams, Roger D., and Richard Bauman. "Ranges of Festival Behavior". In *The Reversible World: Symbolic Inversion in Art and Society*, edited Barbara Babock, 193–208. Cornell University Press, 1978.
Abramitzky, Ran. *The Mystery of the Kibbutz: Egalitarian Principles in a Capitalist World*. Princeton University Press, 2018.
Abrams, Philip, and Andrew McCulloch, with Sheila Adams and Pat Gore. *Communes, Sociology and Society*. Cambridge University Press, 1976.
Ackroyd, Peter. *The Life of Thomas More*. Random House, 1998.
Adams, Percy G. *Travelers and Travel Liars: 1600–1800*. University of California Press, 1962.
Adas, Michael. *Prophets of Rebellion: Millenarian Protest Movements against the European Colonial Order*. University of North Carolina Press, 1979.
Albinski, Nan Bowman. *Women's Utopias in British and American Fiction*. Routledge, 1988.
Albritton, Vicky, and Fredrik Albritton Jonsson. *Green Victorians: The Simple Life in John Ruskin's Lake District*. University of Chicago Press, 2016.
Alcott, Louisa May. *Transcendental Wild Oats and Excerpts from the Fruitlands Diary*. Harvard Common Press, 1975.
Ali, Tariq. *Street Fighting Years: An Autobiography of the Sixties*. Verso, 2005.
Allen, Kelly-Ann. *The Psychology of Belonging*. Routledge, 2021.
Althusser, Louis. *For Marx*. Penguin Books, 1969.
Ames, Russell. *Citizen Thomas More and His Utopia*. Princeton University Press, 1949.
Anderson, Benedict. *Imagined Communities: Reflections on the Origin and Spread of Nationalism*. Verso, 1991.
Andrews, Cecile. *Slow Is Beautiful: New Visions of Community, Leisure and Joi de Vivre*. New Society, 2006.
Applebaum, Anne. *Iron Curtain: The Crushing of Eastern Europe, 1945–56*. Allen Lane, 2012.
Appelbaum, Robert. "Utopia and Utopianism". In *The Oxford Handbook of English Prose, 1500–1640*. Edited by Andrew Hadfield, 253–66. Oxford University Press, 2013.
Appleby, Joyce Oldham. *Economic Thought and Ideology in Seventeenth-Century England*. Princeton University Press, 1978.
———. *Liberalism and Republicanism in the Historical Imagination*. Harvard University Press, 1992.
Arendt, Hannah. *Between Past and Future*. Faber and Faber, 1961.
———. *The Human Condition*. University of Chicago Press, 1958.

Argyle, Michael. *The Psychology of Happiness* [1987]. 2nd ed. Routledge, 2001.

Armogathe, Jean-Robert. "*Per Annos Mille*: Cornelius a Lapide and the Interpretation of Revelation 20:2–8". In *Catholic Millenarianism: From Savonarola to the Abbé Gregoire,* edited by Karl A. Kottman, 45–52. Kluwer, 2001.

Armytage, W.H.G. *Heaven's Below: Utopian Experiments in England, 1560–1960.* Routledge and Kegan Paul, 1961.

———. *Yesterday's Tomorrows: A Historical Survey of Future Societies.* Routledge and Kegan Paul, 1968.

Arndt, Karl. *George Rapp's Harmony Society, 1785–1847.* University of Pennsylvania Press, 1965.

Auerbach, Nina. *Communities of Women: An Idea in Fiction.* Harvard University Press, 1998.

Avineri, Shlomo. *Moses Hess: Prophet of Communism and Zionism.* New York University Press, 1985.

Baczko, Bronislaw. *Utopian Lights: The Evolution of the Idea of Social Progress.* Paragon House, 1989.

Bahro, Rudolf. *Avoiding Social and Ecological Disaster: The Politics of World Transformation.* Gateway Books, 1994.

———. *Building the Green Movement.* Heretic Books, 1986.

Baillie, John. *The Belief in Progress.* Oxford University Press, 1950.

Baker, David Weil. *Divulging Utopia: Radical Humanism in Sixteenth-Century England.* University of Massachusetts Press, 1999.

Bakhtin, Mikhail. *Rabelais and His World.* MIT Press, 1968.

Baldwin, Francis Elizabeth. *Sumptuary Legislation and Personal Regulation in England* [1923]. General Books, 2010.

Ball, Alan M. *Russia's Last Capitalists: The Nepmen, 1921–1929.* University of California Press, 1987.

Bammer, Angelika. *Partial Visions: Feminism and Utopianism in the 1970s.* Routledge, 1991.

Baranowski, Shelley. *Strength through Joy: Consumerism and Mass Tourism in the Third Reich.* Cambridge University Press, 2004.

Barber, Benjamin R. *Consumed: How Markets Corrupt Children, Infantilize Adults, and Swallow Citizens Whole.* W. W. Norton, 2007.

Barker, Adele Marie, ed. *Consuming Russia: Popular Culture, Sex, and Society since Gorbachev.* Duke University Press, 1999.

Bartlett, Djurdja. *Fashion East: The Spectre That Haunted Communism.* MIT Press, 2010.

Baudet, Henri. *Paradise on Earth: Some Thoughts on European Images of Non-European Man.* Yale University Press, 1965.

Baudrillard, Jean. *The Consumer Society: Myths and Structures.* Sage, 1998.

Bauman, Zygmunt. *Collateral Damage: Social Inequalities in a Global Age.* Polity, 2011.

———. *Community.* Polity, 2001.

———. *Consuming Life.* Polity, 2007.

———. *Freedom.* Open University Press, 1988.

———. *The Individualized Society.* Polity, 2001.

———. *Liquid Life.* Polity, 2005.

———. *Liquid Love: On the Frailty of Human Bonds.* Polity, 2003.

———. *Wasted Lives: Modernity and Its Outcasts.* Polity, 2004.

Baumeister, R. F., and M. R. Leary. "The Need to Belong". *Psychological Bulletin* 117 (1995): 497–529.

Baumgartner, Frederic J. *Longing for the End: A History of Millennialism in Western Civilization.* Macmillan, 1999.

Bayertz, Kurt, ed. *Solidarity*. Kluwer Academic, 1999.

Beaumont, Matthew. *Utopia Ltd: Ideologies of Social Dreaming in England, 1870–1900*. Brill, 2005.

Beecher, Jonathan. *Charles Fourier: The Visionary and His World.* University of California Press, 1986.

Beer, Max. *History of British Socialism.* 2 vols. G. Bell and Sons, 1929.

———. *An Inquiry into Physiocracy*. George Allen and Unwin, 1939.

———. *Social Struggles and Socialist Forerunners*. Leonard Parsons, 1924.

Belasco, Warren J. *Appetite for Change: How the Counterculture Took On the Food Industry, 1966–1968*. Pantheon Books, 1989.

Bell, Bernard Iddings. *Crowd Culture: An Examination of the American Way of Life*. Gateway, 1956.

Bell, Daniel. *The Cultural Contradictions of Capitalism*. Heinemann, 1976.

———. *The End of Ideology: On the Exhaustion of Political Ideas in the Fifties*. Free Press, 1960.

Bell, David M. *Rethinking Utopia: Power, Place, Affect*. Routledge, 2017.

Bell, Duncan. *Dreamworlds of Race: Empire and the Utopian Destiny of Anglo-America*. Princeton University Press, 2020.

Bell, Robert R. *Worlds of Friendship*. Sage, 1981.

Ben-Rafael, Eliezer, Yaacov Oved, and Menachem Topel, eds. *The Communal Idea in the 21st Century*. Brill, 2013.

Bercé, Yves-Marie. *History of Peasant Revolts*. Cornell University Press, 1990.

Berdyaev, Nicolas. *The Destiny of Man*. Centenary, 1937.

———. *The End of Our Time*. Sheed and Ward, 1933.

———. *The Meaning of History*. Centenary, 1945.

———. *The Realm of Spirit and the Realm of Caesar*. Victor Gollancz, 1952.

———. *Slavery and Freedom*. Centenary, 1944.

Beresford, Maurice. *The Lost Villages of England*. Lutterworth, 1954.

Berg, Maxine. *Luxury & Pleasure in Eighteenth-Century Britain*. Oxford University Press, 2005.

Berger, Peter L., Brigitte Berger, and Hansfried Kellner. *The Homeless Mind: Modernization and Consciousness*. Pelican Books, 1974.

Berghoff, Hartmut. "Enticement and Deprivation: The Regulation of Consumption in Pre-war Nazi Germany". In Daunton and Hilton, *Politics of Consumption*, 165–184.

Berghoff, Hartmut, and Uwe Spiekermann, eds. *Decoding Modern Consumer Societies*. Palgrave Macmillan, 2012.

Berkhofer, Robert F. *The White Man's Indian: Images of the American Indian from Columbus to the Present*. Vintage Books, 1979.

Berlin, Isaiah. *The Crooked Timber of Humanity* [1959]. John Murray, 1990.

———. *Freedom and Its Betrayal: Six Enemies of Human Liberty*. Chatto and Windus, 2002.

Berman, Marshall. *The Politics of Authenticity: Radical Individualism and the Emergence of Modern Society*. Atheneum, 1970.

Berneri, Marie Louise. *Journey through Utopia*. Routledge and Kegan Paul, 1950.

Berners-Lee, Mike. *How Bad Are Bananas? The Carbon Footprint of Everything*. 2nd ed. Profile Books, 2020.

Berry, Brian J. L. *America's Utopian Experiments*. University Press of New England, 1992.

Berry, Christopher J. *The Idea of Commercial Society in the Scottish Enlightenment*. Edinburgh University Press, 2013.

———. *The Idea of Luxury: A Conceptual and Historical Investigation*. Cambridge University Press, 1994.

Bestor, A. E. *Backwoods Utopias: The Sectarian Origins and Owenite Phase of Communitarian Socialism in America, 1663–1829* [1950]. 2nd ed. University of Pennsylvania Press, 1970.

Betts, Paul. *Within Walls: Private Life in the German Democratic Republic*. Oxford University Press, 2010.

Billington, Sandra. *A Social History of the Fool*. Harvester, 1984.

Bissell, Benjamin. *The American Indian in English Literature of the Eighteenth Century*. Yale University Press, 1925.

Blaim, Artur. *Failed Dynamics: The English Robinsonade of the Eighteenth Century*. Lublin, 1987.

———. *Gazing in Useless Wonder: English Utopian Fictions, 1516–1800*. Peter Lang, 2013.

Blatterer, Harry, Pauline Johnson, and Maria M. Markus, eds. *Modern Privacy: Shifting Boundaries, New Forms*. Palgrave Macmillan, 2010.

Bloch, Ruth H. *Visionary Republic: Millennial Themes in American Thought, 1756–1800*. Cambridge University Press, 1985.

Bloomfield, Paul. *Imaginary Worlds; or, The Evolution of Utopia*. Hamish Hamilton, 1932.

Blumenberg, Hans. *The Legitimacy of the Modern Age*. MIT Press, 1983.

Bockelson, Friedrich Reck-Malleczewen. *A History of the Münster Anabaptists*. Edited by George B. von der Lippe and Viktoria M. Reck-Malleczewen. Palgrave-Macmillan, 2008.

Boissevain, Jeremy. *Friends of Friends: Networks, Manipulators and Coalitions*. Basil Blackwell, 1974.

Bonnett, Alastair, and Keith Armstrong, eds. *Thomas Spence: The Poor Man's Revolutionary*. Breviary Stuff, 2014.

Bookchin, Murray. *Toward an Ecological Society*. Black Rose Books, 1980.

Borch, Christian. *The Politics of Crowds: An Alternative History of Sociology*. Cambridge University Press, 2012.

Bourboulis, Photeine P. *Ancient Festivals of the "Saturnalia" Type*. Spoudon, 1964.

Bowlby, Rachel. *Carried Away: The Invention of Modern Shopping*. Faber and Faber, 2000.

Bowler, Peter J. *A History of the Future: Prophets of Progress from H. G. Wells to Isaac Asimov*. Cambridge University Press, 2017.

Bowman, Sylvia E. *Edward Bellamy*. Twayne, 1986.

Bowman, Sylvia E., Alexander Nikoljukin, Peter Marshall, Robin Gollan, W. R. Fraser, Franz X. Riederer, K. Zysltra, J. Bogaard, Herbert Roth, Lars Almebrink, George Levin, Pierre Michel, and Guido Fink. *Edward Bellamy Abroad. An American Prophet's Influence*. Twayne, 1962.

Boyd, Emily, and Emma L. Tompkins. *Climate Change*. Oneworld, 2009.

Boyle, David. *Authenticity: Brands, Fakes, Spin and the Lust for Real Life*. Flamingo, 2003.

Bradley, Karin, and Johan Hedrén, eds. *Green Utopianism: Perspectives, Politics and Micro-practices*. Routledge, 2014.

Braunthal, Alfred. *Salvation and the Perfect Society: The Eternal Quest*. University of Massachusetts Press, 1979.

Breckman, Warren. "Disciplining Consumption: The Debate about Luxury in Wilhelmine Germany, 1890–1914". *Journal of Social History* 24 (1990–91): 485–505.

Bregman, Rutger. *Utopia for Realists*. Bloomsbury, 2017.

Bren, Paulina, and Mary Neuburger, eds. *Communism Unwrapped: Consumption in Cold War Eastern Europe*. Oxford University Press, 2012.

Brewer, Priscilla J. *Shaker Communities, Shaker Lives*. University Press of New England, 1986.

Brooke, Christopher. *Philosophic Pride: Stoicism and Political Thought from Lipsius to Rousseau*. Princeton University Press, 2012.

Brooks, David. *Bobos in Paradise: The New Upper Class and How They Got There*. Simon and Schuster, 2000.

Brooks, John. *Showing Off in America: From Conspicuous Consumption to Parody Display*. Little, Brown, 1979.

Brown, Ford K. *Fathers of the Victorians: The Age of Wilberforce*. Cambridge University Press, 1961.

Brown, Harrison, James Bonner, and John Weir. *The Next Hundred Years*. Weidenfeld and Nicolson, 1957.

Brown, Peter. *The Cult of the Saints: Its Rise and Function in Latin Christianity*. SCM, 1981.

Buber, Martin. *Paths in Utopia*. Beacon, 1958.

Buckle, Stephen. *Natural Law and the Theory of Property: Grotius to Hume*. Clarendon, 1991.

Bull, Malcolm, ed. *Apocalypse Theory and the End of the World*. Basil Blackwell, 1995.

Bultmann, D. Rudolf. *History and Eschatology*. Edinburgh University Press, 1957.

Burke, Peter. *Popular Culture in Early Modern Europe*. Temple Smith, 1978.

Bury, J. B. *The Idea of Progress*. Macmillan, 1921.

Bushman, Richard L. *The Refinement of America: Persons, Houses, Cities*. Alfred A. Knopf, 1992.

Byrne, Aisling. *Otherworlds: Fantasy and History in Medieval Literature*. Oxford University Press, 2016.

Camic, Charles. *Veblen: The Making of an Economist Who Unmade Economics*. Harvard University Press, 2020.

Campion, Nicolas. *The Great Year: Astrology, Millenarianism and History in the Western Tradition*. Penguin Books, 1994.

Capp, B. S. "Extreme Millenarianism". In *Puritans, the Millennium and the Future of Israel: Puritan Eschatology, 1600 and 1660; A Collection of Essays*, edited by Peter Toon, 66–90. James Clarke, 1970.

Caraman, Philip. *The Lost Paradise: An Account of the Jesuits in Paraguay, 1607–1768*. Sidwick and Jackson, 1975.

Carden, Maren Lockwood. *Oneida: Utopian Community to Modern Corporation*. Johns Hopkins Press, 1969.

Cartledge, Paul. *Spartan Reflections*. Duckworth, 2001.

———. *The Spartans*. Macmillan, 2002.

Cassirer, Ernst. *The Philosophy of the Enlightenment*. Princeton University Press, 1951.

Castle, Terry. *Masquerade and Civilization: The Carnivalesque in Eighteenth-Century English Culture and Fiction*. Stanford University Press, 1986.

Castronova, Edward. *Exodus to the Virtual World: How Online Fun Is Changing Reality*. Palgrave Macmillan, 2007.

Caudwell, Christopher. *Studies in a Dying Culture*. Bodley Head, 1949.

Caute, David. *The Year of the Barricades: A Journey through 1968*. Harper and Row, 1988.

Cave, Terence, ed. *Thomas More's Utopia in Early Modern Europe*. Manchester University Press, 2008.

Chase, Stuart. *The Tragedy of Waste*. Macmillan, 1942.

———. *The Tyranny of Words*. Methuen, 1938.

Chayko, Mary. *Connecting: How We Form Social Bonds and Communities in the Internet Age*. State University of New York Press, 2002.

Cioran, E. M. *History and Utopia* [1960]. Quartet Books, 1996.

Claeys, Gregory. *Dystopia: A Natural History*. Oxford University Press, 2016.

———. "The Five Languages of Utopia". In *Spectres of Utopia*, edited by Artur Blaim and Ludmilla Gruszewska-Blaim, 26–31. Peter Lang, 2012.

———. "From True Virtue to Benevolent Politeness: Godwin and Godwinism Revisited". In *Empire and Revolutions: Papers Presented at the Folger Institute Seminar "Political Thought in the English-Speaking Atlantic, 1760–1800"*, edited by Gordon Schochet, 187–226. Folger Library, 1993.

———. *Imperial Sceptics: British Critics of Empire, 1850–1920*. Cambridge University Press, 2010.

———. *John Stuart Mill: A Very Short Introduction*. Oxford University Press, 2022.

———. *Machinery, Money and the Millennium: From Moral Economy to Socialism, 1815–60*. Princeton University Press, 1987.

———. *Marx and Marxism*. Penguin Books, 2018.

———. *Mill and Paternalism*. Cambridge University Press, 2013.

———. "Mill, Moral Suasion, and Coercion". In *Ethical Citizenship: British Idealism and the Politics of Recognition*, edited by Thom Brooks, 79–104. Palgrave-Macmillan, 2014.

———. "News from Somewhere: Enhanced Sociability and the Composite Definition of Utopia and Dystopia". *History* 98 (2013): 145–73.

———. "'The Only Man of Nature That Ever Appeared in the World': 'Walking' John Stewart and the Trajectories of Social Radicalism, 1790–1822". *Journal of British Studies* 53 (2014): 1–24.

———. "The Origins and Development of Social Darwinism". In *The Cambridge Companion to Nineteenth-Century Thought*, edited by Claeys, 165–83. Cambridge University Press, 2019.

———. *Searching for Utopia*. Thames and Hudson, 2011. 2nd ed. published as *Utopia: The History of an Idea* (Thames and Hudson, 2020).

———. "Socialism and the Language of Rights". In *Revisiting the Origins of Human Rights: Genealogy of a European Idea*, edited by Miia Halme-Tuomisaari and Pamela Slotte, 206–36. Cambridge University Press, 2015.

———. *Thomas Paine: Social and Political Thought*. Unwin Hyman, 1989.

———. "Why Are Utopias Important for Human Mankind?" Talk presented at TedxLinz. TED, February 2019. https://www.ted.com/talks/gregory_claeys_why_are_utopias_important_for_human_mankind.

Claeys, Gregory, and Christine Lattek. "Radicalism, Republicanism, and Revolutionism: From the Principles of '89 to Modern Terrorism". In *The Cambridge History of Nineteenth-Century Political Thought*, edited by Gareth Stedman Jones and Claeys, 200–254. Cambridge University Press, 2011.

Clark, Eric. *The Want Makers: Lifting the Lid off the World Advertising Industry*. Guild, 1988.

Clark, Henry C. *Compass of Society: Commerce and Absolutism in Old-Regime France*. Rowman and Littlefield, 2010.

Clark, Stuart. *Thinking with Demons: The Idea of Witchcraft in Early Modern Europe*. Oxford University Press, 1997.

Clarke, I. F. *The Pattern of Expectation, 1644–2001*. Jonathan Cape, 1979.

Clasen, Claus-Peter. *Anabaptism: A Social History, 1525–1618*. Cornell University Press, 1972.

Coates, Chris. *Utopia Britannica: British Utopian Experiments, 1325–1945*. D&D, 2001.

Cohen, Lizabeth. *A Consumers' Republic: The Politics of Mass Consumption in Postwar America*. Vintage Books, 2004.

Cohn, Norman. "Medieval Millenarism: Its Bearing on the Comparative Study of Millenarian Movements". In Thrupp, *Millennial Dreams in Action*, 31–43.

———. *The Pursuit of the Millennium*. Secker and Warburg, 1947.

Cole, Charles Woolsey. *French Mercantilist Doctrines before Colbert*. R. R. Smith, 1931.

Coleman, Nathaniel. *Lefebvre for Architects*. Routledge, 2015.

———. *Utopias and Architecture*. Routledge, 2005.

Cook, Philip L. *Zion City, Illinois: Twentieth Century Utopia*. Syracuse University Press, 1996.

Cooper, Davina. *Everyday Utopias: The Conceptual Life of Promising Spaces*. Duke University Press, 2014.

Cox, Harvey. *The Feast of Fools: A Theological Essay on Festivity and Fantasy*. Harvard University Press, 1969.

Cranston, Maurice. *Jean-Jacques: The Early Life and Work of Jean-Jacques Rousseau, 1712–1754*. Norton, 1982.

———. *The Noble Savage: Jean-Jacques Rousseau, 1754–1762*. University of Chicago Press, 1991.

Cro, Stelio. *The Noble Savage: An Allegory of Freedom*. Wilfrid Laurier University Press, 1990.

Crocker, David A., and Toby Linden, eds. *Ethics of Consumption: The Good Life, Justice, and Global Stewardship*. Rowman and Littlefield, 1998.

Crowley, David. "Warsaw's Shops, Stalinism and the Thaw". In Reid and Crowley, *Style and Socialism*, 25–48.

Crowley, David, and Susan E. Reid, eds. *Pleasures in Socialism: Leisure and Luxury in the Eastern Bloc*. Northwestern University Press, 2010.

Csikszentmihalyi, Mihalyn, and Eugene Rochberg-Halton. *The Meaning of Things: Domestic Symbols and the Self*. Cambridge University Press, 1981.

Dahrendorf, Ralph. *Essays in the Theory of Society*. Routledge and Kegan Paul, 1968.

Dalrymple, William. *The Anarchy: The Relentless Rise of the East India Company*. Bloomsbury, 2019.

D'Angour, Armand. *The Greeks and the New: Novelty in Ancient Greek Imagination and Experience*. Cambridge University Press, 2011.

Darnton, Robert. *The Forbidden Best-Sellers of Pre-revolutionary France*. HarperCollins, 1996.

Daunton, Martin, and Matthew Hilton, eds. *The Politics of Consumption: Material Culture and Citizenship in Europe and America*. Berg, 2001.

Davis, Fred. *Yearning for Yesterday: A Sociology of Nostalgia*. Free Press, 1979.

Davis, J. C. *Utopia and the Ideal Society: A Study of English Utopian Writing, 1516–1700*. Cambridge University Press, 1981.

Davis, Mike, and Daniel Bertrand Hunt, eds. *Evil Paradises: Dreamworlds of Neoliberalism*. New Press, 2007.

Davis, Natalie Zemon. *The Gift in Sixteenth-Century France*. Oxford University Press, 2000.

Dawson, Christopher. *Progress and Religion: An Historical Enquiry*. Sheed and Ward, 1929.

Dawson, Doyne. *Cities of the Gods: Communist Utopias in Greek Thought*. Oxford University Press, 1992.

Day, Graham. *Community and Everyday Life*. Routledge, 2006.

Degroot, Gerard. *The 60s Unplugged: A Kaleidoscope History of a Disorderly Decade*. Macmillan, 2008.

Dehaene, Michiel, and Lieven De Cauter, eds. *Heterotopia and the City: Public Space in a Postcivil Society*. Routledge, 2008.

De la Mare, Walter. *Desert Islands and Robinson Crusoe*. Faber and Faber, 1930.

Delano, Sterling F. *Brook Farm: The Dark Side of Utopia*. Harvard University Press, 2004.

Derrida, Jacques. *Politics of Friendship*. Verso, 1997.

Desmond, William D. *The Greek Praise of Poverty: Origins of Ancient Cynicism*. University of Notre Dame Press, 2006.

Dichter, Ernest. *The Strategy of Desire*. T. V. Boardman, 1960.

Dickson, Gary. *The Children's Crusade: Medieval History, Modern Mythistory*. Palgrave Macmillan, 2007.

Dietz, Maribel. *Wandering Monks, Virgins, and Pilgrims: Ascetic Travel in the Mediterranean World, A.D. 300–800*. Pennsylvania State University Press, 2005.

Diggins, John P. *The Bard of Savagery: Thorstein Veblen and Modern Social Theory*. Harvester, 1978.

Dillon, Matthew. *Pilgrims and Pilgrimage in Ancient Greece*. Routledge, 1997.

D'Monté, Rebecca, and Nicole Pohl, eds. *Female Communities, 1600–1800*. Macmillan, 2000.

Dobb, Maurice. *Russia Today and Tomorrow*. Hogarth, 1930.

Doherty, Daniel, and Amitai Etzioni, eds. *Voluntary Simplicity: Responding to Consumer Culture*. Rowman and Littlefield, 2003.

Donato, Antonio. *Italian Renaissance Utopias: Doni, Patrizi, and Zuccolo*. Palgrave Macmillan, 2019.

Donner, H. W. *Introduction to Utopia*. Sidgwick and Jackson, 1945.

Douglas, Mary. *In the Active Voice*. Routledge and Kegan Paul, 1984.

Duck, Steve. *Friends, for Life: The Psychology of Personal Relationships*. 2nd ed. Harvester Wheatsheaf, 1991.

Dudley, Donald R. *A History of Cynicism*. Methuen, 1937.

Dudok, Gerard. *Sir Thomas More and His Utopia*. A. H. Kruyt, 1923.

Dunbar, Robin. *Friends: Understanding the Power of Our Most Important Relationships*. Little, Brown, 2020.

———. *How Many Friends Does One Person Need?* Faber and Faber, 2010.

Dunn, Alastair. *The Great Rising of 1381*. Tempus, 2002.

Durning, Alan Thein. *How Much Is Enough? The Consumer Society and the Future of the Earth*. Earthscan, 1992.

Eaton, Ruth. *Ideal Cities: Utopianism and the (Un)built Environment*. Thames and Hudson, 2002.

Edgerton, Robert B. *Sick Societies: Challenging the Myth of Primitive Harmony*. Free Press, 1992.

Ehrenreich, Barbara. *Dancing in the Streets: A History of Collective Joy*. Grant Books, 2007.

Eisenstein, Elizabeth L. *The First Professional Revolutionist: Filippo Michele Buonarroti (1761–1837)*. Harvard University Press, 1959.

Ekström, Karin M., and Helene Brembeck, eds. *Elusive Consumption*. Berg, 2004.

Elgin, Duane. *Voluntary Simplicity: Toward a Way of Life That Is Outwardly Simple, Inwardly Rich*. William Morrow, 1981.

Elgin, Duane, and Arnold Mitchell. "A Movement Emerges". In Doherty and Etzioni, *Voluntary Simplicity*, 145–74.

Eliade, Mircea. *The Myth of the Eternal Return*. Princeton University Press, 1974.

———. *The Sacred and the Profane: The Nature of Religion*. Harper and Row, 1961.

Elias, Norbert. *The Court Society*. Pantheon Books, 1983.

———. *Essays I: On the Sociology of Knowledge and the Sciences*. University College Dublin Press, 2009.

Eliav-Feldon, Miriam. *Realistic Utopias: The Ideal Imaginary Societies of the Renaissance, 1516–1630*. Clarendon, 1982.

Erdman, Carl. *The Origin of the Idea of Crusade*. Princeton University Press, 1977.

Etzioni, Amitai. *The Third Way to a Good Society*. Demos, 2000.

Eurich, Nell. *Science in Utopia: A Mighty Design*. Harvard University Press, 1967.

Evans, Rhiannon. *Utopia Antiqua: Readings of the Golden Age and Decline at Rome*. Routledge, 2008.

Ewen, Stuart. *All Consuming Images: The Politics of Style in Contemporary Culture*. Basic Books, 1988.

———. *Captains of Consciousness: Advertising and the Social Roots of Consumer Culture*. McGraw-Hill, 1976.

Faderman, Lillian. *Surpassing the Love of Men: Romantic Friendship and Love between Women from the Renaissance to the Present*. William Morrow, 1981.

Fairchild, Hoxie Neale. *The Noble Savage: A Study in Romantic Naturalism*. Cornell University Press, 1928.

Featherstone, Mark. *Tocqueville's Virus: Utopia and Dystopia in Western Social and Political Thought*. Routledge, 2008.

Fehér, Ferenc, Agnes Heller, and György Márkus. *Dictatorship over Needs*. Basil Blackwell, 1983.

Ferguson, John. *Utopias of the Classical World*. Thames and Hudson, 1975.

Fink, Z. S. *The Classical Republicans: An Essay in the Recovery of a Pattern of Thought in Seventeenth-Century England*. Northwestern University Press, 1962.

Finucare, Ronald C. *Miracles and Pilgrims: Popular Beliefs in Medieval England*. St. Martin's, 1995.

Fiorino, Daniel J. *A Good Life on a Finite Earth: The Political Economy of Green Growth*. Oxford University Press, 2018.

Firth, Rhiannon. *Utopian Politics: Citizenship and Practice*. Routledge, 2012.

Fischer, Claude S. *To Dwell among Friends: Personal Networks in Town and City*. University of Chicago Press, 1982.

Fishman, Robert. *Bourgeois Utopias: The Rise and Fall of Suburbia*. Basic Books, 1987.

———. *Urban Utopias in the Twentieth Century*. Basic Books, 1977.

Fitting, Peter. "A Short History of Utopian Studies". *Science Fiction Studies* 36 (2009): 121–31.

Fitzpatrick, Sheila, ed. *Cultural Revolution in Russia, 1928–1931*. Indiana University Press, 1978.

Flügel, J. C. *The Psychology of Clothes*. Hogarth, 1930.

Fogarty, Robert S. *All Things New: American Communes and Utopian Movements*. University of Chicago Press, 1990.

———. *American Utopianism*. F. E. Peacock, 1972.

———, ed. *Special Love/Special Sex: An Oneida Community Diary*. Syracuse University Press, 1994.

Foster, Lawrence. *Religion and Sexuality: Three American Communal Experiments of the Nineteenth Century*. Oxford University Press, 1981.

———. *Women, Family, and Utopia: Communal Experiments of the Shakers, the Oneida Community, and the Mormons*. Syracuse University Press, 1991.

Foucault, Michel. *Aesthetics, Method, and Epistemology*. Translated by Robert Hurley et al. Allen Lane, 1998.

Francis, Richard. *Fruitlands: The Alcott Family and Their Search for Utopia*. Yale University Press, 2010.

Frank, Robert H. *Luxury Fever: Money and Happiness in an Age of Excess*. Princeton University Press, 1999.

Frank, Waldo. *The Rediscovery of America*. Charles Scribner's Sons, 1929.

Frankel, Boris. *The Post-industrial Utopians*. Polity, 1987.

Freedman, Paul. "Luxury Dining in the Later Years of the German Democratic Republic". In *Becoming East German: Social Structures and Sensibilities after Hitler*, edited by Mary Fulbrook and Andrew L. Port, 179–200. Berghahn, 2013.

Freud, Sigmund. *Group Psychology and the Analysis of the Ego* [1921]. Bantam Books, 1960.

Friedman, Milton, and Rose Friedman. *Free to Choose*. Secker and Warburg, 1979.

Friesen, Abraham. *Reformation and Utopia: The Marxist Interpretation of the Reformation and Its Antecedents*. Franz Steiner Verlag, 1974.

Frisby, David. *Fragments of Modernity: Theories of Modernity in the Work of Simmel, Kracauer and Benjamin*. Polity, 1985.

Fromm, Erich. *The Anatomy of Human Destructiveness*. Fawcett Crest, 1973.

Fruchtman, Jack, Jr. *The Apocalyptic Politics of Richard Price and Joseph Priestley: A Study in Late Eighteenth Century English Republican Millennialism*. American Philosophical Society, 1983.

Funke, Hans-Günter. "Utopie, Utopiste". In *Handbuch politisch-sozialer Grundbegriffe in Frankreich, 1680–1820*, edited by Rolf Reichardt and Hans-Jürgen Lüsebrink, bk 11, 5–104. R. Oldebourg Verlag, 1991.

Furnham, Adrian. "Friendship and Personal Development". In *The Dialectics of Friendship*, edited by Roy Porter and Sylvana Tomaselli, 92–110. Routledge, 1989.

Fuz, J. K. *Welfare Economics in English Utopias from Francis Bacon to Adam Smith*. Martinus Nijhoff, 1962.

Gabriel, Viannis, and Tim Lang. *The Unmanageable Consumer: Contemporary Consumption and Its Fragmentation*. Sage, 1995.

Gagnier, Regenia. *The Insatiability of Human Wants: Economics and Aesthetics in Market Society*. University of Chicago Press, 2000.

Gans, Herbert J. *The Urban Villagers: Group and Class in the Life of Italian-Americans*. Free Press, 1962.

Garforth, Lisa. *Green Utopias: Environmental Hope before and after Nature*. Polity, 2018.

Garnett, R. G. *Co-operation and the Owenite Socialism Communities in Britain, 1825–45*. Manchester University Press, 1972.

Garno, Diana M. *Citoyennes and Icaria*. University Press of America, 2005.

Garnsey, Peter. *Thinking about Property: From Antiquity to the Age of Revolution*. Cambridge University Press, 2007.

Garrett, Clarke. *Respectable Folly: Millenarians and the French Revolution in France and England*. Johns Hopkins University Press, 1975.

Gaster, Theodor. *New Year: Its History, Customs and Superstitions*. Abelard-Schuman, 1955.

Gates, Bill. *How to Avoid a Climate Disaster: The Solutions We Have and the Breakthroughs We Need*. Allen Lane, 2021.

Gennep, Arnold van. *The Rites of Passage*. Routledge and Kegan Paul, 1960.

Geoghegan, Vincent. *Ernst Bloch*. Routledge, 1996.

Geus, Marius de. *Ecological Utopias: Envisioning the Sustainable Society*. International Books, 1999.

———. *The End of Over-Consumption: Towards a Lifestyle of Moderation and Self-Restraint*. International Books, 2003.

Gibbins, H. de B. *English Social Reformers*. Methuen, 1902.

Gierke, Otto. *Natural Law and the Theory of Society, 1500 to 1800*. 2 vols. Cambridge University Press, 1934.

Giesecke, Annette Lucia. *The Epic City: Urbanism, Utopia, and the Garden in Ancient Greece and Rome*. Harvard University Press, 2007.

Gill, Roger. "In England's Green and Pleasant Land". In *Utopias*, edited by Peter Alexander and Gill, 109–18. Gerald Duckworth, 1984.

Girard, René. *Violence and the Sacred*. Johns Hopkins University Press, 1977.

Goldsmith, M. M. *Private Vices, Public Benefits: Bernard Mandeville's Social and Political Thought*. Cambridge University Press, 1985.

Goldthwaite, R. "The Economy of Renaissance Italy: The Preconditions for Luxury Consumption". *I Tatti Studies: Essays in the Renaissance* 2 (1987): 15–39.

Gomez, Fernando. *Good Places and Non-places in Colonial Mexico: The Figure of Vasco de Quiroga (1470–1565)*. University Press of America, 2001.

Goode, Erich. *Collective Behavior*. Saunders College Publishing, 1992.

Goodman, Paul. *Growing Up Absurd: Problems of Youth in the Organized Society* [1956]. Victor Gollancz, 1961.

Goodman, Percival, and Paul Goodman. *Communitas: Means of Livelihood and Ways of Life* [1947]. Vintage Books, 1960.

Goodwin, Barbara. *Social Science and Utopia: Nineteenth Century Models of Social Harmony*. Harvester, 1978.

Goodwin, Barbara, and Keith Taylor. *The Politics of Utopia*. Hutchinson, 1982.

Goody, Jack. *The Domestication of the Savage Mind*. Cambridge University Press, 1977.

Gorsuch, Anne E. *Youth in Revolutionary Russia: Enthusiasts, Bohemians, Delinquents*. Indiana University Press, 2000.

Gottschalk, Louis R. *Jean Paul Marat: A Study in Radicalism*. University of Chicago Press, 1967.

Graaf, John de, David Wann, and Thomas H. Naylor. *Affluenza: How Overconsumption is Killing Us—and How to Fight Back* [2001]. 3rd ed. BK Currents Books, 2014.

Gray, George Zabriskie. *The Children's Crusade*. Sampson Low, Son, and Marston, 1871.
Gray, John. *The Silence of Animals: On Progress and Other Modern Myths*. Allen Lane, 2013.
Graziano, Frank. *The Millennial New World*. Oxford University Press, 1999.
Greeley, Andrew M. *The Friendship Game*. Doubleday, 1971.
Green, Toby. *Thomas More's Magician: A Novel Account of Utopia in Mexico*. Weidenfeld and Nicolson, 2004.
Greenfield, Kent Roberts. *Sumptuary Law in Nürnberg: A Study in Paternal Government*. Johns Hopkins Press, 1918.
Gregg, Richard. *The Value of Voluntary Simplicity*. Pendle Hill, 1936.
Grendler, Paul F. *Critics of the Italian World: Anton Francesco Doni, Nicolò Franco & Ortensio Lando*. University of Wisconsin Press, 1969.
Groningen, B. A. van. *In the Grip of the Past: Essay on an Aspect of Greek Thought*. Brill, 1953.
Gronow, Jukka. *Caviar with Champagne: Common Luxury and the Ideals of the Good Life in Stalin's Russia*. Berg, 2003.
Gross, David. "Ernst Bloch: The Dialectics of Hope". In *The Unknown Dimension: European Marxism since Lenin*, edited by Dick Howard and Karl. E. Klare, 107–30. Basic Books, 1972.
Guarneri, Carl J. *The Utopian Alternative: Fourierism in Nineteenth-Century America*. Cornell University Press, 1991.
Guenther, Irene. *Nazi Chic? Fashioning Women in the Third Reich*. Berg, 2004.
Guicciardini, Francesco. *The History of Florence* [1510]. Harper Torchbooks, 1970.
Gurova, Olga. "The Ideology of Consumption in the Soviet Union". In Vihavainen and Bogdanova, *Communism and Consumerism*, 68–84.
Habermas, Jürgen. *Philosophical-Political Profiles*. Heinemann, 1983.
Hahn, Cynthia. *Strange Beauty: Issues in the Making and Meaning of Reliquaries, 400–circa 1204*. Pennsylvania State Press, 2012.
Haldane, Charlotte. *Tempest over Tahiti*. Constable, 1963.
Hall, John R. *The Ways Out: Utopian Communal Groups in an Age of Babylon*. Routledge and Kegan Paul, 1978.
Hall, Peter, and Colin Ward. *Sociable Cities: The Legacy of Ebenezer Howard*. John Wiley and Sons, 1998.
Halliday, W. M. *Potlatch and Totem*. J. M. Dent and Sons, 1935.
Hammersley, Rachel. *The English Republican Tradition and Eighteenth-Century France*. Manchester University Press, 2010.
———. *James Harrington: An Intellectual Biography*. Oxford University Press, 2019.
Hampson, Norman. *Saint-Just*. Basil Blackwell, 1991.
Hanley, Ryan Patrick. *The Political Philosophy of Fénelon*. Oxford University Press, 2020.
Hanson, Lawrence, and Elisabeth Hanson. *The Noble Savage: A Life of Paul Gauguin*. Chatto and Windus, 1954.
Hanson, Philip. *Advertising and Socialism*. Macmillan, 1974.
———. *The Consumer in the Soviet Economy*. Macmillan, 1968.
Hardy, Dennis. *Alternative Communities in Nineteenth Century England*. Longman, 1979.
Harris, Max. *Sacred Folly: A New History of the Feast of Fools*. Cornell University Press, 2011.
Harrison, J.F.C. "Millennium and Utopia". in *Utopias*, edited by Peter Alexander and Roger Gill, 61–68. Duckworth, 1984.

———. *Quest for the New Moral World: Robert Owen and the Owenites in Britain and America.* Charles Scribner's Sons, 1969.

Harvey, David. *Rebel Cities: From the Right to the City to the Urban Revolution.* Verso, 2012.

———. *Spaces of Hope.* Edinburgh University Press, 2000.

Hatch, Nathan O. *The Sacred Cause of Liberty: Republican Thought and the Millennium in Revolutionary New England.* Yale University Press, 1977.

Haybron, Daniel M. *Happiness: A Very Short Introduction.* Oxford University Press, 2013.

Hayden, Brian. *The Power of Feasts: From Prehistory to the Present.* Cambridge University Press, 2014.

Hayden, Dolores. *Seven American Utopias: The Architecture of Communitarian Socialism, 1790–1975.* MIT Press, 1976.

Haydn, Hiram. *The Counter-Renaissance.* Charles Scribner's Sons, 1950.

Hayek, Friedrich. *The Road to Serfdom.* University of Chicago Press, 1945.

Haynes, R. D. *H. G. Wells: Discoverer of the Future.* Macmillan, 1980.

Hayward, J.E.S. "Solidarity: The Social History of an Idea in Nineteenth-Century France". *International Review of Social History* 4 (1959): 261–84.

Heath, Joseph, and Andrew Potter, *The Rebel Sell: Why the Culture Can't Be Jammed.* Capstone, 2005.

Heath, Sidney. *Pilgrim Life in the Middle Ages.* T. Fisher Unwin, 1911. Reprinted as *In the Steps of the Pilgrims* (Rich and Cowan, 1950).

Hecht, J. Jean. *The Domestic Servant Class in Eighteenth-Century England.* Routledge and Kegan Paul, 1956.

Hechter, Michael. *Principles of Group Solidarity.* University of California Press, 1987.

Heilbroner, Robert. *An Inquiry into the Human Prospect.* Calder and Boyars, 1975.

———. *The Quest for Wealth: A Study of Acquisitive Man.* Eyre and Spottiswoode, 1958.

Helm, Dieter. *Green and Prosperous Land: A Blueprint for Rescuing the British Countryside.* William Collins, 2019.

Henderson, Fred. *The Economic Consequences of Power Production.* George Allen and Unwin, 1931.

Herman, Gabriel. *Ritualised Friendship and the Greek City.* Cambridge University Press, 1987.

Hertzler, Joyce Oramel. *The History of Utopian Thought.* George Allen and Unwin, 1923.

Hessler, Julie. *A Social History of Soviet Trade: Trade Policy, Retail Practices, and Consumption, 1917–1953.* Princeton University Press, 2004.

Hexter, J. H. *More's Utopia: The Biography of an Idea.* Princeton University Press, 1952.

Higonnet, Patrice. *Goodness beyond Virtue: Jacobins during the French Revolution.* Harvard University Press, 1998.

Hill, Christopher. *Some Intellectual Consequences of the English Revolution.* Weidenfeld and Nicolson, 1980.

———. *The World Turned Upside Down: Radical Ideas during the English Revolution.* Temple Smith, 1972.

Hilton, Marjorie L. *Selling to the Masses: Retailing in Russia, 1880–1930.* University of Pittsburgh Press, 2012.

Hinds, William Alfred. *American Communities and Co-operative Colonies* [1878]. Porcupine, 1975.

Hindus, Maurice. *Under Moscow Skies.* Victor Gollancz, 1936.

Hine, Robert V. *California's Utopian Colonies*. Yale University Press, 1969.
Hirsch, Fred. *Social Limits to Growth*. Routledge and Kegan Paul, 1976.
Hobsbawm, Eric. *Primitive Rebels: Studies in Archaic Forms of Social Movement in the 19th and 20th Centuries*. Manchester University Press, 1959.
Hodkinson, Stephen, and Ian Macgregor Morris, eds. *Sparta in Modern Thought*. Classical Press of Wales, 2012.
Hogg, Michael A. *The Social Psychology of Group Cohesiveness*. Harvester Wheatsheaf, 1992.
Holloway, Mark. *Heavens on Earth: Utopian Communities in America, 1680–1880*. Dover, 1966.
Hölscher, Lucien. "Utopie". *Utopian Studies* 7 (1996): 1–65.
Holstun, James. *A Rational Millennium: Puritan Utopias of Seventeenth-Century England and America*. Oxford University Press, 1987.
Holtoon, F. L. van. *The Road to Utopia: A Study in John Stuart Mill's Social Thought*. Van Gorcum, 1971.
Holyoake, George Jacob. *The History of Co-operation in England*. 2 vols. Trubner, 1875.
Honoré, Carl. *In Praise of Slowness: Challenging the Cult of Speed*. HarperOne, 2004.
Honour, Hugh. *The New Golden Land: European Images of America from the Discoveries to the Present Time*. Allen Lane, 1975.
Hont, Istvan. "The Early Modern Debate on Commerce and Luxury". In *The Cambridge History of Eighteenth-Century Political Thought*, edited by Mark Goldie and Robert Wokler, 379–418. Cambridge University Press, 2006.
———. *Jealousy of Trade: International Competition and the Nation-State in Historical Perspective*. Harvard University Press, 2005.
———. "The Language of Sociability and Commerce: Samuel Pufendorf and the Theoretical Foundations of the 'Four-Stages Theory'". In *The Languages of Political Theory in Early Modern Europe*, edited by Anthony Pagden, 253–76. Cambridge University Press, 1987.
———. *Politics in Commercial Society: Jean-Jacques Rousseau and Adam Smith*. Harvard University Press, 2015.
Horne, Thomas. *The Social Thought of Bernard Mandeville*. Columbia University Press, 1978.
Horowitz, Daniel. *The Anxieties of Affluence: Critiques of American Consumer Culture, 1939–1979*. University of Massachusetts Press, 2004.
Hostetler, John A. *Amish Society*. Johns Hopkins Press, 1963.
Houston, Chloë. *The Renaissance Utopia: Dialogue, Travel and the Ideal Society*. Ashgate, 2014.
Howard, Donald R. *Writers and Pilgrims: Medieval Pilgrimage Narratives and Their Posterity*. University of California Press, 1980.
Hudson, Wayne. "Bloch and a Philosophy of the Proterior". In Thompson and Žižek, *Privatization of Hope*, 21–36.
———. *The Marxist Philosophy of Ernst Bloch*. Macmillan, 1982.
———. *The Reform of Utopia*. Ashgate, 2003.
Huehns, Gertrude. *Antinomianism in English History*. Cresset, 1951.
Hulliung, Mark. *The Autocritique of Enlightenment: Rousseau and the Philosophes*. Harvard University Press, 1994.
Humphery, Kim. *Excess: Anti-consumerism in the West*. Polity, 2010.
Humphrey, Chris. *The Politics of Carnival: Festive Misrule in Medieval England*. Manchester University Press, 2001.

Hundert, Edward J. *The Enlightenment's Fable: Bernard Mandeville and the Discovery of Society*. Cambridge University Press, 1994.

———. "Mandeville, Rousseau, and the Political Economy of Fantasy". In *Luxury in the Eighteenth Century*, edited by Maxine Berg and Elizabeth Eger, 28–40. Palgrave, 2003.

Hunt, Alan. *Governance of the Consuming Passions: A History of Sumptuary Law*. Macmillan, 1996.

Hurlock, Elizabeth B. "Sumptuary Law". In Roach and Eicher, *Dress, Adornment*, 295–302.

Hutter, Horst. *Politics as Friendship: The Origins of Classical Notions of Politics in the Theory and Practice of Friendship*. Wilfrid Laurier University Press, 1978.

Infield, Henrik F. *Co-operative Communities at Work*. Kegan Paul, Trench, Trubner, 1947.

Jack, Malcolm. *Corruption & Progress: The Eighteenth-Century Debate*. AMS, 1989.

Jackson, Tim. *Prosperity without Growth: Economics for a Finite Planet*. Earthscan, 2009.

Jacobs, Jane. *The Life and Death of Great American Cities*. Penguin Books, 1974.

James, E. O. *Seasonal Feasts and Festivals*. Thames and Hudson, 1961.

James, Oliver. *Affluenza: How to Be Successful and Stay Sane*. Vermilion, 2007.

Jameson, Fredric. *Archaeologies of the Future: The Desire Called Utopia and Other Science Fictions*. Verso, 2005.

Jameson, Fredric, Jodi Dean, Saroj Giri, Agon Hamza, Kojin Karatani, Kim Stanley Robinson, Frank Ruda, Alberto Toscano, and Kathi Weeks. *An American Utopia*. Edited by Slavo Žižek. Verso, 2016.

Jamieson, Dale. *Reason in a Dark Time: Why the Struggle against Climate Change Failed—and What It Means for Our Future*. Oxford University Press, 2014.

Jamieson, Lynn. *Intimacy: Personal Relationships in Modern Society*. Polity, 1988.

Jastrow, Morris. "Adam and Eve in Babylonian Literature". *American Journal of Semitic Languages and Literatures* 15 (1899): 193–214.

Jendrysik, Mark Stephen. *Utopia*. Polity, 2020.

Jennings, Chris. *Paradise Now: The Story of American Utopianism*. Random House, 2016.

Johns, Alessa. "Feminism and Utopianism". In *The Cambridge Companion to Utopian Literature*, edited by G. Claeys, 174–99. Cambridge University Press, 2010.

———. *Women's Utopias of the Eighteenth Century*. University of Illinois Press, 2003.

Johnson, Christopher H. *Utopian Communism in France: Cabet and the Icarians, 1839–1851*. Cornell University Press, 1974.

Jonas, Hans. *The Imperative of Responsibility: In Search of an Ethics for the Technological Age*. University of Chicago Press, 1984.

Jones, Howard Mumford. *O Strange New World: American Culture; The Formative Years*. Chatto and Windus, 1965.

Jones, James F., Jr. *La Nouvelle Héloïse: Rousseau and Utopia*. Librarie Droz, 1977.

Jones, Jennifer M. *Sexing* La Mode*: Gender, Fashion and Commercial Culture in Old Regime France*. Berg, 2004.

Jones, Peter M. "The 'Agrarian Law': Schemes for Land Redistribution during the French Revolution". *Past and Present* 133 (1991): 96–133.

Jue, Jeffrey K. *Heaven upon Earth: Joseph Mede (1586–1638) and the Legacy of Millenarianism*. Springer, 2006.

Kallen, Horace M. *The Decline and Rise of the Consumer*. Packard, 1936.

Kaminsky, Howard. "The Free Spirit in the Hussite Rebellion". In Thrupp, *Millennial Dreams in Action*, 166–86.

———. *A History of the Hussite Rebellion*. University of California Press, 1967.

Kanter, Rosabeth Moss. *Commitment and Community: Communes and Utopias in Sociological Perspective*. Harvard University Press, 1972.

Kargon, Robert H., and Arthur P. Molella. *Invented Edens: Techno-cities of the Twentieth Century*. MIT Press, 2008.

Karsenti, Bruno. "Imitation: Returning to the Tarde-Durkheim Debate". In *The Social after Gabriel Tarde*, edited by Matei Candea, 44–61. Routledge, 2010.

Kateb, George. *Utopia and Its Enemies*. 2nd ed. Schocken Books, 1976.

Kater, Michael H. *Hitler Youth*. Harvard University Press, 2004.

Kaufmann, David. "Thanks for the Memory: Bloch, Benjamin and Philosophy of History". *Yale Journal of Criticism* 6 (1993): 143–62.

Kautsky, Karl. *Communism in Central Europe in the Time of the Reformation*. T. Fisher Unwin, 1897.

———. *Thomas More and His Utopia* [1888]. A. and C. Black, 1927.

Keay, John. *The Honourable Company: A History of the East India Company*. HarperCollins, 1991.

Kendrick, Christopher. *Utopia, Carnival, and Commonwealth in Renaissance England*. University of Toronto Press, 2004.

Kenner, Dario. *Carbon Inequality: The Role of the Richest in Climate Change*. Earthscan, 2019.

Kenyon, Timothy. *Utopian Communism and Political Thought in Early Modern England*. Pinter, 1989.

Kesten, Seymour R. *Utopian Episodes: Daily Life in Experimental Communities Dedicated to Changing the World*. Syracuse University Press, 1993.

Kharkhordin, Oleg. *The Collective and the Individual in Russia: A Study of Practices*. University of California Press, 1999.

Kiefer, Otto. *Sexual Life in Ancient Rome*. George Routledge and Sons, 1934.

Killerby, Catherine Kovesi. *Sumptuary Law in Italy, 1200–1500*. Clarendon, 2002.

Kinane, Ian. *Theorising Literary Islands: The Island Trope in Contemporary Robinsonade Narratives*. Rowman and Littlefield, 2017.

Kirk, Andrew G. *Counterculture Green: The Whole Earth Catalog and American Environmentalism*. University Press of Kansas, 2007.

Klausner, Joseph. *The Messianic Idea in Israel*. George Allen and Unwin, 1956.

Klaw, Spencer. *Without Sin: The Life and Death of the Oneida Community*. Penguin Books, 1993.

Klein, Naomi. *The (Burning) Case for a Green New Deal*. Simon and Schuster, 2019.

———. *No Logo*. Picador, 2000.

———. *This Changes Everything: Capitalism vs. the Climate*. Allen Lane, 2014.

Knight, Katherine. *Rationing in the Second World War*. Tempus, 2007.

Knowles, Rob. *Political Economy from Below: Economic Thought in Communitarian Anarchism, 1840–1914*. Routledge, 2004.

Kohn, Margaret. *Brave New Neighborhoods: The Privatization of Public Space*. Routledge, 2004.

Kolakowski, Leszek. *Main Currents of Marxism*. 3 vols. Oxford University Press, 1978.

———. *Modernity on Endless Trial*. University of Chicago Press, 1990.

———. *My Correct Views on Everything*. St. Augustine's, 2005.

———. "Need of Utopia, Fear of Utopia". In *Radicalism in the Contemporary Age*, edited by Seweryn Bialer, vol. 2, *Radical Visions of the Future*, 3–12. Westview, 1977.
Kolbert, Elizabeth. *The Sixth Extinction: An Unnatural History*. Bloomsbury, 2014.
Kolmerten, Carol. *Women in Utopia: The Ideology of Gender in the American Owenite Communities*. Indiana University Press. 1990.
Kolnai, Aurel. *The Utopian Mind*. Athlone, 1995.
König, René. *The Restless Image: A Sociology of Fashion*. George Allen and Unwin, 1973.
Koselleck, Reinhart. *Futures Past: On the Semantics of Historical Time*. Translated by Keith Tribe. Columbia University Press, 2004.
———. *The Practice of Conceptual History*. Translated by Todd Samuel Presner et al. Stanford University Press, 2002.
Kotin, Joshua. *Utopias of One*. Princeton University Press, 2018.
Kowinski, William Severini. *The Malling of America: Travels in the United States of Shopping*. Xlibris, 2002.
Kraybill, Donald B. *The Riddle of Amish Culture*. Johns Hopkins University Press, 1989.
Kraybill, Donald B., and Carl F. Bowman. *On the Backroad to Heaven: Old Order Hutterites, Mennonites, Amish, and Brethren*. Johns Hopkins University Press, 2001.
Krylova, Anna. "Saying 'Lenin' and Meaning 'Party': Subversion and Laughter in Soviet and Post-Soviet Society". In Barker, *Consuming Russia*, 243–65.
Kuchta, David. "The Making of the Self-Made Man: Class, Clothing, and English Masculinity, 1688–1732". In *The Sex of Things: Gender and Consumption in Historical Perspective*, edited by Victoria de Grazia, with Ellen Furlough, 54–78. University of California Press, 1996.
Kumar, Krishan. "Religion and Utopia". In *The Canterbury Papers: Essays on Religion and Society*, edited by Dan Cohn-Sherbok, 69–79. Bellew, 1990.
———. *Utopia and Anti-utopia in Modern Times*. Basil Blackwell, 1987.
———. *Utopianism*. Open University Press, 1991.
Kuper, Adam. *The Invention of Primitive Society: Transformations of an Illusion*. Routledge, 1988.
Ladurie, Emmanuel Le Roy. *Carnival: A People's Uprising at Romans, 1579–80*. Scholar Press, 1980.
Laing, R. D. *The Politics of Experience and The Bird of Paradise*. Penguin Books, 1967.
Lamb, Jonathan. "Fantasies of Paradise". In *The Enlightenment World*, edited by Martin Fitzpatrick, Peter Jones, Christa Knellwolf, and Iain McCalman, 521–35. Routledge, 2004.
Lamont, William M. *Godly Rule: Politics and Religion, 1603–60*. Macmillan, 1969.
Landauer, Gustav. *Revolution and Other Political Writings*. PM Press, 2010.
Landes, Richard. *Heaven on Earth: The Varieties of the Millennial Experience*. Oxford University Press, 2011.
Landsman, Mark. *Dictatorship and Demand: The Politics of Consumerism in East Germany*. Harvard University Press, 2005.
Lane, Robert E. *The Loss of Happiness in Market Democracies*. Yale University Press, 2000.
———. "The Road Not Taken". In Crocker and Linden, *Ethics of Consumption*, 218–48.
Lang, Andrew. *Magic and Religion*. Longmans, Green, 1901.
Lantenari, Vittorio. *The Religions of the Oppressed: A Study of Modern Messianic Cults*. Macgibbon and Kee, 1963.
Lasch, Christopher. *The Culture of Narcissism: American Life in an Age of Diminishing Expectations*. W. W. Norton, 1978.

———. *The True and Only Heaven: Progress and Its Critics*. W. W. Norton, 1991.

Lasky, Melvin J. *Utopia and Revolution*. Macmillan, 1977.

Laslett, Peter. *The World We Have Lost*. Methuen, 1965.

Lasn, Kalle. *Culture Jam: The Uncooling of America*. William Morrow, 1999.

Latham, J.E.M. *Search for a New Eden: James Pierrepont Greaves (1777–1842): The Sacred Socialist and His Followers*. Associated University Presses, 1999.

Laveleye, Emile de. *Luxury*. Swan Sonnenschein, 1891.

Lebergott, Stanley. *Pursuing Happiness: American Consumers in the Twentieth Century*. Princeton University Press, 1993.

Leech, Kenneth. *Youthquake: The Growth of a Counter-culture through Two Decades*. Sheldon, 1973.

Lefebvre, Henri. *The Urban Revolution*. University of Minnesota Press, 2003.

Leiss, William. *The Limits to Satisfaction: On Needs and Commodities*. Marion Boyers, 1978.

Lenček, Lena, and Gideon Bosker. *The Beach: The History of Paradise on Earth*. Secker and Warburg, 1998.

Leopold, David. "The Structure of Marx and Engels' Considered Account of Utopian Socialism". *History of Political Thought* 26 (2005): 443–66.

Lerner, Robert. *The Heresy of the Free Spirit in the Later Middle Ages*. University of California Press, 1972.

Leroux, Robert. "Tarde and Durkheimian Sociology". In *The Anthem Companion to Gabriel Tarde*, edited by Leroux, 119–34. Anthem, 2018.

Levin, Harry. *The Myth of the Golden Age in the Renaissance*. Oxford University Press, 1969.

Levitas, Ruth. *The Concept of Utopia*. Syracuse University Press, 1990.

———. *Utopia as Method: The Imaginary Reconstitution of Society*. Palgrave Macmillan, 2013.

Lewes, Darby. *Dream Revisionaries: Gender and Genre in Women's Utopian Fiction, 1870–1920*. University of Alabama Press, 1995.

Lewis, C. S. *English Literature of the Sixteenth Century*. Clarendon, 1954.

Licht, Hans. *Sexual Life in Ancient Greece*. George Routledge and Sons, 1932.

Lichtenberg, Judith. "Consuming because Others Consume". In Crocker and Linden, *Ethics of Consumption*, 155–75.

Limits to Growth: A Report for the Club of Rome's Project on the Predicament of Mankind. Potomac Associates, 1972.

Lipovetsky, Gilles. *The Empire of Fashion: Dressing Modern Democracy*. Princeton University Press, 1994.

Lipton, Lawrence. *The Holy Barbarians*. W. H. Allen, 1960.

Lockwood, George Browning. *The New Harmony Communities*. Chronicle, 1902.

Logan, George M., ed. *The Cambridge Companion to Thomas More*. Cambridge University Press, 2011.

Lord, Arthur. *Plymouth and the Pilgrims*. Houghton Mifflin, 1920.

Lovejoy, Arthur O. *Essays in the History of Ideas*. Johns Hopkins University Press, 1948.

Lovejoy, Arthur O., and George Boas. *Primitivism and Related Ideas in Antiquity*. Johns Hopkins Press, 1935.

Lovell, Julia. *Maoism: A Global History*. Bodley Head, 2019.

Lovelock, James. *The Revenge of Gaia*. Allen Lane, 2006.

Löwy, Michael. *Redemption and Utopia: Jewish Libertarian Thought in Central Europe*. Athlone, 1992.
Lubac, Henri de. *The Un-Marxian Socialist: A Study of Proudhon*. Sheed and Ward, 1948.
Lukes, Steven. *Individualism*. Basil Blackwell, 1973.
Lunbeck, Elizabeth. *The Americanization of Narcissism*. Harvard University Press, 2014.
Lunt, Peter K., and Sonia M. Livingstone. *Mass Consumption and Personal Identity*. Open University Press, 1992.
Lynas, Mark. *Six Degrees: Our Future on a Hotter Planet*. Harper Perennial, 2008.
Lynd, Robert S., and Helen Merrell Lynd. *Middletown in Transition: A Study in Cultural Conflicts*. Constable, 1937.
Macfadyen, Dugald. *Sir Ebenezer Howard and the Town Planning Movement*. Manchester University Press, 1933.
MacGillvray, J. R. "The Pantisocracy Scheme and Its Immediate Background". In *Studies in English*, edited by Malcolm W. Williams, 131–69. Oxford University Press, 1931.
Maffesoli, Michel. *The Shadow of Dionysus: A Contribution to the Sociology of the Orgy*. State University of New York Press, 1992.
Maitland, Sara, ed. *Very Heaven: Looking Back at the 1960s*. Virago Books, 1988.
Malcolm, Henry. *Generation of Narcissus*. Little, Brown, 1971.
Malm, Andreas. *Corona, Climate, Chronic Emergency: War Communism in the Twenty-First Century*. Verso Books, 2020.
Mann, Michael. *The New Climate War: The Fight to Take Back Our Planet*. Scribe, 2021.
Mannheim, Karl. *Ideology and Utopia: An Introduction to the Sociology of Knowledge* [1929]. Kegan Paul, Trench, Trubner, 1936.
Mansfield, Andrew. *Ideas of Monarchical Reform: Fénelon, Jacobitism and the Political Works of the Chevalier Ramsay*. Manchester University Press, 2015.
Manuel, Frank E. *The New World of Henri Saint-Simon*. Harvard University Press, 1956.
———. *Shapes of Philosophical History*. George Allen and Unwin, 1965.
Manuel, Frank E., and Fritzie P. Manuel. *Utopian Thought in the Western World*. Basil Blackwell, 1979.
Marchant, Ian. *A Hero for High Times*. Jonathan Cape, 2018.
Marc'Hadour, Germain. "Utopia and Martyrdom". In Olin, *Interpreting Thomas More's "Utopia"*, 61–76.
Margolis, Jonathan. *A Brief History of Tomorrow*. Bloomsbury, 2000.
Marsh, Jan. *Back to the Land: The Pastoral Impulse in England, from 1880 to 1914*. Quartet Books, 1982.
Martin, Everett Dean. *The Behavior of Crowds*. Harper and Bros, 1920.
Maslow, Abraham H. *Motivation and Personality*. 3rd ed. Longman, 1987.
Massingham, H. J. *The Golden Age: The Story of Human Nature*. Gerald Howe, 1927.
Mathews, Nancy Mowll. *Paul Gauguin: An Erotic Life*. Yale University Press, 2001.
Mauss, Marcel. *The Gift: Forms and Functions of Exchange in Archaic Societies*. Cohen and West, 1954.
Mazlish, Bruce. *The Revolutionary Ascetic: Evolution of a Political Type*. Basic Books, 1976.
McAllister, Matthew P. *The Commercialization of American Culture: New Advertising, Control and Democracy*. Sage, 1996.

McCall, George, Michal M. McCall, Norman K. Denzin, Gerald D. Suttles, and Suzanne B. Kurth. *Social Relationships*. Aldine, 1970.

McCarthy, Fiona. *The Simple Life: C. R. Ashbee in the Cotswolds*. Lund Humphries, 1981.

McDaniel, Ian. *Adam Ferguson in the Scottish Enlightenment*. Harvard University Press, 2013.

McDannell, Colleen, and Bernhard Lang. *Heaven: A History*. Vintage Books, 1990.

McGrady, Thomas. *Beyond the Black Ocean*. C. H. Kerr, 1901.

McKendrick, Neil, John Brewer, and J. H. Plumb. *The Birth of a Consumer Society: The Commercialization of Eighteenth-Century England*. Hutchinson, 1983.

McKenna, Erin. *The Task of Utopia: A Pragmatist and Feminist Perspective*. Rowman and Littlefield, 2001.

McKibben, Bill. *The End of Nature*. Viking, 1990.

McNeil, Peter, and Giorgio Riello. *Luxury: A Rich History*. Oxford University Press, 2016.

McWilliam, Neil. *Dreams of Happiness: Social Art and the French Left, 1830–1850*. Princeton University Press, 1993.

Mead, Margaret, ed. *Cooperation and Competition among Primitive Peoples*. Beacon, 1937.

Meadows, Donella. *Beyond the Limits: Global Collapse or a Sustainable Future*. Earthscan, 1992.

Meadows, Donella, Jørgen Randers, and Dennis Meadows. *Limits to Growth: The 30-Year Update*. Earthscan, 2005.

Meek, Ronald. *The Economics of Physiocracy*. George Allen and Unwin, 1962.

Meltzer, Graham. "Contemporary Communalism at a Time of Crisis". In Ben-Rafael, Oved, and Topel, *Communal Idea*, 73–90.

Merkel, Ina. *Utopie und Bedürfnis: Die Geschichte der Konsumkultur in der DDR*. Böhlau Verlag, 1999.

Metcalf, Bill. *From Utopian Dreaming to Communal Reality: Cooperative Lifestyles in Australia*. University of New South Wales Press, 1995.

———. *Shared Visions, Shared Lives: Communal Living around the Globe*. Findhorn, 1996.

Milbrath, Lester W. *Envisioning a Sustainable Society: Learning Our Way Out*. State University of New York Press, 1989.

Miles, Clement A. *Christmas in Ritual and Tradition, Christian and Pagan*. T. Fisher Unwin, 1912.

Miles, Malcolm. *Urban Utopias: The Built and Social Architectures of Alternative Settlements*. Routledge, 2008.

Miller, Daniel. *Consumption and Its Consequences*. Polity, 2012.

———. *Stuff*. Polity, 2010.

Miller, Timothy. *Communes in America, 1975–2000*. Syracuse University Press, 2019.

———. *The Encyclopedic Guide to American Intentional Communities*. 2nd ed. Richard W. Couper, 2015.

———. "A Matter of Definition: Just What Is an Intentional Community?". *Communal Societies* 30 (2010): 1–15.

———. *The Quest for Utopia in Twentieth-Century Communes*. Syracuse University Press, 1998.

———. *The 60s Communes: Hippies and Beyond*. Syracuse University Press, 1999.

Minton, Anna. *Ground Control: Fear and Happiness in the Twenty-First-Century City*. Penguin Books, 2009.

Mises, Ludwig von. *Socialism: An Economic and Sociological Analysis*. Jonathan Cape, 1936.

Monbiot, George. *Heat: How to Stop the Planet Burning*. Allen Lane, 2006.

———. *Out of the Wreckage: A New Politics for an Age of Crisis*. Verso, 2017.

Monro, Hector. *The Ambivalence of Bernard Mandeville*. Clarendon, 1975.
Monter, E. William. *Calvin's Geneva*. John Wiley and Sons, 1967.
Moos, Rudolf, and Robert Brownstein. *Environment and Utopia: A Synthesis*. Plenum, 1977.
Morgan, Arthur E. *The Small Community: Foundation of Democratic Life*. Harper and Bros, 1942.
Morton, A. L. *The World of the Ranters: Social Radicalism in the English Revolution*. Lawrence and Wishart, 1970.
Moyn, Samuel. *The Last Utopia: Human Rights in History*. Harvard University Press, 2010.
Muggeridge, Malcolm. *Winter in Moscow*. Eyre and Spottiswoode, 1934.
Mumford, Lewis. *The City in History*. Secker and Warburg, 1961.
———. *The Culture of Cities*. Secker and Warburg, 1940.
———. *The Story of Utopias* [1922]. Viking, 1962.
———. *The Urban Prospect*. Secker and Warburg, 1968.
Musgrove, Frank. *Ecstasy and Holiness: The Counter Culture and the Open Society*. Methuen, 1974.
Nash, James A. "On the Subversive Virtue: Frugality". In Crocker and Linden, *Ethics of Consumption*, 416–36.
Near, Henry. *The Kibbutz Movement: A History*. 2 vols. Oxford University Press, 1992.
———. "Utopian and Post-utopian Thought: The Kibbutz as Model". *Communal Societies* 5 (1985): 41–58.
Nelson, Anitra. *Small Is Necessary: Shared Living on a Shared Planet*. Pluto, 2018.
Nelson, Eric. *The Greek Tradition in Republican Thought*. Cambridge University Press, 2004.
Neuburger, Mary. "Inhaling Luxury: Smoking and Anti-smoking in Socialist Bulgaria, 1947–1989". In Crowley and Reid, *Pleasures in Socialism*, 239–58.
Neumann, Matthias. *The Communist Youth League and the Transformation of the Soviet Union, 1917–1932*. Routledge, 2011.
Newman, E. M. *Seeing Russia*. Funk and Wagnalls, 1928.
Niebuhr, Reinhold. *Faith and History: A Comparison of Christian and Modern Views of History*. Nisbet, 1949.
———. *The Kingdom of God in America*. Harper and Bros, 1937.
Niman, Michael I. *People of the Rainbow: A Nomadic Utopia*. University of Tennessee Press, 1997.
Nisbet, Robert. *History of the Idea of Progress*. Heinemann, 1980.
———. *The Quest for Community*. Oxford University Press, 1953.
———. *The Sociological Tradition*. Heinemann, 1966.
Nordhoff, Charles. *The Communistic Societies of the United States* [1875]. Dover, 1966.
North, Michael. *Novelty: A History of the New*. University of Chicago Press, 2013.
Oakeshott, Michael. *Rationalism in Politics and Other Essays* [1962]. Liberty, 1991.
Offer, Avner. *The Challenge of Affluence: Self-Control and Well-Being in the United States and Britain since 1950*. Oxford University Press, 2006.
O'Gorman, Edmundo. *The Invention of America*. Indiana University Press, 1961.
Olin, John C., ed. *Interpreting Thomas More's "Utopia"*. Fordham University Press, 1989.
Oliver, W. H. "Owen in 1817: The Millennialist Moment". In *Robert Owen: Prophet of the Poor*, edited by Sidney Pollard and John Salt, 166–87. Macmillan, 1971.
Olson, Theodore. *Millennialism, Utopianism, and Progress*. University of Toronto Press, 1982.
Ophuls, William. *Ecology and the Politics of Scarcity: Prologue to a Political Theory of the Steady State*. W. H. Freeman, 1977.

Osborn, Fairfield. *Our Plundered Planet*. Faber and Faber, 1948.
Osokina, Elena. *Our Daily Bread: Socialist Distribution and the Art of Survival in Stalin's Russia, 1927–1941*. M. E. Sharpe, 2001.
Oved, Yaacov. *Two Hundred Years of American Communities*. Transaction, 1988.
Ozouf, Mona. *Festivals and the French Revolution*. Harvard University Press, 1988.
Packard, Vance. *The Hidden Persuaders*. Longmans, Green, 1957.
———. *The Waste Makers*. Longmans, 1961.
Pajur, Astrid. "The Fabric of a Corporate Society: Sumptuary Laws, Social Order and Propriety in Early Modern Tallinn". In *A Taste for Luxury in Early Modern Europe*, edited by Johanna Ilmakunnas and Jon Stobart, 21–38. Bloomsbury, 2017.
Pangle, Lorraine Smith. *Aristotle and the Philosophy of Friendship*. Cambridge University Press, 2003.
Park, Julie. *The Self and It: Novel Objects in Eighteenth-Century England*. Stanford University Press, 2010.
Park, Robert E. *The Crowd and the Public and Other Essays*. University of Chicago Press, 1972.
Parker, Harold T. *The Cult of Antiquity and the French Revolution*. University of Chicago Press, 1937.
Parker, Robert Allerton. *A Yankee Saint: John Humphrey Noyes and the Oneida Community*. G. Putnam's Sons, 1935.
Partington, John S. *Building Cosmopolis: The Political Thought of H. G. Wells*. Ashgate, 2003.
Passmore, John. *The Perfectibility of Man*. Duckworth, 1970.
Paulicelli, Eugenia. *Fashion under Fascism: Beyond the Black Shirt*. Berg, 2004.
Pearl, Jason H. *Utopian Geographies and the Early English Novel*. University of Virginia Press, 2014.
Peck, Linda Levy. *Consuming Splendor: Society and Culture in Seventeenth-Century England*. Cambridge University Press, 2005.
Pepper, David. *Communes and the Green Vision: Counterculture, Lifestyle and the New Age*. Green Print, 1991.
———. *The Roots of Modern Environmentalism*. Routledge, 1984.
Perkin, Harold. *The Origins of Modern English Society, 1780–1880*. Routledge and Kegan Paul, 1969.
Perrot, Michel, ed. *A History of Private Life*. 5 vols. Harvard University Press, 1990.
Pethybridge, Roger. *One Step Backwards, Two Steps Forward: Soviet Society and Politics in the New Economic Policy*. Clarendon, 1990.
Pettifor, Ann. *The Case for the Green New Deal*. Verso, 2019.
Phelan, John Leddy. *The Millennial Kingdom of the Franciscans in the New World*. University of California Press, 1956.
Philp, Mark. *Godwin's Political Justice*. Duckworth, 1986.
Pilbeam, Pamela. *Saint-Simonians in Nineteenth-Century France: From Free Love to Algeria*. Palgrave Macmillan, 2014.
Pinker, Susan. *The Village Effect: Why Face-to-Face Contact Matters*. Atlantic Books, 2014.
Piotrowski, Sylvester A. *Etienne Cabet and the Voyage en Icarie*. Hyperion, 1975. Originally PhD dissertation, Catholic University of America, 1935.
Pitzer, Donald, ed. *America's Communal Utopias*. University of North Carolina Press, 1997.

Platt, Charles. *The Psychology of Social Life*. George Allen and Unwin, 1922.
Pleij, Herman. *Dreaming of Cockaigne: Medieval Fantasies of the Perfect Life*. Columbia University Press, 2001.
Pocock, J.G.A. *The Machiavellian Moment: Florentine Political Thought and the Atlantic Republican Tradition*. Princeton University Press, 1975.
Pohl, Nicole. *Women, Space and Utopia, 1600–1800*. Ashgate, 2003.
Pohl, Nicole, and Brenda Tooley, eds. *Gender and Utopia in the Eighteenth Century*. Ashgate, 2007.
Polak, Fred. L. *The Image of the Future*. 2 vols. Oceana, 1961.
———. "Utopia and Cultural Revival". In *Utopias and Utopian Thought*, edited by Frank E. Manuel, 281–95. Beacon, 1967.
Polanyi, Karl. *Origins of Our Time: The Great Transformation*. Victor Gollancz, 1945.
Popper, Karl. *The Open Society and Its Enemies*. 2 vols. Routledge and Kegan Paul, 1945.
Porter, Roy. *English Society in the Eighteenth Century*. Allen Lane, 1982.
Poster, Mark. *The Utopian Thought of Restif de la Bretonne*. New York University Press, 1971.
Potter, David M. *People of Plenty: Economic Abundance and the American Character*. University of Chicago Press, 1968.
Preece, Rod. *Sins of the Flesh: A History of Ethical Vegetarian Thought*. UBC, 2008.
Price, A. W. *Love and Friendship in Plato and Aristotle*. Clarendon, 1989.
Pusey, W. W. *Louis-Sébastien Mercier in Germany*. Columbia University Press, 1939.
Putnam, Robert D. *Bowling Alone: The Collapse and Revival of American Community*. Simon and Schuster, 2000.
Quesnay, François. *Quesnay's Tableau Économique*. Macmillan, 1972.
Raby, Peter. *Samuel Butler: A Biography*. Hogarth, 1991.
Rae, John. *Life of Adam Smith*. Macmillan, 1895.
Randall, Amy E. *The Soviet Dream World of Retail Trade and Consumption*. Palgrave Macmillan, 2008.
Raworth, Kate. *Doughnut Economics: Seven Ways to Think like a 21st-Century Economist*. Business Books, 2017.
Rawson, Elizabeth. *The Spartan Tradition in European Thought*. Clarendon, 1960.
Raymond, Janice. *A Passion for Friends: Towards a Philosophy of Female Affections*. Women's Press, 1991.
Rees, Christine. *Utopian Imagination and Eighteenth Century Fiction*. Longman, 1996.
Rees, Martin. *On the Future: Prospects for Humanity*. Princeton University Press, 2018.
Reich, Charles. *The Greening of America*. Allen Lane, 1971.
Reid, Susan E., and David Crowley, eds. *Style and Socialism: Modernity and Material Culture in Post-war Eastern Europe*. Berg, 2000.
Reisman, John M. *Anatomy of Friendship*. Irvington, 1979.
Reynolds, E. E. *The Field Is Won: The Life and Death of Saint Thomas More*. Burns and Oates, 1968.
Ribeiro, Aileen. *Fashion in the French Revolution*. B. T. Batsford, 1988.
Rich, Nathaniel. *Losing Earth: The Decade We Could Have Stopped Climate Change*. Picador, 2017.
Riesman, David. *The Lonely Crowd: A Study of the Changing American Character*. Yale University Press, 1952.

———. *Thorstein Veblen*. Charles Scribner's Sons, 1953.

Ritter, Alan. *The Political Thought of Pierre-Joseph Proudhon*. Princeton University Press, 1969.

Ritter, Roland, and Bernd Knaller-Vlay, eds. *Other Spaces: The Affair of the Heterotopia*. Haus der Architektur, 1998.

Ritzer, George. *Enchanting a Disenchanted World: Continuity and Change in the Cathedrals of Consumption*. Sage, 2010.

———. *The McDonaldization of Society*. 9th ed. Sage, 2019.

Roach, Mary Ellen, and Joanne Bubolz Eicher, eds. *Dress, Adornment, and the Social Order*. John Wiley and Sons, 1965.

Robins, Robert S., and Jerrold M. Post. *Political Paranoia: The Psychopolitics of Hatred*. Yale University Press, 1997.

Robinson, Kim Stanley. *The Ministry for the Future*. Orbit Books, 2020.

Robinson, Roxy. *Music Festivals and the Politics of Participation*. Ashgate, 2015.

Roche, Daniel. *The Culture of Clothing: Dress and Fashion in the "Ancien Régime"*. Cambridge University Press, 1996.

Rogers, Richard, and Anne Power. *Cities for a Small Country*. Faber and Faber, 2000.

Rohling, Eelco J. *The Climate Question: Natural Cycles, Human Impact, Future Outlook*. Oxford University Press, 2019.

Rose, R. B. *The Enragés: Socialists of the French Revolution?* Melbourne University Press, 1965.

———. *Gracchus Babeuf: The First Revolutionary Communist*. Edward Arnold, 1978.

Rosen, Michael. *Dignity: Its History and Meaning*. Harvard University Press, 2012.

Ross, Harry. *Utopias Old and New*. Nicholson and Watson, 1938.

Ross, Kristin. *Communal Luxury: The Political Imaginary of the Paris Commune*. Verso, 2015.

Roszak, Theodore. *The Making of a Counter Culture*. Faber and Faber, 1970.

———. *Where the Wasteland Ends*. Faber and Faber, 1973.

Rothstein, Bo, and Eric M. Ushaner. "All for All: Equality, Corruption, and Social Trust". *World Politics* 58 (2005): 41–72.

Rowbotham, Sheila. *Promise of a Dream: Remembering the Sixties*. Allen Lane, 2000.

Rowbotham, Sheila, and Jeffrey Weeks. *Socialism and the New Life: The Personal and Sexual Politics of Edward Carpenter and Havelock Ellis*. Pluto, 1977.

Royle, Edward. *Robert Owen and the Commencement of the Millennium: A Study of the Harmony Community*. Manchester University Press, 1998.

Russell, Bertrand. *In Praise of Idleness and Other Essays*. George Allen and Unwin, 1936.

Russell, Frances Theresa. *Touring Utopia*. Dial, 1932.

Ryback, Timothy W. *Rock around the Bloc: A History of Rock Music in Eastern Europe and the Soviet Union*. Oxford University Press, 1990.

Sahlins, Marshall. *Stone-Age Economics*. Tavistock, 1974.

Saito, Kohei. *Karl Marx's Ecosocialism: Capital, Nature, and the Unfinished Critique of Political Economy*. Monthly Review, 2017.

Sale, Kirkpatrick. *The Conquest of Paradise: Christopher Columbus and the Columbian Legacy*. Hodder and Stoughton, 1991.

Salmond, Anne. *Aphrodite's Island: The European Discovery of Tahiti*. University of California Press, 2010.

Salt, Henry S. *The Creed of Kinship*. Constable, 1935.

———. *Seventy Years among Savages.* George Allen and Unwin, 1921.

Sanford, Charles L. *The Quest for Paradise: Europe and the American Moral Imagination.* University of Illinois Press, 1961.

Sargant, William Lucas. *Social Innovators and Their Schemes.* Smith, Elder, 1858.

Sargent, Lyman Tower. *British and American Utopian Literature, 1516–1985.* Garland, 1988. Updated as *Utopian Literature in English: An Annotated Bibliography from 1516 to the Present* (database; Penn State Libraries Open Publishing, 2016 and continuing), doi:10.18113/P8WC77).

———. "Ideology and Utopia". in *The Oxford Handbook of Political Ideologies,* edited by Michael Freeden, Sargent, and Marc Stears, 439–51. Oxford University Press, 2013.

———. "The Three Faces of Utopianism Revisited". *Utopian Studies* 5 (1994): 1–37.

———. "Utopianism". In *Routledge Encyclopedia of Philosophy* (online). 1998. https://www.rep.routledge.com/articles/thematic/utopianism/v-1.

———. *Utopianism: A Very Short Introduction.* Oxford University Press, 2010.

Sargisson, Lucy. *Fool's Gold? Utopianism in the Twenty-First Century?* Palgrave Macmillan, 2012.

Sargisson, Lucy, and Lyman Tower Sargent. *Living in Utopia: New Zealand's Intentional Communities.* Ashgate, 2004.

Sassatelli, Roberta. *Consumer Culture: History, Theory and Politics.* Sage, 2007.

Schivelbusch, Wolfgang. *Tastes of Paradise: A Social History of Spices, Stimulants, and Intoxicants.* Vintage, 1992.

Schlink, Bernhard. *Heimat als Utopie.* Suhrkamp, 2000.

Schor, Juliet B. *The Overspent American: Upscaling, Downshifting, and the New Consumer.* Basic Books, 1998.

———. *The Overworked American: The Unexpected Decline of Leisure.* Basic Books, 1991.

Schudson, Michael. "Delectable Materialism". In Crocker and Linden, *Ethics of Consumption,* 249–68.

Schumacher, E. F. *Small Is Beautiful: A Study of Economics as if People Mattered.* Abacus, 1974.

Schur, Edwin. *The Awareness Trap: Self-Absorption instead of Social Change.* Quadrangle, 1976.

Scitovsky, Tibor. *The Joyless Economy: An Inquiry into Human Satisfaction and Consumer Dissatisfaction.* Oxford University Press, 1976.

Scott, Jonathan. *Commonwealth Principles: Republican Writing of the English Revolution.* Cambridge University Press, 2004.

Scott, Robert A. *Miracle Cures: Saints, Pilgrimage, and the Healing Powers of Belief.* University of California Press, 2010.

Scudder, Vida D. *Social Ideals in English Letters.* Houghton, Mifflin, 1898.

Seabrook, John. *Nobrow: The Culture of Marketing, the Marketing of Culture.* Methuen, 2000.

Sekora, John. *Luxury: The Concept in Western Thought, Eden to Smollett.* Johns Hopkins University Press, 1977.

Sen, S. R. *The Economics of Sir James Steuart.* G. Bell and Sons, 1957.

Sennett, Richard. *The Corrosion of Character: The Personal Consequences of Work in the New Capitalism.* W. W. Norton, 1998.

———. "Destructive Gemeinschaft". In *Beyond the Crisis,* edited by Norman Birnbaum, 171–200. Oxford University Press, 1977.

———. *The Fall of Public Man.* Cambridge University Press, 1977.

———. *Families against the City: Middle Class Homes of Industrial Chicago, 1872–1890*. Harvard University Press, 1970.

———. *The Uses of Disorder: Personal Identity and City Life*. Faber and Faber, 1996.

Sennett, Richard, and Jonathan Cobb. *The Hidden Injuries of Class*. Cambridge University Press, 1972.

Shadurski, Maxim. *The Nationality of Utopia: H. G. Wells, England, and the World-State*. Routledge, 2020.

Shi, David E. *The Simple Life: Plain Living and High Thinking in American Culture*. University of Georgia Press, 2007.

Shklar, Judith. *After Utopia: The Decline of Political Faith*. Princeton University Press, 1957.

———. *Men and Citizens: A Study of Rousseau's Social Theory*. Cambridge University Press, 1969.

———. *Political Thought and Political Thinkers*. University of Chicago Press, 1998.

Shovlin, John. *The Political Economy of Virtue: Luxury, Patriotism, and the Origins of the French Revolution*. Cornell University Press, 2006.

Shrubsole, Guy. *Who Owns England? How We Lost Our Green and Pleasant Land & How to Take It Back*. William Collins, 2019.

Siebers, Tobin, ed. *Heterotopia: Post-modern Utopia and the Body Politic*. University of Michigan Press, 1994.

Silverstone, Roger, ed. *Visions of Suburbia*. Routledge, 1997.

Siméon, Ophélie. *Robert Owen's Experiment at New Lanark*. Palgrave Macmillan, 2017.

Simmel, Georg. *On Individuality and Social Forms*. University of Chicago Press, 1971.

———. *The Sociology of Georg Simmel*. Free Press, 1950.

Skinner, B. F. *Beyond Freedom and Dignity*. Penguin Books, 1973.

Skinner, Quentin. "Thomas More's *Utopia* and the Virtue of True Nobility". In *Visions of Politics*, vol. 2, *Renaissance Virtues*, 213–31. Cambridge University Press, 2002.

Slater, Don. *Consumer Culture and Modernity*. Polity, 1997.

Smith, Eric C. *Foucault's Heterotopia in Christian Catacombs*. Palgrave Macmillan, 2014.

Smith, Woodruff D. *Consumption and the Making of Respectability, 1600–1800*. Routledge, 2002.

Sear, Joanne, and Ken Sneath. *The Origins of the Consumer Revolution in England*. Routledge, 2020.

Snell, K.D.M. *Spirits of Community: English Senses of Belonging and Loss, 1750–1900*. Bloomsbury, 2016.

Soja, Edward W. *Postmodern Geographies: The Reassertion of Space in Critical Social Theory*. Verso, 1989.

———. *Thirdspace: Journeys to Los Angeles and Other Real-and-Imagined Spaces*. Blackwell, 1996.

Solnit, Rebecca. *A Paradise Built in Hell: The Extraordinary Communities That Arise in Disaster*. Viking, 2009.

Sombart, Werner. *Luxury and Capitalism*. University of Michigan Press, 1967.

———. *Socialism and the Social Movement* [1896]. J. M. Dent, 1909.

Sonenscher, Michael. *Before the Deluge: Public Debt, Inequality, and the Intellectual Origins of the French Revolution*. Princeton University Press, 2007.

———. *Jean-Jacques Rousseau: The Division of Labour, the Politics of the Imagination and the Concept of Federal Government*. Brill, 2020.

———. *Sans-Culottes: An Eighteenth Century Emblem in the French Revolution*. Princeton University Press, 2008.

Soper, Kate. *Post-growth Living: For an Alternative Hedonism*. Verso, 2020.
Souhami, Diana. *Selkirk's Island*. Weidenfeld and Nicolson, 2001.
Souter, Gavin. *A Peculiar People: The Australians in Paraguay*. Sydney University Press, 1981.
Spadafora, David. *The Idea of Progress in Eighteenth-Century Britain*. Yale University Press, 1990.
Spang, Rebecca L. "What Is Rum? The Politics of Consumption in the French Revolution". In Daunton and Hilton, *Politics of Consumption*, 33–50.
Spear, Jeffrey R. *Dreams of an English Eden: Ruskin and His Tradition in Social Criticism*. Columbia University Press, 1984.
Spencer, Colin. *Vegetarianism: A History*. Grub Street, 2000.
Spengler, Joseph J. *French Predecessors of Malthus*. Duke University Press, 1942.
Spiro, Melford E. *Children of the Kibbutz*. Harvard University Press, 1958.
———. *Kibbutz: Venture in Utopia*. Harvard University Press, 1956.
Stayer, James M. *The German Peasants' War and Anabaptist Community of Goods*. McGill-Queen's University Press, 1991.
Stearns, Peter. *Consumerism in World History: The Global Transformation of Desire*. Routledge, 2001.
Stein, Lorenz von. *The History of the Social Movement in France, 1789–1850*. Bedminster, 1964.
Stein, Stephen J. *The Shaker Experience in America*. Yale University Press, 1992.
Stewart, Susan. *On Longing*. Johns Hopkins University Press, 1984.
Stitziel, Judd. *Fashioning Socialism: Clothing, Politics, and Consumer Culture in East Germany*. Berg, 2005.
Stjerno, Steinar. *Solidarity in Europe: The History of an Idea*. Cambridge University Press, 2005.
Stobbart, Lorainne. *Utopia: Fact or Fiction? The Evidence from the Americas*. Alan Sutton, 1992.
Strathern, Paul. *Death in Florence: The Medici, Savonarola and the Battle for the Soul of the Renaissance City*. Jonathan Cape, 2011.
Streeck, Wolfgang. *How Will Capitalism End? Essays on a Failing System*. Verso, 2016.
Strong, Roy. *Feast: A History of Grand Eating*. Jonathan Cape, 2002.
Strugnell, Anthony. *Diderot's Politics*. Martinus Nijhoff, 1973.
Sutton, Robert P. *Communal Utopias and the American Experience: Religious Communities, 1732–2000*. Praeger, 2003.
———. *Communal Utopias and the American Experience: Secular Communities, 1824–2000*. Praeger, 2004.
———. *Heartland Utopias*. Northern Illinois University Press, 2009.
———. *Les Icariens: The Utopian Dream in Europe and America*. University of Illinois Press, 1994.
Svede, Mark Allen. "All You Need Is Lovebeads: Latvia's Hippies Undress for Success". In Reid and Crowley, *Style and Socialism*, 189–208.
Talmon, J. L. *Political Messianism: The Romantic Phase*. Secker and Warburg, 1960.
Talmon, Yonina. *Family and Community in the Kibbutz*. Harvard University Press, 1972.
Tarde, Gabriel. *The Laws of Imitation*. Henry Holt, 1903.
———. *Penal Philosophy*. William Heinemann, 1912.
Thomas, Dana. *Deluxe: How Luxury Lost Its Lustre*. Allen Lane, 2007.
Thomas, Keith. "The Utopian Impulse in Seventeenth-Century England". In *Between Dream and Nature: Essays on Utopia and Dystopia*, edited by Dominic Baker-Smith and C. C. Barfoot, 20–46. Rodopi, 1987.

Thompson, Damian. *The End of Time: Faith and Fear in the Shadow of the Millennium*. Sinclair-Stevenson, 1996.

Thompson, E. P. *William Morris: From Romantic to Revolutionary* [1955]. 2nd ed. Merlin, 1977.

Thompson, Noel. *The Market and Its Critics: Socialist Political Economy in Nineteenth Century Britain*. Routledge, 1988.

———. "Owen and the Owenites: Consumer and Consumption in the New Moral World". In *Robert Owen and His Legacy*, edited by Noel Thompson and Chris Williams, 113–28. University of Wales Press, 2011.

———. *Social Opulence and Private Restraint: The Consumer in British Socialist Thought since 1800*. Oxford University Press, 2015.

———. "Social Opulence, Private Asceticism: Ideas of Consumption in Early Socialist Thought". In Daunton and Hilton, *Politics of Consumption*, 51–68.

Thompson, Peter. "Religion, Utopia, and the Metaphysics of Contingency". In Thompson and Žižek, *Privatization of Hope*, 82–105.

Thompson, Peter, and Slavoj Žižek, eds. *The Privatization of Hope: Ernst Bloch and the Future of Utopia*. Duke University Press, 2013.

Thomson, David. *The Babeuf Plot: The Making of a Republican Legend*. Kegan Paul, Trench, Trubner, 1947.

Thrupp, Sylvia, ed. *Millennial Dreams in Action: Studies in Revolutionary Religious Movements*. Schocken Books, 1970.

Tilman, Rick. *Thorstein Veblen and His Critics, 1891–1963*. Princeton University Press, 1992.

Timasheff, Nicholas. *The Great Retreat: The Growth and Decline of Communism in Russia*. E. P. Dutton, 1946.

Tismaneanu, Vladimir, ed. *Promises of 1968: Crisis, Illusion, and Utopia*. Central European University Press, 2011.

Todorov, Tzvetan. *On Human Diversity: Nationalism, Racism, and Exoticism in French Thought*. Harvard University Press, 1993.

Tönnies, Ferdinand. *Community and Civil Society* [1887]. Edited by José Harris. Cambridge University Press, 2001.

Touraine, Alain. *The May Movement: Revolt and Reform; May 1968—the Student Rebellion and Workers' Strikes—the Birth of a Social Movement*. Random House, 1971.

Turkle, Sherry. *Alone Together: Why We Expect More from Technology and Less from Each Other*. Basic Books, 2011.

———. *Life on the Screen: Identity in the Age of the Internet*. Weidenfeld and Nicolson, 1996.

Turner, Bryan S., and Chris Rojek. *Society & Culture: Principles of Scarcity and Solidarity*. Sage, 2001.

Turner, Victor, ed. *Celebration: Studies in Festivity and Ritual*. Smithsonian Institution Press, 1982.

———. *Dramas, Fields, and Metaphors: Symbolic Action in Human Society*. Cornell University Press, 1974.

———. *From Ritual to Theatre: The Human Seriousness of Play*. Performing Arts Journal, 1982.

———. *The Ritual Process: Structure and Anti-structure*. Transaction, 2008.

Turner, Victor, and Edith Turner. *Image and Pilgrimage in Christian Culture*. Basil Blackwell, 1978.

Tuveson, Ernest Lee. *Millennium and Utopia: A Study in the Background of the Idea of Progress*. University of California Press, 1949.

———. *Redeemer Nation: The Idea of America's Millennial Role*. University of Chicago Press, 1968.
Twitchell, James B. *Lead Us into Temptation: The Triumph of American Materialism*. Columbia University Press, 1999.
———. *Living It Up: Our Love Affair with Luxury*. Columbia University Press, 2002.
Van den Berg, Axel. *The Immanent Utopia: From Marxism on the State to the State of Marxism*. Transaction, 2003.
Veblen, Thorstein. *The Theory of the Leisure Class* [1899]. Allen and Unwin, 1924.
Venturi, Franco. *Utopia and Reform in the Enlightenment*. Cambridge University Press, 1971.
Verrall, A. W. *Collected Literary Essays*. Cambridge University Press, 1913.
Vichert, Gordon. "The Theory of Conspicuous Consumption in the Eighteenth Century". In *The Varied Pattern: Studies in the Eighteenth Century*, edited by Peter Hughes and David Williams, 251–67. A. M. Hakkert, 1971.
Vihavainen, Timo. "Consumerism and the Soviet Project". In Vihavainen and Bogdanova, *Communism and Consumerism*, 113–38.
———. "The Spirit of Consumerism in Russia and the West". In Vihavainen and Bogdanova, *Communism and Consumerism*, 1–27.
Vihavainen, Timo, and Elena Bogdanova, eds. *Communism and Consumerism: The Soviet Alternative to the Affluent Society*. Brill, 2016.
Vincent, John Martin. *Costume and Conduct in the Laws of Basel, Bern, and Zurich, 1370–1800*. Johns Hopkins Press, 1935.
Vincent, K. Steven. *Pierre-Joseph Proudhon and the Rise of French Republican Socialism*. Oxford University Press, 1984.
Vondung, Klaus. *The Apocalypse in Germany*. University of Missouri Press, 2000.
Wachtel, Paul L. "Alternatives to the Consumer Society". In Crocker and Linden, *Ethics of Consumption*, 198–217.
———. *The Poverty of Affluence: A Psychological Portrait of the American Way of Life*. Free Press, 1983.
Wagar, W. Warren. *H. G. Wells and the World State*. Yale University Press, 1961.
———. *Terminal Visions: The Literature of Last Things*. Indiana University Press, 1982.
Wahl, Elizabeth Susan. *Invisible Relations: Representations of Female Intimacy in the Age of Enlightenment*. Stanford University Press, 1999.
Wallace-Wells, David. *The Uninhabitable Earth: A Story of the Future*. Allen Lane, 2019.
Wallas, Graham. *Human Nature in Politics*. Constable, 1919.
Wallerstein, Immanuel. *Utopistics; or, Historical Choices of the Twenty-First Century*. New Press, 1998.
Walzer, Michael. *The Revolution of the Saints: A Study in the Origins of Radical Politics*. Weidenfeld and Nicolson, 1966.
Wayland-Smith, Ellen. *Oneida: From Free Love Utopia to the Well-Set Table*. Picador, 2016.
Weatherill, Lorna. *Consumer Behaviour and Material Culture in Britain, 1660–1760*. Routledge, 1988.
Webb, Diana. *Medieval European Pilgrimage, c.700–c.1500*. Palgrave, 2002.
Weber, Max. *Economy and Society*. University of California Press, 1978.
Weinstein, Donald. "Millenarianism in a Civic Setting: The Savonarola Movement in Florence". In Thrupp, *Millennial Dreams in Action*, 187–206.

———. *Savonarola: The Rise and Fall of a Renaissance Prophet*. Yale University Press, 2011.

Weisman, Alan. *The World without Us*. Thomas Dunne, 2007.

Weiss, Robert S. *Loneliness: The Experience of Emotional and Social Isolation*. MIT Press, 1973.

Wernick, Andrew. *Promotional Culture: Advertising, Ideology and Symbolic Expression*. Sage, 1991.

Whatmore, Richard. *Republicanism and the French Revolution: An Intellectual History of Jean-Baptiste Say's Political Economy*. Oxford University Press, 2000.

Wheeler, Brannon. *Mecca and Eden: Ritual, Relics, and Territory in Islam*. University of Chicago Press, 2006.

Whitehead, Anne. *Paradise Mislaid: In Search of the Australian Tribe of Paraguay*. University of Queensland Press, 1997.

Whitney, Lois. *Primitivism and the Idea of Progress in English Popular Literature of the Eighteenth Century*. Johns Hopkins Press, 1934.

Whitson, Robley Edward, ed. *The Shakers*. Paulist, 1983.

Whyte, William H. *City: Rediscovering the Center*. Doubleday, 1988.

———. *The Organization Man*. Jonathan Cape, 1957.

———. *The Social Life of Small Urban Spaces*. Conservation Foundation, 1980.

Widdicombe, Toby. "Early Histories of Utopian Thought (to 1950)". *Utopian Studies* 3 (1992): 1–38.

Wightman, Richard, and T. J. Jackson Lears, eds. *The Culture of Consumption: Critical Essays in American History*. Pantheon Books, 1983.

Wilkie, Everett C., Jr. "Mercier's *L'An 2440*: Its Publishing History during the Author's Lifetime". *Harvard Library Bulletin* 32 (1984): 5–35.

Wilkinson, Richard G. *The Impact of Inequality: How to Make Sick Societies Healthier*. Routledge, 2005.

Wilkinson, Richard, and Kate Pickett. *The Inner Level: How More Equal Societies Reduce Stress, Restore Sanity and Improve Everyone's Well-Being*. Allen Lane, 2018.

———. *The Spirit Level: Why More Equal Societies Almost Always Do Better*. Allen Lane, 2009.

Williams, Albert Rhys. *The Russian Land*. Geoffrey Bles, 1929.

Williams, George Huntston. *The Radical Reformation*. Weidenfeld and Nicolson, 1962.

———, ed. *Spiritual and Anabaptist Writers*. SGM, 1957.

———. *Wilderness and Paradise in Christian Thought: The Biblical Experience of the Desert in the History of Christianity & the Paradise Theme in the Theological Idea of the University*. Harper and Bros, 1962.

Williams, John Alexander. *Turning to Nature in Germany: Hiking, Nudism, and Conservation, 1900–1940*. Stanford University Press, 2007.

Williams, Raymond. *The Country and the City*. Chatto and Windus, 1973.

Willimott, Andy. *Living the Revolution: Urban Communes & the Soviet Socialism, 1917–1932*. Oxford University Press, 2012.

Wilson, Matthew. *Moralising Space: The Utopian Urbanism of the British Positivists, 1855–1920*. Routledge, 2018.

Wirth, Louis. *Louis Wirth on Cities and Social Life*. University of Chicago Press, 1964.

Witeze, Geraldo, Jr. "Vasco de Quiroga Rewrites *Utopia*". In *Utopias in Latin America Past and Present*, edited by Juan Pro, 53–75. Sussex Academic, 2018.

Wittke, Carl. *The Utopian Communist: A Biography of Wilhelm Weitling*. Louisiana State University Press, 1950.

Wokler, Robert. *Rousseau, the Age of Enlightenment, and Their Legacies*. Princeton University Press, 2012.

Wolfe, Thomas. *You Can't Go Home Again*. Harper and Row, 1940.

Wolfe, Tom. *The Electric Kool-Aid Acid Test*. Farrar Straus and Giroux, 1968.

Wootton, David. "Friendship Portrayed: A New Account of Utopia". *History Workshop Journal* 45 (1998): 28–47.

Worden, Harriet M. *Old Mansion House Memories*. Kenwood, 1950.

Wright, Erik Olin. *Envisioning Real Utopias*. Verso, 2010.

Wright, Johnson Kent. *A Classical Republican in Eighteenth Century France: The Political Thought of Mably*. Stanford University Press, 1997.

Wright, Louis B. *The Colonial Search for a Southern Eden*. University of Alabama Press, 1953.

Wrigley, Richard. *The Politics of Appearances: Representations of Dress in Revolutionary France*. Berg, 2002.

Wroth, Warwick. *The London Pleasure Gardens of the Eighteenth Century*. Macmillan, 1979.

Wunderlich, Roger. *Low Living and High Thinking at Modern Times, New York*. Syracuse University Press, 1992.

Yar, Majid. *The Cultural Imaginary of the Internet: Virtual Utopias and Dystopias*. Palgrave Macmillan, 2014.

Yinger, J. Milton. *Countercultures: The Promise and Peril of a World Turned Upside Down*. Free Press, 1982.

Zablocki, Benjamin David. *The Joyful Community: An Account of the Bruderhof*. Penguin Books, 1971.

Zakharova, Larissa. "How and What to Consume: Patterns of Soviet Clothing Consumption in the 1950s and 1960s". In Vihavainen and Bogdanova, *Communism and Consumerism*, 85–112.

Zanda, Emanuela. *Fighting Hydra-like Luxury: Sumptuary Regulation in the Roman Republic*. Bristol Classical, 2011.

Żeromski, Stefan. *The Coming Spring* [1924]. Central European University Press, 2007.

Zimbardo, Philip G. "Mind Control in Orwell's *Nineteen Eighty-Four*: Fictional Concepts Become Operational Realities in Jim Jones's Jungle Experiment". In *On Nineteen Eighty-Four*, edited by Abbott Gleason et al., 127–54. Princeton University Press, 2005.

Zimmermann, Rainer. "Transforming Utopian into Metopian Systems: Bloch's *Principle of Hope* Revisited". In Thompson and Žižek, *Privatization of Hope*, 246–68.

Zipes, Jack. *Ernst Bloch: The Pugnacious Philosopher of Hope*. Palgrave-Macmillan, 2019.

Zukin, Sharon. *Naked City: The Death and Life of Authentic Spaces*. Oxford University Press, 2010.

Zweiniger-Bargielowska, Ina. *Austerity in Britain: Rationing, Controls, and Consumption, 1939–1945*. Oxford University Press, 2000.

INDEX

A NOTE ON THE TYPE

This book has been composed in Arno, an Old-style serif typeface in the classic Venetian tradition, designed by Robert Slimbach at Adobe.